T0326355

Reaction Engineering

Reaction Engineering

Shaofen Li
Professor, Tianjin University, Tianjin, China

Translated and updated by
Lin Li and Feng Xin

Butterworth-Heinemann
An imprint of Elsevier

CIP

Butterworth-Heinemann is an imprint of Elsevier
The Boulevard, Langford Lane, Kidlington, Oxford OX5 1GB, United Kingdom
50 Hampshire Street, 5th Floor, Cambridge, MA 02139, United States

Notices

Knowledge and best practice in this field are constantly changing. As new research and
experience broaden our understanding, changes in research methods, professional practices, or
medical treatment may become necessary.

Practitioners and researchers must always rely on their own experience and knowledge in
evaluating and using any information, methods, compounds, or experiments described herein.
In using such information or methods they should be mindful of their own safety and the safety
of others, including parties for whom they have a professional responsibility.

To the fullest extent of the law, neither the Publisher nor the authors, contributors, or editors,
assume any liability for any injury and/or damage to persons or property as a matter of products
liability, negligence or otherwise, or from any use or operation of any methods, products,
instructions, or ideas contained in the material herein.

British Library Cataloguing-in-Publication Data
A catalogue record for this book is available from the British Library

Library of Congress Cataloging-in-Publication Data
A catalog record for this book is available from the Library of Congress

ISBN: 978-0-12-410416-7

For Information on all Butterworth-Heinemann publications
visit our website at https://www.elsevier.com

Working together
to grow libraries in
developing countries

www.elsevier.com • www.bookaid.org

Publisher: Jonathan Simpson
Acquisition Editor: Glyn Jones
Editorial Project Manager: Naomi Robertson
Production Project Manager: Mohanapriyan Rajendran
Designer: Mark Rogers

Typeset by MPS Limited, Chennai, India

Contents

11. Fundamentals of Biochemical Reaction Engineering

12. Fundamentals of Polymerization Reaction Engineering

List of Contributors

Qingfeng Ge Southern Illinois University, Carbondale, IL, United States

Xudong Ge University of Maryland, Baltimore, MD, United States

Lin Li Chevron Energy Technology Company, Richmond, CA, United States

Chuan Lu University of Utah, Salt Lake City, UT, United States

Xiaoxia Lu Texas Commission on Environmental Quality, Austin TX, United States

Fumin Wang Tianjin University, Tianjin, China

Tiefeng Wang Tsinghua University, Beijing, China

Feng Xin Tianjin University, Tianjin, China

Preface

As a major branch of Chemical Engineering, Chemical Reaction Engineering studies the modeling and scaling up of the reaction process and has long been a pillar for process research and engineering design in the field of Chemical Engineering. The course of Reaction Engineering is built primarily on the foundation of mathematics and fundamental sciences such as physics and chemistry, relies on mole and energy balance, and teaches reaction kinetics, reactor design, and safe and reliable operation and optimization of different types of reactors.

This book has a straightforward structure from reaction kinetics to reactor design and operation. It starts with an introduction of basic definitions of conversion, selectivity, and yield in chapter 1, and then move on to discuss homogeneous reaction kinetics and intrinsic kinetics of gas-solid catalytic reactions in chapter 2 and macro kinetics of gas-solid catalytic reactions in chapter 6. After presenting reactor design equations for three ideal reactors: batch reactors (BR), continuous stirred tank reactors (CSTR) in chapter 3 and plug flow reactors (PFR) in chapter 4, and flow models for continuous reactors in chapter 5, more practical reactor models for fixed bed chapter 7, fluidized bed in chapter 8, multiple-phase reactors in chapter 9 and fluid-solid non-catalytic reactors in chapter 10 are introduced. Along with reactor model and analysis, the impact of different operation modes on reactor performance are examined, such as batch versus continuous reactors, constant volume versus variable-volume processes, adiabatic versus heat-exchange operations etc. In addition, multiple steady states and reactor operation stability are also discussed. Separate chapters are devoted to applications for biochemical reactions in chapter 11, polymerizations in chapter 12, and electrochemical processes in chapter 13 with an emphasis on unique characters of each application. Among the thirteen chapters, the first seven chapters can be used for teaching Chemical Reaction Engineering to undergraduate students while the latter six chapters are suitable for a graduate course.

This book uses concise and accurate language to teach the fundamental concepts and principles, presents details of model development, and provides a comprehensive coverage of the subject of reaction engineering. Plenty of detailed worked examples and end-of-chapter exercise problems are provided to help the students understand the complex concepts and cultivate the capability to establish reactor models and perform engineering design and analysis.

This book is adapted and translated from the Chinese textbook written by the late Professor Shaofen Li, originally published in 1990. In over two decades, the original Chinese version has been widely used in many universities. In addition, due to its comprehensive nature, it is also a highly valued professional reference. The book also received the highest college textbook award in China. Revised throughout, this English version is developed to reach broader readers around the world. We cherish the outstanding contribution to reaction engineering education by Professor Li, and at the same time would like to thank all the contributors to the original Chinese version: Bangrong Liu, Lu Huang, Ying Zhang, Hui Liao, Haojiang Zhang, Xueming Zhao, Yanxi Chen, Tongyu Cao, Jingwu Sun and Hongwu Ma, all from Tianjin University.

Feng Xin and Lin Li organized the overall efforts of this version and many people contributed.

Chapters 1, 2, 3, 4, and 7 were translated by Lin Li.
Chapters 5 and 11 were translated by Xiaoxia Lu.
Chapters 6 and 13 were translated by Chuan Lu.
Chapter 8 was written by Tiefeng Wang.
Chapter 9 was translated by Qingfeng Ge.
Chapter 10 was written by Feng Xin in Chinese and translated by Xudong Ge.
Chapter 12 was translated by Fumin Wang.

Lin Li conducted the overall review and copyediting of the entire first draft. We also want to thank Chumeng Li for reading and polishing the manuscript with his linguistic skills.

The publication of this English version reflects continued efforts of three generations. It is our sincere hope that it will benefit readers from both academia and industry, and we would appreciate any feedback and suggestions for improvements.

Chapter 1

Introduction

Chapter Outline

Converting feedstock into useful products through chemical processing is widely used in the chemical industry as well as other process industries, such as metallurgy, petroleum refining, energy, and consumer products. A chemical process to make a useful product typically involves three major steps: (1) feedstock preparation; (2) chemical conversion; and (3) product separation and purification. The second step is the centerpiece of a chemical production process and the other two steps are complementary to the core chemical conversion step. From a Chemical Engineering education perspective, the first and third steps are covered by Unit Operations and other courses related to transport phenomena, while the second step is covered by Chemical Reaction Engineering.

1.1 CHEMICAL REACTION ENGINEERING

Chemical Reaction Engineering is a subset of Chemical Engineering, and it is often simply called Reaction Engineering. Its content can be roughly divided into two parts: Reaction Kinetics and Reactor Design and Analysis.

Reaction Kinetics is mainly concerned with mechanism and the rate of chemical reactions. In order to obtain the reaction rate information needed for the design and operation of an industrial reactor, it is necessary to investigate reaction kinetics, including reaction mode and mechanism, rate equation, reaction activation energy, etc. Most chemical reaction processes are very complex; therefore, a certain degree of simplification is usually needed

Reaction Engineering. DOI: http://dx.doi.org/10.1016/B978-0-12-410416-7.00001-X

in order to obtain a useful reaction rate equation that can quantify the reaction progress. Of course, the key for simplification is to capture the main characters of the reaction process while ignoring the factors that are not critical. For a given reaction system (if a catalyst or a solvent is used, the catalyst or solvent is also fixed), the reaction rate depends on temperature, concentrations, and pressure of the reaction system. The kinetic equation is a quantitative description of these variables. However, with the progress of the chemical reactions, the composition, temperature, and pressure of the reaction system will change with time or location or both. As a result, reaction rate will change during the course of the reaction process.

The other main task of Reaction Engineering is to study the changes of the variables inside the reactor so that the conditions and operations of the reactor can be optimized to achieve the best economics. This is reactor design and analysis. We can say the reaction kinetics concerns with a specific location or a "point" inside a reactor, and reactor design and analysis utilizes that localized information to optimize the whole reactor.

There are many different types of reactors, but they can be divided into a few categories. One commonly used approach in reaction engineering study is to categorize the reactors based on fluid phases involved in the reactor. Chemical reactions can be divided into two main categories: homogeneous and heterogeneous. Each category can be further divided into subcategories. For homogeneous reactions, they can be divided into three types: gas, liquid, and solid homogeneous reactions. Heterogeneous reactions can be divided into six major types: gas−solid, gas−liquid, liquid−liquid, liquid−solid, solid−solid, and gas−liquid−solid. On the other hand, based on whether a catalyst is used, chemical reactions can be divided into catalytic and noncatalytic reactions. Most reactions using a solid catalyst are considered heterogeneous reactions. Vanadium catalyst, e.g., is commonly used for oxidation of sulfur dioxide to make sulfur trioxide. Although both the reaction feed (SO_2) and product (SO_3) are gases, this reaction is not considered as a homogeneous reaction. Instead, it is treated as a gas−solid catalytic reaction.

Chemical reaction processes involve not only chemical reactions but also physical phenomena, i.e., transport processes, including momentum, heat, and mass transfer. Transport phenomena along with chemical reactions constitute the core of chemical engineering science. Next, we will use ammonia oxidation on platinum−rhodium wire as an example to illustrate the chemical and transport phenomena during a chemical reaction process:

$$4NH_3 + 5O_2 \rightarrow 4NO + 6H_2O - \Delta Hr$$

Since the chemical reaction takes place on the catalyst surface, the reactants, ammonia, and oxygen, must first transfer from bulk gas phase onto the surface of platinum−rhodium wire. On the other hand, the products, nitrogen monoxide and water will have to transfer from platinum−rhodium wire surface into bulk gas phase. We can see that mass transfer between the gas and solid phases must accompany this chemical reaction. In addition, this reaction is exothermic. As a result of the chemical reaction, the surface temperature of the platinum—rhodium wire will increase, creating a temperature difference between the catalyst surface and gas phase which will in turn lead to heat transfer.

The phenomena discussed above are localized events, i.e., we only examine the reaction and transport phenomena for a specific location. When a reaction process is taking place inside a reactor, various transport processes will also take place. Therefore, we must study the reaction and transport phenomena together, understand the behaviors of various reaction and transport steps involved, their relationships, and their impact on the overall reaction process so that we can adequately identify key issues and find innovative solutions.

Chemical reaction engineering plays a critical role in chemical product development, process design, and reactor scale-up. By using chemical reaction engineering knowledge, a larger scale-up factor can be achieved and less pilot study would be required for the development of a new chemical manufacturing process. All of these lead to a much shorter development time, which translates to better economics and competitive advantages. For existing plants, chemical reaction engineering knowledge is also essential for improving reactor performance and optimizing its operations. Therefore, for both product development, which require scale-up from lab study to building a new chemical plant, and improvement of existing facilities, chemical reaction engineering knowledge is required. In addition to the conventional chemical industry, chemical reaction engineering can also play critical roles in many other areas, such as environmental protection, fuel combustion, and artificial organs.

Chemical reaction engineering, as a branch of chemical engineering, is an applied science that relies on math and fundamental sciences such as physics and chemistry. Its roots stem from the 1930s. Damköhler, with very limited experimental data available at that time, systematically described the impact of diffusion, fluid flow, and heat transfer on reactor yield. This laid a foundation for chemical reaction engineering. At almost the same time, Thiele and Zel'dovich pioneered the study on diffusion reaction problems. In the late 1940s, Hougen and Waston published *Principles of Chemical Process* and Frank-Kamenetskii published *Diffusion and Heat Transfer in Chemical Kinetics*. These books summarized the relationship between chemical reactions and transport processes and explored reactor design, which were considered the core contents of chemical reaction engineering science. However, the chemical reaction engineering conference, attended by scholars from several European countries and held in Amsterdam in 1957, is usually considered as the formal formation of chemical reaction engineering science. The term Chemical Reaction Engineering was first used and defined at this meeting. Since then, the Chemical Reaction Engineering science has been growing with ever increasing number of people studying in this field. Like any other science and technology fields, the development of chemical reaction engineering has been driven by the demands of the society and the industries it serves. In the 1960s, the petrochemical industry experienced significant growth. The tremendous increase of production scale, broadening of feedstock, and improvement of both quantity and quality of the products provided many challenges and opportunities for the growth and development of chemical reaction engineering. The advance of computer technology enabled reaction engineers to quantitatively solve complex problems for reactor design and control. After over a half-century of development there are already fairly matured theories and methods at the core of chemical reaction engineering. Since the 1980s, the high tech industries, such as microelectronics, optical fiber, new materials, and biotechnology, have provided new challenges and opportunities in chemical reaction engineering, and led to new research areas, such as bioreaction engineering, polymerization reaction engineering, and electrochemical reaction engineering.

1.2 CONVERSION AND YIELD OF CHEMICAL REACTIONS

1.2.1 Extent of Reaction

For a chemical reaction, there is a quantitative relationship between the amounts of reactant consumed and product formed known as stoichiometry. For example,

$$\nu_A A + \nu_B B \rightarrow \nu_R A \tag{1.1}$$

In the equation above, ν_A, ν_B, and ν_R are stoichiometric coefficients for components A, B, and R, correspondingly. If the amount of A reacted is ν_A moles, then the amount of B reacted must be ν_B and the amount of R formed must be ν_R. In other words, the ratio of amount of reactant converted to product formed is equals to the ratio of chemical stoichiometric coefficients. If at the beginning of the reaction there are n_{A0} moles of A, n_{B0} moles of B, and n_{R0} moles of R in the reaction system, and after a certain period of time reaction the amount of A, B, and R in the system becomes n_A, n_B, and n_C, respectively, then the amounts of conversion can be calculated by subtracting final quantities from initial values, and the following relationship must hold:

$$(n_A - n_{A0}) : (n_B - n_{B0}) : (n_R - n_{R0}) = \nu_A : \nu_B : \nu_R$$

It is obvious that $(n_A - n_{A0}) < 0$ and $(n_B - n_{B0}) < 0$, and therefore ν_A and ν_B must be negative. This indicates that the amounts of reactants decrease as the reaction progresses. On the other hand, the amount of product increases, reflected by the positive value of ν_R. In this book the stoichiometric coefficients for reactants are defined as negative and the stoichiometric coefficients for products as positive. This will make the description of multi-reaction systems easier. The above equation can be rearranged as:

$$\frac{(n_A - n_{A0})}{\nu_A} = \frac{(n_B - n_{B0})}{\nu_B} = \frac{(n_R - n_{R0})}{\nu_R} = \xi \qquad (1.2)$$

i.e., for each reaction component the ratio of amount of conversion to stoichiometric coefficients is a constant, and ξ is called the extent of reaction. Eq. (1.2) can be used for any reactions:

$$n_i - n_{i0} = \nu_i \xi \qquad (1.3)$$

We can see that only one parameter is needed to describe extent of reaction. If ξ is known, then the amount of each component reacted can be easily calculated. Based on the definition of Eq. (1.2), ξ is always a positive number. Based on the convention used in this book for stoichiometric coefficients the amount of reactants converted calculated by Eq. (1.3) will be negative, reflecting that the reactants are consumed, while the amount of products formed are positive values, because there are net increases of products in the system.

It should be noted that the extent of reaction is an extensive parameter. Sometimes it is more convenient to use an intensive parameter to define the extent of reaction, such as extent of reaction per unit volume or unit mass of the reaction system. If there are multiple reactions taking place, each reaction has its own extent of reaction. If ξ_j is the extent of reaction for reaction j, then the total amount of component i reacted is equal to the sum of the amount of this component converted (or formed) from each reaction:

$$n_i - n_{i0} = \sum_{j=1}^{M} \nu_{ij} \xi_j \qquad (1.4)$$

where ν_{ij} is the stoichiometric coefficient of component i in reaction j, and M is the total number of the reactions.

1.2.2 Conversion

Conversion is a widely used parameter to describe the extent of a reaction, and it is defined as the percentage of a reactant converted:

$$X = \text{amount of a reactant converted/initial amount of the reactant} \qquad (1.5)$$

From the above definition we can see conversion is defined based on a specific reactant. For a reaction with multiple reactants, conversions for different reactants may be different, but they reflect the extent of the same reaction. Therefore strictly speaking, it doesn't matter which reactant is selected to define the conversion. However, in practice we should try to select a reactant that would make the calculation more convenient and would allow us to gain more insight about the reaction. For most industrial applications, the feed composition usually is not equal to the reaction stiochiometric ratio, which means for some reactants excessive amounts are fed to the reactor, and for the components that are excessive, their maximum conversions will be less than 100%. It is a common practice to select a nonexcessive component, called a key component, to define reaction conversion so that the maximum conversion is 100%. If the feed composition corresponds to the stoichiometric ratio of the reaction, it doesn't matter which component is selected to calculate the reaction conversion, since all will have the same value.

In most cases the key component is the most expensive one. Other components are cheaper and also excessive. Therefore the conversion of the key component has direct impact on process economics and can serve as a good measurement of reaction performance.

For conversion calculation it is also important to select an adequate starting point. For continuous flow reactors it is commonplace to choose reactor inlet as the initial state to calculate the reaction conversion. For batch reactors the initial state is defined at the start of the reaction. For some applications multiple reactors in series are used. In this situation reaction conversion is normally calculated based on the feed composition to the first reactor, not the inlet composition of each reactor, since this is more convenient, especially for comparing different designs.

Due to equilibrium or other limitations, the reaction conversion for some systems can be very low. In order to improve feed utilization, which is usually critical for reducing cost, it is a common practice to separate feed components from product stream from reactor outlet and recycle them back to reactor inlet to mix with fresh feed. For example, ammonia synthesis from hydrogen and nitrogen, methanol synthesis from carbon monoxide and hydrogen, and ethanol synthesis from ethylene and steam all use such feed recycle schemes to achieve high utilization of the feedstocks. For such reactors there will be two definitions for reaction conversion. One is called single pass conversion, which is the conversion achieved from single pass through the reactor, and can also be understood as the conversion calculated based on reactor inlet composition. Another is known as overall conversion, which is calculated based on fresh feedstock entering the reactor system and final product stream leaving the reactor system. Obviously, overall conversion is always higher than single pass conversion since recycling increases reactant conversion.

If the conversion of the key component is known, then the amounts of other reactants that are converted can be calculated based on feed composition and reaction stoichiometry. The relationship between conversion and extent of reaction can be obtained by combining Eqs. (1.3) and (1.5):

$$X_i = -\frac{\nu_i \xi}{n_{i0}} \tag{1.6}$$

Example 1.1
Chloroethylene, the monomer for Poly(chloroethene), usually known as polyvinyl chloride (PVC), is typically synthesized through reaction between acetylene and hydrogen chloride on mercury chloride catalyst:

$$C_2H_2 + HCl \rightarrow CH_2 = CHCl$$

Since acetylene is more expensive than HCl, excessive HCl is usually added to the feed stream. If HCl is 10% more than stoichiometrically required and at reactor outlet the chloroethylene molar fraction is 90%, what is conversion for acetylene and HCl?

Solution
The stoichiometric ratio of HCl to acetylene is 1. Since HCl is 10% more than stoichiometric ratio, the molar ratio of acetylene to HCl in the feed is 1:1.1. Assuming that for every 1 mole of acetylene entering the reactor x mol will be converted, then we can write the following mol balance for the reactor:

	Inlet	Outlet
C_2H_2	1	$1 - x$
HCl	1.1	$1.1 - x$
$CH_2 = CHCl$	0	x
Total =	2.1	$2.1 - x$

chloroethylene molar fraction at reactor outlet is 90%, so

$$x/(2.1 - x) = 0.9$$
$$\text{Therefore} \quad x = 0.9947 \text{ mol}$$

The above calculation is based on 1 mole of C_2H_2, so the conversion of C_2H_2 is

$$X_{C_2H_2} = 0.9947/1 = 99.47\%$$

The amount of HCl reacted is the same as that of C_2H_2, so the conversion of HCl is:

$$X_{HCl} = 0.9947/1.1 = 0.9043 = 90.43\%$$

The above calculation clearly shows that conversion calculated based on different reactants is different. If there is no excessive HCl in the feed then the conversion values for these two reactants will be the same.

1.2.3   Yield and Selectivity

Reaction conversion reflects how much reactant is converted, and the yield defines the amount of product formed:

$$Y_R = \left| \frac{\nu_A}{\nu_B} \right| \frac{\text{Amount of product formed}}{\text{Initial quantity of key component}} \qquad (1.7)$$

ν_A and ν_R are stoichiometric coefficients for key component A and product R, respectively. The reason that we have to include the stoichiometric coefficient in the above definition is to ensure the maximum yield is 100%. Eq. (1.7) can be rewritten as:

$$Y = \frac{\text{Amount of key component consumed to make the product}}{\text{Initial quantity of the key component}} \qquad (1.8)$$

By comparing Eqs. (1.5) and (1.8) it is obvious that for single reactions the value of conversion and yield is the same and the yields for different products are also the same. However, this is not the case if more than one reaction takes place in the reactor. For example, when ethylene is oxidized on a silver catalyst two reactions will take place:

$$C_2H_4 + 1/2O_2 \rightarrow \underset{\underset{O}{\diagdown \diagup}}{H_2C{-}CH_2} \qquad (1.9)$$

$$C_2H_4 + 3O_2 \rightarrow 2CO_2 + 2H_2O \qquad (1.10)$$

Ethylene can be converted to either ethylene oxide or carbon dioxide. As a result, the conversion of ethylene does not equal the yield of either ethylene oxide or carbon dioxide.

For industrial applications mass yield is also commonly used. The definition of mass yield is similar to Eq. (1.7), but is defined in mass, instead of moles. In addition, stoichiometric coefficients are not included in the definition. Therefore, the maximum value of mass yield can be higher than 100%. For a reaction system involving recycling, similar to conversion, there will also be a single pass yield and an overall yield. The single pass yield reflects the product yield achieved when the feed stream passes through the reactor only once, while the overall yield measures the total product produced after recycles. Similarly, the difference between single pass and overall yield can also be understood as the product yield calculated with different bases, i.e., the single pass yield is calculated based on reactor inlet and overall yield is based on inlet of whole reactor system. Overall yield is always higher than single pass conversion.

One reaction variable, extent of reaction, conversion, or yield, can only describe the extent of a single reaction. For multiple reaction systems, more reaction variables are needed. For the ethylene oxidation example given above, if only conversion is known, then we will only know the total amount of ethylene converted, or the overall result of two reactions taking place in the reactor. We won't know how much ethylene is converted into ethylene oxide and how much is converted to carbon dioxide. If we add additional reaction variables, such as the yield of ethylene oxide, then we will be able to fully describe the reaction system and calculate compositions.

For a multiple reaction system where multiple reactions take place, in addition to conversions and yields, selectivity is another important concept, and is defined as:

$$S = \frac{\text{Amount of key component consumed to make desired product}}{\text{Total amount of key component converted}} \quad (1.11)$$

Due to the side reactions not all the reactant converted will form desired products, and from Eq. (1.11) we can see that selectivity is always less than 1. Reaction selectivity reflects the relative reaction extent between main or desired reactions and the side reactions. From Eqs. (1.5), (1.6), and (1.11) we can obtain the following relationships among conversion, yield, and selectivity:

$$Y = SX \quad (1.12)$$

It should be noted that the definitions of reaction yield and selectivity may not be the same in different books. For example, the selectivity defined in this book is known as yield in other books. Therefore extra caution needs to be paid when using these concepts.

Example 1.2

Ethylene can be oxidized on a silver catalyst to make ethylene oxide. If at the inlet of the catalytic reactor the molar percentages of C_2H_4 is 15%, O_2 7%, CO_2 10%, and Ar 12%, and the rest is nitrogen. At the reactor outlet the molar percentages for C_2H_4 and O_2 are 13.1% and 4.8%, respectively. Please calculate conversion of ethylene, the yield of ethylene oxide, and reaction selectivity.

Solution

We will conduct the calculation based on 100 moles of feed. Assume x and y are the amounts of ethylene oxide and CO_2 formed, respectively. Based on the feed composition and stoichiometry of this reaction system (see Eqs. (1.9) and (1.10)), we can make the following table:

	Reactor inlet	Reactor outlet
C_2H_4	15	$15 - x - y/2$
O_2	7	$7 - x/2 - 3y/2$
C_2H_4O	0	x
CO_2	10	$10 + y$
H_2O	0	y
Ar	12	12
N_2	56	56
Total	100	$100 - x/2$

The molar percentages of ethylene oxide and CO_2 are 13.1% and 4.8%, respectively, therefore:

$$(15 - x - y/2)/(100 - x/2) = 0.131$$
$$(7 - x/2 - 3y/2)/(100 - x/2) = 0.048$$

(Continued)

(Continued)

By solving the above two equations we can obtain:

$$x = 1.504 \text{ mol}$$
$$y = 0.989 \text{ mol}$$

The amount of ethylene converted is:

$$1.504 + 0.989/2 = 1.999 \text{ mol}$$

Therefore the conversion of ethylene is:

$$1.999/15 = 0.1333 \quad \text{or} \quad 13.33\%$$

The yield of ethylene oxide is

$$1.504/15 = 0.1003 \quad \text{or} \quad 10.03\%$$

Using Eq. (1.12) we can calculate reaction selectivity:

$$S = 0.1003/0.1333 = 0.7524 \quad \text{or} \quad 75.24\%$$

From the above calculations we can see that both ethylene conversion and ethylene oxide yields are very low. In industrial production the reactor effluent is first washed by water to remove ethylene oxide and then washed by caustic solution to remove CO_2. The rest of the gas is compressed and then mixed with fresh ethylene and oxygen feeding into the reactor to form a recycle loop. It is due to such recycling schemes that higher conversion of ethylene and high yield of ethylene oxide are achieved.

1.3 CLASSIFICATIONS OF CHEMICAL REACTORS

Reactor design and analysis is one of the core elements of chemical reaction engineering. There are many different types of reactors that can be categorized based on different aspects of reactor characters. Based on structure characters, reactors can be categorized into the following types, as shown in Fig. 1.1:

1. Tubular reactors. Their main character is its length is much larger than its diameter, and empty inside, i.e., does not contain any internal structure, as shown in Fig. 1.1(A). A tubular reactor is suitable for homogeneous reactions, such as a cracker to make ethylene by thermally cracking hydrocarbon feedstocks, such as ethane or naphtha.
2. Tank reactors, sometimes also called stirred tank reactor, as shown in Fig. 1.1(B). Typically its height is a little larger than its diameter, with height to diameter ratio of 2 to 3. There are stir and baffles inside the tank, and in many cases heat exchangers are also installed inside the reactor to maintain reaction temperature. Another option to control the reactor temperature is to have an external heat exchanger, and to use a pump to circulate reaction fluid through the heat exchanger. Obviously,

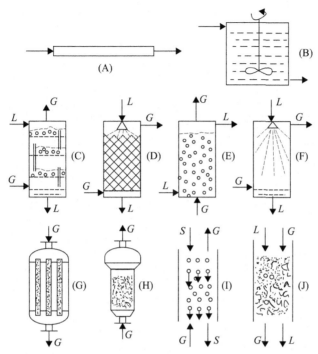

FIGURE 1.1   Illustration of different types of reactors. (A) Tubular reactor; (B) Tank reactor; (C) Tray column; (D) Packed column; (E) Bubble column; (F) Spray column; (G) Fixed bed reactor; (H) Fluidized bed reactor; (I) Moving bed reactor; (J) Trickle bed reactor. *G*, Gas; *L*, Liquid; *S*, Solid.

for reactions with low reaction heat it may not be necessary to have heat exchange equipment. Tank reactors are widely used for both homogeneous (most liquid phase) and heterogeneous reactions, such as gas−liquid reactions, liquid−liquid reactions, liquid−solid reactions, or gas−liquid−solid reactions. Many organic synthesis processes, such as esterification, nitration, sulfonation, and chlorination, use stirred tank reactors.

3. Column (tower) reactors. Column reactors usually have a height to diameter ratio greater than 10, and typically some type of internals such as packing or sieve plate are installed inside the column to enhance interphase contact. Fig. 1.1(C) illustrates a tray column, and Fig. 1.1(D) illustrates a packed column. Column reactors are mainly used for reactions between two fluids, such as gas−liquid or liquid−liquid reactions. Bubble columns (Fig. 1.1(E)) are also column reactors but without any internals, where gas bubbles through the liquid phase. Another type of column reactor is a spray tower (Fig. 1.1(F)), commonly used for gas−liquid reactions with liquid dispersed in the gas phase as droplets,

the opposite of the behavior observed in a bubble column. For any type of column reactor, the two fluids can either flow cocurrently or counter-currently, depending on which flow pattern gives a better performance for the specific reactions taking place inside the reactor.

4. Fixed bed reactors. The key characteristic of a fixed bed reactor is that it is packed with solid pellets, typically a catalyst, but in some case the solid could also be a reactant participating in the reactions. Fixed bed reactors are one of the most widely used reactors for heterogeneous catalytic reactions, and have been used in ammonia synthesis, methanol synthesis, oxidation of p-xylene, etc. Fig. 1.1(G) is an illustration of a parallel tube reactor. The solid catalyst is packed inside each tube, and reaction feed flows downward through the catalyst bed. The space between the tubes is filled with a heat transfer medium to maintain reaction temperature. For exothermal reactions, a common practice is to use the cold feed as a heat exchange medium to absorb reaction heat since it also will heat up the feed to reaction temperature so that the feed can be fed to the catalyst bed. This type of reactor is called an autothermal reactor. Some fixed reactors are operated under adiabatic conditions, i.e., no heat exchange between the reactor and its environment. In addition to heterogeneous catalytic reactions, fixed bed reactors are also used for gas−solid and liquid−solid noncatalytic reactions.

5. Fluidized bed reactors. This is also a reactor used for reactions involving solid particles in the system. However, unlike in the fixed bed reactors where solid particles are packed into a bed, in a fluidized bed reactor the solid particles are floating and moving in all directions inside the reactor. There are two major types of fluidized bed reactors. In one type, the solid particles will be carried away by the fluid, and after being separated from the fluid they will be recycled back to the reactor. This type of reactor is also called a circulating fluidized bed reactor. Another major type is the bubbling bed fluidized bed reactor. In this type of reactor, the solid particles move around but stay in the reactor. They are then dispersed in the fluid to form a fluidized bed which behaves like a bubbling liquid. The fluidized bed shows properties similar to that of a liquid, so it is also called a pseudoliquid bed. Fig. 1.1(H) shows a schematic of a bubbling bed reactor. At the bottom of the reactor there is a distributing plate to hold the solid particles. Fluid will enter the reactor from the bottom and flow upward through the distributing plate and penetrate through the particle bed. When fluid velocity reaches a certain value the particle bed will become loose, and further increases of fluid velocity will fluidize the bed. Typically the reactor will have various internals such as baffles, heat exchangers, and fluid−solid separation equipment to ensure uniform fluidization, maintain reaction temperature, separate the solid from the fluid, etc. Fluidized bed reactors can be used for gas−solid, liquid−solid and gas−liquid−solid reactions, either catalytic or noncatalytic, and are

one of the most widely used reactor types. One typical example is Fluidized Catalytic Cracking (FCC), a process which uses a circulating fluidized bed reactor. The examples of using bubbling fluidized bed reactors include oxidation of naphthalene, ammonia-oxidization of propylene, and oxidative dehydrogenation of butane. Fluidized bed reactors are also widely used for processing solid feedstock, such as the roasting of yellow iron ore and sphalerite, as well as calcinations of limestone.

6. Moving bed reactors. Similar to fixed bed reactors, this is also a reactor involving solid particles. However, the solid is continuously fed into the reactor from the top. The solid will move downward and eventually be removed from the bottom. If the solid consists of catalyst particles, then they will be lifted in a riser to the top and recycled back to the reactor. Inside the reactor, fluid flows upward to form a countercurrent flow pattern with the solids. This type of reactor is most suitable for catalytic reactions where the catalyst needs to be regenerated continuously, or a solid processing process. Fig. 1.1(I) is a schematic of a moving bed reactor.

7. Trickle bed reactors, as illustrated in Fig. 1.1(J). Strictly speaking it is also a fixed bed reactor. A catalyst bed is used for gas−liquid reactions. A typical application is hydrodesulfurization in refining industry. Typically gas and liquid will cocurrently flow downward. There are also applications where gas and liquid flow countercurrently.

Obviously, there are many different types of reactors and it is not possible to summarize all of them in few short paragraphs. For example, we did not mention rotating reactors used for gas−solid or solid-only reactions. It has its own characteristics, such as relying on reactor rotation to continuously move solid materials from one side of the reaction to the other side.

1.4 OPERATION MODES OF CHEMICAL REACTORS

There are three operation modes for industrial reactors: (1) batch, (2) continuous, and (3) semibatch (or semicontinuous).

1. Batch operation. Reactors operated under batch mode are also called batch reactors. Its main character is that all the feeds are loaded into the reactor at the beginning, and after a certain period of reaction time when the desired conversion is achieved, the whole content in the reactor, which includes mainly reaction products and small quantity of unconverted feeds, will be unloaded. Then the reactor needs to be cleaned and prepared for the next loading, reaction, and unloading cycle.

   A batch reaction is an unsteady process, and the composition inside the reactor changes with time. This is the most fundamental character of a batch reactor. Fig. 1.2 illustrates concentrations inside the reactor as a function of time. With an increase of time, the concentration of reactant

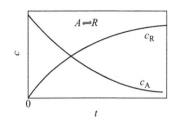

FIGURE 1.2   Illustration of concentrations as a function of time inside a batch reactor.

A will decrease from an initial concentration of C_{A0} to zero (for irreversible reactions) or equilibrium concentration (for reversible reactions). For a single reaction, product concentration will increase with time. If there are multiple reactions taking place, product concentration may not monolithically increase with time. For the series reaction system $A \rightarrow R \rightarrow Q$, the concentration of R will initially increase with time, reach a maximum, and then decrease with time. So for this type of reaction, a longer reaction time may not be beneficial for achieving a high yield of R.

During the reaction process in a batch reactor, there is no input or output of reaction materials, i.e., there is no material flow. In addition, the reaction takes place under constant volume conditions. If reacting materials are gases, they will fill the whole space inside the reactor; therefore it is obvious that the reacting volume is constant, i.e., equal to the reactor volume. If reacting materials are liquid, they will not fill the whole reactor volume. However, since the impact of pressure changes on liquid volume is very small, the reacting volume can also be treated as a constant without introducing much error.

Almost all the batch reactors are tank reactors. Batch reactors are suitable for reactions with slow reaction rates and processes requiring low production rate. It is most suitable for industries that will produce small quantities but a high variety of products. A typical example is the pharmaceutical industry.

2. Continuous operation. For this mode the feedstock is continuously fed into the reactor and, simultaneously, the product is also continuously removed from the reactor. A reactor operated under continuous mode is called a continuous reactor or a flow reactor. All the reactors reviewed in the previous section can be operated under continuous mode, and for some reactors used in industrial applications continuous operation is the only mode under which they can be operated.

Most reactors that operate under a continuous mode are under steady state. At any specific location in the reaction system parameters such as concentration and temperature do not change with time. However, those parameters are functions of locations. Fig. 1.3 illustrates concentrations inside a flow reactor as functions of axial location. The concentration of reactant A decreases along flow directions, changing from inlet concentration of C_{A0} to outlet concentration of C_{AL}. On the other hand, the

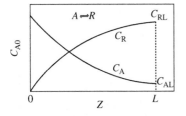

concentration of product R will increase from inlet concentration (usually equal to zero) to outlet concentration of C_{RL}. For a reversible reaction, the limit for both C_{AL} and C_{RL} will be their corresponding equilibrium values, and obviously it would take an infinite length to achieve the equilibrium limits. For irreversible reactions, reactant A, if not in excessive amount, can be completely consumed, but for some reactions it will need a reactor with infinite length. The shape of the concentration curves in Fig. 1.3 and in Fig. 1.2 are similar. The difference is that one is a function of time and the other is a function of location, which is the key difference between these two operation modes.

Most large-scale applications use continuous reactors since continuous operation, compared with batch operation, has the advantages of steady product quality, high productivity, and can easily achieve automated operation. However, when a continuous operation system is built, it is very difficult if not impossible to produce other products, and in some cases even significantly adjusting productivity is also not an easy task. In contrast, batch operation is much more flexible.

3. Semibatch operation. If a reactor is operated with at least one reactant continuously fed into the reactor or one product continuously leaving the reactor while other components are added to or removed from the reactor in batch, then this reactor is under semibatch operation and called a semibatch reactor. One example of semibatch operation is the production of chloride benzene by reaction between benzene and chlorine. In this process, liquid benzene is loaded into the reactor at the beginning. Chlorine gas is continuously fed into the reactor and unreacted chlorine continuously leaves the reactor. When the composition inside the reactor reaches the desired value, chlorine gas flow will stop, and product will be unloaded from the reactor.

Semibatch operations possess some characteristics of both batch and continuous operations. Some components continuously flow through the reactor, just like those in a continuous operation. Other components are loaded and unloaded in batches so the whole production process is still in batch mode. Therefore, composition inside a semicontinuous reactor will be functions of both reaction time and location. Tubular reactors, tank reactors, column reactors, and fixed bed reactors can all be operated under semicontinuous mode.

1.5 MODELS IN REACTOR DESIGN

The tasks for reactor design include: (1) select an adequate reactor type; (2) determine optimal operation conditions; and (3) for a given productivity requirement calculate reaction volume based on the reactor type and operation conditions selected. The reaction volume is the space that allows chemical reactions to take place and reactor size is actually determined based on this value. These three tasks are the main contents of chemical reaction engineering. They are not independent of each other, but instead are related. It is necessary to compare many options to eventually make a final decision. For example, a reactor with a simple structure may require a large volume to achieve the desired results. Therefore, a decision has to be made based on careful evaluations between simple or complex structures and large or small volumes. In most situations, the decision criterion is process economics, and the reactor design and optimization has to also meet the requirement of optimization of the whole production process.

At the core of reactor design is the reaction volume, i.e., calculating reaction volume for given reactor types and operation conditions. Reaction volume depends on the consumption rate of reacting components, which in turn depends on the composition, pressure, and temperature of the reaction system. For most reactors, the composition, temperature, and pressure of the reaction system inside the reactor changes with location, time, or both. Therefore, the reaction rate inside the reactor is not a constant. In order to calculate the reaction volume needed, a set mathematical equations that describe the relationships between those variables is needed. These will be the base design equations for reactor design.

There are three types of equations for reactor design: (1) mass balance equations that describe concentration changes, which are also called as continuity equations; (2) energy balance equations that describe temperature changes, also called energy equations; and (3) momentum balance equations that describe pressure changes. And the fundamental principles for establishing these three types of equations are conservation laws for mass, energy, and momentum.

Before deriving these equations, we need first to define the variables and control volume. The variables can be divided into dependent variables, which are also called state variables, and independent variables. For reactor design and analysis, it is common to select concentrations of key components, c_i, as dependent variables to establish mass balance equations. In some situations, it might be more convenient to select conversions X_i, or yields Y_i, or even partial pressures p_i as dependent variables. For energy and momentum balance equations, temperature and pressure are normally selected as dependent variables, respectively.

There are two types of independent variables, i.e., temporal and spatial variables. For steady state processes, the dependent variables do not change with time, therefore there is no need to consider temporal variables when deriving the conservative equations. For unsteady state process, both temporal and spatial variables have to be considered. The number of spatial variables depends on the dimensions of the reaction space. In this book we will only discuss one dimension system, i.e., we will only consider one spatial variable, because this will make mathematical derivation much simpler. In most cases reactor axial distance is selected as the independent variable.

The control volume is the space where the conservation equations will be established, and usually the maximum volume inside which the reaction rates can be considered as constants is selected as the control volume. For example, if both concentration and temperature inside a tank reactor are uniform, then the whole reaction volume can be selected as the control volume, since reaction rate at any point inside the reaction volume is the same. If this is not the case, i.e., reaction rate changes with location, then we have to choose a microvolume as the control volume.

With variables and control volume being selected, conservation equations can be established. We will start with material balance equations, which are mass conservation equations for key components over the control volume and can be generally expressed as:

$$(\text{Input rate for key component } i) = (\text{Output rate for key component } i) +$$
$$(\text{Consumption rate for key component } i) + (\text{Accumulation rate for key component } i)$$

$$(1.13)$$

In the above equation, key component i is a reactant. If the key component is a product, then the second item on the right hand side of Eq. (1.13) will be the key component formation rate and should be moved to the left hand side of the equation. If there is only one reaction taking place in the reactor, then the conversion or yield of the key component can describe the reaction progress, and concentrations of other nonkey components can be calculated. If there are multiple reactions taking place, we need to first determine how many key components are needed to describe the reaction system, and then select key components to establish material balance equations similar to Eq. (1.13). Therefore, for multiple reaction systems, multiple material balance equations are needed. For a multiphase reaction system, material balance equations have to be established for each phase, i.e., even more material balance equations are needed to describe the reaction system. Generally speaking, the number of material balance equations will be doubled for each additional phase.

Next we need to set up energy balance equations. For a majority of reactors, the change of potential energy, kinetic energy, and work can be ignored, and the energy balance equation is actually a heat balance equation, which in general can be expressed as

$$(\text{Heat input per unit time}) = (\text{Heat output per unit time}) +$$
$$(\text{Reaction heat per unit time}) + (\text{Heat accumulated per unit time})$$

$$(1.14)$$

In contrast to material balance equations which have to be established for each individual key components, the heat balance equation is established for the whole reaction mixture. Therefore, it does not matter how many reactions are taking place in the reactor, only one heat balance equation is needed. Of course, if there are multiple phases in the reactor and the temperatures in each phase are different, the heat balance for each phase needs to be established and the heat transfer between phases has to be considered. The second term in Eq. (1.14) is a sum of heat release from all the reactions. The reaction heat is positive for endothermic reactions and negative for exothermic reactions.

Similar to material and heat balance equations, momentum balance equation can be expressed as:

$$(\text{Momentum input}) = (\text{Momentum output}) + (\text{Momentum consumption})$$
$$+ (\text{Momentum accumulation})$$
$$(1.15)$$

This equation is based on Newton's Second Law for flowing fluid. For the momentum balance of a flow reactor, only pressure and friction force need to be considered. If the pressure drop is low then the reactor can be treated as being operated at constant pressure, then momentum balance equation is not needed. Most ambient pressure reactors can be treated as constant pressure systems.

From the above discussion, all three balance equations follow the same format:

$$(\text{Input}) = (\text{Output}) + (\text{Consumption}) + (\text{Accumulation}) \qquad (1.16)$$

All the balance equations contain four terms, but can be simplified based on specific situations. Some terms may not be needed for a specific reactor and in some situations one term may include many subterms.

These three types of balance equations are coupled, and have to be solved simultaneously. In addition, adequate initial and boundary conditions have to be established.

For a specific reactor, in order to establish balance equations that adequately describe the reactor behavior, we need not only to understand reactor structure and characters, but also solid knowledge of reaction kinetics, thermodynamics, transport phenomena, etc.

We have discussed general principles and foundations for establishing conservation equations. The reaction process is typically very complex. It is extremely difficult if not impossible to write a set of equations that can fully and accurately describe all the details of a reaction system. In fact, it is not necessary to do that. What we need is a set of equations that can capture key characters of the reaction system. Therefore, we need first to analyze the reaction system so that we can establish a physical model that captures

key characters while ignoring minor details. Based on this simplified physical model we can write conservation equations that serve as a mathematical
model for the reactor. In the following chapters we will introduce
many models for different types of reactors.

1.6 SCALE-UP OF INDUSTRIAL REACTORS

The commercialization of a new chemical product, from lab research to
industrial scale production, requires multiple steps of technology development, i.e., the technology has to be tested and demonstrated at different
scales. Obviously, with increase of production scale, the reactor size will
also increase accordingly. The scale-up of an industrial reactor is very important and also a very difficult task for chemical engineers. Chemical processing is different from physical processing and the increase of scale will lead
not only to quantitative changes but also to qualitative changes.

Similarity method is a scale-up method based on similarity theory and
dimension analysis has been successfully used in many industries such as
shipbuilding, airplane manufacture, and dam construction. However, such a
method is not adequate for reactor scale-up. This is because it is impossible
to achieve similarity simultaneously in diffusion, fluid dynamics, heat transfer, and chemical reactions. For example, in order to maintain similarity for
chemical reactions, the ratio of reactor length to linear velocity of the reaction mixture must be kept constant. On the other hand, similarity in fluid
dynamics and diffusion requires the production of reactor length and linear
velocity of reaction mixture to be a constant. It is obvious that it is impossible to meet these two conditions simultaneously. There are other examples
that illustrate the incompatibility of similarity criteria. Therefore, reactor
scale-up has long been reliant on multistep empirical process.

The multistep empirical process starts with a small-scale reactor for process development study to find optimal operation conditions and
suitable reactor types to meet both technical and economic objectives. Based
on data from small-scale tests, a larger scale equipment will be designed and
constructed to conduct the so-called model test. Based on the model test, a
pilot plant will be designed and tested and eventually an industrial reactor
will be designed based on the pilot test results. In many situations, to ensure
success, multiple pilot plant tests are conducted so that for each step the
scale-up factor is not too big. Multistep scale-up processes are very time-
consuming and expensive, and obviously not an ideal approach. The foundation of such an approach is the test data obtained at different scales, not the
fundamental physics and chemistry involved in the reaction process. So it is
an empirical approach and usually cannot achieve a largescale-up factor.

Since the 1960s, a method based on a mathematical model of the reactor
has been gradually developed, and is considered as a more ideal approach. It
uses mathematical model equations to describe and design the reactor,

predict reactor performance, and optimize operation conditions at different scales. Whether the mathematical model can adequately describe the reactor behaviors depends on the understanding of the reaction process inside the reactor, and such understanding has to rely on experimental study. Therefore, the foundation of a modeling approach for reactor scale-up is still an experimental study. However, the experimental method and objective for model approaches is very different from that of empirical approach.

A modeling approach for reactor scale-up typically includes the following steps: (1) Lab-scale experiments. This includes; new product synthesis, new catalyst development, reaction kinetic study, etc. At this stage the focus is the fundamentals of process chemistry. (2) Small-scale tests. It is still a lab-scale study but larger than previous stages, and the structure of the reactor should be similar to what will be used at industrial scale. For example, if a multitube fixed bed reactor will be used in the commercial unit, a single-tube fixed bed reactor should be used in the lab. The main objective of this stage is to examine the impact of physical process on the chemical reaction and the impact of industrial feedstock. (3) Large cold model test. The main objective of this stage is to study transport phenomena inside the reactor. As pointed out before, chemical reaction processes are always influenced by various transport processes, and their effects are highly dependent on equipment scale. (4) Pilot plant test. At this stage, not only does the scale become larger, but more importantly, the whole process and equipment used will be very similar to those for commercial units. One objective for this stage is to validate and modify the mathematical model and provide the data needed for commercial unit design. Another main objective is to examine catalyst life and activity changes over a long period of time. In addition, corrosion of the material used for equipment construction has also to be examined at this stage. (5) Computer simulation. This will be used in all of the four steps listed above. Computer simulation is used to analyze test results, optimize operation conditions, validate and improve reactor models, predict reactor performance, develop a mathematical model that can predict the operation and performance of a large-scale industrial reactor, and eventual design the commercial reactor.

At the core of the model approach is to develop mathematical model. In fact, the ultimate objective of all the experimental efforts in each scale-up step is to develop a mathematical model that can be used for industrial reactor design. Fig. 1.4 illustrates the procedure for reactor model development. The information obtained from lab-scale experiments will be used to develop chemical and physical models for the reaction process. The principles outlined in previous sections will then be used to set up a mathematical model for the reactor. This reactor model will be validated, modified, and improved based on small-scale tests to get a new model. This new model will be further improved based on pilot plant data and eventually the mathematical model for a commercial reactor design will be established.

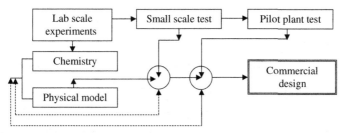

FIGURE 1.4 Procedure for reactor model development.

A model approach is a reactor scale-up method based on extensive experimental data. Without solid test data we cannot understand the key characters of the reaction system. On the other hand, solid knowledge and skills of chemical reaction engineering are also critical since these will guide the experimental study to gain the most useful data. In addition, computer simulation is a key enabler since it will allow extensive analysis and optimization. Therefore, the modeling approach is a combination of experiments, theoretical analysis, and computer stimulations.

The experimental steps outlined above are a general description of a typical procedure for reactor scale-up. For some reaction systems it may not necessary to go through all the steps. There are successful stories where a commercial reactor was successfully designed based on small-scale test results, i.e., without a pilot test, and a very large-scale-up factor has been achieved. For example, dimerization of propylene to isoprene was successfully designed with a scale-up factor of 17,000. However, those successes are exceptions rather than the norm. Most chemical reaction processes are very complex. Even with modern test facilities, reaction engineering knowledge, and computer capability it is still very challenging to scale-up a reaction process and design a commercial scale reactor.

Finally, it should be noted that the ultimate standard for industrial reactor design is its economic and social benefit. The investment−benefit analysis can be conducted by establishing an economic balance equation, the same format as Eq. (1.16), to calculate investment, returns, feed cost, operation cost, product price profit, etc. Although economic return is critical, it can only be pursued without compromising social good. For chemical product manufacture, the most important aspect related to social consequences is the environmental impact, such as pollution and noise during the production process. The design process must adopt adequate strategies to control environmental impact, at least to ensure all emissions will meet government regulations. In addition, operation safety is the most important factor that needs to be properly addressed. The designer must choose an operation condition that is safe. Adequate safeguards such as fire and explosion prevention must be included in the design. In summary, there are many factors that need to be considered during an actual reactor design. In this book we will focus on technical aspects.

FURTHER READING

Weterterp KR, Van Swaaij WPM, Beenackers AACM. Chemical reactor design and operation. New York: John Wiley & Sons; 1984.

Chen M, Yuan W. Development of industrial reaction process. Beijing: Chemical Industry Press; 1985.

Franks RGE. Modeling and simulation in chemical engineering. New York: John Wiley & Sons; 1972.

Levenspiel O. Chem Eng Sci 1980;35(6):4.

Dankohler G. Int Chem Eng 1988;28(1):432.

PROBLEMS

1.1 Methanol can be oxidized to formaldehyde on a silver catalyst:

$$2CH_3OH + O_2 \rightarrow 2HCHO + 2H_2O$$
$$2CH_3OH + 3O_2 \rightarrow 2CO_2 + 4H_2O$$

At reactor inlet the molar ratio of methanol:air:steam in the feed is 2:4:1.3. Methanol conversion is 72%, and formaldehyde yield is 69.2%. Please calculate:

1. Reaction selectivity

2. Gas composition at reactor outlet

1.2 Methanol is commercially manufactured by reaction between carbon monoxide and hydrogen on a copper—zinc—aluminum catalyst. The reaction system includes the following reactions:

$$CO + 2H_2 \leftrightharpoons CH_3OH$$
$$2CO + 4H_2 \leftrightharpoons (CH_3)_2O + H_2O$$
$$CO + 3H_2 \leftrightharpoons CH_4 + H_2O$$
$$4CO + 8H_2 \leftrightharpoons C_4H_9OH + 3H_2O$$
$$CO + H_2O \leftrightharpoons CO_2 + H_2$$

Due to equilibrium limitation carbon monoxide cannot be fully converted into methanol. In order to increase feed utilization unreacted feed components are recycled, i.e., after reaction the gas stream is cooled down to separate condensable components (crude methanol). Majority of uncondensable hydrogen and carbon monoxide are compressed and mixed with fresh feed stream to feed into the synthesis column while a small slip stream is vented. The overall flow diagram is given below:

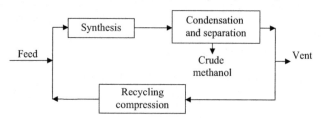

The molar fraction of each component in fresh feed and recycle stream (after condensation) is given below:

Component	Fresh feed (%)	After condensation (%)
CO	26.82	15.49
H_2	68.25	69.78
CO_2	1.46	0.82
CH_4	0.55	3.62
N_2	2.92	10.29

Composition of crude methanol is:

CH_3OH = 89.15 wt.%

$(CH_2)_2O$ = 3.55 wt.%

C_4H_9OH = 1.1 wt.%

H_2O = 6.2 wt.%

At operation temperature and pressure all other components can be considered as uncondensable but can partially dissolved in crude methanol. The amount of dissolved per kg crude methanol is given below:

CO_2: 9.82 g

H_2: 1.76 g

CH_4: 2.14 g

N_2: 5.38 g.

If the molar ratio of recycle gas to fresh feed is 7.2, please calculate:

1. Carbon monoxide single pass conversion and overall conversion

2. Single pass and overall yield of methanol

Chapter 2

Fundamentals of Reaction Kinetics

Chapter Outline

The main objective of chemical reaction engineering research is the design and operation of an industrial reactor to conduct chemical reactions more effectively at an industrial scale. Such efforts require knowledge from multiple disciplines and reaction kinetics is one of the most fundamental knowledge needed. In this chapter, based on the knowledge of Physical Chemistry, we will review some fundamental concepts of chemical reaction kinetics and discuss methods and procedures for kinetic analysis. Both homogeneous and heterogeneous catalytic kinetics will be discussed.

Reaction Engineering. DOI: http://dx.doi.org/10.1016/B978-0-12-410416-7.00002-1
25

2.1 REACTION RATE

Any chemical reaction will proceed at a certain rate, which is typically defined as the amount of reaction of a specific component in the reaction system per unit time and volume. For reaction:

$$\nu_A A + \nu_B B \rightarrow \nu_R R \tag{2.1}$$

the reaction rate defined based on reaction component A, B, and R can be expressed as:

$$r_A = -\frac{1}{V}\frac{dn_A}{dt}, \; r_B = -\frac{1}{V}\frac{dn_B}{dt}, \; r_R = \frac{1}{V}\frac{dn_R}{dt} \tag{2.2}$$

Since A and B are reactants, their quantities will decrease with time, and therefore their derivative with respect to time will be negative, i.e., $dn_A/dt < 0$, $dn_B/dt < 0$. R is a product, and therefore $dn_R/dt > 0$. In order to maintain reaction rate as a positive value, a minus sign is added in front of the derivative term in the definition of reaction rate based on a reactant. Obviously, the values of reaction rates defined based on different reacting components are different, i.e., $r_A \neq r_B \neq r_R$, unless the stoichiometric coefficients for each component is the same. When discussing reaction rate, we must specify which component we are talking about. Although the values of reaction rates for different components are different, all of them describe the same reaction.

From chemical stoichiometry we know that the ratio of amount of reactants converted to that of product formed must follow stoichiometry, i.e.:

$$dn_A : dn_B : dn_R = \nu_A : \nu_B : \nu_R$$

Therefore;

$$(-r_A):(-r_B):r_R = \nu_A : \nu_B : \nu_R$$

Or:

$$\frac{-r_A}{\nu_A} = \frac{-r_B}{\nu_B} = \frac{r_R}{\nu_R} = \bar{r} = \text{Constant} \tag{2.3}$$

Eq. (2.3) indicates that no matter which component is chosen to define the reaction rate, the value to its stoichiometric coefficient ratio is a constant. Therefore the reaction rate can also be defined as:

$$\bar{r} = \frac{1}{\nu_i V}\frac{dn_i}{dt} \tag{2.4a}$$

Based on the definition of extent of reaction discussed in Chapter 1, Introduction, Eq. (2.4a) can be expressed as:

$$\bar{r} = \frac{1}{V}\frac{d\xi}{dt} \tag{2.4b}$$

Eq. (2.4a) or (2.4b) is a generalized reaction rate definition, and its value does not depend on which component is used. As long as $\bar{r}$ is known, reaction rate for any component can be easily calculated by multiplying $\bar{r}$ with its corresponding stoichiometric coefficient. Such treatment is most convenient for reaction kinetic analysis for a multiple reaction system.

Since:

$$n_A = Vc_A$$

Substitute it into Eq. (2.2)

$$r = -\frac{1}{V}\frac{d(c_A V)}{dt} = -\frac{dc_A}{dt} - \frac{c_A}{V}\frac{dV}{dt} \tag{2.5}$$

For constant volume process, V is a constant, then Eq. (2.5) is reduced to conventional reaction rate definition widely used in chemical reaction kinetic studies:

$$r_A = -\frac{dc_A}{dt} \tag{2.6}$$

In Eq. (2.6) reaction rate is expressed as concentration changes as a function of time. For a volume changing process, the second term in Eq. (2.5) is not zero, hence Eq. (2.6) is not valid. Under this situation the concentration change is caused by both the reaction itself and the volume change of the reaction mixture.

Fig. 2.1 illustrates a flow reactor. Reactant A is fed into the reactor continuously at molar flow rate of F_{A0}. At steady state, all the parameters at any specific point inside the reactor will not change with time. In this situation, neither Eq. (2.2) nor (2.5) can be used to describe reaction rate, since both of them explicitly use time as a variable to define the reaction rate. Instead, we need to use a reaction rate definition that does not explicitly involve time as a variable. Let's pick any point M inside the reactor, and the volume of this "point" is dV_r, where V_r is reaction volume, representing the space for reaction to take place. Since dV_r is a microvolume element, all the parameters inside this volume can be considered uniform and treated as constants. If the molar flow rate of reactant A entering this volume is F_A, and after passing through this volume becomes $F_A - dF_A$ due to the chemical reaction taking place, then the reaction rate, defined

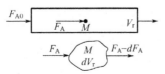

FIGURE 2.1   Reaction rate for a flow system.

as the amount of reactant A converted per unit volume in unit time, can be expressed as:

$$r_A = -\frac{dF_A}{dV_r} \tag{2.7}$$

Eq. (2.7) has the same meaning as Eq. (2.2), and the only difference is that it does not use time explicitly to describe the reaction progress.

For multiphase reactions, instead of reaction volume V_r, interphase area, a, can also be used to define reaction rate:

$$r'_A = -\frac{dF_A}{da} \tag{2.8}$$

For reaction systems involving solids, especially solid catalysts, it is usually more convenient to define reaction rate based on solid weight:

$$r''_A = -\frac{dF_A}{dW} \tag{2.9}$$

For homogeneous systems the reaction rate is normally defined based on reaction volume. For heterogeneous reactions any of the three types of reaction rate expressions introduced above can be used. Therefore, when discussing reaction rate for a heterogeneous reaction system we need to be careful to distinguish which definition is used. When using reaction volume to define the reaction rate of a heterogeneous system, we must be especially careful to specify whether the reaction volume is the volume of a specific phase or the total volume of all phases. The reaction rates defined based on the above three definitions can be easily converted to each other. For example, for a reaction using a solid catalyst, if V_r is the bulk volume of the catalyst and ρ_b is its bulk density, and a_V is the external specific surface area, then:

$$r'_A = a_V r'_A = \rho_b r''_A \tag{2.10}$$

Example 2.1
Dimerization of butadiene is carried out at an isothermal condition at 350°C and constant volume. The total pressure of the reaction system was measured as a function of time:

t/min:	0	6	12	26	38	60
p/kPa:	66.7	62.3	58.9	53.5	50.4	46.7

Please calculate reaction rate at 26 min.

Solution
If A and R represent butadiene and its dimer, respectively, then the dimerization reaction can be expressed as:

$$2A \rightarrow R$$

(Continued)

(Continued)

Since the reaction takes place under isothermal and constant volume conditions and the total numbers of moles in the reaction system changes, the total pressure changes with time reflect the reaction progress. When $t = 0$ the butadiene concentration is c_{A0}. At time t butadiene concentration is c_A, and based on stoichiometry the corresponding dimer concentration is $(c_{A0} - c_A)/2$. Then the total reacting component concentration is $(c_{A0} + c_A)/2$. Based on ideal gas law:

$$\frac{c_{A0}}{(C_{A0} + C_A)/2} = \frac{p_0}{p} \tag{A}$$

where p_0 is the total pressure at $t = 0$.

Eq. (A) can be expressed as:

$$c_A = c_{A0}\left(2\frac{p}{p_0} - 1\right) \tag{B}$$

Since the reaction proceeds under constant volume conditions, we can use Eq. (2.6) to describe reaction rate. Substitute Eq. (B) into Eq. (2.6):

$$r_A = -\frac{dc_A}{dt} = -\frac{2c_{A0}}{p_0}\frac{dp}{dt} \tag{C}$$

Again based on ideal gas law:

$$c_{A0} = \frac{p_0}{RT}$$

Then Eq. (C) can be rearranged as

$$r_A = -\frac{2}{RT}\frac{dp}{dt} \tag{D}$$

A plot of p versus t can be generated based on the data provided, as shown in Fig. 2A. A tangent line can be drawn at $t = 26$ min, and the slope is dp/dt, which is -1.11 kPa/min. Then from Eq. (D) the reaction rate expressed as the amount of butadiene converted is:

$$r_A = -\frac{2 \times (-1.11)}{8.314(350 + 273)} = 4.29 \times 10^{-4} \text{kmol}/(m^3 \cdot \text{min})$$

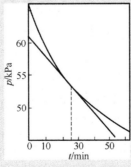

FIGURE 2A Total pressure as a function of reaction time.

(Continued)

(Continued)

If the reaction rate is expressed as dimer formed, then the reaction rate is 2.15×10^{-4} kmol/(m$^3 \cdot$ min).

In this example, a graphic method is used to estimate the derivative value. Graphic methods are good for illustrating the concept involved in the calculation, but are not very accurate. It is much easier and more accurate to derive a correlation and calculate the derivative using widely available computer programs.

2.2 REACTION RATE EQUATIONS

There are many factors that can affect the rate of a chemical reaction. The important ones include temperature, concentration, pressure, solvent, catalyst used, etc. Among these factors, temperature and concentration are the two parameters that will have an impact on all the chemical reactions. Not all the reactions use solvent or catalyst and similarly pressure can only influence certain types of reactions. Typically, a quantitative correlation between reaction rate and temperature and concentration is used to describe the reaction progress under constant pressure and for given solvent and catalyst (if used), which is called the rate equation or kinetic equation:

$$R = f(c, T) \tag{2.11}$$

where c is the concentration vector, i.e., there may be more than one concentration that will have an impact on the reaction rate. Obviously, different reactions will have different rate equations. At the very least, parameters will be different even if the formats are the same. Rate equations are extremely important for chemical reaction engineering because they are the foundation for reactor design and analysis.

If the chemical reaction is an elementary reaction, it is straightforward to write the rate equation based on mass action law. For example, for an elementary reaction (2.1) the rate equation is:

$$r_A = k c_A^{\nu_A} c_B^{\nu_B} \tag{2.12}$$

where k is the reaction rate constant which is a function of temperature. In fact most reactions are not elementary reactions. Therefore, it is not practical to establish reaction rate equations based on mass action law. However, a nonelementary reaction can be treated with a combination of several elementary reaction steps that constitute the reaction mechanism. Therefore, based on the reaction mechanism a rate equation can be derived.

Assume reaction $A \rightarrow P + D$ is based on the following elementary reaction steps:

$$A \leftrightharpoons A^* + P \tag{2.13}$$

$$A^* \rightarrow D \tag{2.14}$$

$$\overline{A \rightarrow P + D} \tag{2.15}$$

Both Eqs. (2.13) and (2.14) are elementary reactions, and A^* is a reaction intermediate complex. A commonly used approach to derive reaction rate equations is to assume one of the elementary steps is the rate-determining step while all the other steps are at equilibrium. The rate-determining step is the slowest reaction step and as a result its rate will determine the overall reaction rate. If Eq. (2.14) is the rate determining step, then based on mass action law, the rate equation for this step can be expressed as:

$$r_A = r_A^* = k_2 c_A^* \tag{2.16}$$

Reaction Eq. (2.13) is at equilibrium, therefore:

$$\frac{c_A^* c_P}{c_A} = K_1$$

or:

$$c_A^* = K_1 c_A / c_P$$

where K_1 is a equilibrium constant. Substitute the above equation into Eq. (2.16):

$$r_A = k_2 K_1 c_A / c_P = k c_A / c_P \tag{2.17}$$

Both k_2 and K_1 are functions of temperature and can be combined into a single constant, which is the reaction rate constant for this reaction. Eq. (2.17) is the rate equation of the reaction. From this example we can see that for a nonelementary reaction its rate equation cannot be directly derived based on mass action law. If we write the rate equation based on mass action law, we will get a first order reaction equation since this is a one-molecule reaction. However, when we derive the rate equation based on the reaction mechanism as shown in Eq. (2.17) we see that the reaction rate also depends on the concentration of product P.

On the other hand, are there any nonelementary reactions whose rate equation follows mass action law? The answer is yes. For example, for oxidation of nitrogen monoxide:

$$2NO + O_2 \rightarrow 2NO_2$$

Its rate equation is:

$$r = k c_{NO}^2 c_{O_2} \tag{2.18a}$$

So the rate equation is the same as those directly derived based on mass action law. But in reality this reaction is not an elementary reaction, and it is commonly believed that the reaction includes the following two elementary steps:

$$NO + NO \rightleftharpoons (NO)_2$$

$$(NO)_2 + O_2 \rightarrow 2NO_2$$

And the second step is the rate determining step, and hence:

$$r = k_2 c_{NO_2} c_{O_2}$$

And the first step is at equilibrium:

$$c_{NO_2} = K_1 c_{NO}^2$$

Substituted into the previous equation:

$$r = k_2 K_1 c_{NO}^2 c_{O_2} = k c_{NO}^2 c_{O_2} \qquad (2.18b)$$

It must be noted that if the rate equation has the same format as that derived based on mass action law it does not mean the reaction is an elementary reaction. On the other hand, if a reaction is an elementary reaction then its rate equation can definitely be derived based on mass action law.

However, if the experimental data is consistent with the rate equation derived based on the proposed reaction mechanism, does that mean the proposed mechanism is correct? The answer is no. What is proved is the proposed mechanism did not contradict experimental result and therefore is a possible mechanism. The reality is that different mechanisms can lead to the exact same rate equation. We can still use the nitrogen monoxide oxidation reaction as an example. If the reaction mechanism is:

$$NO + O_2 \rightleftharpoons NO_3$$

$$NO_3 + NO \rightarrow 2NO_2$$

And the second is the rate-determining step; we can derive a rate equation that is exactly the same as Eq. (2.18b), which was derived from a different mechanism. Therefore, being consistent with experimental data is only a necessary but not a sufficient condition for the proposed mechanism to be correct. In order to prove whether a reaction mechanism is correct or not, additional experimental data is required. For example, the above two mechanisms for nitrogen monoxide oxidation reaction assume $(NO)_2$ and NO_3 as intermediate complexes, respectively. In order to figure out which mechanism is correct, we need additional experimental studies to investigate which intermediate complex exists in the reaction system. Otherwise, either of the two mechanisms discussed above is only a possibility.

Unfortunately, so far for a majority of chemical reactions their mechanisms are unclear and we still have to rely on experimental data to validate reaction rate equation. For power law rate equations, the impact of temperature and concentrations on reaction rates are typically treated separately:

$$r = f_1(T)f_2(\mathbf{c})$$

At any given temperature, the temperature function $f_1(T)$ is a constant, expressed as reaction rate constant k, and the concentration function is expressed as an exponential function of each component's concentrations:

$$f_2(\mathbf{c}) = c_A^{\alpha_A} c_B^{\alpha_B} \dots$$

Then the reaction rate equation is:

$$r = kc_A^{\alpha_A} c_B^{\alpha_B} \dots = k \prod_{i=1}^{N} c_i^{\alpha_i} \tag{2.19}$$

A rate equation with this format is called the power law rate equation. α_A and α_B are the reaction orders for reactants A and B, respectively. If the reaction is an elementary reaction, then α_A and α_B will be equal to n_A and n_B, correspondingly. Kinetic parameters such as k, α_A, and α_B have to be measured through experimental studies. It should be noted that even for nonreversible reactions the product concentrations could also have an impact on the reaction rate. Hence Eq. (2.19) includes concentration terms for all components. The actual impact of each component on the reaction will have to be determined through experimental measurement.

Strictly speaking every chemical reaction is reversible, and the net reaction rate is the difference between the rate of forward reaction and that of backward reactions. The so-called irreversible reactions can be considered as a special situation where the backward reaction rate is negligible. Reaction rate for a reversible reaction can also be expressed in the power law equation:

$$r = \overrightarrow{k} \prod_{i=1}^{N} c_i^{\alpha_i} - \overleftarrow{k} \prod_{i=1}^{N} c_i^{\beta_i} \tag{2.20}$$

where $\overrightarrow{k}$ and $\overleftarrow{k}$ are the reaction rate constants for forward and backward reaction, respectively; α_i and β_i are reaction orders for forward and backward reactions, respectively. There are relationships between α_i and β_i, as well as between rate constants $\overrightarrow{k}$ and $\overleftarrow{k}$ and the reaction equilibrium constant K_c.

For a reversible reaction:

$$\nu_A A + \nu_B B \rightleftharpoons \nu_R R$$

its rate equation is:

$$r_A = \overrightarrow{k} c_A^{\alpha_A} c_B^{\alpha_B} c_R^{\alpha_R} - \overleftarrow{k} c_A^{\beta_A} c_B^{\beta_B} c_R^{\beta_R}$$

At equilibrium $r_A = 0$, therefore:

$$\overrightarrow{k} c_A^{\alpha_A} c_B^{\alpha_B} c_R^{\alpha_R} = \overleftarrow{k} c_A^{\beta_A} c_B^{\beta_B} c_R^{\beta_R}$$

or

$$\frac{c_R^{\beta_R - \alpha_R}}{c_A^{\alpha_A - \beta_A} c_B^{\alpha_B - \beta_B}} = \frac{\overrightarrow{k}}{\overleftarrow{k}} \tag{2.21}$$

Assuming A, B, and R all obey ideal gas law, and when the reaction reaches its equilibrium:

$$c_A^{\nu_A} c_B^{\nu_B} c_R^{\nu_R} = K_c \tag{2.22}$$

Both A and B are reactants, and their stoichiometric coefficients ν_A and ν_B are negative.

Assuming ν is a positive number, the above equation can be expressed as:

$$c_A^{\nu_A/\nu} c_B^{\nu_B/\nu} c_R^{\nu_R/\nu} = K_c^{1/\nu} \tag{2.23}$$

Eqs. (2.21–2.23) all describe the same fact, i.e., the reaction is at equilibrium, therefore by comparing these three equations we can obtain:

$$\beta_A - \alpha_A = \nu_A/\nu$$

$$\beta_B - \alpha_B = \nu_B/\nu$$

$$\beta_R - \alpha_R = \nu_R/\nu$$

or

$$\frac{\beta_A - \alpha_A}{\nu_A} = \frac{\beta_B - \alpha_B}{\nu_B} = \frac{\beta_R - \alpha_R}{\nu_R} = \frac{1}{\nu} \tag{2.24}$$

And:

$$\overrightarrow{k}/\overleftarrow{k} = K_c^{1/\nu} \tag{2.25}$$

Eq. (2.24) indicates that the ratio of the difference between forward and backward reaction orders to the stoichiometric coefficient is a constant. This conclusion can be used to check whether a reaction rate is correct. On the other hand, using this property can reduce the number of kinetic parameters that have to be measured. Eq. (2.25) gives the relationship between the reaction rate constant and equilibrium constant. Unless $\nu = 1$, equilibrium constant is generally not equal to the ratio of forward to backward rate constants.

Both Eqs. (2.24) and (2.25) involve a parameter of ν. What does this parameter mean physically? It represents how many times the rate-determining step has to occur. For example, if the mechanism for reaction $2A + B = R$ is the following:

1. $A \rightleftharpoons A^*$
2. $A^* + B \rightleftharpoons X$
3. $A^* + X \rightleftharpoons R$

Then A^* and X are the reaction intermediates. To produce 1 mol R, the first step has to take place two times, and the other two steps only need to happen once. If the first step is the rate determining step, then $v = 2$. This is called the stoichiometric number. Be careful to not confuse it with stoichiometric coefficients.

Example 2.2

The esterification reaction between acetic acid and butanol takes place under isothermal conditions:

$$CH_3COOH + C_4H_9OH \rightleftharpoons CH_3COOC_4H_9 + H_2O$$

The initial concentrations of acetic acid and butanol are 0.2332 kmol/m³ and 1.16 kmol/m³, respectively. The amount of acetic acid converted as a function of time is measured as shown in the following table:

Time (h)	Acetic acid converted (kmol/m³)	Time (h)	Acetic acid converted (kmol/m³)	Time (h)	Acetic acid converted (kmol/m³)
0	0	3	0.03662	6	0.06086
1	0.01636	4	0.04525	7	0.06833
2	0.02732	5	0.05405	8	0.07398

Please derive the rate equation for this reaction.

Solution

Since the acetic acid conversion is very low and there is excessive butanol available, the impact of the backward reaction can be ignored. In addition, we do not need to consider the influence of butanol concentration on the reaction rate. Therefore the reaction rate equation can be expressed as:

$$r_A = -\frac{dc_A}{dt} = kc_A^\alpha \tag{A}$$

Integrating Eq. (A):

$$(\alpha - 1)kt = \frac{1}{c_A^{\alpha-1}} - \frac{1}{c_{A0}^{\alpha-1}} \tag{B}$$

We can use the trial-and-error method to obtain α and k in Eq. (B), i.e., assuming a value of α and then examining whether the value of right-hand side

(Continued)

(Continued)

of Eq. (B) will be a linear function of time. For this specific example, if we assume $\alpha = 2$, then Eq. (B) can be simplified as:

$$kt = \frac{1}{c_A} - \frac{1}{c_{A0}} \tag{C}$$

Based on the experimental data of acetic acid converted over different times given in the table above, the acetic acid concentration at a given time can be calculated by subtracting the amount of acetic acid converted from initial concentrations. The value of $\frac{1}{c_A} - \frac{1}{c_{A0}}$ can be calculated, as shown in Table 2A. Fig. 2B shows a plot of $\frac{1}{c_A} - \frac{1}{c_{A0}}$ as a function of time which shows great linearity. Therefore, we know that the reaction is second order with respect to acetic acid.

TABLE 2A Concentration of acetic acid (c_A) at different time

t/h	c_A	$\dfrac{1}{c_A} - \dfrac{1}{c_{A0}}$
0	0.2332	0
1	0.2168	0.3244
2	0.2059	0.5686
3	0.1966	0.7983
4	0.1879	1.0337
5	0.1792	1.2922
6	0.1723	1.5157
7	0.1649	1.7761
8	0.1592	1.9932

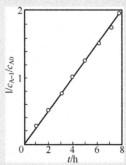

FIGURE 2B $\frac{1}{c_A} - \frac{1}{c_{A0}}$ dependence on t.

(Continued)

(Continued)

The data given in this example was obtained with excessive butanol, i.e., during the experiment the concentration of butanol is maintained at a high value. As a result, we cannot examine the influence of butanol on the reaction rate. Whether the butanol will have an impact on the reaction needs to be further studied under different test conditions.

2.3 EFFECT OF TEMPERATURE ON REACTION RATE

In the power law rate equation the impact of temperature on reaction rate is expressed by temperature function $f_1(T)$. At a given temperature $f_1(T)$ is a constant, and expressed by rate constant k. Therefore $f_1(T)$ is a correlation between k and temperature. The relationship between reaction rate constant k and temperature T is usually described using the Arrhenius equation:

$$k = A \exp(-E/RT) \tag{2.26}$$

where A is the preexponential factor, which has the same unit as k. E is reaction activation energy and R is the gas constant.

Reaction rate constant k is also called the specific reaction rate. It represents the reaction rate when the concentration of every reacting component equals 1. The unit of the reaction rate constant depends on the format of the reaction rate equation and the unit used to describe the composition of the reaction mixture. For example, if the reaction rate of a first order reaction is expressed as $kmol/(m^3 \cdot s)$, and the concentration is expressed as $kmol/m^3$, then the unit for k is $1/s$. But if the reaction rate is expressed as $kmol/kg \cdot s$, then the unit of k is $m^3/(kg \cdot s)$.

For gas phase reactions, the reaction composition can be expressed as partial pressure p_i, concentration c_i, or mol fraction y_i. If the corresponding rate constant is expressed as k_p, k_c, and k_y, respectively, then those rate constants follow the following relationship:

$$k_c = (RT)^\alpha k_p = (RT/p)^\alpha k_y \tag{2.27}$$

where p is the total pressure, α is the total reaction order. Of course the above equation is only valid for a gas system obeying the ideal gas law and power law reaction rate equation.

Taking logarithm on both sides of Eq. (2.26):

$$\ln k = \ln A - E/RT \tag{2.28}$$

Therefore when plotting $\ln(k)$ against $1/T$, a straight line will be obtained, and the slope of this line can be used to calculate reaction activation energy. It should be noted that the Arrhenius equation (i.e., Eq. (2.26)) is only valid in a given temperature range, and it normally cannot be used outside the valid temperature range. Although in most cases the Arrhenius equation can satisfactorily describe the relationship between reaction rate constant and temperature, there are situations where $\ln(k)$ is not a linear function of $1/T$. This may be caused by the fact that the rate equation used is not adequate or in the temperature range of the reaction kinetic being investigated or the reaction mechanism has changed. In some situations the preexponential factor A is a function of temperature, and as a result the correlation between $\ln(k)$ and $1/T$ is not linear.

Eq. (2.26) can also be written as:

$$\frac{d \ln k}{dT} = \frac{E}{RT^2} \tag{2.29}$$

If both the forward and backward reactions follow the Arrhenius equation, then:

$$\frac{d \ln \vec{k}}{dT} = \frac{\vec{E}}{RT^2} \quad \text{and} \quad \frac{d \ln \overleftarrow{k}}{dT} = \frac{\overleftarrow{E}}{RT^2} \tag{2.30}$$

where $\vec{E}$ and $\overleftarrow{E}$ is the activation energy for forward and backward reaction, respectively. Taking logarithm on both sides of Eq. (2.25):

$$\ln \vec{k} - \ln \overleftarrow{k} = \frac{1}{\nu} \ln K_p$$

Taking derivative with respect to temperature:

$$\frac{d \ln \vec{k}}{dT} - \frac{d \ln \overleftarrow{k}}{dT} = \frac{1}{\nu} \frac{d \ln K_p}{dT} \tag{2.31}$$

Based on thermodynamics for a constant pressure process:

$$\frac{d \ln K_p}{dT} = \frac{\Delta H_r}{RT^2} \tag{2.32}$$

Using Eqs. (2.32) and (2.30) in Eq. (2.31), we can obtain:

$$\vec{E} - \overleftarrow{E} = \frac{1}{\nu} \Delta H_r \tag{2.33}$$

This equation gives a relationship between activation energy of forward and backward reactions. For endothermic reactions $\Delta H_r > 0$, therefore $\vec{E} > \overleftarrow{E}$. For exothermic reactions, $\Delta H_r < 0$, and $\vec{E} < \overleftarrow{E}$.

Based on Eq. (2.26), the reaction rate always increases with temperature (except for some very rare systems), and the temperature dependence shows very strong nonlinearity, i.e., a small change of temperature will lead to a significant change of reaction rate. Therefore, temperature is the most important factor that has to be carefully considered during reactor design and analysis. Temperature control is also very critical for reactor operation.

For reversible reactions, the reaction rate is the difference between forward and backward reactions. When temperature increases, the reaction rate of both forward and backward reactions increase, but will the overall reaction rate also increase? To answer this question we can rearrange Eq. (2.20) as:

$$r = \overrightarrow{k} f(X_A) - \overleftarrow{k} g(X_A) \tag{2.34}$$

For a given initial reaction mixture composition, if conversion of component A is X_A, then concentrations of all other components can be expressed as functions of X_A. In Eq. (2.34) $f(X_A)$ is the concentration term, expressed as a function of X_A, for the forward reaction rate equation. Similarly, $g(X_A)$ is the concentration term, also expressed as a function of X_A, for the backward reaction rate. Taking the derivative of Eq. (2.34) with respect to T:

$$\left(\frac{\partial r}{\partial T}\right)_{X_A} = f(X_A)\frac{d\overrightarrow{k}}{dT} - g(X_A)\frac{d\overleftarrow{k}}{dT} \tag{2.35}$$

If reaction constants for both forward and backward reactions follow the Arrhenius equation, then:

$$\frac{d\overrightarrow{k}}{dT} = \frac{\overrightarrow{k}\overrightarrow{E}}{RT^2}$$

And

$$\frac{d\overleftarrow{k}}{dT} = \frac{\overleftarrow{k}\overleftarrow{E}}{RT^2}$$

where $\overrightarrow{E}$ and $\overleftarrow{E}$ are the activation energies for forward and backward reactions, respectively. Using the above two equations in Eq. (2.35):

$$\left(\frac{\partial r}{\partial T}\right)_{X_A} = \frac{\overrightarrow{E}}{RT^2}\overrightarrow{k}f(X_A) - \frac{\overleftarrow{E}}{RT^2}\overleftarrow{k}g(X_A) \tag{2.36}$$

Since $r \geq 0$, therefore $\overrightarrow{k}f(X_A) \geq \overleftarrow{k}g(X_A)$, and for endothermic reactions, $\overrightarrow{E} > \overleftarrow{E}$, therefore from Eq. (2.36):

$$\frac{\overrightarrow{E}}{RT^2}\overrightarrow{k}f(X_A) > \frac{\overleftarrow{E}}{RT^2}\overleftarrow{k}g(X_A)$$

i.e.:

$$\left(\frac{\partial r}{\partial T}\right)_{X_A} > 0$$

This tells us that for reversible endothermic reactions the overall reaction rate always increases with temperature. Fig. 2.2 illustrates the relationship between reaction rate, temperature, and conversion for a reversible endothermic reaction. The curves in the figure are called constant-rate curves, i.e., the reaction rate is the same at any point on the line. The curve at $r = 0$ is

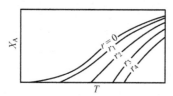

FIGURE 2.2 Relationship between reaction rate, temperature, and conversion for a reversible endothermic reaction.

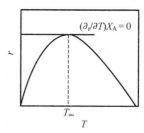

FIGURE 2.3 Relationship between reaction rate and temperature for a reversible exothermic reaction.

called the equilibrium curve and the corresponding conversion is called the equilibrium conversion, which represents the equilibrium limit for the reaction. For a reversible endothermic reaction the equilibrium constant increases with temperature, and hence the equilibrium conversion also increases with temperature. All other curves that are under the equilibrium curve are constant-rate lines with a reaction rate larger than zero:

$$r_4 > r_3 > r_2 > r_1$$

From this figure we can see that at a given temperature, reaction rate decreases with increase of conversion; at a given conversion, the reaction rate increases with temperature.

For reversible exothermic reactions, since $\vec{E} < \overleftarrow{E}$, but $\vec{k}f(X_A) < \overleftarrow{k}g(X_A)$, from Eq. (2.36):

$$\left(\frac{\partial r}{\partial T}\right)_{X_A} \gtrless 0$$

Therefore, for a reversible exothermic reaction, when temperature increases the reaction rate can either increase or decrease, as shown in Fig. 2.3. The curve in the figure is obtained at constant conversion, so it is called a constant-conversion curve. At lower temperature, the reaction rate increases with temperature and gradually reaches a maximum. After that, a further increase of temperature will lead to a decrease of the reaction rate. This is because at low temperature the reaction is far away from its equilibrium, the overall reaction rate is mainly determined by the kinetics, and $\left(\frac{\partial r}{\partial T}\right)_{X_A} > 0$. At higher temperature, the equilibrium constant decreases and has stronger constraints on the overall reaction rate. And eventually the reaction will become equilibrium limited, reflected by the fact of $\left(\frac{\partial r}{\partial T}\right)_{X_A} < 0$. At the maximum point of the reaction, the rate-temperature curve, $\left(\frac{\partial r}{\partial T}\right)_{X_A} = 0$, which corresponds to the highest reaction rate, and the corresponding temperature is known as the optimal reaction temperature.

The optimal temperature can be determined by using extremum method. Let the right-hand side of Eq. (2.36) equal zero:

$$\vec{E}\vec{k}f(X_A) - \overleftarrow{E}\overleftarrow{k}g(X_A) = 0$$

Or:

$$\frac{\vec{E}\vec{A}\exp\left(-\dfrac{\vec{E}}{RT_{op}}\right)}{\overleftarrow{E}\overleftarrow{A}\exp\left(-\dfrac{\overleftarrow{E}}{RT_{op}}\right)} = \frac{g(X_A)}{f(X_A)} \tag{2.37}$$

where $\vec{A}$ and $\overleftarrow{A}$ are the preexponential factors for forward and backward reactions, respectively. When the reaction reaches its equilibrium, $r = 0$, and from Eq. (2.34)

$$\frac{g(X_A)}{f(X_A)} = \frac{\vec{k}}{\overleftarrow{k}} = \frac{\vec{A}\,\exp(-\vec{E}/RT_e)}{\overleftarrow{A}\,\exp(-\overleftarrow{E}/RT_e)} \tag{2.38}$$

In the above equation T_e is the equilibrium temperature when conversion is X_A. Substitute Eq. (2.38) into Eq. (2.37):

$$\frac{\vec{E}\vec{A}\,\exp\left(-\dfrac{\vec{E}}{RT_{op}}\right)}{\overleftarrow{E}\overleftarrow{A}\,\exp\left(-\dfrac{\overleftarrow{E}}{RT_{op}}\right)} = \frac{\vec{A}\,\exp\left(-\dfrac{\vec{E}}{RT_e}\right)}{\overleftarrow{A}\,\exp\left(-\dfrac{\overleftarrow{E}}{RT_e}\right)}$$

Taking logarithm on both sides of the above equation and simplifying, we can obtain the optimal temperature as:

$$T_{op} = \frac{T_e}{1 + \dfrac{RT_e}{\overleftarrow{E} - \vec{E}}\ln\dfrac{\overleftarrow{E}}{\vec{E}}} \tag{2.39}$$

Eq. (2.39) does not explicitly show the relationship between conversion and the optimal temperature. However, the equilibrium temperature is a function of conversion; therefore the optimal temperature is an implicit function of conversion. For any given conversion, there must correspondingly be an equilibrium temperature and an optimal temperature.

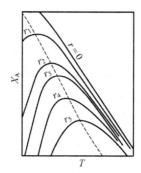

FIGURE 2.4   Relationship between reaction rate, temperature, and conversion for a reversible exothermic reaction.

For a reversible exothermic reaction, the relationship between reaction rate, conversion, and temperature is shown in Fig. 2.4, which is usually called a $T{-}X_A$ plot. The curve with $r = 0$ is the equilibrium curve, which represents the limit of the reaction. All other curves are constant-rate curves with the following order:

$$r_4 > r_3 > r_2 > r_1$$

From Fig. 2.4, every constant-rate curve shows a maximum, i.e., the highest conversion. The corresponding temperature is the optimal temperature. Connecting all the maximum points, we will obtain an optimal temperature curve, illustrated by the dashed line in Fig. 2.4. This curve is a geometric illustration of Eq. (2.39). For a reversible exothermic reaction, if the reaction process follows the optimal curve then the reaction always proceeds at maximum rate. Obviously, for an industrial reactor it is not realistic to fully achieve such optimal operation. But when designing an industrial reactor, we should pursue such optimal operation as often as possible.

In summary, for both reversible and irreversible reactions the reaction rates decrease with the increase of conversion. For irreversible and endothermic reversible reactions, reaction rate always increases with temperature. For reversible exothermic reactions, reaction rate is at a maximum when the reaction temperature follows the optimal temperature profile.

Example 2.3
Ammonia is commercially manufactured by synthesis reaction between nitrogen and hydrogen at high temperature and pressure over an iron catalyst:

$$N_2 + 3H_2 \rightleftharpoons 2NH_3 \tag{A}$$

(Continued)

(Continued)

The ammonia synthesis reaction is reversible and exothermic, therefore it is desirable to operate the reactor following the optimal temperature profile. Please calculate the optimal temperatures for the following conditions:

1. Reaction pressure = 25.33 MPa, feed mixture with hydrogen to nitrogen ratio = 3:1, and ammonia content is 17%;
2. The same conditions as in (1) but ammonia content is 12%;
3. Reaction pressure = 32.42 MPa, other conditions are the same as in (1)

The activation energies for the forward and backward reaction on this catalyst are 58.168×10^3 J/mol and 167.48×10^3 J/mol, respectively. The equilibrium constant is a function of temperature T (K) and total pressure (MPa):

$$\log(K_p) = \frac{2172.26 + 19.6478p}{T} - (4.2405 + 0.02149p) \tag{B}$$

Solution

1. First we need to calculate the mixture composition when ammonia is 17%, and then determine K_p, which in turn can be used to calculate equilibrium temperature. The optimal temperature then can be calculated using Eq. (2.39).

    For ammonia synthesis reaction the stoichiometric hydrogen to nitrogen ratio is 3:1, and the hydrogen to nitrogen ratio in the feed mixture is also 3:1, therefore when the ammonia content is 17% the mixture composition is:

    NH_3: 17%
    N_2: $\frac{1}{4} \times (1 - 0.17) \times 100 = 20.75\%$
    H_2: $\frac{3}{4} \times (1 - 0.17) \times 100 = 62.25\%$

    Based on Eq. (A) the equilibrium constant is:

    $$K_p = \frac{p_{NH_3}}{p_{H_2}^{1.5} p_{N_2}^{0.5}} = \frac{0.17 \times 25.33}{(0.6225 \times 25.33)^{1.5}(0.2075 \times 25.33)^{0.5}} = 3.000 \times 10^{-2} (MPa)^{-1}$$

    Substitute the value of K_p into Eq. (B):

    $$\log(3.00 \times 10^{-2}) = \frac{2172.26 + 19.6478 \times 25.33}{T_e} - (4.2405 + 0.02149 \times 25.33)$$

    So the equilibrium temperature $T_e = 818.5$K
    Then the optimal temperature can be calculated by using Eq. (2.39)

    $$T_{op} = \frac{818.5}{1 + \frac{8.3144 \times 818.5}{(167.48 - 58.618) \times 10^3} \ln \frac{167.48 \times 10^3}{58.618 \times 10^3}} = 768K$$

2. Due to the change in ammonia content, the gas mixture composition will change as will T_e and T_{op}. The calculation procedure is the same as above, and the final results are:

    $$T_e = 872.5 \text{ K}$$
    $$T_{op} = 815.3 \text{ K}$$

(Continued)

(Continued)

3. Although the gas composition is the same as that in (1), the increase of total pressure will lead to a decrease in K_p. The final results are:

$$T_e = 849.8 \text{ K}$$
$$T_{op} = 795.6 \text{ K}$$

From the above calculations we can see that with the increase of ammonia content (i.e., increase of conversion), the optimal temperature decreases, following the normal trend as illustrated in Fig. 2.4. If the pressure of the reaction system increases while other conditions are the same, the optimal temperature will increase.

2.4 MULTIPLE REACTIONS

When multiple reactions take place simultaneously in a reaction system, it is usually called a multiple reaction system. Most of the real reaction systems belong to this category. Since multiple reactions take place simultaneously, any component in the reaction system may participate either in one reaction, in multiple, or even in all the reactions. When one component participates in more than one reaction, it may be a reactant for one reaction but a product for another reaction. In this situation, with the progress of the reactions, the change in the amount of specific component is the overall result of all the reactions it participates in. The change in the amount of specific component in the reaction mixture during unit time and in unit volume is called the consumption rate (if component i is a reactant) or formation rate (if component i is a product) for this specific component, which is usually expressed as $\mathscr{R}_i$. Next we will first discuss how to calculate $\mathscr{R}_i$ and then review the different types of multiple reactions. And finally we will briefly introduce the concept of the reaction network.

2.4.1 Consumption Rate and Formation Rate

Based on the definition given above, the reaction rate for a given component is a net result of contributions from all the reactions in which this component participates. Therefore $\mathscr{R}_i$ is a sum of reaction rates of component i in each reaction:

$$\mathscr{R}_i = \sum_{j=1}^{M} \nu_{ij} \bar{r}_i \tag{2.40}$$

$\bar{r}_j$ is the reaction rate of reaction j defined by Eq. (2.4a), and ν_{ij} is the stoichiometric coefficient of component i in reaction j. So $\nu_{ij}\bar{r}_j$ is the rate for reaction j calculated based on component i. For a given reaction component i could be either a reactant or a product. If component i is a reactant, ν_{ij} is negative; if component i is a product, then v_{ij} is positive. $\mathscr{R}_i$ could be either positive or negative. If $\mathscr{R}_i$ is positive, the amount of this component will increase with progress of the reactions, and $\mathscr{R}_i$ is called the formation rate. On the other hand, if $\mathscr{R}_i$ is negative it is called the consumption rate since this component is consumed during the reaction process. The difference between conversion or formation rate $\mathscr{R}_i$ and reaction rate r_i is that the former reflects a results of multiple reactions and the later is for specific reaction. Obviously if there is only one reaction taking place then $\mathscr{R}_i$ and r_i will be the same, i.e., $r_i = |\mathscr{R}_i|$. The reason that an absolute value has to be taken for $\mathscr{R}_i$ is that $\mathscr{R}_i$ could be either positive or negative but r_i is always positive.

If we know the reaction rates of individual reactions we can calculate the conversion or formation rate by using Eq. (2.40). However, when investigating kinetics of a multiple reaction system, what is measured is the overall reaction results, i.e., the conversion or formation rates of different components. For kinetic studies it is critical to know the rates of individual reactions, and therefore we need to calculate $\bar{r}_j$ based on R_i. In order to do so we first have to identify how many reactions are taking place. In many situations it is necessary to focus on the most important reactions while ignoring the less important ones. If the number of reactions that have to be considered is M, then we have to measure conversion or formation rates of at least M components. Substituting $\mathscr{R}_i$ of M components into Eq. (2.40) we will have M linear algebra equations, which can be solved to obtain $\bar{r}_j$. It should be noted that those M reactions must be independent; otherwise the algebra equations obtained would be underspecified and won't have a unique solution. An independent reaction is defined as a reaction that cannot be obtained by a linear combination of other reactions in the reaction system. For example, for methane steam reforming the following reactions can take place:

$$CH_4 + H_2O \rightleftharpoons CO + 3H_2 \tag{2.i}$$

$$CH_4 + 2H_2O \rightleftharpoons CO_2 + 4H_2 \tag{2.ii}$$

$$CO + H_2O \rightleftharpoons CO_2 + H_2 \tag{2.iii}$$

However, only two of the above reactions are independent since one of them can be obtained by a linear combination of two other reactions. For example, subtracting Eq. (2.i) from Eq. (2.ii) will lead to Eq. (2.iii), or combining Eq. (2.i) and Eq. (2.ii) will lead to Eq. (2.iii).

2.4.2 Basic Types of Multiple Reactions

There are three basic types of multiple reactions, i.e., side-by-side reactions, parallel reactions, and consecutive or series reactions. Reversible reactions involve both forward and backward reactions and should be considered as a multiple reaction system. However, reversible reactions can be treated as a single reaction, and hence we will not discuss them here.

1. Side-by-side reactions
   If in a reaction system each reaction involves different reactants, then these reactions are called side-by-side reactions. For example:

$$A \rightarrow P$$

$$B \rightarrow Q$$

   These two reactions take place side-by-side. Each reaction can be treated independently to calculate reaction rates. If many of these reactions take place, the reaction rate for any specific reaction will not be affected by the concentrations of other reactions' reactants, except for some heterogeneous catalytic reactions. Therefore, the fact that each reaction proceeds independently is the kinetic character of side-by-side reactions. However, if reaction volume changes, the rate of one reaction may influence the rates of other reactions. For example, if gas phase reactions A → P and 2 B → Q take place simultaneously, with the progress of the second reaction the reaction volume will change, leading to concentration changes of A and P, which will in turn have an impact on the reaction rate of the first reaction.

2. Parallel reactions
   If several reactions use the same reactants but lead to different, or at least partially different products, those reactions are parallel reactions. The following reactions are parallel reactions:

$$A \rightarrow P, r_P = k_1 c_A^{\alpha}$$

$$\nu_A A \rightarrow Q, r_Q = k_2 c_A^{\beta}$$

   Reactant A can be converted to either P or Q, so this type of reaction is also called a competitive reaction. If our goal is to produce P, P is the objective product, and the first reaction is the main reaction while the second is the side reaction. Maximizing the reaction rate of the main reaction while minimizing that of the side reaction to improve the yield of objective product is the key task for multiple reaction analysis.

   Instantaneous selectivity is commonly used to evaluate the relative reaction rates of main and side reactions. Assuming the reaction order with respect to reactant A for the first reaction is α and for the second reaction is β, respectively, using Eq. (2.40) the consumption rate for A is:

$$\mathscr{R}_A = -k_1 c_A^{\alpha} + \nu_A k_2 c_A^{\beta} \tag{2.41}$$

Since $\nu_A < 0$, $\mathscr{R}_A < 0$, reflecting the fact that A is a reactant. Based on reaction selectivity defined in Chapter 1, Introduction, the instantaneous selectivity can be expressed as:

$$S = \mu_{PA} \frac{\mathscr{R}_P}{|\mathscr{R}_A|} \tag{2.42}$$

where μ_{PA} is the moles of A consumed to make 1 mole of product P. P is a product, therefore $\mathscr{R}_P$ is positive. To make the instantaneous selectivity always positive, the absolute value of $\mathscr{R}_A$ is used to define the instantaneous selectivity.

$$\mathscr{R}_P = r_P = k_1 c_A^\alpha$$

And

$$\mu_{PA} = 1$$

Substituting the above equation as well as Eq. (2.41) into Eq. (2.42):

$$S = \frac{k_1 c_A^\alpha}{k_1 c_A^\alpha + |\nu_A| k_2 c_A^\beta} = \frac{1}{1 + \frac{k_2}{k_1} |\nu_A| c_A^{\beta - \alpha}} \tag{2.43}$$

We can examine the impact of concentration and temperature on instantaneous selectivity using this equation. At a given temperature, k_1 and k_2 are constants. The effect of reaction concentration c_A on instantaneous selectivity depends on reaction orders of the main and side reactions. When $\alpha = \beta$, Eq. (2.43) can be simplified as:

$$S = \frac{k_1}{k_1 + |\nu_A| k_2}$$

So the instantaneous selectivity is independent of concentration and is only a function of reaction temperature. If the reaction order of the main reaction is larger than that of the side reaction, i.e., $\alpha > \beta$, the instantaneous selectivity increases with reactant concentration. On the other hand, if $\beta > \alpha$, a lower reactant concentration will lead to a higher instantaneous selectivity.

Next we can examine the impact of reaction temperature on instantaneous selectivity. Assuming reaction activation energy for the main and side reaction is E_1 and E_2, respectively, and the influence of temperature on reaction rate constant k_1 and k_2 follow the Arrhenius equations, then Eq. (2.43) can be expressed as:

$$S = \frac{1}{1 + \frac{A_2}{A_1} \exp\left(\frac{E_1 - E_2}{RT}\right) |\nu_A| c_A^{\beta - \alpha}}$$

At given concentrations, the influence of temperature on instantaneous selectivity depends on the relative values of the activation energy of the main and side reactions. When the activation energy of the main reaction E_1 is higher than that of the side reaction E_2, higher temperature will lead to higher instantaneous selectivity. On the other hand, if $E_2 > E_1$, the increased temperature will cause instantaneous selectivity to decrease.

In addition to selecting adequate reactant concentration and temperature, using a catalyst is another way to improve reaction selectivity. A good catalyst can not only accelerate reaction rates, but also promote the reaction proceeding in a desired direction. Therefore, it can enhance the main reaction while inhibiting the undesired side reaction.

The above analysis focuses on a reaction system with two parallel reactions. In fact, the principle and approach can be extended to a reaction system involving multiple reactions and more multiple reaction networks, i.e., Eq. (2.42) is valid for any reaction system. In order to better understand its meaning Eq. (2.42) can be rewritten as:

$$S = \mu_{PA} \frac{\text{Formation rate of desired product}}{|\text{Consumption rate of key reactant}|} = \mu_{PA} \frac{\mathscr{R}_P}{|\mathscr{R}_A|} = \mu_{PA} \frac{\sum_j \nu_{Pj} \bar{r}_j}{\sum_j \nu_{Aj} \bar{r}_j}$$

$$(2.44)$$

So the definition of reaction selectivity is fundamentally the same as the one used in Chapter 1, Introduction. The only difference is that here, the selectivity is defined based on consumption rate and formation rate, while in Chapter 1, Introduction, the selectivity was defined based on the amount of product formed and reactant converted. The reason that selectivity defined in Eq. (2.42) is called instantaneous selectivity is that it will change with reaction composition and temperature, and is an instantaneous value. The instantaneous selectivity is also called point selectivity or microselectivity, since its value will change at different locations inside a flow reactor.

3. Consecutive reactions

When a product of one reaction is a reactant of another reaction, these two reactions are consecutive reactions. For example, nitrogen monoxide can be oxidized into nitrogen dioxide. In addition, two nitrogen dioxide molecules can combine to make dinitrogen tetroxide:

$$2NO + O_2 + \leftrightarrows 2NO_2$$
$$2NO_2 \leftrightarrows N_2O_4$$

Another example is chlorination of methane, which can make chloromethane, dichloromethane, trichloromethane, and carbon tetrachloride:

$$CH_4 \xrightarrow{Cl_2} CH_3Cl \xrightarrow{Cl_2} CH_2Cl_2 \xrightarrow{Cl_2} CHCl_3 \xrightarrow{Cl_2} CCl_4$$

For our discussions we will use the following consecutive reactions:

$$A \rightarrow P \rightarrow Q$$

No matter whether P or Q is the desired product, it is always beneficial to increase the conversion of reactant A. This is the character of a consecutive reaction system. If Q is the desired product, increasing reaction rates of either or both of these two reactions is beneficial, and this can be achieved by, e.g., increasing reaction temperature. If P is the desired product, in order to produce more P, the second reaction should be as slow as possible. Since P can be further converted into Q, it is an intermediate product, and must have a maximum value, which means that there will be an optimal value of extent of reaction. For methane chlorination, chloromethane, dichloromethane, and trichloromethane are all intermediate products and will have maximum yields. If the objective is to make chloromethane, the reaction time should be kept short; otherwise more polychloride will be produced. In extreme situations, no chloromethane will be left. The maximum yield of the desired product depends on the reactor type and will be discussed in the next two chapters.

Similar to parallel reactions, using a catalyst that enhances the first reaction while inhibiting the second is an effective approach to improve the yield of product P. On the other hand, the yield of P can also be improved by properly selecting reaction conditions and reactor types. For example, if the activation energy of the first reaction is higher than that of the second reaction, higher reaction temperatures will favor the yield of P; the opposite is also true, i.e., a lower temperature should be used if the activation energy of the first reaction is lower than that of the second reaction.

Example 2.4
Ethylene is typically manufactured by the thermal cracking of ethane at a high temperature, and the process involves the following reactions:

$$C_2H_6 \rightleftharpoons C_2H_4 + H_2, \quad \overline{r_1} = \overrightarrow{k_1} c_A - \overleftarrow{k_1} c_E c_H$$
$$2C_2H_6 \rightarrow C_3H_8 + CH_4, \quad \overline{r_2} = k_2 c_A$$
$$C_3H_6 \rightleftharpoons C_2H_2 + CH_4, \quad \overline{r_3} = \overrightarrow{k_3} c_P - \overleftarrow{k_3} c_R c_M$$
$$C_2H_2 + C_2H_4 \rightarrow C_4H_6, \quad \overline{r_4} = k_4 c_R c_E$$
$$C_2H_4 + C_2H_6 \rightarrow C_3H_6 + CH_4 \quad \overline{r_5} = k_5 c_E c_A$$

The subscript A, R, E, P, M, and H represent C_2H_6, C_2H_2, C_2H_4, C_3H_6, CH_4, and H_2. Please derive equations for instantaneous selectivity of ethylene.

Solution
Ethylene is the desired product and ethane is the key reactant, from Eq. (2.40) the ethane consumption rate can be expressed as:

$$\mathscr{R}_A = -\overline{r_1} - 2\overline{r_2} - \overline{r_5} = -\overrightarrow{k_1} c_A + \overleftarrow{k_1} c_E c_H - 2k_2 c_A - k_5 c_E c_A$$
$$= \overleftarrow{k_1} c_E c_H - (\overrightarrow{k_1} + 2k_2 + k_5 c_E)c_A$$

(Continued)

(Continued)

And the ethylene formation rate is:

$$\mathscr{R}_E = \overrightarrow{r_1} - \overleftarrow{r_4} - \overleftarrow{r_5} = \vec{k}_1\,c_A - \overleftarrow{k}_1\,c_E c_H - k_4\,c_R c_E - k_5\,c_E c_A = \vec{k}_1\,c_A - (\overleftarrow{k}_1\,c_E - k_4\,c_R + k_5\,c_A)c_E$$

Based on reaction stoichiometry, 1 mole of ethane will be consumed to make 1 mole ethylene, i.e., $\mu_{EA} = -1$.

Substituting the above equations into Eq. (2.44), the instantaneous selectivity of ethylene is:

$$S = \frac{\vec{k}_1\,c_A - (\overleftarrow{k}_1\,c_H + k_4\,c_R + k_5\,c_A)c_E}{\left(\vec{k}_1 + 2k_2 + k_5\,c_E\right)c_A - \overleftarrow{k}_1\,c_E c_H}$$

2.4.3   Reaction Network

The previous section introduced three basic types of multiple reaction systems and their main characters. In reality it is common that all three types of reactions exist and form a complex network that is called a reaction network. For example, the naphthene oxidation on vanadium catalyst can be described by the following reaction network:

Naphthalene can be oxidized to either phthalic anhydride or naphthoquinone, i.e., reaction 1 and 2 are parallel to each other. On the other hand, naphthoquinone can be further oxidized into phthalic anhydride that can be fully oxidized into carbon dioxide and water. Therefore reaction 1, 3, and 4 are consecutive reactions, and reaction 2 and 4 are also consecutive reactions.

Another example is the reactions between ethylene expoxide and water that includes both parallel and consecutive reactions:

$$H_2O + (CH_2)_2O \xrightarrow{1} CH_2OHCH_2OH$$

$$CH_2OHCH_2OH + (CH_2)_2O \xrightarrow{2} CH_2OHCH_2OCH_2CH_2OH$$

$$CH_2OHCH_2OCH_2CH_2OH + (CH_2)_2O \xrightarrow{3} CH_2OHCH_2OCH_2CH_2OCH_2CH_2OH$$

If we use A, B, P, Q, and R to represent ethylene epoxide, water, glycol, diethylene glycol, and triethylene glycol, the above three reactions can be rewritten as the following three formats:

$$A + B \xrightarrow{1} P \qquad A \xrightarrow[1]{+B} P$$

$$A + P \xrightarrow{2} Q \qquad A \xrightarrow[2]{+P} Q \qquad B \xrightarrow[1]{+A} P \xrightarrow[2]{+A} Q \xrightarrow[3]{+A} R$$

$$A + Q \xrightarrow{3} R \qquad A \xrightarrow[3]{+Q} P$$

(a)   (b)   (c)

Format (b) is similar to a parallel reaction system, and format (c) is a series of consecutive reactions. The selection of proper reaction conditions for this reaction system depends on the desired product, relative cost of reactants, the cost of product separation, etc. For hydration of ethylene epoxide if glycol is the desired product, the initial concentration of ethylene epoxide should not be very high to improve reaction selectivity. The reactions between ammonia and ethylene, and methanol and ethylene also show similar reaction behaviors.

With an increase in the number of reactions, the kinetic behavior of the reaction system becomes more complex and the reaction system simulation will be more difficult. It is common practice to simplify the reaction network. For example, the reaction network presented above for naphthalene oxidation is already a simplified description of the real reaction system. However, it can be further simplified by only considering reaction 2 and 4, i.e., using two consecutive reactions to describe the oxidation of naphthalene to capture the key character of the reaction system. Of course, whether such simplification, i.e., ignoring some reactions or combining several reactions together, is acceptable has to be verified through theoretical analysis and comparison with experimental data.

However, for a reaction system using a feedstock with complex composition, many chemical reactions take place simultaneously, and even after simplification there are still many reactions. For example, for heavy oil catalytic cracking, catalytic reforming of naphtha and other reaction processes used in the refining and petrochemical industry, the feedstock contains hundreds of components. It is almost impossible to list all the reactions and corresponding reaction rate equations. A practical approach is to combine reaction components that have similar reaction properties into a single reaction entity, and then study the reaction behavior of this pseudo component. This approach is called lumping. For example, a simple reaction model for catalytic cracking reactions contains four pseudo components:

Heavy oil → gasoline   → coke + gases

This reaction network is similar to that describing the naphthalene oxidation reactions, and the only difference is that pseudo components, instead of individual compounds, are used here. This model is very simple but still can capture basic reaction behavior.

2.5   TRANSFORMATION AND INTEGRATION OF REACTION RATE EQUATIONS

As mentioned above, for a given reaction system, reaction rate is a function of temperature and concentrations of reaction components, and the rate equation is the mathematical equation that describes this relationship. However, for solving practical problems such rate equations are not convenient to use, because even under isothermal conditions, the concentrations of reaction components are variables. Fortunately, for a given feed composition not all of the concentrations are independent variables. Therefore, it is possible to transform the reaction rate equation and use one variable to describe the entire reaction. As discussed in previous section, during the reaction process the changes in concentrations of reaction components must follow the stoichiometry of the reaction, and based on this relationship the reaction progress can be described using the conversion of the key component or the yield of the desired product.

2.5.1   Single Reaction

For gas phase reaction

$$\nu_A A + \nu_B B \rightarrow \nu_P P$$

the rate equation is:

$$r_A = k c_A^\alpha c_B^\beta \tag{2.45}$$

At the beginning of the reaction there is no P in the reaction mixture, and the concentration of components A and B are c_{A0} and c_{B0}, respectively. If the reaction takes place under isothermal conditions, k is a constant. Now we want to transform Eq. (2.45) so that the reaction rate is a function of the conversion of component A (X_A), and there are two situations that need to be treated separately.

1. Constant-volume process
   Based on concentration definition:

$$c_A = \frac{n_A}{V}$$

Generally speaking, during the reaction process both n_A and V are variables, i.e., both are functions of conversion. But for constant−volume processes, since they are already defined that the volume of reaction mixture is a constant, we only need to consider n_A as a function of conversion X_A. Based on conversion definition:

$$n_A = n_{A0}(1 - X_A) \tag{2.46}$$

Thus:

$$c_A = \frac{n_{A0}(1 - X_A)}{V} = c_{A0}(1 - X_A) \tag{2.47}$$

From stoichiometry we know that to convert ν_A moles of A ν_B moles of B will be consumed,

$$n_B = n_{B0} - \frac{\nu_B}{\nu_A} n_{A0} X_A$$

then

$$c_B = \frac{n_{B0} - \frac{\nu_B}{\nu_A} n_{A0} X_A}{V} = c_{B0} - \frac{\nu_B}{\nu_A} c_{A0} X_A \tag{2.48}$$

Substituting Eqs. (2.47) and (2.48) into Eq. (2.45):

$$r_A = k c_{A0}^{\alpha} (1 - X_A)^{\alpha} \left(C_{B0} - \frac{\nu_B}{\nu_A} c_{A0} X_A \right)^{\beta} \tag{2.49}$$

Now the right-hand side of Eq. (2.45) is a function of X_A. Similarly, the left-hand side can also be expressed as a function of X_A by substituting Eq. (2.46) into reaction rate definition:

$$r_A = -\frac{1}{V} \frac{d n_{A0}(1 - X_A)}{dt} = \frac{n_{A0}}{V} \frac{dX_A}{dt} = c_{A0} \frac{dX_A}{dt} \tag{2.50}$$

Eq. (2.49) can also be written as:

$$c_{A0} \frac{dX_A}{dt} = k c_{A0}^{\alpha} (1 - X_A)^{\alpha} \left(c_{B0} - \frac{\nu_B}{\nu_A} c_{A0} X_A \right)^{\beta} \tag{2.51}$$

2. Variable-volume process

If the reaction does not take place under constant-volume conditions, the volume of reaction mixture changes with conversion. The amount of each component before and after the reaction can be summarized as follows:

Component	Before the reaction	At conversion $= X_A$
A	n_{A0}	$n_{A0} - n_{A0} X_A$
B	n_{B0}	$n_B = n_{B0} - \dfrac{\nu_B}{\nu_A} n_{A0} X_A$
P	0	$\dfrac{\nu_P}{|\nu_A|} n_{A0} X_A$
Total	$n_{t0} = n_{A0} + n_{B0}$	$n_t = n_{t0} + n_{A0} X_A \delta_A$

where

$$\delta_A = \sum \nu_i / |\nu_A| \tag{2.52}$$

δ_A represents the changes of total number of moles of reaction mixture when 1 mole of A is converted. $\sum \nu_i$ is the sum of stoichiometric coefficients of reactants and products. If $\delta_A > 0$, the total number of moles in the reaction mixture increases as the reaction progresses, i.e., $n_t > n_{t0}$. If $\delta_A < 0$, the total number of moles in the reaction mixture decreases, i.e., $n_t < n_{t0}$. $\delta_A = 0$ means no change of total number of moles in the reaction mixture as the reaction progresses. If the gas mixture follows the ideal gas law, under isothermal and constant pressure conditions:

$$\frac{V_0}{V} = \frac{n_{t0}}{n_t}$$

Or:

$$\frac{V_0}{V} = \frac{n_{t0}}{n_{t0} + n_{A0}X_A\delta_A}$$

The above equation can be rearranged as:

$$V = V_0(1 + y_{A0}X_A\delta_A) \tag{2.53}$$

This equation gives the relationship between reaction mixture volume at conversion X_A and its initial value. y_{A0} is the initial molar fraction of component A and equals n_{A0}/n_{t0}. From Eq. (2.53) we can see that when $\delta_A > 0$, $V > V_0$; when $\delta_A < 0$, $V < V_0$; and when $\delta_A = 0$, $V = V_0$.

Substituting Eq. (2.53) into Eq. (2.47) and Eq. (2.48):

$$c_A = \frac{c_{A0} - c_{A0}X_A}{1 + y_{A0}X_A\delta_A}$$

and

$$c_B = \frac{c_{B0} - \frac{\nu_B}{\nu_A}c_{A0}X_A}{1 + y_{A0}X_A\delta_A}$$

Substituting the above two equations into Eq. (2.45) we can obtain the relationship between the reaction rate and conversion under variable-volume conditions:

$$r_A = k\frac{c_{A0}^\alpha(1-X_A)^\alpha\left(c_{B0} - \frac{\nu_B}{\nu_A}c_{A0}X_A\right)^\beta}{(1 + y_{A0}X_A\delta_A)^{\alpha+\beta}} \tag{2.54}$$

The difference between this equation and the one for constant-volume conditions is that in Eq. (2.54) there is a volume correction factor

$(1 + y_{A0}X_A\delta_A)^m$. The exponential term m depends on reaction order, and for the reaction discussed above $m = \alpha + \beta$. If $\delta_A = 0$, i.e., the total number of moles in the reaction mixture does not change, Eq. (2.54) will be reduced to Eq. (2.49).

To express reaction rate under variable-volume condition as a function of conversion, we can substitute Eqs. (2.46) and (2.53) into Eq. (2.1):

$$r_A = \frac{-1}{V_0(1 + y_{A0}X_A\delta_A)}\frac{d(n_{A0} - n_{A0}X_A)}{dt} = \frac{c_{A0}}{1 + y_{A0}X_A\delta_A}\frac{dX_A}{dt}$$

Compared to the equation under constant-volume condition, there is a volume correction factor in this equation.

The above discussions use concentrations to describe the reaction composition, and the relationship between concentration and conversion can be in general expressed as:

$$c_i = \frac{c_{i0} - \frac{\nu_i}{\nu_A}c_{A0}X_A}{1 + y_{A0}\delta_A X_A} \tag{2.55}$$

This equation is valid for both reactants and products, and under either constant or variable volume conditions. Similar equations can be derived when partial pressure or mole fraction is used to describe reaction composition:

$$p_i = \frac{p_{i0} - \frac{\nu_i}{\nu_A}p_{A0}X_A}{1 + y_{A0}\delta_A X_A} \tag{2.56}$$

$$y_i = \frac{y_{i0} - \frac{\nu_i}{\nu_A}y_{A0}X_A}{1 + y_{A0}\delta_A X_A} \tag{2.57}$$

It should be noted that the above discussion on variable-volume reaction processes is only applicable to gas phase reactions. For liquid phase reaction processes, whether the total number of moles changes or not, it usually can be treated as a constant-volume system without introducing much error. A reactor operated under batch mode, either gas phase or liquid phase, can be treated as a constant-volume process. For isothermal flow reactors, in additional to liquid reaction, gas reactions with no change in the total number of moles can be treated using constant-volume processes.

All the above transformations are based on given feed composition. If feed composition changes, the relationship between reaction rate and conversion will also change.

After being expressed as a function of conversion X_A and time t, the rate equation can be integrated to obtain an algebra equation. This can be illustrated using the reaction discussed above as an example. Assuming

$|\nu_A| = |\nu_B| = |\nu_P| = 1$ and the reaction is first order with respect to both A and B, Eq. (2.54) becomes:

$$\frac{c_{A0}}{1 - y_{A0}X_A}\frac{dX_A}{dt} = \frac{kc_{A0}(1 - X_A)(c_{B0} - c_{A0}X_A)}{(1 - y_{A0}X_A)^2}$$

After simplification:

$$\frac{dX_A}{dt} = \frac{k(1 - X_A)(c_{B0} - c_{A0}X_A)}{1 - y_{A0}X_A}$$

$$k\int_0^t dt = \int_0^{X_A} \frac{1 - y_{A0}X_A}{(1 - X_A)(c_{B0} - c_{A0}X_A)}dX_A$$

$$t = \frac{1}{k}\left[\frac{1 - c_{B0}y_{A0}/c_{A0}}{c_{B0} - c_{A0}}\ln\left(1 - \frac{c_{A0}X_A}{c_{B0}}\right) + \frac{1 - y_{A0}}{c_{B0} - c_{A0}}\ln\frac{1}{1 - X_A}\right]$$

This is the correlation between conversion and reaction time for a second order reaction. If the reaction takes place under constant-volume conditions, integrating Eq. (2.51):

$$t = \frac{1}{k(c_{B0} - c_{A0})}\ln\frac{1 - c_{A0}X_A/c_{B0}}{1 - X_A}$$

Obviously, analytical solutions can only be obtained for simple reaction rate equations. For most situations, numerical methods have to be used to describe conversion changes with time.

Example 2.5

For gas phase hydrogenation of benzene on nickel catalyst:

$$C_6H_6 \text{ (B)} + 3H_2 \text{ (H)} \rightarrow C_6H_{12} \text{ (C)}$$

The rate equation is:

$$r_B = \frac{kp_B p_H^{0.5}}{1 + Kp_B} \tag{A}$$

where p_B and p_H are the partial pressures of benzene and hydrogen, respectively, and k and K are constants. If there is no cyclohexane in the feed, the molar fractions of benzene and hydrogen are y_{B0} and y_{H0}, respectively, and the total pressure of the reaction system is p. Please transform Eq. (A) so that the reaction rate is a function of conversion X_B.

Solution

For this reaction the total number of moles decreases as the reaction progresses

$$\delta_B = \frac{\nu_C + \nu_B + \nu_H}{|\nu_B|} = \frac{1 - 1 - 3}{1} = -3$$

(Continued)

(Continued)

From Eq. (2.57)

$$y_B = \frac{y_{B0} - y_{B0}X_B}{1 - 3y_{B0}X_B}$$

and

$$y_H = \frac{y_{H0} - 3y_{B0}X_B}{1 - 3y_{B0}X_B}$$

Since $p_B = py_B$ and $p_H = py_H$, substitute the above equations into Eq. (A). We can obtain the reaction rate r_B as a function of conversion X_B:

$$r_B = \frac{k\left(p\frac{y_{B0} - y_{B0}X_B}{1 - 3y_{B0}X_B}\right)\left(p\frac{y_{H0} - 3y_{B0}X_B}{1 - 3y_{B0}X_B}\right)^{0.5}}{1 + Kp\frac{y_{B0} - y_{B0}X_B}{1 - 3y_{B0}X_B}} = \frac{kp^{1.5}y_{B0}(1 - X_B)(y_{H0} - 3y_{B0}X_B)^{0.5}}{(1 - 3y_{B0}X_B)^{1.5} + Kpy_{B0}(1 - X_B)(1 - 3y_{B0}X_B)^{0.5}}$$

2.5.2 Multiple Reactions

For a reaction system that includes multiple chemical reactions, similar procedures can be used for rate equation transformation. For each reaction, a reaction variable, which is typically reaction conversion or yield, needs to be selected. The format of the reaction rate equation after transformation depends on the reaction variable selected.

For a reaction system that includes N reaction components and M chemical reactions, M reaction equations will be needed. For the convenience of mathematical treatment those M reactions can be expressed as:

$$\nu_{11}A_1 + \nu_{21}A_2 + \ldots + \nu_{N1}A_N = 0$$
$$\nu_{12}A_1 + \nu_{22}A_2 + \ldots + \nu_{N2}A_N = 0$$
$$\ldots\ldots$$
$$\nu_{1M}A_1 + \nu_{2M}A_2 + \ldots + \nu_{NM}A_N = 0$$

or

$$\sum_{i=1}^{N} \nu_{ij}A_i = 0, j = 1, 2, \ldots M \tag{2.58}$$

For the transformation of reaction equations in a multiple reaction system, it is most convenient to choose extent of reaction as the reaction variable, since for any component i its stoichiometric coefficient for reaction j, ν_{ij}, times extent of reaction ξ_j, is equal to the amount of i reacted in reaction j. $\nu_{ij}\xi_j$ could be either positive or negative, depending on whether component i is a reactant or product for that reaction. If it is a reactant, $\nu_{ij}\xi_j$ is negative; if it is a product, $\nu_{ij}\xi_j$ is positive. We can add the amount of component i reacted in each reaction to obtain total amount of component i reacted:

$$n_i - n_{i0} = \sum_{j=1}^{M} \nu_{ij} \xi_j \tag{2.59}$$

Adding reacted amounts of all components, we can obtain total molar change in the reaction system:

$$\sum_{i=1}^{N} (n_i - n_{i0}) = \sum_{i=1}^{N} \sum_{j=1}^{M} \nu_{ij} \xi_j$$

Or

$$n_t - n_{t0} = \sum_{i=1}^{N} \sum_{j=1}^{M} \nu_{ij} \xi_j \tag{2.60}$$

where n_{t0} and n_t are the total moles in the reaction system at the beginning and any reaction time t, respectively. At given pressure and temperature:

$$V/V_0 = n_t/n_{t0}$$

Rearranging Eq. (2.60) we can obtain volume of reaction system at any given time:

$$V = V_0 \left(1 + \frac{1}{n_{t0}} \sum_{i=1}^{N} \sum_{j=1}^{M} \nu_{ij} \xi_j \right) \tag{2.61}$$

$$n_i = n_{i0} + \sum_{j=1}^{M} \nu_{ij} \xi_j$$

Therefore:

$$c_i = \frac{n_i}{V} = \frac{n_{i0} + \sum_{j=1}^{M} \nu_{ij} \xi_j}{V_0 \left(1 + \frac{1}{n_{t0}} \sum_{i=1}^{N} \sum_{j=1}^{M} \nu_{ij} \xi_j \right)} \tag{2.62a}$$

For constant volume process, $V = V_0$,

$$c_i = \frac{1}{V_0} \left(n_{i0} + \sum_{j=1}^{M} \nu_{ij} \xi_j \right) \tag{2.62b}$$

Eq. (2.62a) is a general equation applicable to any component in a reaction system that follows the ideal gas law. It is obvious that the equation for a single reaction system, Eq. (2.55), is only a simplified version of Eq. (2.62a). Liquid reaction systems usually can be considered as constant volume systems, therefore Eq. (2.62b) is a general equation applicable to a liquid reaction system.

After transformation, the rate equations for multiple reactions are expressed as functions of reaction variables, which are typically a group of ordinary differential equations coupled to each other and hence have to be solved numerically.

Example 2.6

Catalytic dehydrogenation of ethyl benzene takes place at 898K and ambient pressure:

1. $C_6H_5C_2H_5 \rightleftharpoons C_6H_5 - CH = CH_2 + H_2$
2. $C_6H_5C_2H_5 \rightarrow C_6H_6 + C_2H_4$
3. $C_6H_5C_2H_5 + H_2 \rightarrow C_6H_5 - CH_3 + CH_4$

Using subscripts A, B, T, H, and S to represent ethyl benzene, benzene, toluene, hydrogen, and styrene, respectively, under reaction temperature and pressure the reaction rate equations for the above reactions are:

$$r_1 = 0.1283 \, (p_A - p_S p_H/0.04052)$$
$$r_2 = 5.745 \times 10^{-3} p_A$$
$$r_3 = 0.2904 p_A p_H$$

In the above equations the unit for reaction rate is kmol/(kg·h) and pressure unit is MPa. The feed contains 10% ethyl benzene and 90% water. Please calculate the consumption rate of ethyl benzene when the yield of styrene, benzene, and toluene is 60%, 0.5%, and 1%, respectively.

Solution

Use 10 kmol of ethyl benzene as the base for the calculation, then the amount of steam in the feed stream is 90 kmol. Since styrene, benzene, and toluene are formed from reactions 1, 2 and 3, respectively, the extent of reaction for these three reactions are, $\xi_1 = 6$ kmol, $\xi_2 = 0.05$ kmol, and $\xi_3 = 0.1$ kmol. This is a variable-volume reaction; therefore Eq. (2.62a) has to be used to calculate reaction mixture composition. The reacted amounts for all the components in the reaction system are listed in Table 2B.

TABLE 2B Amount of Each Component During Ethyl Benzene Dehydrogenation

Component	n_{i0}	ν_{i1}	ν_{i2}	ν_{i3}	$\nu_{i1}\xi_1$	$\nu_{i2}\xi_2$	$\nu_{i3}\xi_3$	$\sum_{j}^{3} \nu_{ij}\xi_j$
Ethyl benzene	10	−1	−1	−1	−6	−0.05	−0.1	−6.15
Styrene	0	1	0	0	6	0	0	6
Toluene	0	0	0	1	0	0	0.1	0.1
Benzene	0	0	1	0	0	0.05	0	0.05
Hydrogen	0	1	0	−1	6	0	−0.1	5.9
Methane	0	0	0	1	0	0	0.1	0.1
Ethylene	0	0	1	0	0	0.05	0	0.05
Steam	90	0	0	0	0	0	0	0
$\sum_{i=1}^{8} n_{i0} = 100$							$\sum_{i=1}^{8} \sum_{j}^{3} \nu_{ij}\xi_j = 6.05$	

(Continued)

(Continued)

Since the rate equations are functions of partial pressures, Eq. (2.62a) has to be converted into an equation based on partial pressures. From ideal gas law:

$$c_i = p_i/(RT) \quad \text{and} \quad V_0 = n_{t0}RT/p$$

Substitute the above two correlations into Eq. (2.62a) and we can obtain:

$$p_i = \frac{\left(n_{i0} + \sum_j^M \nu_{ij}\xi_j\right)p}{n_{t0} + \sum_i^N \sum_j^M \nu_{ij}\xi_j} \tag{A}$$

Using the values in Table 2B, the partial pressures can be calculated from Eq. (A)

$$p_A = \frac{(10 - 6.15) \times 0.1013}{100 + 6.05} = 3.677 \times 10^{-3} \text{ MPa}$$

$$p_H = \frac{5.9 \times 0.1013}{100 + 6.05} = 5.635 \times 10^{-3} \text{ MPa}$$

$$p_S = \frac{6 \times 0.1013}{100 + 6.05} = 5.731 \times 10^{-3} \text{ MPa}$$

The consumption rate for ethyl benzene:

$$|\mathscr{R}_A| = |-r_1 - r_2 - r_3| = 0.1238\left(p_A - \frac{p_S p_H}{0.04052}\right) + 5.745 \times 10^{-3} p_A + 0.2904 p_A p_H$$

$$= 0.1283(3.67 \times 10^{-3} - 5.731 \times 10^{-3} \times \frac{5.635 \times 10^{-3}}{0.040512} + 5.745$$

$$\times 10^{-3} \times 3.67 \times 10^{-3} + 0.2904 \times 3.67 \times 10^{-3} \times 5.635 \times 10^{-3}$$

$$= 3.95710^{-4} \text{kmol}/(\text{kg·h})$$

Since a majority of the feed is steam that does not participate in the reaction, the reaction volume does not change much during the reaction process although ethyl benzene dehydrogenation is a variable-volume reaction. In this situation, the reaction can be treated as a constant-volume process, and error caused by the approximation is about 6%.

2.6 HETEROGENEOUS CATALYSIS AND ADSORPTION

To make chemical reactions commercially feasible, a catalyst is usually required. There are two major types of catalytic processes: homogeneous and heterogeneous catalysis. The heterogeneous process uses a solid catalyst to catalyze chemical reactions that take place on the surface of the solid catalyst. Therefore, it is extremely important to study the adsorption of reaction components on the catalyst surface. Next we will first provide a brief introduction of heterogeneous catalysis and then discuss adsorptions.

2.6.1   Heterogeneous Catalysis

Based on transition state theory, chemical reaction rate depends on the free energy of the intermediate complex formed during the transition from reactant to product. The function of a catalyst is to reduce the free energy. As a result, the reaction rate of a catalytic reaction will be much faster than that of noncatalytic reactions. In addition, for multiple reaction systems the catalyst can also help direct the reaction proceeding toward to the desired product, i.e., accelerate the main reaction to improve the yield and selectivity of the desired product. For example, for decomposition of ethanol, when an acid catalyst such as γ-alumina is used, ethanol will be dehydrated:

$$C_2H_5OH \rightarrow C_2H_4 + H_2O$$

If a metal catalyst is used, such as copper, the product will not be ethylene. Instead ethanol will dehydrogenate to form acetaldehyde:

$$C_2H_5OH \rightarrow CH_3CHO + H_2$$

Most solid catalysts are solid particles with various shapes. The sizes can be as large as over 10 mm, or as small as micrometers. Of course, the size of the catalysts needs to be carefully selected based on the reaction involved and reactor used. Typically a solid catalyst includes three parts, active component, promoter, and support. The active component and promoter are distributed on the support.

The function of a support is to increase surface area since the reaction takes place on the surface. The most commonly used support is oxides such as alumina and silica. High specific surface area, i.e., surface area per unit volume, is usually desired. Hence porous materials are widely used. Some porous materials, such as γ-alumina, can have a very high specific surface area of $100-300$ m^2/g. On the other hand, α-alumina has a much lower specific surface area of $1-10$ m^2/g. Different reactions have different requirements for specific surface area. Usually high surface area means smaller pore size, which will increase resistance for diffusion. Diffusion inside a catalyst particle is a very important issue, which will be discussed in Chapter 6, Chemical Reaction and Transport Phenomena in Heterogeneous System. In addition to providing surface area, the support also helps to improve physical properties, such as enhance mechanical strength, improve heat transfer, reduce deactivation, etc.

The catalytic function is provided by the active component. Commonly used active components include metals and metal oxides. Most metal catalysts require a support, such as silver catalyst used for the epoxidation of ethylene. There are also metal catalysts that are not supported. One example is the platinum–rhodium catalyst for oxidation of ammonia, which is simply a metal mesh. Some oxide catalysts such as γ-alumina for dehydration of ethanol and SiO_2-Al_2O_3 catalyst for heavy oil hydrocracking are not supported. But there are also metal oxide catalysts that are supported. For example, vanadium oxide supported on corundum is used as catalyst for benzene oxidation.

The content of the promoter in a catalyst is typically very low, and its function is to enhance the activity, selectivity, and stability of the main active component. There are two major types of promoters; one is to maintain the structure and another is to adjust the property of the main active component. For example, Al_2O_3 is added into iron catalyst for ammonia synthesis to prevent the loss of activity caused by growth of microcrystals of iron. On the other hand, K_2O is also added to the iron catalyst, and the function of K_2O is to enhance the intrinsic activity of iron. From this example we can see that there might be more than one promoter in a catalyst. A promoter itself usually has no or very low activity for the reaction.

Since the reaction takes place on the catalyst surface, heterogeneous catalytic reactions include many steps such as adsorption, surface reaction, and desorption. Assuming a heterogeneous catalytic reaction:

$$A + B \rightleftharpoons R \tag{2.63}$$

include the following steps:

1. Adsorption: $A + \sigma \rightleftharpoons A\sigma$ $\tag{2.64a}$

2. Adsorption: $B + \sigma \rightleftharpoons B\sigma$ $\tag{2.64b}$

3. Reaction: $A\sigma + B\sigma \rightleftharpoons R\sigma + \sigma$ $\tag{2.64c}$

4. Desorption: $R\sigma \rightleftharpoons R + \sigma$ $\tag{2.64d}$

where σ represents the adsorption sites, and $A\sigma$, $B\sigma$, and $R\sigma$ represent A, B, and R adsorbed on the surface. Adding all these four steps together will give the overall reaction equation, i.e., Eq. (2.63). Obviously the above equations are used as an example to illustrate typical steps involved in a heterogeneous catalytic reaction. Each reaction may involve different steps. Even for the same reaction, the steps involved may also depend on what catalyst is used. Some steps may not be necessary for certain reactions. For example, step 3 may form R directly without involving adsorbed R, i.e., $R\sigma$ and the final desorption step. For the two reactants A and B, it is also possible that only one of them is adsorbed, then there will be only one adsorption step, and the third step will be the reaction between adsorbed reactant with another reactant that is not adsorbed. However, at least one reactant has to be adsorbed, otherwise the catalyst cannot function. Currently there are still no comprehensive procedures to determine reaction steps involved in a heterogeneous catalytic reaction. As a result, an empirical approach has to be used.

2.6.2 Adsorption and Desorption

As discussed above, adsorption is an indispensable step for heterogeneous catalytic reactions. Here we will first introduce two types of adsorption, i.e., physical adsorption and chemical adsorption, then discuss equilibrium and rate of chemical adsorption process.

1. Physical and chemical adsorptions
   The driving force for physical adsorption is Van der Waals force. Any gas can absorb on solid surfaces, but the strength and amount of adsorption depends on the properties of the gas molecule. Physical adsorption typically occurs at low temperature, and the adsorption decreases quickly with an increase in temperature. For physical adsorption, gas molecules can form multilayer adsorption, and adsorption rate is fast and reversible. In addition the adsorption heat is low, typically in the range of 5−25 kJ/mol, which is similar to gas liquidation heat, and therefore it is easy to reach equilibrium. Most catalytic reactions take place at high temperature (higher than critical temperature), and under such conditions there is very little physical adsorption. As a result, the physical adsorption is not important to a catalytic reaction and is negligible.

   Chemical adsorption is characterized by the chemical bond formed between adsorbing molecules and the solid surface. It is selective, meaning that not all gas molecules can be chemically adsorbed. Chemical adsorption usually takes place at high temperature, and adsorption rate increases with temperature. In addition, it is monolayer adsorption with large adsorption heat, typically in the range $40 \sim 200$ kJ/mol which is in the same order of reaction heat. During chemical adsorption the structure of the adsorbed molecule changes. It is due to such molecule structure change that the activation energy is lowered and reaction rate is accelerated or catalyzed. Therefore, chemical adsorption on solid surfaces is a very important foundation for the study of heterogeneous catalytic kinetics.

   In summary, at low temperatures, the rate for chemical adsorption is low and physical adsorption will dominate and quickly reach equilibrium. The equilibrium adsorption decreases with increase of temperature. At high temperatures, physical adsorption is very weak, and chemical adsorption will dominate. Therefore, chemical adsorption is the main character of heterogeneous reactions. Since chemical adsorption forms a monolayer, when the active surface is fully covered by the gas molecules, the amount of adsorption cannot increase anymore.

2. Ideal adsorption
   In order to quantitatively describe the chemical adsorption process, an adequate adsorption model is needed. The simplest and also most widely used model is the ideal adsorption model. The basic assumptions for the ideal adsorption model include: (1) The energy of the adsorbing surface is uniform, in other words, every adsorption site is identical and has the same energy; (2) The interaction between adsorbed molecules is negligible and (3) Monolayer adsorption. The adsorbing surface meeting the above conditions is called an ideal surface. The ideal adsorption model is also called the Langmuir model.

Due to gas dynamics, the gas molecules constantly collide with the catalyst surface. However, not every molecule that collides with the surface can be adsorbed on the active site. Instead, only those molecules with enough dynamic energy can be adsorbed. The adsorbed molecules can also leave the surface (desorption). Therefore, the adsorption process is a dynamic equilibrium. The rate of gas molecules adsorption onto the catalyst surface is proportional to the numbers of molecules that collide with the catalyst surface, and the number of collisions is proportional to gas partial pressure. Therefore, gas adsorption rate is proportional to gas partial pressure. Since the adsorption takes place on a free surface (not covered by other adsorbed molecules), the adsorption rate is also proportional to free surface area. Assuming that gas A will adsorb on to an adsorption site σ:

$$A + \sigma \underset{k_d}{\overset{k_a}{\rightleftharpoons}} A\sigma$$

The adsorption rate is:

$$r_a = k_a p_A (1 - \theta_A) \tag{2.65}$$

where k_a is the adsorption rate constant that depends on temperature and can be expressed as an exponential function of temperature. θ_A is surface coverage, representing the fraction of surface covered by adsorbed A. Chemical adsorption is monolayer adsorption, the area that is not covered by adsorbed molecules is called free surface, and the uncovered fraction is also called the vacant fraction $\theta_V = 1 - \theta_A$

The desorption rate r_d is proportional to the number of adsorbed molecules:

$$r_d = k_d \theta_A \tag{2.66}$$

where k_d is the desorption rate constant which changes with temperature exponentially.

The net rate for adsorption is:

$$r = k_a p_A (1 - \theta_A) - k_d \theta_A$$

At adsorption equilibrium, the net rate is zero, i.e., $r = 0$, then

$$k_a p_A (1 - \theta_A) = k_d \theta_A$$

Or:

$$\theta_A = \frac{K_A p_A}{1 + K_A p_A} \tag{2.67}$$

where $K_A = k_a / k_d$, the equilibrium constant. Eq. (2.67) is the Langmuir adsorption isotherm. The adsorption constant reflects the strength of gas molecule adsorption on the surface. For very weak adsorption, $K_A p_A \ll 1$, Eq. (2.67) can be simplified as;

$$\theta_A = K_A p_A \tag{2.68}$$

For very strong adsorption, $K_A p_A >> 1$, Eq. (2.67) becomes $\theta_A = 1$.

The dependence of adsorption constant on temperature can be expressed as:

$$K_A = K_{A0} \exp(q/RT) \tag{2.69}$$

where q is heat of adsorption. From Eq. (2.69), when temperature increases, K_A will become smaller, which means with an increase of temperature, surface coverage will decrease.

After discussing monomolecular adsorption, we will analyze bimolecular adsorption. Assuming both A and B molecules can be adsorbed by the catalyst surface, and bimolecular adsorption can be described as:

$$A + B + 2\sigma \overset{k_a}{\underset{k_d}{\rightleftharpoons}} A\sigma + B\sigma$$

Since A and B are adsorbed simultaneously, the vacant fraction is $\theta_V = 1 - \theta_A - \theta_B$. If the adsorption rate for molecule A is r_{aA} and desorption rate is r_{dA}:

$$r_{aA} = k_{aA} p_A (1 - \theta_A - \theta_B) \tag{2.70}$$

$$r_{dA} = k_{dA} \theta_A \tag{2.71}$$

Similarly adsorption and desorption rates for molecule B can be expressed as:

$$r_{aB} = k_{aB} p_B (1 - \theta_A - \theta_B) \tag{2.72}$$

$$r_{dB} = k_{dB} \theta_B \tag{2.73}$$

At equilibrium, the net rates are zero:

$$r_{aA} = r_{dA}, r_{aB} = r_{dB} \tag{2.74}$$

Solving Eqs. (2.70) to (2.74) simultaneously, we obtain:

$$\theta_A = \frac{K_A p_A}{1 + K_A p_A + K_B p_B}$$

$$\theta_B = \frac{K_B p_B}{1 + K_A p_A + K_B p_B}$$

The vacant coverage:

$$\theta_V = 1 - \theta_A - \theta_B = \frac{1}{1 + K_A p_A + K_B p_B}$$

If there are m different molecules that can be simultaneously adsorbed, then the coverage rate for molecule i is:

$$\theta_i = \frac{K_i p_i}{1 + \sum_{i=1}^{m} K_i p_i} \tag{2.75}$$

And the vacant fraction is:

$$\theta_v = \frac{1}{1 + \sum\limits_{i=1}^{m} K_i p_i} \qquad (2.76)$$

If during the adsorption process the molecule dissociates into atoms, each atom will occupy an active site:

$$A_2 + 2\sigma \underset{k_d}{\overset{k_a}{\rightleftharpoons}} 2A\sigma$$

The adsorption and desorption rates can be expressed as:

$$r_a = k_a p_A (1 - \theta_A)^2$$

$$r_d = k_d \theta_A^2$$

The net adsorption rate is:

$$r = k_a p_A (1 - \theta_A)^2 - k_d \theta_A^2$$

At equilibrium the net rate is zero, then:

$$\theta_A = \frac{\sqrt{K_A p_A}}{1 + \sqrt{K_A p_A}} \qquad (2.77)$$

Eq. (2.77) is the Langmuir isotherm for dissociative adsorption.

3. Real adsorption

   If an adsorption process does not meet the assumptions of an ideal adsorption, it is called real adsorption. In reality, the catalyst surface is not uniform and the adsorption energy changes. At the beginning, the gas will first be adsorbed on the most active sites, and the adsorption is strong and will release a lot of heat. With more active sites covered, the adsorption will become weaker and the heat release also decreases. As a result, the activation energy required for the adsorption process increases. Therefore, the adsorption rate depends on surface coverage.

   On the other hand, for ideal adsorption it is assumed that there is no interaction between adsorbed species, and this is obviously not realistic. If the adsorbed species interact with each other, the abovementioned phenomena will occur, then the activation energy, heat effect, as well as rate of adsorption will depend on surface coverage. However, it is difficult to distinguish whether such dependence is caused by nonuniformity of the surface or the interaction between adsorbed species. From a practical perspective, it is not necessary to distinguish, and we just need to consider such nonuniform phenomenon.

   Due to either surface nonuniformity or adsorbed species interactions, activation energies of adsorption and desorption as well as the heat of adsorption change with surface coverage. If we assume the activation energies are linear functions of surface coverage:

$$E_a = E_a^0 + \alpha \theta_A \qquad (2.78)$$

$$E_d = E_d^0 + \beta \theta_A \qquad (2.79)$$

E_a^0 and E_d^0 represent the activation energy for adsorption and desorption at zero coverage, respectively. α and β are constants. From Eqs. (2.78) and (2.79), we can see that with increasing coverage the activation energy for adsorption increases while the activation energy for desorption decreases. Since

$$q = E_d - E_a$$

And using Eqs. (2.78) and (2.79) we obtain:

$$q = (E_d^0 - E_a^0) - (\alpha + \beta)\theta_A = q_0 - \gamma \theta_A \qquad (2.80)$$

where $q_0 = E_{d0} - E_{A0}$, and $\gamma = \alpha + \beta$. Based on the above equations, the rate equations for adsorption and desorption can be derived:

$$r_a = k_{a0} p_A \exp\left(-\frac{\alpha \theta_A}{RT}\right)$$

$$r_d = k_{d0} \exp\left(\frac{\beta \theta_A}{RT}\right)$$

At equilibrium, $r_a - r_d = 0$, therefore:

$$\theta_A = \frac{RT}{\alpha + \beta} \ln (k_{a0}/k_{d0} p_A) = \frac{RT}{\alpha + \beta} \ln (K_0 p_A) \qquad (2.81)$$

where $K_0 = k_{a0} / k_{d0}$. Eq. (2.81) is the isotherm for real adsorption, which is also called the Temkin isotherm.

Another isotherm for real adsorption is:

$$\theta_A = K p_A^m, m < 1 \qquad (2.82)$$

This equation was first proposed as an empirical correlation, and was later derived by assuming the activation energies for adsorption and desorption are linear functions of $\ln(\theta_A)$. Eq. (2.82) is also called the Freundlich isotherm. In fact, both the Temkin and Freundlich isotherm have their own disadvantages. On the one hand, the Temkin isotherm, θ_A is not zero when $p_A = 0$, and this is obviously unrealistic. On the other hand, the Freundlich isotherm meets the requirement of $\theta_A = 0$ when $p_A = 0$, but θ_A could be larger than 1 if p_A is high enough, which is also unrealistic.

2.7   KINETICS OF HETEROGENEOUS CATALYTIC REACTIONS

From a chemical engineering perspective, the main task for heterogeneous kinetic study is to obtain the reaction rate equation. As discussed above, a heterogeneous reaction involves multiple steps including adsorption, desorption, surface reaction, etc. It is not difficult to write a rate equation for each step, and in the previous section we already discussed rate equations for adsorption and desorption. Surface reaction is an elementary step, and hence mass action law can be used. The challenging part is to derive the overall reaction rate equation based on the information of each individual step, and this is the focus of this section.

2.7.1 Steady-State Approximation and Rate-Determining Step

In order to derive the overall reaction rate equation, it is necessary to make some approximations to simplify the problem. The two most widely used approximations are the steady-state and rate-determining step. In fact we have already used such approximations, but here we will discuss them more comprehensively so that their physical meaning can be better understood.

If a reaction process is under steady-state, the concentration of reaction intermediates will not change with time, i.e.:

$$\frac{dc_I}{dt} = 0, I = 1, 2, \ldots N \tag{2.83}$$

where c_I is the concentration of the intermediate compounds. This is the steady-state approximation. This approximation can also be described as the reaction rates for consecutive steps are all the same, since the reaction process reaches a steady-state. Now we will use a simple example to illustrate that these two descriptions are equivalent. Assume a chemical reaction:

$$A \leftrightharpoons R$$

includes the following two steps:

$$A \leftrightharpoons A^*$$

$$A^* \leftrightharpoons R$$

The rates for these two steps are:

$$r_1 = \overrightarrow{k_1} c_A - \overleftarrow{k_1} c_{A^*}$$

$$r_2 = \overrightarrow{k_2} c_{A^*} - \overleftarrow{k_2} c_R$$

The concentration change of the intermediate compound A^* can be expressed as:

$$\frac{dc_{A^*}}{dt} = \left(\overrightarrow{k_1} c_A - \overleftarrow{k_1} c_{A^*} \right) - \left(\overrightarrow{k_2} c_{A^*} - \overleftarrow{k_2} c_R \right) = r_1 - r_2$$

When $\frac{dc_{A^*}}{dt} = 0$

$$r_1 = r_2 \tag{2.84}$$

Therefore these two descriptions are equivalent.

Now, we can use the same example to discuss the concept of the rate-determining step. Both steps are reversible, and the reaction rate r_i is the difference between the forward reaction rate $\vec{r}_i$ and backward reaction rate $\overleftarrow{r}_i (i = 1, 2)$:

$$r_1 = \vec{r}_1 - \overleftarrow{r}_1$$

$$r_2 = \vec{r}_2 - \overleftarrow{r}_2$$

where $\vec{r}_1 = \vec{k}_1 c_A$, $\overleftarrow{r}_1 = \overleftarrow{k}_1 c_{A^*}$, $\vec{r}_2 = \vec{k}_2 c_{A^*}$; $\overleftarrow{r}_2 = \overleftarrow{k}_2 c_R$. The relative magnitudes of reaction rates are illustrated in Fig. 2.5. As shown in the figure, the forward and backward reaction rates for different steps are different, but based on steady-state assumption, the net reaction rates for all steps are the same and equal to the overall reaction rate, i.e.:

$$r_1 = r_2 = r$$

Both reaction steps are reversible. Although the net reaction rates are the same, the distances from their equilibrium for these two reaction steps are different. We can use the following to descript how close each step approaches its equilibrium:

$$\frac{\vec{r}_1 - \overleftarrow{r}_1}{\vec{r}_1}$$

and

$$\frac{\vec{r}_2 - \overleftarrow{r}_2}{\vec{r}_2}$$

If the value is zero, it indicates that the step is at equilibrium. Since $\vec{r}_1 \gg \vec{r}_2$ and $\vec{r}_1 - \overleftarrow{r}_1 = \vec{r}_2 - \overleftarrow{r}_2$, the first step is much closer to its equilibrium than the second step. As a result, the second step is considered to be the rate-determining step, and its rate represents the rate of the overall reaction. All other steps except the rate-determining step can be approximately treated as at equilibrium. Such treatment is actually also a reflection of the fact that the reaction is at steady-state.

There is another definition of the rate-determining step, the slowest step in the reaction sequence is considered as the rate-determining step. But this

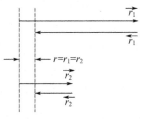

FIGURE 2.5 Illustration of relative magnitude of forward and backward reaction rates.

definition seems to contradict the steady-state assumption. When the reaction is at steady-state, rates of all the steps are equal to each other. Then, how do we understand the slowest step? The reaction steps in a reaction sequence are related to each other, and before the overall reaction achieves its steady state, the rates of different steps are different, i.e., some steps proceed faster or slower than others. Only when the reaction reaches the steady-state will the rates of all steps become equal. Before the reaction reaches steady-state, the slowest step is the rate-determining step. There is another way to understand this concept. If all the steps proceed independently and are not related to other steps, the rate of different steps will be different (of course, in very rare case it is possible that all the steps proceed at the same rate), and the slowest step is the rate-determining step. It does not matter which concept is used for the rate-determining step since they reflect the same fact.

The steady-state and rate-determining steps are two important concepts for deriving rate equations for chemical reactions. Using these two concepts, many kinetic equations can be derived and simplified. However, for some reactions it is not possible to identify a reaction step as the rate-determining step, which means the rates for different steps are close to each other, or the degrees of approaching their equilibrium are similar. On the other hand, it is also possible that there are two rate-determining steps. Finally, it should be noted that the rate-determining step might change due to a change in conditions. For example, at early stages, one step is the rate-determining step, but at a later time another step becomes the rate-determining step.

2.7.2 Rate Equations of Heterogeneous Catalytic Reactions

Using steady-state and rate-determining step concepts, we can derive the reaction rate equation based on the proposed reaction mechanisms. Here we can still use the example of the reaction of A + B = R, and its reaction sequence as given in Eqs. (2.64a–2.64d). We will derive the overall rate equations under different scenarios.

1. Surface reaction control. Now the third step is the rate-determining step, and the overall rate is equal to the rate of this step. Using mass action law for the surface reaction described by Eq. (2.64c):

$$r = \vec{k}_s \theta_A \theta_B - \overleftarrow{k}_s \theta_R \theta_V \qquad (2.85)$$

θ_V is the vacant fraction, equal to $1 - \theta_A - \theta_B - \theta_R$. The other steps are at equilibrium, therefore:

$$k_{aA} p_A \theta_V - k_{dA} \theta_A = 0 \text{ or } \theta_A = K_A p_A \theta_V \qquad (2.86)$$

$$k_{aB} p_B \theta_V - k_{dB} \theta_B = 0 \text{ or } \theta_B = K_B p_B \theta_V \qquad (2.87)$$

$$k_{aR} p_R \theta_V - k_{dR} \theta_R = 0 \text{ or } \theta_R = K_R p_R \theta_V \qquad (2.88)$$

where $K_A = k_{aA}/k_{dA}$, $K_B = k_{aB}/k_{dB}$, and $K_R = k_{aR}/k_{dR}$. Substituting Eqs. (2.86–2.88) into Eq. (2.85):

$$r = \vec{k}_s K_A p_A K_B p_B \theta_V^2 - \overleftarrow{k}_s K_R p_R \theta_V^2 \qquad (2.89)$$

Since $\theta_A + \theta_B + \theta_R + \theta_V = 1$, using this correlation and combining Eqs. (2.86–2.88), we can obtain:

$$\theta_V = \frac{1}{1 + K_A p_A + K_B p_B + K_R p_R}$$

Substituting into Eq. (2.89), the reaction rate equation is:

$$r = \frac{\vec{k}_s K_A K_B p_A p_B - \overleftarrow{k}_s K_R p_R}{(1 + K_A p_A + K_B p_B + K_R p_R)^2} = \frac{k(p_A p_B - p_R/K_P)}{(1 + K_A p_A + K_B p_B + K_R p_R)^2} \qquad (2.90)$$

where k is the rate constant for the forward reaction, equal to $\vec{k}_s K_A K_B$. $K_P = \vec{k}_s K_A K_B / \overleftarrow{k}_s K_R$, which is the equilibrium constant for this reaction. When the reaction reaches its equilibrium, $r = 0$, then from Eq. (2.90) we can obtain the definition equation of the equilibrium constant. Since $k \neq 0$, the item inside the parenthesis in the numerator must be zero.

2. Adsorption control of component A

Now the first step is the rate-determining step, and the reaction rate equals to the rate of adsorption of component A. From Eq. (2.65):

$$r = k_{aA} p_A \theta_V - k_{dA} \theta_A \qquad (2.91)$$

The other three steps are at equilibrium. When the third step of the surface reaction reaches its equilibrium, from Eq. (2.64c)

$$\theta_R \theta_V / (\theta_A \theta_B) = \vec{k}_s / \overleftarrow{k}_s = K_S \qquad (2.92)$$

K_S is the equilibrium constant for the surface reaction. Using θ_B and θ_R given by Eqs. (2.87) and (2.88), respectively:

$$\theta_A = K_R p_R \theta_V / (K_S K_B p_B) \qquad (2.93)$$

Substituting into Eq. (2.91):

$$r = k_{aA} p_A \theta_V - k_{dA} K_R p_R \theta_V / (K_S K_B p_B) \qquad (2.94)$$

Since $\theta_A + \theta_B + \theta_R + \theta_V = 1$, and substituting Eq. (2.93) and Eqs. (2.87) and (2.88) into the above equation:

$$\theta_V = \frac{1}{1 + K_R p_R / K_S K_p p_B + K_B p_B + K_R p_R}$$

Then Eq. (2.94) becomes:

$$r = \frac{k_{aA} p_A - k_{dA} K_R p_R / K_S K_B p_B}{1 + K_R p_R / K_S K_B p_B + K_B p_B + K_R p_R} = \frac{k_{aA}(p_A - p_R / K_P p_B)}{1 + K_A p_R / K_p p_B + K_B p_B + K_R p_R} \qquad (2.95)$$

where K_P is the reaction equilibrium constant, and its relationship with adsorption equilibrium constants and the equilibrium constant for the surface reaction is the same as that in Eq. (2.90).

3. Desorption control of R

Now the last step is the rate-determining step, and from Eq. (2.64d):

$$r = k_{dR}\theta_R - k_{aR}p_R\theta_V \qquad (2.96)$$

The first three steps are at equilibrium, and substituting Eqs. (2.86) and (2.87) into Eq. (2.92):

$$\theta_R = K_S K_A K_B p_A p_B \theta_V$$

$\theta_A + \theta_B + \theta_R + \theta_V = 1$, therefore:

$$\theta_V = \frac{1}{1 + K_A p_A + K_B p_B + K_S K_A K_B p_A p_B}$$

Substituting θ_R and θ_V into Eq. (2.96)

$$r = \frac{k_{dR} K_S K_A K_B p_A p_B - k_{aR} p_R}{1 + K_A p_A + K_B p_B + K_S K_A K_B p_A p_B} = \frac{k(p_A p_B - p_R/K_P)}{1 + K_A p_A + K_B p_B + K_R K_P p_A p_B} \qquad (2.97)$$

where $k = k_{dR} K_S K_A K_B$. Eq. (2.97) is the reaction rate equation when desorption is the rate-determining step.

So far we have derived the rate equations for reaction $A + B = R$ for three scenarios based on the proposed reaction steps. In all three scenarios, we do not consider the presence of inert gas. An inert gas component in the reaction mixture may have an impact on the reaction rate, since it, although not participating in the reactions, may be adsorbed on the catalyst surface. If this is the case, a $K_I p_I$ term needs to be added in the denominator in the rate equation, where p_I is the partial pressure of the inert component, and K_I is its equilibrium constant. For example, Eq. (2.90) will become:

$$r = \frac{k(p_A p_B - p_R/K_P)}{(1 + K_A p_A + K_B p_B + K_R p_R + K_I p_I)^2} \qquad (2.98)$$

All of the three rate equations, i.e., Eqs. (2.90), (2.95), and (2.97), are derived based on ideal surface. They are different, but can be expressed in the same format:

Reaction rate = (Kinetic term) × (Driving force)/(Adsorption term)^n

The kinetic term is the reaction rate constant k, which is a function of temperature. For a reversible reaction, the driving force is the distance from its equilibrium. The farther away the reaction is from its equilibrium state the stronger the driving force is, and the higher the reaction rate will be. All three reaction rate equations contain the term $(p_A p_B - p_R/K_P)$ which is the driving force term. For an irreversible reaction, the driving force term will not be a difference of two terms. Instead the driving force term will only contain reactant pressures, and in this situation the driving force represents the extent of the reaction progress. The adsorption term describes which components are adsorbed on the catalyst surface and how strong those adsorptions are. This type of kinetic equation is called a hyperbolic type rate equation. In fact, as long as we use the ideal surface assumption, it does not matter which step is the rate-determining step the overall reaction rate equation will always be the hyperbolic type.

We can summarize the procedure of deriving the rate equation for a heterogeneous catalytic reaction as follows:

1. Propose the reaction steps;
2. Determine the rate-determining step and write the rate equation for this step, which also is the rate for the overall reaction;
3. Consider all other steps are at equilibrium and write equilibrium equations for them. Express surface coverage of each reacting component as a function of partial pressures;
4. The sum of coverages must be equal to one. Use this equation and combine it with surface coverage equations obtained from step (3) to obtain the vacant fraction as a function of partial pressures;
5. Substitute equations obtained from step (3) and (4) into the rate equation derived from step (2), rearrange and simplify to obtain the rate equation.

The above discussions are based on the fact that all the chemical adsorptions take place on the same type of adsorption site. In other words, on the catalyst surface there is only one type of adsorption site that can adsorb all or part of the components in the reaction system. Some catalysts can have two different active sites. Example 2.7 will show how to derive a reaction rate equation in this situation.

As pointed out before, it is extremely rare that an adsorption is ideal. However, hyperbolic type reaction rate equations are still widely used in the literature. In many situations the equilibrium constant obtained from regression of the reaction rate equation is different from the value measured through independent adsorption test. Therefore it is questionable whether the reaction rate equation based on ideal surface would be correct. But how can this type of rate equation still correlate with laboratory data reasonably well? The main reason is that this type of rate equation is a multiparameter model that is very flexible mathematically. In the rate equation there are many parameters that can be adjusted to make the rate equation in good agreement with the experimental data. However, such an agreement only demonstrates that the rate equation derived based on proposed reaction steps can fit the experimental data. It does not prove that the proposed reaction steps make up the real reaction mechanism. In fact, in many cases the same rate equation can be obtained based on different reaction steps.

The procedure to derive the reaction rate equation based on real adsorption models is the same as that based on the ideal adsorption model, i.e., the steps summarized above can be used. The difference is that the adsorption rate equation and adsorption isotherm will be different. Example 2.8 uses ammonia synthesis reaction on iron catalyst as an example to illustrate the derivation of the reaction rate equation based on a real adsorption model. When the real adsorption model is used, the reaction rate equation is usually a power law function, although for some situations it could also be hyperbolic.

For some gas–solid catalytic reactions both power law and hyperbolic rate equations have been used and compared with kinetic data, and the conclusion was that the accuracy of these two types of rate equations was similar, and it is hard to say which is better. Therefore, unless there is enough experimental data to prove the proposed mechanism is correct, the power law rate equation is not necessarily better even though it is derived based on a real adsorption model. One should be very careful to use the term of mechanistic model. However, from a reactor design and analysis perspective, the power law rate equation is simpler, since, unlike the hyperbolic rate equation, it contains less temperature dependent parameters. For either lab data regression or reactor design applications, the power law rate equation is simpler and more convenient to use. Of course, the application of the rate equation should be within the condition range of the experimental data from which the rate equation is derived.

Example 2.7
Cyclohexane is an important feedstock in the petrochemical industry, and is manufactured commercially by benzene hydrogenation on nickel catalyst:

$$C_6H_6 + 3H_2 \rightleftharpoons C_6H_{12}$$

The reaction temperature is below 200°C, and the reaction is irreversible and exothermic. Assume there are two types of active sites on the nickel catalyst. One adsorbs benzene and the intermediate compounds; another adsorbs hydrogen. Cyclohexane is not adsorbed. The reaction sequence is:

(1) $H_2 + 2\sigma_2 \rightleftharpoons 2H\sigma_2$
(2) $C_6H_6 + \sigma_1 \rightleftharpoons C_6H_6\sigma_1$
(3) $C_6H_6\sigma_1 + H\sigma_2 \rightleftharpoons C_6H_7\sigma_1 + \sigma_2$
(4) $C_6H_7\sigma_1 + H\sigma_2 \rightleftharpoons C_6H_8\sigma_1 + \sigma_2$
(5) $C_6H_8\sigma_1 + H\sigma_2 \rightleftharpoons C_6H_9\sigma_1 + \sigma_2$
(6) $C_6H_9\sigma_1 + H\sigma_2 \rightleftharpoons C_6H_{10}\sigma_1 + \sigma_2$
(7) $C_6H_{10}\sigma_1 + H\sigma_2 \rightleftharpoons C_6H_{11}\sigma_1 + \sigma_2$
(8) $C_6H_{11}\sigma_1 + H\sigma_2 \rightleftharpoons C_6H_{12} + \sigma_1 + \sigma_2$

(Continued)

(Continued)

If step (3) is the rate-determining step, and except for benzene and hydrogen the adsorptions of all intermediate compounds are very weak, please derive the reaction rate equation.

Solution

Since the step (3) is the rate-determining step, under steady-state the reaction rate is:

$$r = k_S \theta_{1B} \theta_{2H} \tag{A}$$

where θ_{1B} and θ_{2H} are surface coverage fractions of benzene and hydrogen, respectively. Steps (1) and (2) are at equilibrium and except for benzene and hydrogen the adsorptions of other intermediate compounds are very weak, therefore:

$$\theta_{1B} = \frac{K_B p_B}{1 + K_B p_B} \tag{B}$$

$$\theta_{2H} = \frac{(K_H p_H)^{0.5}}{1 + (K_H p_H)^{0.5}} \tag{C}$$

Substitute Eqs. (B) and (C) into (A):

$$r = k_S \frac{\sqrt{K_H} K_B \sqrt{p_H} p_B}{(1 + \sqrt{K_H p_H})(1 + K_B p_B)} \tag{D}$$

In the temperature range 90–180°C, $K_H p_H \ll 1$, so Eq. (D) can be simplified as:

$$r = \frac{k p_B p_H^{0.5}}{1 + K_B p_B} \tag{E}$$

where $k = k_S K_B K_H^{0.5}$.

Example 2.8

The reaction steps for ammonia synthesis reaction are proposed as follows:

1. $N_2 + 2\sigma \leftrightarrows 2N\sigma$
2. $H_2 + 2\sigma \leftrightarrows 2H\sigma$
3. $N\sigma + H\sigma \leftrightarrows NH\sigma + \sigma$
4. $NH\sigma + H\sigma \leftrightarrows NH_2\sigma + \sigma$
5. $NH_2\sigma + H\sigma \leftrightarrows NH_3\sigma + \sigma$
6. $NH_3\sigma \leftrightarrows NH_3 + \sigma$

The nitrogen adsorption is the rate-determining step. Please derive the reaction rate equation using the Temkin isotherm model.

(*Continued*)

(Continued)

Solution

Using the Temkin isotherm model, the rate equation for the first step can be expressed as:

$$r = k_a p_N \exp\left(-\frac{\alpha \theta_N}{RT}\right) - k_d \exp\left(\frac{\beta \theta_N}{RT}\right) \tag{A}$$

Since the first step is the rate-determining step, the rest of the steps are at equilibrium and can be combined together. Multiply Eq. (2) by 3, Eq. (3) and (6) by 2, and add Eq. (2) to (6);

$$2N\sigma + 3H_2 \rightleftharpoons 2NH_3 + 2\sigma \tag{B}$$

Since steps (2) to (5) are at equilibrium, Eq. (B) must also be at equilibrium, therefore:

$$K_p^2 = \frac{p_A^2}{p_H^3 p_{N^*}} \tag{C}$$

where p_A and p_H are partial pressures of ammonia and hydrogen, respectively. p_{N^*} is not nitrogen partial pressure in the gas phase; instead, it is the nitrogen partial pressure that is at equilibrium with p_A and p_H. From Temkin isotherm Eq. (2.81)

$$\theta_N = \frac{RT}{\alpha + \beta} \ln(K_0 p_N^*)$$

Using Eq. (C) we can obtain:

$$\theta_N = \frac{RT}{\alpha + \beta} \ln\left(K_0 \frac{p_A^2}{p_H^3 K_p^2}\right)$$

Substituting this equation into Eq. (A), and rearranging to obtain the rate equation for ammonia synthesis reaction:

$$r = \vec{k} p_N \left(\frac{p_A^2}{p_H^3}\right)^{-a} - \overleftarrow{k} \left(\frac{p_A^2}{p_H^3}\right)^{b} \tag{D}$$

where

$$a = \alpha/(\alpha + \beta)$$

$$b = \beta/(\alpha + \beta)$$

$$\vec{k} = k_a \left(K_0/K_p^2\right)^{-a}$$

$$\overleftarrow{k} = k_d \left(K_0/K_p^2\right)^{b}$$

We can see a power law rate equation is obtained when the real adsorption model is used. Of course, this is not always the case. For iron catalyst, the experimental measured values of a and b are $a = b = 0.5$, then Eq. (D) becomes:

$$r = \vec{k} p_N \frac{p_H^{1.5}}{p_A} - \overleftarrow{k} \frac{p_A}{p_H^{1.5}}$$

2.8 DETERMINATION OF KINETIC PARAMETERS

Kinetic parameters are parameters included in the reaction rate equations, such as adsorption equilibrium constants, rate constants, and reaction order. Both adsorption equilibrium constants and rate constants depend on temperature, and the dependence normally follows the Arrhenius equation. The parameters included in the Arrhenius equation, activation energy, preexponential factor, heat of adsorption, etc., are also part of the kinetic parameters, and can be obtained as long as the rate constant and adsorption equilibrium constants are determined at different temperatures. Therefore, for hyperbolic type reaction equations, the key step is to determine the rate constant and adsorption equilibrium constants. For power law rate equations, it is the reaction order and rate constant that need to be determined.

Fundamentally, the kinetic parameters have to be obtained based on kinetic data from experimental study. Obvious the accuracy of the kinetic parameters depends on the quality of the experimental data. For a given reaction rate equation, there are two approaches to determine kinetic parameters: one is the integration method and the other is the differential method.

2.8.1 Integration Method

The first step for the integration method is to integrate the reaction rate equation. For example, if a reaction takes place under constant volume conditions, the power law rate equation can be expressed as:

$$r_A = -\frac{dc_A}{dt} = kc_A^{\alpha} \tag{2.99}$$

In order to obtain reaction order α and rate constant k using the integration method, first we need to integrate Eq. (2.99). After integration:

$$\ln\frac{c_{A0}}{c_A} = kt \quad \alpha = 1$$

$$\frac{1}{c_A^{\alpha-1}} - \frac{1}{c_{A0}^{\alpha-1}} = (\alpha - 1)kt \quad \alpha \neq 1 \tag{2.100}$$

where c_{A0} is the initial concentration of component A. From Eq. (2.100) we know that if we plot $1/c_A^{\alpha-1}$ against time t we should get a straight line (see Fig. 2.6, and also reference to Example 2.2 with slope of $k(\alpha - 1)$ and a intercept of $1/c_{A0}^{\alpha-1}$. But both α and k are parameters to be determined, and we need to first assume a value of α, and make the plot using experimental data to see whether it is a straight line. A straight line indicates that the value of α is correct. Otherwise another value of α has to be tried until a satisfactory value is obtained. It should be noted that the experiments must be conducted under isothermal conditions to maintain k as a constant, otherwise Eq. (2.100) is not valid.

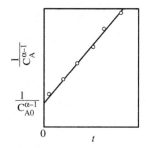

FIGURE 2.6 Estimation of reaction order and rate constant by integration method.

If the rate equation is:

$$r_A = \frac{kK_A p_A}{1 + K_A p_A} \tag{2.101a}$$

kinetic parameters k and K_A can also be determined using the integration method. But first we need to decide how to express r_A. For gas solid catalytic reactions Eq. (2.9) is commonly used. Therefore:

$$-\frac{dF_A}{dW} = \frac{kK_A p_A}{1 + K_A p_A} \tag{2.101b}$$

Before integration, F_A and p_A have to be expressed as a function of conversion X_A. Assuming F_{A0} is the initial molar flux of component A and the initial molar fraction of A is y_{A0}, for constant volume process:

$$F_A = F_{A0}(1 - X_A)$$

$$p_A = P\, y_{A0}(1 - X_A)$$

where P is the total pressure of the reaction system. Using the above two equations in Eq. (2.101b), and after simplification:

$$F_{A0} \frac{dX_A}{dW} = \frac{kK_A P y_{A0}(1 - X_A)}{1 + K_A P y_{A0}(1 - X_A)}$$

Integrating the above equation:

$$\int_0^W \frac{dW}{F_{A0}} = \int_0^{X_A} \frac{1 + K_A P y_{A0}(1 - X_A)}{kK_A P y_{A0}(1 - X_A)} dX_A$$

$$\frac{W}{F_{A0}} = \frac{X_A}{k} - \frac{\ln(1 - X_A)}{kK_A P y_{A0}}$$

Rearranging:

$$\frac{W}{F_{A0} X_A} = \frac{1}{k} - \frac{\ln(1 - X_A)}{kK_A P y_{A0} X_A}$$

Therefore, at a given pressure, temperature, initial gas composition, and catalyst quantity, measuring the reaction conversion at different F_{A0}, and plotting $\ln(1 - X_A)/X_A$ against $W/(F_{A0} \times X_A)$ will form a straight line with a slope of $1/(kK_A Py_{A0})$ and an intercept of $1/k$. Then k and K_A can be determined.

The graphic method was used in both examples to determine the kinetic parameters. When the number of parameters to be determined is no more than two, the graphic method is very convenient.

However, when the number of parameters is more than two, it will be difficult to use the graphic method unless some special design of the kinetic experiments allows for simplification. For example, the power law rate equation below contains three parameters (k, α, and β):

$$r = kc_A^{\alpha} c_B^{\beta}$$

If we design the experiments so that at the beginning concentration of component B is much higher than that of component A, concentration of component B can be considered as constant, and hence can be combined with k as a new parameter. Based on this approximation, the reaction rate equation will be equivalent to Eq. (2.99), and the reaction order for component A, α, and rate constant can be determined. Similarly, if we keep component A excessive, by following the same procedure the reaction order for component B and the rate constant can be determined.

However, in most situations, it is not practical to simplify the reaction rate equation through experimental design due to various reasons, such as the limit of experimental conditions or the reaction rate equation is too complicated. Optimization is a more general and powerful approach to parameter determination, which will be discussed in the next section.

2.8.2 Differential Method

The differential method directly uses the reaction rate equation to determine kinetic parameters based on the reaction rate data measured at different conditions. Using Eq. (2.99) as an example, take the logarithm on both sides:

$$\ln r_A = \alpha \ln c_A + \ln k$$

Based on experimental data, a straight line can be obtained when plotting $\ln r_A$ against $\ln c_A$. The slope is α, and the intercept is $\ln k$.

For Eq. (2.101a), it can be rearranged as:

$$\frac{p_A}{r_A} = \frac{1}{k}p_A + \frac{1}{kK_A} \tag{2.101c}$$

Plot p_A / r_A against p_A, the slope of the straight line obtained is $1/k$, and the intercept is $1/kK_A$. Generally speaking, the graphic method is effective if the kinetic parameters are no more than two and the rate equation can be linearized.

Similar to the integration method, when there are more than two parameters in the rate equation, for the differential method it is not practical to use the graphic method for parameter estimation unless some special experimental design would allow for simplification. Therefore, for kinetic data analysis using either integration or differential methods the most commonly used and versatile approach is regression of experimental data based on statistic principle. Such an approach does not have a limit on the number of parameters, the format of the kinetic equations, or the experimental design, and is very versatile. Next we will discuss the fundamental principles of this approach.

Theoretically speaking, the number of experimental data required are equal to the number of parameters to be determined. For example, Eq. (2.101a) has two parameters, i.e., k and K_A. If two reaction rates, r_{A1} and r_{A2}, were measured experimentally at two partial pressures of A, p_{A1} and p_{A2}. Using these two sets of data in Eq. (2.101a) we will get two equations of k and K_A, and by solving these two equations we can determine k and K_A. However, since it is unavoidable that there will be some experimental error, it is not reliable to determine two parameters based on just two sets of data. In reality, the number of experimental data sets must be more than the number of parameters to be determined. More experimental data would make parameter estimation more reliable. Therefore, there will be more equations than the number of parameters to be determined and the least square method can be used to determine the values of the parameters.

The fundamental principle for the least square method is to seek parameter values that will make the residual sum of square to be at minimum. The residual is the difference between experimental value η and the value predicted by the model $\hat{\eta}$, and therefore the residual sum of square:

$$\Phi = \sum_{i=1}^{M} (\eta_i - \hat{\eta}_i)^2 = \min \tag{2.102}$$

where M is the number of experiments, η is the physical value to be measured. For example, when the differential method is used for kinetic study, typically the reaction rate is measured. Under this situation Eq. (2.102) becomes:

$$\Phi = \sum_{i=1}^{M} (r_i - \hat{r}_i)^2 = \min \tag{2.103}$$

Therefore the goal is to minimize the residual sum of square between the experimentally measured rate and the value calculated by the reaction rate equation. For Eq. (2.101a), the above equation becomes:

$$\Phi = \sum_{i=1}^{M} \left(r_i - \frac{kK_A p_{Ai}}{1 + K_A p_{Ai}} \right)^2 = \min \tag{2.104}$$

where p_{Ai} is the partial pressure of component A for ith experiment, and r_i is the measured reaction rate at this partial pressure. Now the residual sum of square Φ is a function of k and K_A. What we need to do next is to find the values of k and K_A to minimize Φ, and this can be achieved by using the extremum seeking method. Taking the derivative of Eq. (2.104) with respect to k and K_A, and setting $d\Phi/dk = 0$ and $d\Phi/dK_A = 0$, we will get two equations. By solving these two equations we will obtain k and K_A. The same procedure can be used for more parameters. The only difference is the number of equations will increase.

Based on the procedure described above, to obtain parameters using the least square method we need to solve a group of equations as follows:

$$\frac{\partial \Phi}{\partial k_i} = 0, \quad i = 1, 2, 3, \ldots, N$$

k_i are the kinetic parameters and N is the number of kinetic parameters to be determined. This is a group of algebra equations. In most situations the kinetic equations are nonlinear, therefore this is a group of nonlinear algebra equations. As a result, this procedure is often called a nonlinear least square method or nonlinear regression.

Solving a group of nonlinear algebra equations is not easy, especially when there are many equations. If the rate equations can be transformed into a linear format, it will make the parameter estimation much easier mathematically. For example, Eq. (2.101a) can be rewritten in a linear format as Eq. (2.101c). If $p_A/r_A = \eta$, $1/k = \alpha$, and $1/(kK_A) = \beta$, Eq. (2.101a) can be expressed as:

$$\eta = \alpha p_A + \beta \tag{2.105}$$

After transformation, the residual sum of square becomes:

$$\Phi = \sum_{i=1}^{M} (\eta_i - \alpha p_{Ai} - \beta)^2 = \min$$

Taking the derivative with respect to α and β and letting them equal zero, two linear algebra equations of α and β can be obtained, and then solved to get α and β. Based on the definition of α and β given above, k and K_A can be calculated. Regression after linearization is called the linear regression method, or linear least square method. The obvious advantage of the linear least square method is a group of linear algebra equations needs to be solved, and solving linear algebra equations is a lot easier and simpler than solving nonlinear algebra equations. However, the transformation may introduce additional errors for parameter estimation. Parameter estimation is very rich in content, and here we can only give a brief introduction of its fundamental principles.

Example 2.9

For benzene hydrogenation on nickel catalyst, the reaction rate equation has already been derived in Example 2.7

$$r_B = \frac{k p_B p_H^{0.5}}{1 + K_B p_B}$$

where p_B and p_H are the partial pressure of benzene and hydrogen, respectively; k is the rate constant and K_B is the adsorption equilibrium constant of benzene. At 423K the reaction rates at various gas compositions were measured in the lab and the data are given in Table 2C. Please calculate the rate constant and benzene adsorption equilibrium constant.

TABLE 2C Reaction Rate for Benzene Hydrogenation Reaction at 423K

Partial Pressure $p_i \times 10^3$ MPa			Reaction Rate $r_B \times 10^3 / [\text{mol}/(\text{g} \cdot \text{h})]$
Benzene	Hydrogen	Cyclohexane	
2.13	93.0	4.29	18.1
2.42	85.5	11.50	19.0
3.81	78.0	17.20	27.0
5.02	86.8	7.65	30.9
5.80	79.6	13.90	35.2
13.9	84.0	1.92	42.4
10.7	80.6	7.47	39.6
9.58	89.3	1.93	40.8
9.02	88.1	2.78	36.5
7.95	86.9	4.24	35.7
6.46	86.3	6.48	33.8
4.73	92.4	4.6	30.4
4.01	92.5	2.34	29.7
3.30	92.2	3.20	26.3

Solution

First we transform the rate equation into linear form:

$$\frac{p_B p_H^{0.5}}{r_B} = \frac{1}{k} + \frac{K_B}{k} p_B \qquad (A)$$

(Continued)

(Continued)

If we plot $\frac{p_B p_H^{0.5}}{r_B}$ against p_B we will get a straight line, and the slope and intercept can be used to calculate k and K_B. For convenience, let $\frac{p_B p_H^{0.5}}{r_B} = y$, $1/k = b$, and $\frac{K_B}{k} = a$, then Eq. (A) becomes:

$$y = ap_B + b \tag{B}$$

Using the data given in Table 2C, the values of y can be calculated, as shown in Table 2D, and then plot y against p_B, as shown in Fig. 2C. The slope of the straight line is 4.763 and the intercept is 0.0215, i.e.:

TABLE 2D Relationship Between y and p_B

$p_B \times 10^3$	$y \times 10^2$	$p_B^2 \times 10^6$	$p_B y \times 10^5$
2.13	3.589	4.537	7.645
2.42	3.742	5.856	9.012
3.81	3.941	14.52	15.015
5.02	4.786	25.20	24.026
5.80	4.649	33.64	26.964
13.90	9.592	193.20	133.329
10.70	7.671	114.50	82.080
9.58	7.017	91.78	67.223
9.02	7.335	81.36	66.162
7.95	6.565	63.20	52.192
5.46	5.615	41.73	36.273
4.73	4.730	22.7	22.273
4.01	4.106	19.08	16.465
3.30	3.810	10.89	12.573
Σ 0.08883	0.7713	7.189×10^{-4}	5.713×10^{-3}

Unit: p_i: MPa, r_B: mol/(g·h).

$b = 0.0215 = 1/k$
$k = 46.51 \text{ mol/(g·h·MPa}^{1.5})$
$a = 4.763 = K_B/k$
$K_B = 4.763 \times 46.51 = 221.51/\text{MPa}$

For comparisons' sake, a and b are estimated by using the least square method. By seeking the minimum of the residual sum of square the following bivariate linear regression equations can be obtained:

(Continued)

(Continued)

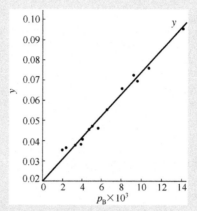

FIGURE 2C   Relationship between y and p_B.

$$a = \frac{\sum_M p_B \sum_M y - M \sum_M p_B y}{\left(\sum_M p_B\right)^2 - M \sum_M p_B^2} \qquad (C)$$

$$b = \frac{1}{M}\left[\sum_M y - a \sum_M p_B\right] \qquad (D)$$

where M is the number of experimental data. In this example, $M = 14$. In order to calculate a and b using Eqs. (C) and (D), p_B^2 and $p_B y$ need to be calculated, which are also shown in Table 2D. Then using Eqs. (C) and (D)

$$a = \frac{0.08883 \times 0.7713 - 14 \times 5.713 \times 10^{-3}}{0.08883^2 - 14 \times 7.189 \times 1^{-4}} = 5.2716$$

$$b = \frac{0.7713 - 5.2716 \times 0.08883}{14} = 0.02166$$

Therefore

$k = 1/b = 46.16 \text{ mol/(g·h·MPa}^{1.5})$

$K_B = ak = 5.2716 \times 46.17 = 243.41/\text{MPa}$

We can see the values of k and K_B calculated from these two approaches are very close. Generally speaking, the values obtained from the linear least square method are more accurate.

Example 2.10

For gas phase hydrogenation of benzene on nickel catalyst, the rate constant and benzene adsorption equilibrium constant are experimentally measured at different temperatures:

(Continued)

(Continued)

T/K	363	393	423	453
k/[mol/(g·h·MPa$^{1.5}$)]	14.52	25.96	45.07	66.03
K_B/(1/MPa)	1495	537.3	237.9	99.41

Please calculate reaction activation energy and heat of adsorption for benzene.

Solution

The temperature dependences of the reaction rate constant and benzene adsorption equilibrium constant on temperature are expressed below:

$$k = A \exp\left(-\frac{E}{RT}\right) \tag{A}$$

$$K_B = K_{B0} \exp\left(\frac{q_B}{RT}\right) \tag{B}$$

Take logarithm on both sides of Eqs. (A) and (B)

$$\ln(k) = \ln(A) - E/(RT) \tag{C}$$

$$\ln(K_B) = \ln K_{B0} + q_B/(RT) \tag{D}$$

Plotting $\ln(k)$ and $\ln(K_B)$ against $1/T$ will give straight lines and the slopes can be used to calculate activation and energy and heat of adsorption. Using the data given above, $1/T$, $\ln(k)$, and $\ln(K_B)$ can be calculated:

$1/T \times 10^3$	2.755	2.545	2.364	2.208
$\ln k$	−0.759	−0.178	0.374	0.756
$\ln K_B$	5.02	3.99	3.18	2.31

The plots are shown in Fig. 2D. The solid line is $\ln(k)$ versus $1/T$ and the dashed line is $\ln(K_B)$ versus $1/T$, and the slope for these two lines are -2.84×10^3 and 5.3×10^3, respectively. From Eq. (C), the slope is $-E/R$, therefore:

$$E = 2.84 \times 10^3 \times 8.313 = 2.36 \times 10^4 \text{ J/mol}$$

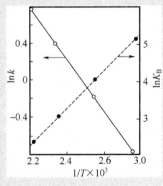

FIGURE 2D Dependence of $\ln(k)$ and $\ln(K_B)$ on $1/T$.

(Continued)

(Continued)

From Eq. (D) we see the slope of the dashed line is q_B/R, therefore:

$$q_B = 5.3 \times 10^3 \times 8.313 = 4.41 \times 10^4 \text{ J/mol}$$

Similarly, since both $\ln(k)$ and $\ln(K_B)$ are linear functions of $1/T$, the linear least square method can also be used to calculate E and q_B, which usually gives more accurate results.

2.9 PROCEDURE FOR DEVELOPING REACTION RATE EQUATION

Studying reaction kinetics is one of the core areas for a reaction engineer, and its main goal is to obtain reaction rate equations. The development of the reaction rate equation must be based on experimental data since so far it is still not possible to derive a reaction rate equation solely based on theoretical considerations. As pointed out above, developing a reaction rate equation usually includes the following steps: (1) propose the reaction mechanism and derive reaction rate equation; (2) conduct an experimental study to obtain kinetic data; and (3) evaluate reaction rate equations and estimate parameters using the experimental data, and eventually select the most suitable rate equation. Obviously, these steps are related to each other, and iterations will be needed to obtain desired results.

For a kinetic study of any reaction system, the first task is to understand what reactions are involved in the system, and which among them is the main reaction and what the side reactions are. It is also critical to understand how these reactions are related to each other, i.e., in series or parallel. In addition to literature research, experimental study is required to obtain such information. The next step is to propose potential reaction models and possible reaction mechanisms based on the information obtained, and then derive reaction rate equations. The derived rate equations must be evaluated based on experimental data to eventually develop the reaction rate equations that can adequately describe the kinetic behavior of the reaction system. For reaction rate equation evaluation first we need to examine how well they simulate the experimental data. In addition, both the values and trend of kinetic parameters must be reasonable. For example, if a power law rate equation is used, and the reaction order estimated based on experimental data is greater than 3, this indicates that the kinetic model is obviously not correct, and therefore, even if the rate equation may simulate the experimental data reasonably well, the rate equation still should not be accepted. Also if the rate constant or adsorption equilibrium constant is negative, the reaction model has to be revised.

A reaction rate equation is a mathematical description of reaction kinetics. As indicated above, both model selection and parameter estimation have to be based on experimental data. Therefore, experimental measurement is at the core of kinetic model development. Only with sufficient and accurate experimental data is it possible to develop a reliable rate equation. The experimental design of a kinetic study includes two parts, i.e., design of experimental reactor and selection of test conditions. It is critical to ensure that the experimental results can capture the key characters of the chemical reactions involved and the data reflects the intrinsic kinetics. To achieve this, the potential impact of physical processes such as mass and heat transfer have to be eliminated. In addition, in most cases the kinetic study needs to be conducted at isothermal conditions, i.e., the effects of concentrations on reaction rate are studied at given temperature, and the impact of temperature is examined by comparing kinetic data obtained at different temperatures. Since reaction rate is very sensitive to temperature, the temperature uniformity of the experimental reactor is very critical for the quality of experimental data. As a result, temperature uniformity is a key criterion for evaluating an experimental reactor. In the following chapter we will have more detailed discussion on experimental reactors and kinetic study.

It is quite common that in order to obtain reaction rate equations for a reaction system tens of reaction models have to be evaluated. It is also possible that several models can simulate the experimental data well. In order to distinguish different reaction kinetic models various statistic analysis and tests can be used. But it is more important to validate the model against experimental results. For example, the models can be examined against data obtained over a broader range of conditions to see which model can predict the behavior better. Another example is to examine the impact of a specific factor on the reaction rate and see which model responds better. In summary, model selection, experimental study, and parameter estimation are related to each other.

It needs to be emphasized that from a practical perspective the reaction rate equations should be as simple as possible, as long as they can adequately describe the kinetic behavior inside the reactor under a certain range of conditions. Therefore, the objective must be clearly defined when developing a kinetic model so that the model will meet the requirements. The kinetic model does not need to be comprehensive and cover all the possible situations, which would only make the model too complex and less usable.

FURTHER READING

Li S. Chemical and catalytic reaction engineering. Beijing: Chemical Industry Press; 1986.

Carberry JJ. Chemical and catalytic reaction engineering. New York: McGraw-Hill; 1976.

Butt JB. Reaction kinetics and reactor design. New Jersey: Prentice-Hall, Eaglewood Cliffs; 1980.

Boudart M. Kinetics of chemical processes. New Jersey: Prentice-Hall, Eaglewood Cliffs; 1968.

PROBLEMS

2.1 Hydrolysis of A takes place in a reactor with constant volume of 4 liters. The initial concentration of A is 12.32% (mass fraction), and the density of the mixture is 1 g/mL. The relative molecular weight of reactant A is 88. The concentration of A as a function of time was measured under isothermal and constant pressure conditions:

Reaction time, h	1.0	2.0	3.0	4.0	5.0	6.0	7.0	8.0	9.0
c_A, mol/L	0.9	0.61	0.42	0.28	0.17	0.12	0.08	0.045	0.03

Calculate the hydrolysis rate of A at reaction time of 3.5 h.

2.2 Methanation reaction takes place in a tubular reactor at ambient pressure and 300°C:

$$CO + 3H_2 \xrightarrow{Ni} CH_4 + H_2O$$

Catalyst volume is 10 mL. The feed gas contains 3% CO, and the rest are N_2 and H_2. CO conversions are measured at different feed flow rates:

$Q_0/(cm^3/min)$	83.3	67.6	50.0	38.5	29.4	22.2
$X/\%$	20	30	40	50	60	70

Find CO consumption rate when feed flow rate $Q_0 = 50$ cm^3/min.

2.3 The forward reaction rate for water gas shift reaction on Fe-Mg catalyst is:

$$r = k_w y_{CO}^{0.85} y_{CO_2}^{-0.4} \text{kmol}/(\text{kg} \cdot \text{h})$$

where y_{CO} and y_{CO_2} are instantaneous molar fractions of CO and CO_2, respectively. Reaction rate constant at 0.1013 MPa and 700K is $k_w = 0.0535$ kmol/(kg · h). If the specific surface area of the catalyst is 30 m^2/g, and the packed density is 1.13 g/cm^3, please calculate:

1. Reaction rate constant based on per unit volume of reactor k_v;
2. Reaction rate constant based on per unit interface area k_s;
3. Reaction rate constant when composition of the reaction system is expressed by partial pressures of reaction components k_p;
4. Reaction rate constant when composition of the reaction system is expressed by molar concentrations of reaction components k_c.

2.4 Liquid phase reaction A + B → C + D takes place under isothermal conditions, and the reaction rate equation is:

$$r_A = 0.8c_A^{1.5}c_B^{0.5} \text{mol}/(\text{L} \cdot \text{min})$$

If equal volume of feed A and B are mixed and both concentrations are 3 mol/L, please find the conversion of A at 4 min.

2.5 The molar fraction of feed stream to an ammonia synthesis reactor is: 3.5% NH_3, 20.87% N_2, 62.6% H_2, 7.08% Ar, and 5.89% CH_4. The operation pressure of the reactor is 30 MPa. At a specific location in the reactor the temperature is 490°C and the ammonia molar fraction of the reacting mixture is 10%, please calculate the reaction rate at this location. The reaction rate equation for ammonia synthesis on iron catalyst is:

$$r = k_1 p_{N_2} \frac{p_{H_2}^{1.5}}{p_{NH_3}} - k_2 \frac{p_{NH_3}}{p_{H_2}^{1.5}} \, \text{kmol}/(\text{m}^3 \cdot \text{h})$$

The activation energy of backward reaction $\overleftarrow{E} = 1.758 \times 10^5 \, \text{J/mol}$. At 450°C $k_2 = 1.02 \times 10^5 \, \text{kmol} \cdot \text{Pa}^{0.5}/(\text{m}^3 \cdot \text{h})$, and $\frac{k_1}{k_2} = K_P^2$. And at 490°C K_P can be calculated by:

$$\log K_P = \frac{2074.8}{T} - 2.4943 \log T - 1.256 \times 10^{-4} T$$
$$+ 1.8546 \times 10^{-7} T^2 + 3.2061/\text{MPa}$$

2.6 The T-X relationship plots for two reactions are shown below, where AB is the equilibrium line, NP is the optimal temperature curve, AP is the isothermal curve, and HB is equal conversion line. Please answer the following questions based on these two plots:

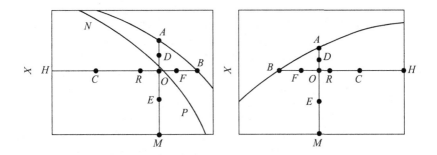

1. Are they reversible or irreversible reactions?
2. Are they endothermic or exothermic reactions?
3. Among points A, D, O, E, and M on the isothermal curve, which one has the fastest reaction rate? Which one has slowest rate?
4. Among points H, C, R, O, F, and B, which one has the fastest reaction rate? Which one has the slowest rate?
5. Which point, C or R, has a faster reaction rate?
6. Among the 10 points on the plot, which one has fastest reaction rate?

2.7 B106 catalyst was used to study reaction kinetics of water gas shift reaction of CO. The activation energy of the forward reaction was measured as 9.629×10^4 J/mol. If the reverse reaction can be ignored, please find the ratio of reaction rate at 550°C to that at 200°C when the reaction mixture composition is the same.

2.8 SO_2 oxidation reaction takes place on vanadium catalyst at 0.103 MPa. The molar composition of the feed stream is 7% SO_2, 11% O_2, and 82% N_2. Please estimate the optimal temperature when conversion is 80%. The activation energy of the forward reaction is 9.211×10^4 J/mol, and the stoichiometric number is ½. Reaction equation is $SO_2 + ½ O_2 \rightleftharpoons SO_3$, and the equilibrium constant as a function of temperature can be described by:

$$\log K_D = \frac{4905.5}{T} - 4.1455$$

Reaction heat $\Delta H_r = -9.629 \times 10^4$ J/mol.

2.9 The following liquid phase reactions take place in a constant volume reactor:

$$A + B \rightarrow R \quad r_R = 1.6c_A$$

$$2A \rightarrow D \quad r_D = 8.2c_A^2$$

where r_R and r_D represents the formation rate, expressed as kmol/ $(m^3 \cdot h)$ of product R and D, respectively. Feed is a mixture of A and B, and the concentration of A is 2 kmol/m^3. Calculate reaction time needed to achieve 95% conversion of A.

2.10 Catalytic hydrogenation of tri-methyl benzene takes place at 0.1 MPa and 523K:

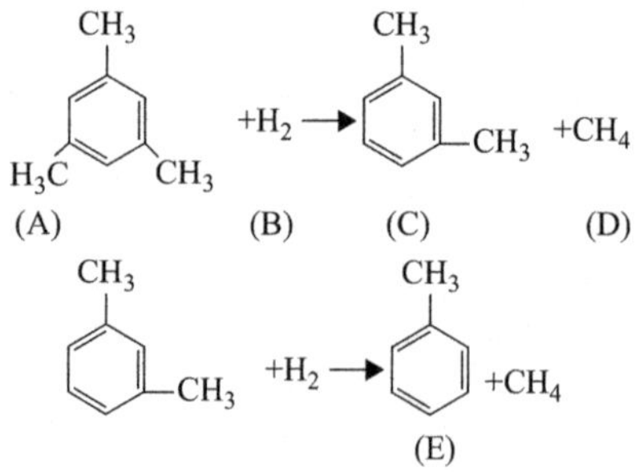

The feed mixture contains 66.67 mol % of hydrogen and 33.33 mol % of tri-methyl benzene. When the conversion of tri-methyl benzene is 80% at the reactor outlet, the hydrogen molar fraction in the mixture is 20%. Please calculate:

1. Gas composition at reactor outlet;
2. Formation rate of di-methyl benzene at reactor outlet if the reaction rate equations for these two reactions are:

$$r_A = 6300 \; c_A c_B^{0.5}, \; kmol/(m^3.h)$$
$$r_E = 3400 \; c_c c_B^{0.5}, \; kmol/(m^3.h)$$

2.11 Gas phase decomposition of ethyl nitrite takes place isothermally under 210°C

$$C_2H_5NO_2 \rightarrow NO + \tfrac{1}{2} CH_3CHO + \tfrac{1}{2} C_2H_5OH$$

This is a first order irreversible reaction, and the reaction rate as a function of temperature can be expressed as $= 1.39 \times 10^{14} \exp\left(-\frac{18973}{T}\right)$, 1/s. The reaction takes place under constant volume condition, the initial pressure is 0.1013 MPa, and the feed is pure ethyl nitrite. Please calculate the decomposition rate of ethyl nitrite and the formation rate of ethanol when the ethyl nitrite conversion is 80%. Recalculate the ethanol formation rate under constant pressure conditions.

2.12 The steam reforming of methane takes place isothermally on a nickel catalyst at 750°C:

$$CH_4 + 2H_2O \rightarrow CO_2 + 4H_2$$

The methane to steam molar ratio in the feed is 1:4. The reaction is first order with respect to all reactants, and the rate constant $k = 2 \; m^3/(mol \cdot s)$.
1. If the reaction takes place under constant volume condition and initial total pressure is 0.1013 MPa, calculate CO_2 and H_2 formation rate when methane conversion is 80%.
2. If the reaction takes place under constant pressure conditions and the other conditions are the same as (1), calculate CO_2 formation rate.

2.13 The following gas phase reactions take place isothermally at 473K and constant pressure:

$$A \rightarrow 3R \quad r_R = 1.2c_A$$

$$A \rightarrow 2S \quad r_S = 0.5c_A$$

$$A \rightarrow T \quad r_T = 2.1c_A$$

where c_A is the molar concentration of reactant A, kmol/m^3, and the unit for reaction rates is kmol/(m$^3 \cdot$ min). In the feed, reactant A to inert gas volume ratio $= 1:1$. Calculate the A consumption rate when its conversion is 85%.

2.14 Cumene decomposition reaction takes place on a Pt catalyst:

$$C_6H_5CH(CH_3)_2 \leftrightarrows C_6H_6 + C_3H_6$$

If cumene, benzene, and propylene are represented by A, B, and R, respectively, the reaction sequence can be expressed as:

$$A + \sigma \leftrightharpoons A\sigma$$
$$A\sigma \leftrightharpoons B\sigma + R$$
$$B\sigma \leftrightharpoons B + \sigma$$

If the surface reaction is the rate-limiting step, please derive the reaction rate equation for cumene decomposition reaction.

2.15 Ethylene oxidation reaction takes place on a silver catalyst:

$$\underset{(A)}{2C_2H_4} + \underset{(B)}{O_2} \rightarrow \underset{(C)}{2C_2H_4O}$$

The reaction sequence can be expressed as:

(1) $A + \sigma \leftrightharpoons A\sigma$

(2) $B_2 + 2\sigma \leftrightharpoons 2B\sigma$

(3) $A\sigma + B\sigma \leftrightharpoons R\sigma + \sigma$

(4) $R\sigma \leftrightharpoons R + \sigma$

If the third step is the rate-determining step, please derive the reaction rate equation.

2.16 Reaction $A \rightarrow B + D$ includes the following steps:

(1) $A + \sigma \leftrightharpoons A\sigma$

(2) $A\sigma \rightarrow B\sigma + D$

(3) $B\sigma \leftrightharpoons B + \sigma$

Please derive the reaction rate equation if the first step is the rate-determining step.

2.17 At relatively low temperature, the reaction rate for the water gas shift reaction:

$$\underset{(A)}{CO} + \underset{(B)}{H_2O} \rightarrow \underset{(C)}{CO_2} + \underset{(D)}{H_2}$$

can be expressed as:

$$r = \frac{kp_A p_B}{1 + K_A p_A + K_C p_C}$$

Could you propose a reaction sequence based on this rate equation?

2.18 Using the data provided in Problem 2.1, derive the kinetic equation using both integrated and differential methods.

2.19 Methanation reaction takes place on a nickel catalyst:

$$CO + 3H_2 \leftrightharpoons CH_4 + H_2O$$

Methane formation rate were measured at 200°C as a function of CO and H_2 partial pressure:

p_{CO}/MPa	0.1013	0.1823	0.4133	0.7294	1.063
p_{H_2}/MPa	0.1013	0.1013	0.1013	0.1013	0.1013

| $r_{CH_4} \times 10^3/[\text{mol}/(\text{g} \cdot \text{min})]$ | 7.33 | 13.2 | 30.0 | 52.8 | 77 |

If a power law rate equation is selected, please estimate the reaction order for carbon monoxide and forward rate constant.

2.20 The reaction rate equation for ethylene oxidation on platinum catalyst can be expressed as:

$$r = \frac{kp_A p_B}{(1 + K_B p_B)^2}$$

where p_A and p_B represent ethylene and oxygen partial pressure, respectively. The reaction rates were measured under isothermal condition at 430K:

No	$p_A \times 10^3/\text{MPa}$	$p_B \times 10^3/\text{MPa}$	$r \times 10^4/[\text{mol}/(\text{g} \cdot \text{min})]$
1	8.99	3.23	0.672
2	14.22	3.00	1.072
3	8.86	4.08	0.598
4	8.32	2.03	0.713
5	4.37	0.89	0.610
6	7.75	1.74	0.834
7	7.75	1.82	0.828
8	6.17	1.73	0.656
9	6.13	1.73	0.694
10	6.98	1.56	0.791
11	2.87	1.06	0.418

Please estimate reaction rate constant k and adsorption equilibrium constant K_B.

Chapter 3

Tank Reactor

Chapter Outline

The tank reactor is one of the widely used reactor types in commercial applications. They can be used for both homogeneous reactions (mostly liquid phase homogeneous reactions) and heterogeneous reactions such as gas−liquid, liquid−solid, liquid−liquid, and gas−liquid−solid reactions, and operated in continuous, batch or semibatch modes. Structure characteristics of tank reactors have been discussed in Chapter 1, Introduction, and this chapter will mainly discuss design, analysis, and optimal operations of tank reactors.

Since most tank reactors have an internal stirring mixing device, reactant concentrations inside a tank reactor can be considered as uniform, and such an assumption is valid for most applications. In fact, this is a very important assumption for the analysis of tank reactors. Similarly, it is also reasonable to assume the temperature inside a tank reactor to be uniform. However, if there are two or more phases in the reactor, then compositions and temperature in each phase may not be the same. As a result, there will be interphase

Reaction Engineering. DOI: http://dx.doi.org/10.1016/B978-0-12-410416 7.00003 3 mass and heat transfer, which must be taken into account when designing and analyzing multiphase reactors and will be discussed in later chapters. This chapter will only discuss homogeneous reactions, or heterogeneous reactions that can be treated using a pseudo homogeneous model.

3.1 MASS BALANCE FOR TANK REACTOR

Assume there are M homogeneous reactions taking place in a tank reactor and the numbers of key components needed to describe this reaction system is K. If all the M reactions are independent, then $K = M$; if not all the reactions are independent, then $K < M$. As mentioned above, concentration and temperature inside the reactor can be considered uniform and equal to those at reactor outlet, so we can use whole reactor volume V_r as control volume for mass balance. Under this situation time is the only independent variable. Assuming that during time interval dt the change of key component i is dn_i, as shown in Fig. 3.1, the volumetric flow rates at reactor inlet and outlet are Q_0 and Q, respectively, then the amounts of key component i flowing in and out the reactor are $Q_0 c_{i0} dt$ and $Q c_i dt$, respectively. The amount of key component i reacted during time period dt is $\mathcal{R}_i V_r dt$. Based on mass balance:

$$Q_0 c_{i0} dt = Q c_i dt - \mathcal{R}_i V_r dt + dn_i$$

If i is a reactant, then $\mathcal{R}_i$ is negative; if i is a product then $\mathcal{R}_i$ is positive. The above equation can be rearranged as:

$$Q_0 c_{i0} = Q c_i - \mathcal{R}_i V_r + \frac{dn_i}{dt}, \quad i = 1, 2, \ldots, K \tag{3.1}$$

These are the mass balance equations for a tank reactor, which is a group of ordinary differential equations. The conversion or formation rate of key component, $\mathcal{R}_i$ has been discussed in Chapter 2, Fundamentals of Reaction Kinetics, and can be expressed as:

$$\mathcal{R}_i = \sum_{j=1}^{M} \nu_{ij} \overline{r}_j \tag{3.2}$$

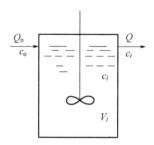

FIGURE 3.1 Notation of a tank reactor.

$\mathscr{R}_i$ calculated from the above equation can be either negative for a reactant or positive for a product. Substitute Eq. (3.2) into Eq. (3.1):

$$Q_0 c_{i0} = Q c_i - V_r \sum_{j=1}^{M} \nu_{ij} \bar{r}_j + \frac{dn_i}{dt}, \quad i = 1, 2, \ldots, K \tag{3.3}$$

The accumulating rate of key component i, dn_i/dt, can be either positive or negative, depending on the actual situations. For a reactor under steady-state, the accumulating rate will be zero, then Eq. (3.3) can be simplified as:

$$Q_0 c_{i0} = Q c_i - V_r \sum_{j=1}^{M} \nu_{ij} \bar{r}_j, \quad i = 1, 2, \ldots, K \tag{3.4}$$

These are mass balance equations for a tank reactor operated under continuous mode at steady-state, which is a group of algebra equations.

For batch operation there is no material input or output, i.e., $Q_0 = Q = 0$, then Eq. (3.3) becomes:

$$-V_r \sum_{j=1}^{M} \nu_{ij} \bar{r}_j + \frac{dn_i}{dt} = 0, \quad i = 1, 2, \ldots, K \tag{3.5}$$

These are mass balance equations for a tank reactor operated under batch mode, which is a group of ordinary differential equations.

Therefore, the general mass balance equations, Eq. (3.3), are only needed for a continuous tank reactor under unsteady-state or a semicontinuous tank reactor. For a semibatch tank reactor with reactants continuously being fed to the reactor and products being removed from the reactor in batch mode, or the other way around, Eq. (3.3) can also be further simplified.

3.2 DESIGN OF ISOTHERMAL BATCH TANK REACTOR (SINGLE REACTION)

The key characteristic of a batch reactor is that both feed loading and product unloading are in batch mode, which offers operation flexibility. Therefore, batch tank reactors can be used for the production of many different products, or the same product with different specifications. It is especially useful for the production of chemicals in small quantities, such as pharmaceuticals, reagents, auxiliary agents, additives, etc. They are also widely used in fine chemical production. In other process industries, batch reactors are also used for small-scale production or processes that need a long reaction time.

Since a batch reactor is operated in batch mode, the operating time includes two parts: reaction time which spans from the end of feed loading to the instant when product yield reaches the target and auxiliary time which includes loading, unloading, cleaning, etc. The main task for batch reactor design is to determine the operation time for each batch, and most importantly the reaction time needed. Auxiliary time is usually estimated based on experience. Next we will discuss reaction time calculation for different types of reactions.

3.2.1 Calculation of Reaction Time and Reaction Volume

Assuming the reaction taking place in the reactor is $A + B \rightarrow R$, and A is chosen as the key component. If at the beginning the amount of A is n_{A0}, then

$$n_A = n_{A0}(1 - X_A)$$

From Eq. (3.5)

$$-V_r \mathcal{R}_A - n_{A0} \frac{dX_A}{dt} = 0 \tag{3.6}$$

When $t = 0$, $X_A = 0$, integrating Eq. (3.6) we can obtain the reaction time needed to achieve conversion X_{Af}

$$t = \int_0^{X_{Af}} \frac{n_{A0} dX_A}{V_r(-\mathcal{R}_A)} \tag{3.7}$$

Eq. (3.7) can be used for any batch reaction process: heterogeneous or homogeneous; isothermal or nonisothermal.

The batch reactor is a closed system, and there is no material or work exchange with the environment, therefore the batch reaction process can be treated as a constant volume process. In other words, the volume of reaction mixture, V_r, can be considered a constant. For homogeneous reactions, $n_{A0}/V_r = c_{A0}$, then Eq. (3.7) can be simplified as:

$$t = c_{A0} \int_0^{X_{Af}} \frac{dX_A}{(-\mathcal{R}_A)} \tag{3.8}$$

For a single reaction, $(-\mathcal{R}_A) = r_A$, therefore as long as the reaction rate equation is given, reaction time needed to reach conversion X_{Af} can be calculated by using Eq. (3.8). From Eq. (3.8) we can see that for single reactions taking place in a batch reactor the reaction time can be calculated by directly integrating the reaction rate equation. Reaction time depends on the initial concentrations of reactants (except for first order reaction) and final conversion required. Of course, this conclusion is valid when reaction temperature is constant. Reaction temperature has a big impact on the reaction time needed. For example, for the ath-order reaction taking place in an isothermal batch reactor:

$$(-\mathcal{R}_A) = r_A = kc_A^a \tag{3.9}$$

or:

$$(-\mathscr{R}_A) = kc_{A0}^a(1-X_A)^a \tag{3.10}$$

Insert Eq. (3.10) into Eq. (3.8):

$$t = \frac{1}{kc_{A0}^{a-1}}\int_0^{X_{Af}}\frac{dX_A}{(1-X_A)^a} = \frac{(1-X_{Af})^{1-a} - 1}{(a-1)kc_{A0}^{a-1}}(a \neq 1) \tag{3.11}$$

Rate constant k is a function of temperature. Under isothermal conditions, k is constant and therefore can be taken out of the integration sign. From Eq. (3.11) we can see that the higher the reaction temperature, the larger the value of k, and the shorter the reaction time will be needed to achieve the same conversion. However, will this conclusion be valid if the reaction is reversible and exothermic?

For a first order irreversible reaction, we can obtain:

$$t = \frac{1}{k}\ln\frac{1}{1 - X_{Af}} \tag{3.12}$$

Comparing Eq. (3.12) with Eq. (3.11) the obvious difference between first and nonfirst order reactions is that for first order reactions the reaction time is independent of the initial concentrations of reactants.

From the design equation for batch reactors we can reach an important conclusion: reaction time needed to achieve a given conversion depends only on reaction rate, i.e., reaction kinetics, and is independent of reactor size. Reactor size is determined by production rate. The reaction time calculation equations derived above can be used for either small or large size reactors. Therefore, data obtained from small-scale reactors in the laboratory can be used for the design of large-scale commercial reactors if the reaction conditions are kept the same. However, it is very hard to maintain the exact same conditions for two very different scale reactors. For example, it is easy to achieve isothermal operation for a small-scale reactor in the laboratory, but it is quite difficult for a large-scale reactor. For small laboratory reactors, concentration uniformity can be easily achieved through stirring, but it is a lot harder to achieve uniform concentration distribution inside a large reactor. Therefore, the performance and behavior of a large commercial batch reactor will always be different in some aspects from those of a small reactor in the laboratory.

Reactor volume for a batch reactor is determined by processing rate Q_0 and operating time. Processing rate Q_0 can be easily calculated based on design requirements. Operating time includes two parts: reaction time t, which needs to be calculated using Eq. (3.8), and auxiliary time t_0, which can only be estimated based on experiences. The reaction volume of a batch reactor is:

$$V_r = Q_0(t + t_0) \tag{3.13}$$

Obviously, Q_0 is the volume of material processed per unit time.

The actual reactor volume V is larger than reaction volume V_r since there must be some space above the reaction mixture. Reactor volume is usually calculated by:

$$V = \frac{V_r}{f} \tag{3.14}$$

where f is the loading factor, typically in the range of 0.4–0.85, and should be selected based on property of reaction mixture. For boiling or foaming liquid a lower value of f, e.g., 0.4–0.6, should be chosen. On the other hand for nonfoaming fluids a loading factor of 0.7–0.85 can be selected.

Example 3.1

A batch reactor is used for production of ethyl acetate through an esterification reaction between ethanol and acetic acid, and required production rate is 12,000 kg per day:

$$\begin{array}{cccc} CH_3COOH + C_2H_5OH = CH_3COOC_2H_5 + H_2O \\ A \qquad\qquad B \qquad\qquad R \qquad\qquad S \end{array}$$

Feed mass ratio is A:B:S = 1:2:1.35, and the density of reaction mixture is 1020 kg/m³ which is assumed to be constant throughout the reaction process. Total auxiliary time including loading, unloading, heating and cleaning is 1 h. The reaction takes place isothermally at 100°C and the rate equation is:

$$r_A = k_1(c_A c_B - c_R c_S/K)$$

At 100°C $k_1 = 4.76 \times 10^{-4}$ L/(mol·min), and equilibrium constant $K = 2.92$. Calculate the reaction volume needed for 35% conversion of acetic acid. Based on the reaction mixture property a loading factor of 0.75 can be used.

Solution

First, the acetic acid feeding can be calculated based on the required ethyl acetate production capacity:

$$F_{A0} = \frac{12000}{88 \times 24 \times 0.35} = 16.23 \text{ kmol/h}$$

In the above equation 88 is the relative molecular weight of ethyl acetate. In the feed solution acetic acid:ethanol:water = 1:2:1.35, so $1 + 2 + 1.35 = 4.35$ kg feed solution contain 1 kg acetic acid. Then volumetric feed flow rate is:

$$\frac{16.23 \times 60 \times 4.35}{1020} = 4.155 \text{ m}^3/\text{h}$$

In above equation, 60 is the relative molecular weight of acetic acid. Feed solution concentration is:

$$c_{A0} = \frac{16.23}{4.155} = 3.908 \text{ mol/L}$$

(Continued)

(Continued)

Relative molecular weights of ethanol and water are 46 and 18, respectively, and based on the initial concentration of acetic acid and mass ratio in the feed solution, the initial concentrations of ethanol and water can be calculated:

$$c_{B0} = \frac{3.908 \times 60 \times 2}{46} = 10.2 \text{ mol/L}$$

$$c_{S0} = \frac{3.908 \times 60 \times 1.35}{18} = 17.59 \text{ mol/L}$$

To use Eq. (3.8) for reaction time calculation the rate equation given in the problem needs to be expressed as a function of conversion, so we first need to express concentrations as functions of conversion:

$$c_A = c_{A0}(1 - X_A)$$
$$c_B = c_{B0} - c_{A0}X_A$$
$$c_R = c_{A0}X_A$$
$$c_S = c_{S0} + c_{A0}X_A$$

Substitute the above equations into Eq. (3.8):

$$r_A = k_1(a + bX_A + cX_A^2)c_{A0}^2 \tag{A}$$

In Eq. (A):

$$a = c_{B0}/c_{A0}$$
$$b = -(1 + c_{B0}/c_{A0} + c_{S0}/c_{A0}K)$$
$$c = 1 - 1/K$$

Substitute Eq. (A) into Eq. (3.8):

$$t = \frac{1}{k_1 c_{A0}} \int_0^{X_{Af}} \frac{dX_A}{a + bX_A + cX_A^2} = \frac{1}{k_1 c_{A0}\sqrt{b^2 - 4ac}} \ln \frac{\left(b + \sqrt{b^2 - 4ac}\right)X_{Af} + 2a}{\left(b - \sqrt{b^2 - 4ac}\right)X_{Af} + 2a} \tag{B}$$

Based on the definition of a, b, and c:

$$a = 10.2/3.908 = 2.61$$
$$b = -[1 + 10.2/3.908 + 17.59/(3.908 \times 2.92)] = -5.15$$
$$c = 1 - 1/2.92 = 0.6575$$
$$\sqrt{b^2 - 4ac} = \sqrt{(-5.15)^2 - 4 \times 2.61 \times 0.6575} = 4.434$$

Then from Eq. (B) we can calculate reaction time:

$$t = \frac{1}{4.76 \times 10^{-4} \times 3.908 \times 4.434} \ln \frac{(-5.15 + 4.434) \times 0.35 + 2 \times 2.61}{(-5.15 - 4.434) \times 0.35 + 2 \times 2.61} = 118.8 \text{ min.}$$

From Eq. (3.13) we can calculate reaction volume:

$$V_r = Q_0(t + t_0) = 4.155 \times (118.8/60 + 1) = 12.38 \text{ m}^3$$

Actual reactor volume is:

$$12.38/0.75 = 16.51 \text{ m}^3$$

3.2.2 Optimal Reaction Time

As mentioned above, for a batch reactor the actual operating time for each batch production includes reaction time and auxiliary time. For given reactions in a specific reactor the auxiliary time is fixed. Reaction concentrations decrease with increasing reaction time, and correspondingly the formation rate of products also decrease with reactant concentration. Therefore with increasing reaction time there will be more product accumulated but product formation per unit time may not increase. As a result, there will be an optimal reaction time for maximizing production per unit time.

For reaction $A \to R$, if product concentration is c_R, then production of R per unit time is:

$$F_R = \frac{V_r c_R}{t + t_0} \tag{3.15}$$

Differentiating Eq. (3.15) with respect to reaction time:

$$\frac{dF_R}{dt} = \frac{V_r \left[(t + t_0) \dfrac{dc_R}{dt} - c_R \right]}{(t + t_0)^2} \tag{3.16}$$

Set $\frac{dF_R}{dt} = 0$
Then from Eq. (3.16):

$$\frac{dc_R}{dt} = \frac{c_R}{t + t_0} \tag{3.17}$$

Eq. (3.17) shows the conditions that must be met in order to achieve maximum production per unit time, and optimal reaction time can be calculated by using this equation. Next we will show how to find the optimal reaction time graphically. First, using Eq. (3.8) we can obtain the relationship between reaction time and reactant concentration (or conversion). Then we can express c_R as a function of reaction time t and plot c_R against t, as shown in curve OMN in Fig. 3.2. Find A on abscissa so that $OA = t_0$, and

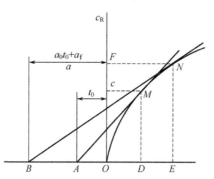

FIGURE 3.2 Optimal reaction time for a batch tank reactor.

then draw a straight line MA from A $(-t_0, 0)$ which is tangent to curve OMN at M, then OD on the abscissa corresponding to point M is the optimal reaction time. From Fig. 3.2 we can see at optimal reaction time the product concentration $c_R = MD$; $AD = t + t_0$; the slope of AM is MD/AD, which is equal to $\frac{dc_R}{dt}$, so Eq. (3.17) is satisfied.

It must be noticed that the optimal reaction time discussed above is to achieve the maximum production per unit time. If the purpose is to minimize feed consumption per unit product, then the longer the reaction time the better. Obviously, different objective functions will lead to different conclusions. If the objective is to minimize operation cost, then we need to use total cost per unit weight product as the objective function. Assuming operating cost per unit reaction time is a, per unit auxiliary time is a_0, and there is a fixed cost of a_f, the total cost per unit weight product is:

$$A_T = \frac{at + a_0 t_0 + a_f}{V_r c_R} \tag{3.18}$$

The optimal reaction time to minimize A_T can be obtained by seeking extremum through differentiating Eq. (3.18) with respect to t:

$$\frac{dA_T}{dt} = \frac{1}{V_r c_R^2}\left[ac_R - (at + a_0 t_0 + a_f)\frac{dc_R}{dt}\right] \tag{3.19}$$

Let $\dfrac{dA_T}{dt} = 0$

Then

$$\frac{dc_R}{dt} = \frac{c_R}{t + (a_0 t_0 + a_f)/a} \tag{3.20}$$

In order to achieve minimum production cost Eq. (3.20) must be satisfied. Similarly, the graphical method can be used to determine the optimal reaction time. On $c_R - t$ plot (Fig. 3.2) from B $[-(a_0 t_0 + a_f)/a,\ 0]$ draw a straight line that is tangent to OMN we obtain line NB, then the abscissa of the point of tangency N gives the optimal reaction time $t = OE$, and the ordinate gives c_R.

3.3 DESIGN OF ISOTHERMAL BATCH TANK REACTOR (MULTIPLE REACTIONS)

3.3.1 Parallel Reactions

Assuming the following parallel reactions take place in an isothermal batch reactor:

$$A \to P \quad r_P = k_1 c_A$$
$$A \to Q \quad r_Q = k_2 c_A$$

P is the desired product, i.e., the first reaction is the main reaction and the second is the side reaction. Using Eq. (3.5) we can write the mass balance for each component:

$$V_r(k_1 + k_2)c_A + \frac{dn_A}{dt} = 0 \tag{3.21}$$

$$-V_r k_1 c_A + \frac{dn_P}{dt} = 0 \tag{3.22}$$

$$-V_r k_2 c_A + \frac{dn_Q}{dt} = 0 \tag{3.23}$$

Since there are only two reactions and each of them is independent, the number of key components is also two, and hence we only need any two of the three equations (from Eqs. (3.21) to (3.23)). For constant volume and a homogeneous system, these three equations can be rewritten by dividing each side with V:

$$(k_1 + k_2)c_A + \frac{dc_A}{dt} = 0 \tag{3.24}$$

$$-k_1 c_A + \frac{dc_P}{dt} = 0 \tag{3.25}$$

$$-k_2 c_A + \frac{dc_Q}{dt} = 0 \tag{3.26}$$

At $t = 0$, $c_A = c_{A0}$, $c_P = 0$, $c_Q = 0$, reaction time can be calculated by integrating Eq. (3.24):

$$t = \frac{1}{k_1 + k_2} \ln \frac{c_{A0}}{c_A} \tag{3.27}$$

or

$$t = \frac{1}{k_1 + k_2} \ln \frac{1}{1 - X_A} \tag{3.28}$$

It seems that only one mass balance equation is needed to calculate the reaction time required, and all the rest of the mass balance equation are superfluous. This is true if we only care about conversion of A. In reality, the purpose of these reactions is to make product, what we are really concerned with is how much product will be produced, but we cannot get this information from Eq. (3.27) or (3.28). We have to use Eq. (3.25) to calculate yield of product P. In order to do so Eq. (3.27) needs to be rewritten as:

$$c_A = c_{A0}\exp[-(k_1 + k_2)t] \tag{3.29}$$

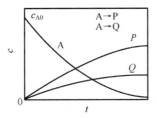

FIGURE 3.3   Concentration–reaction time relationship for parallel reactions.

Substitute Eq. (3.29) into Eq. (3.25) and after integration we can obtain:

$$c_P = \frac{k_1 c_{A0}}{k_1 + k_2}\left\{1 - \exp[-(k_1 + k_2)t]\right\} \tag{3.30}$$

This equation gives the concentration of product P at a given reaction time. From $c_Q = c_{A0} - c_A - c_P$:

$$c_Q = \frac{k_2 c_{A0}}{k_1 + k_2}\left\{1 - \exp[-(k_1 + k_2)t]\right\} \tag{3.31}$$

Eqs. (3.29), (3.30), and (3.31) represent the relationship between reaction system composition and reaction time, which is shown in Fig. 3.3. We can see that the concentration of reactant A always decreases with increasing reaction time, but the concentration of product P and Q increase with reaction time. Divide Eq. (3.30) by Eq. (3.31):

$$\frac{c_P}{c_Q} = \frac{k_1}{k_2}$$

Therefore at any given reaction time the product concentration ratio equals the reaction rate constant ratio. This relationship can also be obtained by dividing Eq. (3.25) by Eq. (3.26) and then integrating. However, it must be noticed that such a conclusion is only valid when the formats of the reaction rate equations for the two reactions are exactly the same. Also for the reaction system discussed above the stoichiometric coefficients for each component are the same. If the stoichiometric coefficients are different, the relationships between concentrations and reaction time need to be modified accordingly, but the function format will be the same as long as the reaction rate equation format are the same.

The above discussion is for two first order reactions. It is pretty straightforward to derive equations for a reaction system with M first order reactions. The concentration of reactant A can be expressed as:

$$c_A = c_{A0}\exp\left(-t\sum_1^M k_i\right) \tag{3.32}$$

Concentration of product i is:

$$c_i = \frac{k_i c_{A0}}{\sum\limits_{1}^{M} k_i} \left[1 - \exp\left(-t \sum\limits_{1}^{M} k_i \right) \right], \quad i = 1, 2, \ldots, M \qquad (3.33)$$

After reaction time is calculated, reaction volume can be easily determined following the same procedure used in the previous section for a single reaction.

Example 3.2

The following liquid phase reactions take place in an isothermal batch tank reactor:

$$A + B \rightarrow P \quad r_P = 2c_A \text{ kmol}/(m^3 \cdot h)$$
$$2A \rightarrow Q \quad r_Q = 0.5c_A^2 \text{ kmol}/(m^3 \cdot h)$$

At beginning the concentrations of both A and B are 2 kmol/m^3. The target product is P, calculate conversion of A and yield of P at reaction time = 3 h.

Solution

$$\mathscr{R}_A = -r_P - 2r_Q = -2c_A - 2 \times 0.5c_A^2 = -2c_A - c_A^2 \qquad (A)$$

From Eq. (3.5):

$$\frac{dn_A}{dt} - V_r \mathscr{R}_A = 0 \qquad (B)$$

For constant volume system, substitute Eq. (A) into Eq. (B) and we can obtain:

$$\frac{dc_A}{dt} + 2c_A + c_A^2 = 0 \qquad (C)$$

Integrating the above equation gives:

$$\int_0^t dt = \int_{c_{A0}}^{c_A} \frac{-dc_A}{2c_A + c_A^2}$$

$$t = \frac{1}{2} \ln \frac{c_{A0}(2 + c_A)}{c_A(2 + c_{A0})} \qquad (D)$$

Initial concentration of component A is $c_{A0} = 2$ kmol/m^3, reaction time $t = 3$ h, then using Eq. (D) we can calculate concentration of component A:

$$3 = \frac{1}{2} \ln \frac{2(2 + c_A)}{c_A(2 + 2)} = \frac{1}{2} \ln \frac{2 + c_A}{2c_A}$$

Therefore:

$$c_A = 2.482 \times 10^{-3} \text{ kmol}/m^3$$

(Continued)

(Continued)

The concentration of component A at reaction time is equal to 3 h. At this time the conversion is:

$$X_A = \frac{2 - 2.482 \times 10^{-3}}{2} = 0.9988 = 99.88\%$$

The amount of P produced cannot be determined by just knowing the conversion of A, since A can be converted to either P or Q. Based on the rate equation given in the problem:

$$\frac{dc_P}{dt} = 2c_A \tag{E}$$

From Eqs. (C) and (E):

$$\frac{dc_A}{dc_P} = -1 - \frac{1}{2}c_A$$

Therefore:

$$\int_0^{c_P} dc_P = -\int_{c_{A0}}^{c_A} \frac{-dc_A}{1 + c_A/2}$$

$$c_P = 2\ln\frac{1 + c_{A0}/2}{1 + c_A/2} \tag{F}$$

$$c_P = 2\ln\frac{1 + 2/2}{1 + 2.482 \times 10^{-3}/2} = 1.3838 \text{ kmol/m}^3$$

The yield of P is:

$$Y_P = \frac{1.3838}{2} = 0.6919 = 69.19\%$$

We can see that although the conversion of A is 99.88% and the yield of P is only 69.19%, the rest (99.88%−69.19% = 30.69%) was converted to Q.

3.3.2   Consecutive Reactions

Now considering that the following first order irreversible reactions take place in an isothermal batch reactor:

$$A \xrightarrow{k_1} P \xrightarrow{k_2} Q$$

Since there are two independent reactions, we need two mass balance equations to describe this reaction system. If A and P are chosen to be key components, using a procedure similar to the one we discussed above for parallel reactions the following two equations can be derived based on Eq. (3.5)

$$-\frac{dc_A}{dt} = k_1 c_A \tag{3.34}$$

$$\frac{dc_P}{dt} = k_1 c_A - k_2 c_P \tag{3.35}$$

If at $t = 0$, $c_A = c_{A0}$, $c_P = 0$, $c_Q = 0$, then integrating Eq. (3.34):

$$c_A = c_{A0} e^{-k_1 t} \tag{3.36}$$

or:

$$t = \frac{1}{k_1} \ln \frac{c_{A0}}{c_A} = \frac{1}{k_1} \ln \frac{1}{1 - X_A} \tag{3.37}$$

Similar to the situation of a parallel reaction system, we can determine the reaction time needed to achieve given conversion by using only one mass balance equation, but we cannot determine the concentration and yield of P. Obviously Eq. (3.37) describes the progress of the first reaction but it cannot tell us the progress of the second reaction.

In order to calculate the concentration of P, we can substitute Eq. (3.36) into Eq. (3.35):

$$\frac{dc_P}{dt} + k_2 c_P = k_1 c_{A0} e^{-k_1 t} \tag{3.38}$$

This is a first order linear ordinary differential equation, which can be solved analytically using initial condition given above:

$$c_P = \frac{k_1 c_{A0}}{k_1 - k_2} (e^{-k_2 t} - e^{-k_1 t}) \tag{3.39}$$

Since $c_A + c_P + c_Q = c_{A0}$,

$$c_Q = c_{A0} - c_A - c_P = c_{A0} \left[1 + \frac{k_2 e^{-k_1 t} - k_1 e^{-k_2 t}}{k_1 - k_2} \right] \tag{3.40}$$

Eqs. (3.36), (3.39), and (3.40) describe the relationships between the concentrations of each component and reaction time, which can be represented by the three curves in Fig. 3.4. From this figure we can see that the

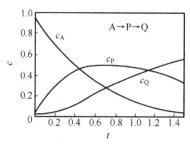

FIGURE 3.4　Concentration—reaction time relationship for consecutive reactions.

concentration of reactant A decreases with increasing reaction time while the concentration of product Q increases. However, the concentration of product P first increases and then decreases with reaction time, exhibiting a maximum. This is a character of consecutive reaction systems due to the fact that P is a product from the conversion of A but at the same time it can be further converted to product Q. At short reaction times the consumption rate of A to P will be higher than the consumption rate of P to Q, the concentration of P increases with the progress of time. At longer reaction times the consumption rate of P to Q will be higher than the consumption rate of A to P, therefore the concentration of P will decrease. From this discussion we can see that if P is the target product then the reaction time needs to be carefully selected to achieve maximum yield of P. The optimal reaction time can be determined by differentiating Eq. (3.39) with respect to reaction t and letting $dc_P/dt = 0$:

$$t_{opt} = \frac{\ln(k_1/k_2)}{k_1 - k_2} \tag{3.41}$$

Using t calculated by Eq. (3.41) in Eq. (3.39), the maximum concentration of P can be determined. If the target product is Q, then the longer the reaction time the higher the yield of Q.

It should be noted that if $k_1 = k_2$ then Eqs. (3.39) to (3.41) would become undefined. Under this situation, will you be able to figure out how to describe the relationship between product concentrations and reaction time? Will there still be a maximum on the concentration curve for the intermediate product P?

All the above discussions are for consecutive reaction systems involving only two first order reactions, but the main conclusions can be expanded to multiple reaction systems. There must be a maximum concentration for intermediate product. For reaction systems involving nonfirst order reactions, the same procedure can be used, but in most situations the balance equations cannot be solved analytically and a numerical method must be used. As long as the reaction time is determined, the reaction volume can be easily calculated by using Eq. (3.13).

Example 3.3

The following reactions take place in an isothermal batch reactor:

$$NH_3 + CH_3OH \xrightarrow{k_1} CH_3NH_2 + H_2O$$

$$CH_3NH_2 + CH_3OH \xrightarrow{k_2} (CH_3)_2NH + H_2O$$

Both of the reactions are first order with respect to corresponding reactants, and at reaction temperature $k_2/k_1 = 0.68$. Calculate the maximum yield of methylamine and the corresponding conversion of ammonia.

(*Continued*)

(Continued)

Solution

The two reaction rate equations are:

$$r_1 = k_1 c_A c_M \tag{A}$$

$$r_2 = k_2 c_B c_M \tag{B}$$

Subscripts A, B, and M represent ammonia, methylamine, and methanol, respectively.

The ammonia consumption rate is:

$$(-\mathscr{R}_A) = r_1 = -\frac{dc_A}{dt} = k_1 c_A c_M$$

Formation rate of methylamine is:

$$\mathscr{R}_B = r_1 - r_2 = \frac{dc_B}{dt} = k_1 c_A c_M - k_2 c_B c_M$$

From the above two equations:

$$-\frac{dc_B}{dc_A} = 1 - \frac{k_2 c_B}{k_1 c_A} \tag{C}$$

$c_A = c_{A0}(1 - X_A)$, $c_B = c_{A0} Y_B$, so Eq. (C) can be written as:

$$\frac{dY_B}{dX_A} = 1 - \frac{k_2 Y_B}{k_1(1 - X_A)} \tag{D}$$

Eq. (D) is a first order linear ordinary differential equation, and its initial conditions are:

$$X_A = 0$$
$$Y_B = 0$$

And the solution of Eq. (D) is:

$$
\begin{aligned}
Y_B &= \exp\left[-\int \frac{k_2 \, dX_A}{k_1(1 - X_A)}\right]\left[\int \exp\left(\int \frac{k_2 \, dX_A}{k_1(1 - X_A)}\right) dX_A + c\right] \\
&= \exp\left[\ln(1 - X_A)^{k_2/k_1}\right]\left\{\int \exp\left[\ln(1 - X_A)^{-k_2/k_1}\right] dX_A + c\right\} \\
&= (1 - X_A)^{k_2/k_1}\left[\int (1 - X_A)^{-k_2/k_1} dX_A + c\right] \\
&= (1 - X_A)^{k_2/k_1}\left[\frac{-(1 - X_A)^{1 - k_2/k_1}}{1 - k_2/k_1} + c\right]
\end{aligned}
\tag{E}
$$

c in Eq. (E) is the integration constant, and can be calculated by applying the initial conditions to Eq. (E)

$$c = 1/(1 - k_2/k_1)$$

(Continued)

(Continued)

Substitute c back into Eq. (E):

$$Y_B = \frac{1}{1 - k_2/k_1}\left[(1 - X_A)^{k_2/k_1} - (1 - X_A)\right] \tag{F}$$

In order to calculate the maximum yield of methylamine we need to differentiate Eq. (F) with respect to X_A:

$$\frac{dY_B}{dX_A} = \frac{1}{1 - k_2/k_1}\left[\frac{-k_2}{k_1}(1 - X_A)^{k_2/k_1 - 1} + 1\right]$$

let $\frac{dY_B}{dX_A} = 0$, then

$$\frac{k_2}{k_1}(1 - X_A)^{k_2/k_1 - 1} = 1$$

Therefore,

$$X_A = 1 - (k_1/k_2)^{1/(k_2/k_1 - 1)} \tag{G}$$

This is the ammonia conversion when the yield of methylamine is at its maximum. It is given that $k_2/k_1 = 0.68$, so from Eq. (G)

$$X_A = 1 - (1/0.68)^{1/(0.68-1)} = 0.7004$$

Using this X_A value in Eq. (F) we can then calculate the maximum yield of methylamine:

$$Y_B = \frac{1}{1 - 0.68}\left[(1 - 0.7004)^{0.68} - (1 - 0.7004)\right] = 0.4406 = 44.06\%$$

In reality, the dimethylamine from the second reaction can further react with methanol to form trimethylamine:

$$(CH_3)_2NH + CH_3OH \xrightarrow{k_3} (CH_3)_3N + H_2O$$

If this reaction needs to be considered, will the maximum yield of methylamine calculated above change? Please also give the reasons that support your conclusion.

3.4 REACTOR VOLUME FOR CONTINUOUS TANK REACTOR (CSTR)

In commercial applications a continuous tank reactor, which is also known as a Continuous Stirred Tank Reactor (CSTR), is usually operated under steady-state. For steady-state operations, all mass flows as well as reactor conditions do not change with time; therefore the reactor design equations are a group of algebra equations. In addition, most tank reactors are used for liquid phase reactions with no significant volume change, and therefore can be considered as constant volume. The potential error caused by the assumption that reactor inlet and outlet volumetric flow rates are the same is very small, so Eq. (3.4) can be simplified as:

$$V_r = \frac{Q_0(c_i - c_{i0})}{\sum_{j=1}^{M} \nu_{ij}\overline{r_j}} \quad i = 1, 2, \ldots, K \tag{3.42}$$

This is the reactor volume calculation equation for a continuous tank reactor. When the feed flow rate, composition, and rate equations are given the reactor volume needed to meet certain conversions can be calculated from Eq. (3.42).

If there is only one reaction taking place in the reactor and component A is the key component, then Eq. (3.42) can be further simplified as:

$$V_r = \frac{Q_0(c_{A0} - c_A)}{(-\mathscr{R}_A)} \tag{3.43}$$

It can also be expressed as a function of conversion:

$$V_r = \frac{Q_0 c_{A0} X_{Af}}{-\mathscr{R}_A(X_{Af})} \tag{3.44}$$

For CSTR under steady-state, both temperature and concentrations in the reactor are constants, as are the reaction rates. Therefore $\mathscr{R}_A$ or $\overline{r_j}$ in Eqs. (3.42) to (3.44) are constant during the reaction process, i.e., uniform throughout the reactor and do not change with time. It is for this reason that the reactor volume can be determined through simple algebraic calculations.

Constant reaction rate is a unique character that differentiates CSTR from other types of reactors. It also needs to be pointed out that $\mathscr{R}_A$ or $\overline{r_j}$ needs to be calculated using composition in the reactor, which is the same as that at reactor outlet.

When multiple reactions take place, a group of algebraic equations need to be solved simultaneously to calculate reactor outlet composition for a given reactor volume. This is because the equations described by Eq. (3.42) are usually coupled to each other. If the reactor inlet and outlet compositions as well as production rate are given, then only one equation is needed to calculate reaction volume no matter how many reactions take place in the reactor.

As discussed in the previous section, the reaction volume for batch tank reactions can be calculated based on reaction time and production rate. For CSTR under steady-state the reactor volume is calculated based on the mass balance equation. In order to compare productivity of CSTR the concept of space-time is usually used:

$$\tau = \frac{V_r}{Q_0} = \frac{\text{Reaction volume}}{\text{Volumetric flow rate of feed}} \tag{3.45}$$

Obviously space-time has the same unit as time. Smaller space-time means higher reactor productivity. For two CSTRs with the same inlet and outlet compositions, the one with smaller space-time has higher productivity. Of course the feed volumetric flow rates for the two reactors must be calculated at same temperature and pressure. For constant volume reaction processes, space-time equals mean residence time in the reactor.

The inverse of space-time is space velocity, which represents the amount of feed processed by unit reactor volume per unit time and has a unit of 1/time. The larger the space velocity, the higher the reactor productivity will be. In order to compare different reactors, the volumetric flow rate is typically defined under standard conditions. For reactions using solid catalysts, it is common to use the catalyst weight to represent reaction volume, and therefore use weight space velocity, which is defined as amount of feed processed by unit catalyst weight per unit time. Of course space velocity can still be defined based on catalyst volume, which is typically the bulk volume of catalyst bed.

In many situations reactor feed is liquid, which is then vaporized and participates gas phase reactions in the reactor. The space velocity calculated based on feed liquid volumetric flow rate is called liquid space velocity. In addition, space velocity can also be defined based on a specific component, such as carbon space velocity, hydrocarbon space velocity, etc. One needs to be very careful when using the space velocity.

Example 3.4
Using the same data and requirements given in Example 3.1, calculate the reaction volume needed if a continuous tank reactor is used to produce ethyl acetate.

Solution
In Example 3.1 we already calculated feed rate and composition:

$$Q_0 = 4.155 \text{ m}^3/\text{h}$$
$$c_{A0} = 3.908 \text{ mol/L}$$
$$c_{B0} = 10.2 \text{ mol/L}$$
$$c_{S0} = 17.59 \text{ mol/L}$$
$$c_{R0} = 0 \text{ mol/L}$$

and

$$k_1 = 4.76 \times 10^{-4} \text{ L/(mol} \cdot \text{min)}$$
$$K = 2.92$$

The acetic acid consumption rate can be expressed as a function of conversion:

$$r_A = k_1[c_{A0} - c_{A0}X_A)(c_{B0} - c_{A0}X_A) - (c_{R0} + c_{A0}X_A)(c_{S0} + c_{A0}X_A)/K]$$

(*Continued*)

(Continued)

Using initial composition and reaction rate constant k_1 and equilibrium constant K listed above:

$$r_A = (18.97 - 37.44X_A + 4.78X_A^2) \times 10^{-3}$$

For CSTR the composition inside the reactor is the same as reactor outlet, so the reaction rate needs to be calculated using conversion at reactor outlet. The final conversion is 35%, and using the above equation we obtain:

$$r_A = (18.97 - 37.44 \times 0.35 + 4.78 \times 0.35^2) \times 10^{-3}$$
$$= 6.452 \times 10^{-3} \text{ mol/(L·min)} = 0.3871 \text{ kmol/(m}^3 \cdot \text{h)}$$

The reaction volume can be calculated by using Eq. (3.44)

$$V_r = \frac{4.155 \times 3.908 \times 0.35}{0.3971} = 14.68 \text{ m}^3$$

In Example 3.1 it was calculated that the reaction volume for a batch reactor was 12.38 m³, which is smaller than the volume of the continuous tank reactor. The reason for this is that for batch reactors the reaction rate changes with time, from higher value at the beginning to lower value at the end. However, for CSTR the reaction rate is constant, and the value corresponds to the lowest reaction rate in a batch reactor. From this aspect the CSTR is inferior to a batch reactor.

Example 3.5
Using a CSTR to conduct the same reactions as those in Example 3.2, If the space-time is 3 h, what will be the final conversion of component A and final yield of P?

Solution
Mass balance equations for A can be written based on Eq. (3.42):

$$V_r = \frac{Q_0(c_{A0} - c_A)}{2c_A + c_A^2}$$

Or

$$\tau = \frac{V_r}{Q_0} = \frac{(c_{A0} - c_A)}{2c_A + c_A^2} \tag{A}$$

Similarly for P:

$$\tau = \frac{c_P}{2c_A} \tag{B}$$

Substitute $\tau = 3$ h and $c_{A0} = 2$ kmol/m³ into Eq. (A):

$$3c_A^2 + 7c_A - 2 = 0$$

Then the concentration of A at reactor outlet can be calculated:

$$c_A = 0.2573 \text{ kmol/}m^3$$

(Continued)

(Continued)

Neglect another root which is negative, so the final conversion of A is:

$$X_{Af} = (2 - 0.2573)/2 = 0.8714$$

Using the value of c_{A0} in Eq. (B) the concentration of P can be calculated:

$$c_P = 2 \times 3 \times 0.2573 = 1.544 \text{ kmol/m}^3$$

Therefore the yield of P is:

$$Y_P = 1.544/2 = 0.722$$

From this example it is clear when space-time of a CSTR equals the reaction time for a batch reactor, both conversion and yield from these two reactors are different. Such a difference must be taken into consideration during reactor scale-up since the reactor for commercial production may be operated differently from lab reactor. For this specific example the yield obtained from CSTR is higher, this is because the lower concentration of A in CSTR favors the formation of desired product P.

3.5 CSTR IN SERIES AND PARALLEL

In commercial applications it is common to use multiple tank reactors to conduct the same reaction. For a given production rate it needs to be decided whether it would be better to use one big reactor or several smaller ones. If multiple reactors are used, it also needs to be decided how to arrange these reactors to achieve optimal performance. In this section we will discuss how to design a multiple reactor system.

3.5.1 Overview

In previous sections we discussed reactor volume calculation for tank reactors. For example, if a CSTR is used for a single reaction then its volume can be calculated by using Eq. (3.43). Next we will use Fig. 3.5 to discuss

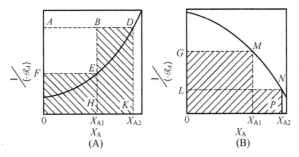

FIGURE 3.5 Graphical representations of reactor volumes for CSTR. (A) Normal kinetics; (B) Abnormal kinetics.

this equation geometrically. Based on the reaction rate equation $1/(-\mathscr{R}_A)$ is plotted as a function of X_A, as shown in Fig. 3.5. In most situations the consumption rate $(-\mathscr{R}_A)$ decreases with an increase of X_A, this is called normal kinetics, so $1/(-\mathscr{R}_A)$ increases with X_A, as shown in Fig. 3.5(A). Fig. 3.5(B) shows $1/(-\mathscr{R}_A) \sim X_A$ plot for abnormal kinetics when reaction order is negative as in the situations of autocatalytic reactions. When reactant A is converted to conversion X_{A2} in a CSTR, from Eq. (3.44) we know the reaction volume is $Q_0 c_{A0} X_{A2}/(-\mathscr{R}_{A2})$. From Fig. 3.5(A) we can see that $X_{A2}/(-\mathscr{R}_{A2})$ equals the area of rectangle OADK. Therefore, for a given feed condition reaction volume is proportional to this rectangular area.

If two CSTRs are used in series, and the conversion reaches to X_{A1} in the first reactor and then to X_{A2} in the second, which is the same as that achieved when using one big CSTR. From Fig. 3.5(A) we know that reaction volume in the first and the second reactor is proportional to the rectangular area OFEH and BHDK, respectively. Since these two reactors are in series, the feed flow rate to both reactors is the same, i.e., Q_0. In addition, both X_{A1} and X_{A2} are defined based on the compositions of feed to the first reactor, i.e., based on c_{A0}. Therefore, Eq. (3.44) needs to be modified to be applied to the second reactor. Specifically, the numerator needs to be changed from X_A to $(X_{A2} - X_{A1})$. This is because Eq. (3.44) is only valid for conversion at inlet is zero. For two reactors operated in series, the inlet of the second reactor is the outlet stream from the first reactor with conversion of X_{A1}. $(X_{A2} - X_{A1})$ is the conversion of component A in the second reactor.

Based on the above analysis and by referring to Fig. 3.5(A) we can see for a single reactor the reaction volume is proportional to the rectangular area OADK and for two reactors in series the total reaction volume is proportional to the sum of the two rectangular areas, OFEH and HBDK. It is obvious that area OADK > (area OFEH + area HBDK). Under this situation, the total reaction volume of a two reactor in series is smaller than that of a single reactor. This conclusion can be expanded to multiple CSTR in series. The more reactors used in a series the smaller the total reaction volume will be. Therefore for normal kinetics it is always advantageous to use multiple CSTRs in series to reduce total reaction volume.

For abnormal kinetics the situation is just the opposite. The total reaction volume for multiple CSTR in series is larger than that of a single CSTR, as shown in Fig. 3.5(B).

For normal kinetics the CSTR is operated at the minimum reaction rate while for abnormal kinetics the reactor is operated at the maximum reaction rate, this is the reason that the optimal arrangement of reactors is totally different for these two situations.

In some commercial applications the required reaction volume is so big that makes it impractical to manufacture a single reactor, then multiple reactors have to be used. Based on the discussions above, for normal kinetics these reactors should be arranged in series. For abnormal kinetics, each reactor

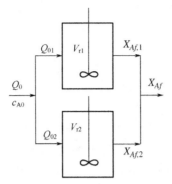

FIGURE 3.6 CSTR in parallel.

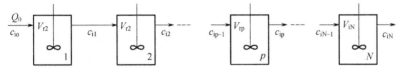

FIGURE 3.7 CSTR in series.

should be operated independently, i.e., the reactors need to be arranged in parallel. Fig. 3.6 shows a configuration of two reactors in parallel. For parallel configuration, the flow rate distribution to each reactor needs to be decided. A general rule is to maintain the same space-time in each tank reactor so that the conversion in each reactor will be the same since such an arrangement will lead to the optimal results. In order to maintain the same space-time the feed flow rate to each reactor should be proportional to the reaction volume.

3.5.2 Calculations for Multiple Reactors in Series

Fig. 3.7 shows N reactors in series, and the volume of each reactor is $V_{r1}, V_{r2}, \ldots V_{rN}$, respectively. It is relatively easy to calculate the final conversion X_{iN} or final yield, Y_{iN}. What we need to do is to write mass balance equations for key component i for each reactor:

$$V_{r_p} = \frac{Q_0(c_{ip} - c_{ip-1})}{\left(\sum_j^M \nu_{ij}\overline{r_j}\right)_p}, \quad p = 1, 2, \ldots, N; \ i = 1, 2, \ldots, K; \ j = 1, 2, \ldots, M$$

$$(3.46)$$

The number of key components is K, therefore for each reactor there will be K equations, and the total number of equations will be KN. If feed flow rate and composition, reaction volume, and temperature in each reactor are known, the outlet composition of the last reactor can be calculated by solving Eq. (3.46). The calculation can be started from the first reactor. The inlet composition for the first reactor is known, so using Eq. (3.46) we can obtain its outlet composition, which will be the inlet composition for the second reactor. This process is continued until we obtain the outlet composition of the last reactor. If there are multiple reactions taking place, each step will involve solving a group of algebraic equations. Such a procedure is called tank-by-tank calculations.

For most design problems, the known aspects are feed flow rate and composition and required product composition; the number of reactors and the reaction volume of each reactor needs to be determined. Generally speaking, the more reactors used the smaller the total reaction volume will be. From this perspective using more reactors seems beneficial. However, more reactors also require more piping, valves, control instruments, etc. as well as a more complicated operation procedure. Therefore, the optimal numbers of reactors need to be carefully selected through economic analysis based on specific situations. In addition, the selection of temperatures and concentrations in each reactor, which will decide the reaction volume ratio, is also an optimization problem and will be discussed later. The common practice is to use the same size reactors, since this will simplify reactor design and fabrication.

For the sake of simplicity, next we will discuss a single reaction with reactant A as key component, then Eq. (3.46) will become:

$$V_{\mathrm{rp}} = \frac{Q_0(c_{\mathrm{Ap}-1} - c_{\mathrm{Ap}})}{(-\mathscr{R}_{\mathrm{Ap}})}, \quad p = 1, 2, \ldots, N \tag{3.47}$$

Or:

$$V_{\mathrm{rp}} = \frac{Q_0 c_{\mathrm{A0}}(X_{\mathrm{Ap}} - X_{\mathrm{Ap}-1})}{[-\mathscr{R}_{\mathrm{Ap}}(X_{\mathrm{Ap}})]}, \quad p = 1, 2, \ldots, N \tag{3.48}$$

For a first order irreversible reaction:

$$V_{\mathrm{rp}} = \frac{Q_0(X_{\mathrm{Ap}} - X_{\mathrm{Ap}-1})}{k(1 - X_{\mathrm{Ap}})}, \quad p = 1, 2, \ldots, N \tag{3.49}$$

Assuming the reaction volume of each reactor is the same as well as the space time τ; and if all the reactors are operated at the same temperature then the reaction rate constant k will also be the same. Based on the definition of space time Eq. (3.49) can be rearranged as:

$$\frac{1 - X_{\mathrm{Ap}-1}}{1 - X_{\mathrm{Ap}}} = 1 + k\tau, \quad p = 1, 2, \ldots, N \tag{3.50}$$

Multiply all these N equations:

$$\frac{1}{1 - X_{\mathrm{AN}}} = (1 + k\tau)^N$$

Or

$$\tau = \frac{1}{k}\left[\left(\frac{1}{1-X_{AN}}\right)^{1/N} - 1\right] \qquad (3.51)$$

If the numbers of the reactors are given, the space time needed to achieve the final conversion X_{AN} can be calculated from Eq. (3.51), then determining reaction volume is straightforward. Please note that the space time calculated is for one reactor. The total space time is $N\tau$, and total reactor volume $Q_0 N\tau$.

For a first order reaction when the reaction volume and operation temperature of each reactor are the same the space time can be calculated by using Eq. (3.51), and it is not necessary to calculate from one reactor to the next. However, for a nonfirst order reaction, even if all the reactors have the same reaction volume and operation temperature the design equations (Eq. (3.48)) will still be a group of nonlinear algebraic equations. For example, for a second order reaction:

$$r_A = kc_A^2 = kc_{A0}^2(1-X_A)^2$$

Substituting this rate equation into Eq. (3.48) and rearranging:

$$\tau = \frac{X_{Ap} - X_{Ap-1}}{kc_{A0}(1-X_{Ap})^2}, \quad p = 1, 2, \dots, N \qquad (3.52)$$

Eq. (3.52) is a group of nonlinear algebraic equations, and it is almost impossible to eliminate all the intermediate conversions, $X_{A1}, X_{A2}, \dots, X_{AN-1}$, to obtain an equation similar to Eq. (3.51). Therefore Eq. (3.52) has to be solved from one reactor to the next. Eq. (3.52) can be rearranged as:

$$k\tau c_{A0}X_{Ap}^2 - (2k\tau c_{A0} + 1)X_{Ap} + k\tau c_{A0} + X_{Ap-1} = 0$$

This is a quadratic equation with X_{Ap} as a variable and easy to solve. For a given reaction volume or τ in each reactor, starting from the first reactor we can calculate outlet conversion for each reactor, and eventually determine numbers of reactor needed, N, to achieve the required final conversion X_{AN}. If the number of reactors is given and the reaction volume of each reactor needs to be determined, we can assume a value of τ and then perform the calculation from one reactor to the next. If the calculated final conversion X_{AN} does not match the required value, it means the assumed τ is not correct and needs to be adjusted. Such a process needs to be repeated until the calculated final conversion matches the required value. Obviously this is a tedious calculation process but fortunately it can be easily done by a computer with simple programming.

For other nonlinear kinetics (other than second order) it is almost impossible to get any analytical solutions, and therefore Eq. (3.52) have to be solved simultaneously through numerical computation.

The calculation process can be illustrated graphically by plotting $(-\mathscr{R}_A)$ against X_A. Since the conversion at reactor outlet has to satisfy both the rate equation and mass balance equation, the intersection of these two curves represents the status of reactor outlet. Rearranging Eq. (3.48):

$$\left[-\mathscr{R}_A(X_{Ap})\right] = \frac{c_{A0}}{\tau_p}X_{Ap} - \frac{c_{A0}}{\tau_p}X_{Ap-1} \tag{3.53}$$

From this equation we can see on the $X_A - [\mathscr{R}_A(X_{Ap})]$ plot that the mass balance equation is a straight line with slope of c_{A0}/τ_p (τ_p is space time of tank p). The mass balance line for reactors 1, 2, 3, and 4 are shown in Fig. 3.8 as OM, NX_{A1}, PX_{A2}, and QX_{A3}, respectively. If all the reactors have the same volume then these lines are parallel to each other. MNPQ is the kinetic curve obtained based on the reaction rate equation. In Fig. 3.8 there is only one kinetic curve, suggesting all the reactors are operated at the same temperature. Otherwise there will be multiple kinetic curves since reaction rate is a function of temperature. In this situation the outlet conversion of a reactor is the intersection point of the mass balance line and the kinetic curve of that specific reactor.

If all the reactors have the same volume and also operate at the same temperature, for a given reaction volume the slope of mass balance line is fixed. The calculation process to determine the number of reactors needed to achieve the final conversion can be illustrated in Fig. 3.8. From point O, draw the mass balance line OM for the first reactor, which intersects the kinetic line at M. The abscissa of M is the conversion of the first reactor X_{A1}, which is also the inlet conversion of the second reactor. From X_{A1} draw a straight line parallel to OM, the intersection with the kinetic curve, N, represents the outlet status of the second reactor. Repeat this process until the final conversion meets the requirement.

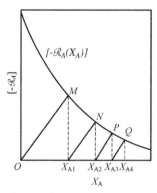

FIGURE 3.8   Conversions of CSTR in series.

Example 3.6
Using the conditions and requirements given in Example 3.4, except using three reactors with equal volume in series, calculate total reaction volume.

Solution
Based on Eq. (3.48) the mass balance equations for the three reactors can be written as:

$$V_{r1} = \frac{Q_0 c_{A0} X_{A1}}{-\mathscr{R}_A(X_{A1})}$$

$$V_{r2} = \frac{Q_0 c_{A0}(X_{A2} - X_{A1})}{-\mathscr{R}_A(X_{A2})}$$

$$V_{r3} = \frac{Q_0 c_{A0}(X_{A3} - X_{A2})}{-\mathscr{R}_A(X_{A3})}$$

Since $V_{r1} = V_{r2} = V_{r3}$:

$$\frac{X_{A1}}{X_{A2} - X_{A1}} = \frac{\mathscr{R}_A(X_{A1})}{\mathscr{R}_A(X_{A2})} \tag{A}$$

$$\frac{X_{A2} - X_{A1}}{X_{A3} - X_{A2}} = \frac{\mathscr{R}_A(X_{A2})}{\mathscr{R}_A(X_{A3})} \tag{B}$$

The relationship between $\mathscr{R}_A$ and X_A has been already derived in Example 3.4:

$$-\mathscr{R}_A(X_A) = (18.97 - 37.44 X_A + 4.78 X_A^2) \times 10^{-3} \tag{C}$$

$X_{A3} = 0.35$, from Eq. (C):

$$-\mathscr{R}_A(X_{A3}) = 6.452 \times 10^{-3} \text{ mol}/(\text{L·min})$$

Substitute this value and Eq. (C) into (A) and (B):

$$\frac{X_{A1}}{X_{A2} - X_{A1}} = \frac{18.97 - 37.44 X_{A1} + 4.78 X_{A1}^2}{18.97 - 37.44 X_{A2} + 4.78 X_{A2}^2} \tag{D}$$

$$\frac{X_{A2} - X_{A1}}{0.35 - X_{A2}} = \frac{18.97 - 37.44 X_{A2} + 4.78 X_{A2}^2}{6.452} \tag{E}$$

Solving Eqs. (D) and (E) simultaneously:

$$X_{A1} = 0.1598$$
$$X_{A2} = 0.2714$$

Using these values in the mass balance equation for the first reactor:

$$V_{r1} = \frac{(4.155/60)(3.908)(0.1598)}{[18.98 - 37.44(0.1598) + 4.78(0.1598)^2] \times 10^{-3}} = 3.299 \text{ m}^3$$

Since $V_{r1} = V_{r2} = V_{r3} = 3.299$ m^3, the total reaction volume = 9.897 m^3

If two reactors are used in the series, then $X_{A2} = 0.35$, and X_{A1} can be calculated from Eq. (A), which is 0.2202. Using the same procedure as above, the
(Continued)

(Continued)

total reaction volume can be calculated, which is 10.88 m³. From Example 3.4 we know if only one reactor is used, then the reaction volume is 14.58 m³. From these calculations we can clearly see that the total reaction volume will be smaller when there are more reactors used. Of course, such a conclusion is not valid for reactions with abnormal kinetics.

3.5.3 Optimal Reaction Volume Ratio for CSTR in Series

In order to minimize the total reaction volume for a fixed number of reactors to achieve a given final conversion, there is an optimal reaction volume ratio. This can also be considered as an optimization of conversions at each reactor outlet except the last one. Again we will discuss this problem for a single reaction system, and the total reaction volume is:

$$V_r = V_{r1} + V_{r2} + \cdots + V_{rN}$$

$$= Q_0 c_{A0} \left[\frac{X_{A1} - X_{A0}}{(-\mathscr{R}_{A1})} + \frac{X_{A2} - X_{A1}}{(-\mathscr{R}_{A2})} + \cdots + \frac{X_{AN} - X_{AN-1}}{(-\mathscr{R}_{AN})} \right]$$

Differentiating the above equation with respect to $X_{Ap}(p = 1, 2, \ldots, N-1)$

$$\frac{\partial V_r}{\partial X_{Ap}} = Q_0 c_{A0} \left[\frac{1}{(-\mathscr{R}_{Ap})} - \frac{1}{(-\mathscr{R}_{Ap+1})} + (X_{Ap} - X_{Ap-1}) \frac{\partial \frac{1}{(-\mathscr{R}_{Ap})}}{\partial X_{Ap}} \right]$$

Let $\frac{\partial V_r}{\partial X_{Ap}} = 0$:

$$\frac{1}{(-\mathscr{R}_{Ap+1})} - \frac{1}{(-\mathscr{R}_{Ap})} = (X_{Ap} - X_{Ap-1}) \frac{\partial \frac{1}{(-\mathscr{R}_{Ap})}}{\partial X_{Ap}} \tag{3.54}$$

These are the conditions that must be met to achieve a minimum total reaction volume. The outlet conversion and reaction volume of each reactor can be calculated by solving Eq. (3.54). For example, for a first order irreversible reaction:

$$(-\mathscr{R}_A) = kc_{A0}(1 - X_A)$$

Therefore:

$$\frac{\partial[1/(-\mathscr{R}_{Ap})]}{\partial X_{Ap}} = \frac{1}{kc_{A0}(1-X_{Ap})^2}$$

If all the reactors are operated at the same temperature, then substitute the above equation into Eq. (3.54):

$$\frac{X_{Ap+1} - X_{Ap}}{1 - X_{Ap+1}} = \frac{X_{Ap} - X_{Ap-1}}{1 - X_{Ap}}, \quad p = 1, 2, \ldots, N - 1$$

Which can be rewritten as:

$$\frac{Q_0 c_{A0}(X_{Ap+1} - X_{Ap})}{k c_{A0}(1 - X_{Ap+1})} = \frac{Q_0 c_{A0}(X_{Ap} - X_{Ap-1})}{k c_{A0}(1 - X_{Ap})}, \quad p = 1, 2, \ldots, N - 1$$

Obviously, the left- and right-hand of the above equation represent reaction volume of the $(p + 1)th$ and pth reactor, respectively, so:

$$V_{rp+1} = V_{rp}, \quad p = 1, 2, \ldots, N - 1$$

From this we can conclude that the total reaction volume of multiple CSTR in series will be the smallest for a first order irreversible reaction when the reaction volumes for all the reactors are the same.

For non-first order reactions, we cannot obtain the optimal reaction volume ratio through analytical solution, and a group of nonlinear algebraic Eq. (3.54) have to be solved to obtain conversions of each reactor, which can then be used to calculate reaction volume.

Generally speaking, when using multiple CSTRs in a series to conduct an α order reaction, if $\alpha > 1$ then the reactor volume should gradually increase along the flow direction, i.e., a small reactor at the front and big one at the end, to achieve the smallest total reaction volume; if $0 < \alpha < 1$ the sequence should be the opposite, i.e., reactor volume should decrease along the flow direction. For $\alpha = 1$, as discussed above, the optimal solution is all the reactors have the same volume. If $\alpha = 0$, then the reaction rate is independent from concentration, and as result the total volume of multiple reactors, no matter how many of them are in series, will equal to that of a single CSTR. And from a reactor volume perspective it is not necessary to use multiple CSTRs in series. If $\alpha < 0$, as discuss before, using single CSTR is better than multiple CSTRs in series.

3.6 YIELD AND SELECTIVITY FOR MULTIPLE REACTIONS IN A TANK REACTOR

For a multiple reaction system we first need to consider the yield of the desired product since it has a direct impact on both the quantity and quality of the product. Selectivity is equally important since it reflects how efficiently the feed is utilized. Yield and selectivity depends on types, operation modes, and conditions of the reactor. In this section we will discuss the impact of those factors on the performance of a tank reactor.

3.6.1 Overall Yield and Overall Selectivity

Assuming that for every 1 mol of desired product, μ_{pA} mol of reactant A will be consumed, from Chapter 2, Fundamentals of Reaction Kinetics, the instantaneous selectivity is:

$$S = \mu_{pA} \frac{\mathscr{R}_p}{(-\mathscr{R}_A)} = \mu_{pA} \frac{dn_p}{(-dn_A)} = \frac{dY_p}{dX_A} \quad (3.55)$$

Since S is an instantaneous value, it is obvious that both X_A and Y_P are also instantaneous variables. For batch reactors, all three will change with time. For continuous reactors under steady-state, they are functions of location in the reactor (except for CSTR). Eq. (3.55) can be rewritten as:

$$dY_P = SdX_A$$

After integrating:

$$Y_{Pf} = \int_0^{X_{Af}} SdX_A \quad (3.56)$$

Here, Y_{Pf} is called an overall yield or final yield, and X_{Af} is the overall conversion or final conversion. Neither are instantaneous values, but rather the final results of the reactions. In other words, they represent the behavior of the whole reactor.

From the definition of conversion, selectivity, and yield:

$$Y_{Pf} = S_O X_{Af} \quad (3.57)$$

S_O is the overall selectivity, which reflects the result of the whole reactor and is also called integrated selectivity. Substituting Eq. (3.56) into Eq. (3.57) we can obtain the relationship between overall selectivity and instantaneous selectivity.

$$S_O = \frac{1}{X_{Af}} \int_0^{X_{Af}} SdX_A \quad (3.58)$$

The relationship between overall selectivity and conversion depends on reaction kinetics, reactor type, and operation mode, and is usually very complicated. For the same tank reactor operated at different modes, the final yields will be different although the final conversions are the same. This can be illustrated by Fig. 3.9.

The curve in Fig. 3.9(A) represents the relationship between instantaneous selectivity and conversion for a batch tank reactor. Based on Eq. (3.56) when final conversion is X_{Af} the area under the curve shaded with vertical lines represents the final yield. If the tank reactor is operated continuously under steady-state (i.e., as a CSTR) to achieve the same final conversion, then the

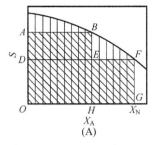 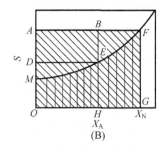

FIGURE 3.9 Final yields of tank reactors. (A) Instantaneous selectivity monotonically decreases with conversion; (B) Instantaneous selectivity monotonically increases with conversion.

final yield is represented by the rectangular area ODFG. This is because a CSTR is operated under constant temperature and concentration conditions and the instantaneous selectivity is also a constant. In this situation, the right-hand side of Eq. (3.56) is represented by the rectangular area of ODFG. In addition, from Eq. (3.58) we know $S_O = S$, i.e., the instantaneous selectivity also equals the final selectivity, and this is a character of this type of reactor. It is obvious that the area under the curve is larger than the rectangular area of ODFG, indicating the final yield of a batch tank reactor is higher than that of a CSTR, although the final conversions for these two reactors are the same. If two CSTRs in series are used to achieve the same final conversion, then the final yield is the sum of areas of rectangle OABH and EFGH, and is higher than that of a single CSTR. Obviously, with an increase in the number of CSTRs in the series, the final yield will gradually approach the value of a batch tank reactor. Based on the above discussion, at the same final conversion the final yield for different tank reactors is in the following order:

Batch tank reactor > Multiple CSTR in series > Single CSTR

However, this conclusion is obtained based on the assumption that instantaneous selectivity will monotonically decrease with an increasing conversion. If the instantaneous selectivity monotonically increases with increasing conversion, as shown in Fig. 3.9(B), then the order of the final yield will be just the opposite, and such a conclusion can be obtained by analyzing Fig. 3.9(B) following the same logic as we used to analyze Fig. 3.9(A).

3.6.2   Parallel Reactions

Assuming the following parallel reactions take place in a tank reactor:

$$A + B \rightarrow P, \quad r_P = k_1 c_A^{\alpha_1} c_B^{\beta_1}$$
$$A + B \rightarrow Q, \quad r_Q = k_2 c_A^{\alpha_2} c_B^{\beta_2}$$

P is the desired product. The instantaneous selectivity based on Eq. (3.55) is:

$$S = \frac{1}{1 + \frac{k_2}{k_1} c_A^{\alpha_2 - \alpha_1} c_B^{\beta_2 - \beta_1}}$$

This equation tells us that concentration and temperature are factors that have direct influence on instantaneous selectivity, which has been discussed in Chapter 2, Fundamentals of Reaction Kinetics. By comparing the reaction order for main and side reactions, the influence of concentration on yields can be analyzed. The impact of reaction temperature depends on the activation energies for main and side reactions.

When designing a reactor the reactant concentrations and reaction temperature should be selected to favor the higher yield of desired product. The concentration condition inside the reactor depends on the type and operation mode of the reactor, feed concentration and ratio, and feeding mode. The straightforward approach is to select feed concentrations that would meet the concentration requirement to achieve a high yield. For example if low concentration favors the yield of the desired product then low concentration feed should be used. However, for the same productivity, lower feed concentration will require larger feed flow rate, which will lead to higher operating costs and higher costs for product separation. Eventually the decision has to be made based on economic analysis. Next we will discuss how to meet the concentration requirement for tank reactors from a technical perspective.

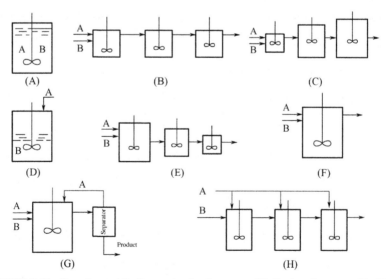

FIGURE 3.10 Operating and feeding mode of tank reactors. (A) Batch tank reactor; (B) CSTR in series; (C) CSTR in series with reactor volume increase; (d) Semibatch tank reactor; (E) CSTR in series with reactor volume decrease; (F) CSTR; (G) CSTR with recycle of reactant A; (H) CSTR in series with reactant A distributed to each reactor.

For the parallel reactions given above, if both A and B need to be at high concentration, then batch operation is preferred (Fig. 3.10(A)). For a continuous operation, using multiple CSTR is better than single reactor, and both A and B should be added into the first reactor (Fig. 3.10(B)). It would be even better if the reaction volumes gradually increase (Fig. 3.10(C)) since the first two reactions can be operated at higher concentrations. If both A and B need to be at low concentration then a single CSTR is preferred (Fig. 3.10(F)); in some situations multiple reactors have to be used. It is then desirable to use reactors with different sizes and gradually decrease reaction volumes so that after the first reactor the reactant concentrations will decrease to a lower level, as shown in Fig. 3.10(E). If one reactant needs to be at high concentration and the other needs to be at low concentration, then the arrangements shown in Fig. 3.10(D), (G), and (H) can be used. Fig. 3.10(D) represents a semibatch operation. At the beginning the reactor is loaded with reactant B, then reactant A is continuously added into the reactor. The addition rate of reactant A should be determined based on reaction extent. Generally the addition rate of A should be high at the beginning and gradually decrease. This is because the reaction is fast at the beginning and slows down gradually. The addition rate of A should match its consumption rate to maintain a low concentration of A and a high concentration of B in the reactor. Fig. 3.10(H) represents a continuous operation. Instead of being added together with B to the first reactor, reactant A is distributed to each reactor to achieve a low concentration of A and a high concentration of B in each reactor.

Another way to control the concentration of the reactants to meet the requirement of the reactions is to adjust feed ratio. For example, if the reactions require a high concentration of A and a low concentration of B, then the excess A can be added in the feed to maintain $c_{A0} > c_{B0}$. In this situation, unreacted A needs to be separated after the reactions and recycled back, as shown in Fig. 3.10(G).

Fig. 3.10 shows several typical modes for reactor arrangement. Obviously other reactor arrangements can be designed based on the characters of the reactions.

As indicated above, for a given conversion, the instantaneous selectivity is a function of temperature and depends on the relative magnitude of activation energies of the main reaction (E_1) and side reaction (E_2). When $E_1 = E_2$, the instantaneous selectivity is independent of temperature, but in practice it is desirable to operate at higher temperatures to increase reactor production intensity. When $E_2 > E_1$, lower temperature is desired to achieve high instantaneous selectivity, but this will also lead to lower production intensity. Therefore, there is optimal reaction temperature to achieve maximum productivity of the desired product. CSTR operated at the optimal temperature will give the highest production rate of the desired product. For a batch reaction there is optimal temperature progression, i.e., reactor should be operated at different temperatures at different times. The optimization discussion above is based on the maximized productivity of the desired product. In reality, it is more common to use the best economics as the objective function for optimization.

Example 3.7

Pure A is used to produce P in a CSTR:

$$A \rightarrow P, \quad r_P = k_1 c_A$$
$$A \rightarrow U, \quad r_U = k_2 c_A$$

The temperature dependence of reaction rate constants, k_1 and k_2 can be described by the Arrhenius equation. The preexponential factor $A_1 = 3.533 \times 10^{18} h^{-1}$, $A_2 = 4.368 \times 10^5 h^{-1}$; Activation energy $E_1 = 41,800$ J/(mol·K), $E_2 = 141,000$ J/(mol·K). If space time is 1 h, at what operation temperature will the yield of P be highest?

Solution

Using Eq. (3.42) the mass balance equations for component A and P can be written as:

$$\frac{V_r}{Q_0} = \frac{c_{A0} - c_A}{r_P + r_U} \tag{A}$$

$$\frac{V_r}{Q_0} = \frac{c_P}{r_P} \tag{B}$$

$V_r/Q_0 = \tau$, $c_A = c_{A0}(1 - X_A)$, and $c_P = c_{A0} Y_P$, substitute these equations and reaction rate equations into (A) and (B):

$$\tau = \frac{X_A}{(k_1 + k_2)(1 - X_A)} \tag{C}$$

$$\tau = \frac{Y_P}{k_1(1 - X_A)} \tag{D}$$

From Eq. (C):

$$X_A = \frac{(k_1 + k_2)\tau}{1 + (k_1 + k_2)\tau} \tag{E}$$

Substitute into Eq. (D) and simplify:

$$Y_P = \frac{k_1 \tau}{1 + (k_1 + k_2)\tau} \tag{F}$$

$$\frac{dY_P}{dT} = \frac{\tau \left[(1 + k_2 \tau)\dfrac{dk_1}{dT} - k_1 \tau \dfrac{dk_2}{dT} \right]}{[1 + (k_1 + k_2)\tau]^2}$$

Let $dY_P/dT = 0$:

$$(1 + k_2 \tau)\frac{dk_1}{dT} - k_1 \tau \frac{dk_2}{dT} = 0 \tag{G}$$

This is the condition to be met to achieve the maximum yield of P.

$$\frac{dk_1}{dT} = \frac{d}{dT} A_1 \exp\left(-\frac{E_1}{RT}\right) = \frac{k_1 E_1}{RT^2}$$

(Continued)

(Continued)

Similarly:

$$\frac{dk_2}{dT} = \frac{k_2 E_2}{RT^2}$$

Substitute these two equations into Eq. (G):

$$(1 + k_2\tau)E_1 = k_2 E_2 \tau$$

Using the Arrhenius equations in above equation:

$$\exp\left(-\frac{E_2}{RT}\right) = A_2\tau\left(\frac{E_2}{E_1} - 1\right)$$

$$T = \frac{E_2}{R\ln\left[A_2\tau\left(\dfrac{E_2}{E_1} - 1\right)\right]} \tag{H}$$

Then the optimal temperature can be calculated:

$$T = \frac{141,000}{8.313\ln\left[3.533 \times 10^{18} \times 1 \times \left(\dfrac{141,000}{41,800} - 1\right)\right]} = 389\ \text{K}$$

It should be noted that this is the optimal temperature for the given space time of 1 h. If space time changes, the optimal temperature will also change. For example, if space time = 0.5 h then the optimal temperature will be 396K.

3.6.3 Consecutive Reactions

For first order irreversible reactions:

$$A \xrightarrow{k_1} P \xrightarrow{k_2} Q$$

We have derived reaction system composition as a function of time and found that there was an optimal yield for product P. If a CSTR is used, will there also be an optimal yield for product P? To answer this question let's first write the mass balance A and P based on Eq. (3.42):

$$V_r = \frac{Q_0(c_{A0} - c_A)}{k_1 c_A} \tag{3.59}$$

$$V_r = \frac{Q_0 c_P}{k_1 c_A - k_2 c_P} \tag{3.60}$$

From Eq. (3.59):

$$c_A = c_{A0}/(1 + k_1\tau) \tag{3.61}$$

Substituting it into Eq. (3.60):

$$Y_{P,M} = \frac{c_P}{c_{A0}} = \frac{k_1\tau}{(1 + k_1\tau)(1 + k_2\tau)} \qquad (3.62)$$

The concentration of Q can be easily calculated by using $c_Q = c_{A0} - c_P - c_A$. Eqs. (3.61) and (3.62) give the relationship between reaction system composition and space-time. Differentiating Eq. (3.62) with respect to τ and set $dY_P/d\tau = 0$,

$$\tau_{op} = \frac{1}{\sqrt{k_1 k_2}} \qquad (3.63)$$

Eq. (3.63) is the optimal space time, and using Eq. (3.62) we can obtain the maximum yield of P:

$$(Y_{P,M})_{max} = \frac{k_1}{\left(\sqrt{k_1} + \sqrt{k_2}\right)^2} \qquad (3.64)$$

Therefore, there is also a maximum yield for consecutive reactions taking place in a CSTR. In order to compare the maximum yields between a CSTR and a batch tank reactor, the maximum yield in a batch tank reactor can be derived by substituting Eq. (3.41) into Eq. (3.39) and using the definition of yield:

$$(Y_{P,B})_{max} = (k_1/k_2)^{k_2/(k_2-k_1)}, \quad k_1 \neq k_2 \qquad (3.65)$$

At the same reaction temperature maximum the yield of P in a batch tank reactor, $(Y_{P,B})_{max}$, is always higher than that in a CSTR, $(Y_{P,M})_{max}$.

When $k_1 = k_2$, Eq. (3.39) is undefined, and from Eq. (3.38):

$$Y_{P,B} = \frac{c_P}{c_{A0}} = k_1 t e^{-k_1 t}, \quad k_1 = k_2 \qquad (3.66)$$

Differentiating with respect to t and letting $dY_P/dt = 0$, we can obtain the optimal reaction time: $\tau_{op} = 1/k_1$. Applying this optimal reaction time in Eq. (3.66), and the maximum yield of P when $k_1 = k_2$ is:

$$(Y_{P,B})_{max} = 1/e = 0.368$$

From Eq. (3.59)

$$\tau = \frac{1 - c_A/c_{A0}}{k_1 c_A/c_{A0}} = \frac{X_A}{k_1(1 - X_A)}$$

Substituting into Eq. (3.62) we can obtain the relationship between conversion and yield in a CSTR:

$$Y_{P,M} = \frac{k_1 X_A(1 - X_A)}{k_2 X_A + k_1(1 - X_A)} \qquad (3.67)$$

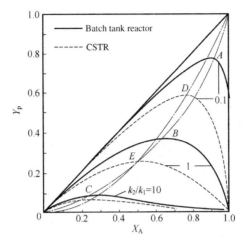

FIGURE 3.11 Conversions and yields of consecutive reactions in batch and continuous tank reactors.

Substituting Eq. (3.37) into Eq. (3.39) and simplifying, we then obtain the relationship between conversion and yield for a batch tank reactor:

$$Y_{P,B} = \frac{k_1}{k_1 - k_2} \left[(1 - X_A)^{k_2/k_1} - (1 - X_A) \right], \quad k_1 \neq k_2 \tag{3.68}$$

If $k_1 = k_2$, then substituting Eq. (3.37) into Eq. (3.66):

$$Y_{P,B} = (X_A - 1) \ln(1 - X_A) \tag{3.69}$$

In order to compare conversion and yield in a batch tank reactor and a CSTR it is more convenient to plot yield Y_P as a function of conversion by using Eqs. (3.67) to (3.69), as shown in Fig. 3.11. The solid and dashed lines represent the relationship between conversion and yield for a batch tank reactor and a CSTR, respectively. In the figure there are three sets of lines, representing the three situations for $k_2/k_1 = 0.1$, 1, and 10. It is clear for any k_2/k_1 value, at the same conversion the yield from a batch tank reactor is always higher than that from a CSTR; also the smaller the value of k_2/k_1 the bigger the difference is. In addition, all the curves show extremum, which we have discussed before. Curve ABC is the trajectory of the maximum yield for a batch tank reactors and curve DEF is the trajectory of maximum yield for a CSTR. Both trajectories depend on k_2/k_1.

From Fig. 3.11 we can also see that for both a batch tank reactor and a CSTR the yield increases with decreasing k_2/k_1. k_2/k_1 are functions of temperature. If both k_1 and k_2 follow the Arrhenius equation then:

$$k_2/k_1 = \frac{A_2}{A_1} \exp\left(\frac{E_1 - E_2}{RT}\right)$$

If $E_1 > E_2$, then the higher the temperature the smaller the k_2/k_1 and hence the higher the yield. The situation is the opposite if $E_1 < E_2$, i.e., lower temperature will lead to higher yield. However, lower temperature will also lead to a slower reaction and lower production intensity of the reactor. Therefore, reaction temperature needs to be carefully selected based on the activation energies of main and side reactions, as well as other considerations. If $E_1 = E_2$, then reaction temperature will not have an influence on k_2/k_1. In this situation it is usually more desirable to operate at higher temperatures to achieve higher productivity.

Another effective way to change k_2/k_1 is to use a catalyst. Catalysts can not only accelerate reaction rates but also change reaction pathways. The yield for product P can be enhanced by using a catalyst that favors the reaction for P formation (increasing k_1) and prevents the reaction for P conversion (decreasing k_2).

All the discussions above focus on the situation when P is the desired product. If the desired product is Q the situation is a lot simpler. For batch reactors, as long as the reaction time is long enough, eventually almost all of A will be converted to Q. Similarly for CSTR the yield of Q will also approach 100% if space time is long enough.

Example 3.8
Methanol reacts with ammonia in a CSTR, and the kinetic data were given in Example 3.3. If the reaction temperature is also the same as that in Example 3.3, what is the maximum yield of methylamine and corresponding conversion of ammonia?

Solution
As in Examples 3.3, let A, B, and M represent ammonia, methylamine, and methanol, respectively. Based on mass balance of A and B:

$$V_r = \frac{Q_0(c_{A0} - c_A)}{k_1 c_A c_M} \tag{A}$$

$$V_r = \frac{Q_0 c_B}{k_1 c_A c_M - k_2 c_B c_M} \tag{B}$$

Divide Eq. (A) by Eq. (B):

$$\frac{c_{A0} - c_A}{c_B}\left(1 - \frac{k_2 c_B}{k_1 c_A}\right) = 1$$

Since $c_A = c_{A0}(1 - X_A)$, $c_B = c_{A0} Y_B$, the above equation can be simplified as:

$$Y_B = \frac{k_1 X_A(1 - X_A)}{k_2 X_A + k_1(1 - X_A)} \tag{C}$$

(Continued)

(Continued)

Eq. (C) looks exactly the same as Eq. (3.67) which is derived for first order consecutive reactions $A \rightarrow P \rightarrow Q$. Although the reaction between ammonia and methanol is second order, the rate equation is first order with respect to each reactant. This is why the yield equation is the same.

In order to calculate the maximum yield of methylamine, differentiate Eq. (C) with respect to X_A:

$$\frac{dY_B}{dX_A} = \frac{1 - 2X_A - (k_2/k_1 - 1)X_A^2}{[1 + (k_2/k_1 - 1)X_A]^2}$$

Let $dY_B/dX_A = 0$:

$$1 - 2X_A - (k_2/k_1 - 1)X_A^2 = 0$$

$k_2/k_1 = 0.68$, X_A can be determined by solving this quadratic equation:

$$X_A = 0.548$$

Substituting this value into Eq. (C), we can obtain the maximum yield of methylamine:

$$Y_B = 0.3004$$

Compared with the results from Example 3.3 we can see that the maximum yield from a CSTR is lower than that from a batch tank reactor.

3.7 SEMIBATCH TANK REACTOR

As mentioned before, if high concentration of one reactant and low concentration of another reactant will favor the yield of the desired product then using a semibatch reactor would be beneficial. For some highly exothermic reactions the reaction temperature of a semibatch reactor can be controlled. In addition to using a cooling medium to remove reaction heat, by adjusting feed flow rate and therefore modulating the reaction rate. One possible way to increase product yield of reversible reactions is to remove the product. Removing product can also increase the reaction rate for a reversible reaction. For example, reaction distillation is a process that combines reaction with distillation. If all the reactants are added to the reactor at the same time but the products are removed from the reactor continuously during the reaction process, then the reactor is operated at semibatch mode. In addition to the examples above, semibatch reactors can also be used in other situations.

A common character of both batch and semibatch operations is the composition of reaction mixture changes with time, therefore time must be an independent variable. Eq. (3.6) is the design equation for a semibatch reactor. Next we will discuss how to use this equation for semibatch reactor calculation.

Assuming the following liquid phase reaction takes place in a semibatch reactor:

$$A + B \rightarrow R, \quad r_A = k'c_A c_B$$

For this reaction Eq. (3.1) can be rewritten as:

$$Q_0 c_{A0} = Q c_A - \mathscr{R}_A V + \frac{d(V c_A)}{dt} \tag{3.70}$$

where V is the volume of reaction mixture inside the reactor, which is a function of time. Assuming the reactor is first loaded with B and its initial volume is V_0, then reactant A with concentration c_{A0} is continuously added to the reactor at a flow rate of Q_0, and no material flows out from the reactor, i.e., $Q = 0$, then Eq. (3.70) becomes:

$$Q_0 c_{A0} = -\mathscr{R}_A V + \frac{d(V c_A)}{dt} \tag{3.71}$$

If B is in excess then the reaction can be treated as first order, i.e., $r_A = k c_A$, then Eq. (3.71) becomes:

$$\frac{d(V c_A)}{dt} + k V c_A = Q_0 c_{A0} \tag{3.72}$$

At any given time the volume of reaction mixture is:

$$V = V_0 + \int_0^t Q_0 dt$$

If reactant A is added at a constant flow rate, then Q_0 is a constant, therefore:

$$V = V_0 + Q_0 t \tag{3.73}$$

If $V c_A$ in Eq. (3.72) is considered as a variable, the Eq. (3.72) is a first order ordinary differential equation, and its initial conditions are $t = 0$, $V c_A = 0$. When Q_0 is a constant, the solution of Eq. (3.72) is:

$$V c_A = \frac{Q_0 c_{A0}}{k} [1 - \exp(-kt)] \tag{3.74}$$

Substituting Eq. (3.73) into Eq. (3.74):

$$\frac{c_A}{c_{A0}} = \frac{1 - \exp(-kt)}{k(t + V_0/Q_0)} \tag{3.75}$$

This is the relationship between reactant A's concentration in the reactor and reaction time. The relationship between the concentration of product R and reaction time is:

$$V c_R = Q_0 c_{A0} t - V c_A \tag{3.76}$$

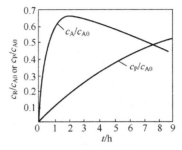

FIGURE 3.12 Concentration—reaction time relationship in a semibatch tank reactor.

Substituting Eqs. (3.73) to (3.75) into (3.76):

$$\frac{c_R}{c_{A0}} = \frac{kt - [1 - \exp(-kt)]}{k(t + V_0/Q_0)} \tag{3.77}$$

Concentrations of A and R at different times can be calculated by using Eqs. (3.75) and (3.77), respectively. Fig. 3.12 shows the results when $k = 0.2 \, \text{h}^{-1}$ and $V_0/Q_0 = 0.5 \, \text{h}$. From this figure we can see there is a maximum on the c_A-time curve. If there is no reaction taking place the concentration of A will increase due to the continuous addition of A. When a reaction takes place, A will be gradually consumed. At early stages the concentration of A is low, as is the consumption rate. Therefore, the concentration of A increases with time. With the continuous addition of A its concentration increases and so does the consumption rate. When the consumption rate is higher than the addition rate, the concentration of A will decrease with time. The concentration of product R always increases with time since it is produced from the reaction.

The above discussion is based on the fact that Q_0 is a constant. If Q_0 is not a constant, then we need to know Q_0 as a function of t to solve Eq. (3.72). If the influence of B concentration cannot be neglected, then Eq. (3.72) will be a partial differential equation, which typically needs to be solved numerically.

Example 3.9
In order to increase the yield of the desired product, the tank reactor discussed in Example 3.2 will be operated in semibatch mode. The reaction temperature is kept constant. First, 1 m^3 of B at 4 mol/m^3 is added to the reactor, then a total 1 m^3 of A at 4 mol/m^3 is continuously added to the reactor at constant flow rate over a 3-h period to react with B. Calculate the concentrations of A and P when the addition is completed, and then compare the results with those in Example 3.2.

(Continued)

(Continued)

Solution

Based on the description this is a semibatch reactor with continuous feed addition but product removal is in batches, so we can use Eq. (3.71) to obtain the relationship between reactant A's concentration and time. From Example 3.2 we know the consumption rate of A is:

$$\mathscr{R}_A = -r_P - 2r_Q = -2c_A - c_A^2$$

Using this equation in Eq. (3.71)

$$Q_0 c_{A0} = (2c_A + c_A^2)V + \frac{d(Vc_A)}{dt} \tag{A}$$

The 1 m^3 A is added continuously at constant flow rate over a 3-h period, $Q_0 = 1/3$ m^3/h. At the beginning there is 1 m^3 of B in the reactor so the initial reaction volume is 1 m^3. At any given time t the reaction volume is:

$$V = 1 + \frac{1}{3}t \tag{B}$$

Eq. (A) can be rewritten as:

$$Q_0 c_{A0} = (2c_A + c_A^2)V + c_A \frac{dV}{dt} + V \frac{dc_A}{dt}$$

Or

$$\frac{dc_A}{dt} = \frac{Q_0 c_{A0}}{V} - 2c_A - c_A^2 - \frac{c_A}{V}\frac{dV}{dt} \tag{C}$$

The concentration of A in the feed is $c_{A0} = 4$ $kmol/m^3$, substitute Eq. (B) as well as the values of Q_0 and c_{A0} into Eq. (C), and after simplifying:

$$\frac{dc_A}{dt} = \frac{4 - c_A}{3 + t} - 2c_A - c_A^2 \tag{D}$$

This is a first order nonlinear differential equation that can be solved using numerical algorithms such as the Runge–Kutta method. Table 3A gives the results from Eq. (D). From this table we can see at $t = 3$ h the concentration of A in the reactor is 0.2886 $kmol/m^3$, and the corresponding conversion is:

$$\frac{4 - 0.2886 \times 2}{4} = 0.8557 = 85.57\%$$

The concentration of P can be calculated through mass balance:

$$\frac{d(Vc_P)}{dt} = \mathscr{R}_P V \tag{E}$$

From Example 3.2 $\mathscr{R}_P = r_P = 2c_A$.
Substitute Eq. (B) and $\mathscr{R}_P$ into Eq. (E), and after simplifying:

$$\frac{d(Vc_P)}{dt} = 2(1 + t/3)c_A \tag{F}$$

(Continued)

(Continued)

TABLE 3A Concentration–Time Relationship in the Reactor

t/h	c_A/(kmol/m^3)
0	0
0.2	0.2030
0.4	0.3103
0.6	0.3608
0.8	0.3802
1.0	0.3833
1.2	0.3783
1.4	0.3694
1.6	0.3508
1.8	0.3477
2.0	0.3367
2.2	0.3261
2.4	0.3160
2.6	0.3063
2.8	0.2972
3.0	0.2886

When $t = 0$, $c_P = 0$, so $Vc_P = 0$. When $t = 3$ h, total 1 m^3 of A was added to the reactor, so $V = 2$ m^3. Eq. (F) can be integrated by treating Vc_P as a variable:

$$c_P = \int_0^3 c_A \, dt + \frac{1}{3} \int_0^3 t c_A \, dt \qquad (G)$$

The data of c_A and t are given in Table 3A, which can be used to calculate the two integrations in Eq. (G) numerically:

$$\int_0^3 c_A \, dt = 0.9637$$

$$\int_0^3 t c_A \, dt = 1.49$$

The concentration of P can be calculated from Eq. (G)

$$c_P = 0.9637 + 1.49/3 = 1.46 \text{ kmol/m}^3$$

So the yield of target product is $1.46/2 = 0.73$.

(Continued)

(Continued)

Compared with the results of Example 3.2, at the same reaction time the conversion for semibatch is 85.57%, which is lower than that from a batch operation. However, the yield of desired product P (73%) is higher than that from batch operation (69.19%). The reason for the higher yield for the desired product is that the concentration of A in the semibatch reactor is lower, which favors the formation of the desired product but also leads to lower conversion. In real situations it is more desirable to achieve higher yield, therefore for this example the semibatch reactor is a better option.

3.8   NONISOTHERMAL BATCH REACTOR

All the discussions about tank reactors in the previous sections are for isothermal processes. For batch tank reactors it is extremely difficult to achieve true isothermal operation. If the reaction heat effect is small the reactor may be approximately considered isothermal, but for reactions with large heat effect it is very hard to achieve isothermal operation. On the other hand, for many reactions nonisothermal operation will lead to better performance. Therefore, studying nonisothermal operations is of great practical importance.

Temperature is a very important factor for reactor operation and has impacts on conversion, yield, and reactor intensity. Temperature also influences physical properties of the reaction mixture, which will in turn have impacts on heat and mass transfer as well as the power consumption of the propeller for mixing. Therefore, for batch reactors it is of critical importance to determine the relationship between reaction temperature and time since this relationship is needed for both reactor design and analysis.

The relationship between reaction temperature and time for a batch reaction process can be derived from energy balance, which is based on Eq. (1.9). A batch tank reactor is a closed system, and if the work from the propeller to the reaction mixture can be neglected, then from first law of thermodynamics:

$$dq = dU \tag{3.78}$$

The equation states the fact that heat exchange between the reaction system and the surrounding environment equals the change of internal energy. For most reaction systems, especially for liquid phase reactions, the difference between internal energy and enthalpy is very small. Therefore it is more convenient and practical to use the following equation instead of Eq. (3.78)

$$dq = dH \tag{3.79}$$

For tank reactors we can choose the whole reaction volume as the control volume, and for batch operations all the parameters will be a function of time, so dq and dH in Eq. (3.79) represent the exchange of heat and enthalpy, respectively, between the reaction system and the surrounding environment during the period of dt.

Assuming at time t the temperature of reaction system is T, and during time period of dt the change of temperature is dT. When the reaction temperature increases from T to $T + dT$, the change of enthalpy will be dH. Since enthalpy change depends only on the initial and final states and is independent of the path of the changes, we can choose a path that is most convenient to calculate the enthalpy changes. If we select T_r as the reference temperature for the calculation, then we first need to calculate the enthalpy change needed to change the reaction system temperature (increase or decrease) to T_r:

$$\Delta H_1 = m_t \int_T^{T_r} C_{pt} dT \approx m_t \overline{C}_{pt} (T_r - T)$$

where m_t is the mass of the reaction system, C_{pt} is the heat capacity of the reaction system. Since C_{pt} is a function of temperature, the symbol of approximately equal in the equation reflects the approximation of using average heat capacity $\overline{C}_{pt}$ in the temperature range between T and T_r.

Next we need to calculate enthalpy change for the reaction taking place at temperature T_r. Here we assume there is only one reaction, then:

$$dH_2 = \Delta H_r(-\mathscr{R}_A) V_r dt$$

Here ΔH_r is the reaction heat, which is negative for an exothermic reaction, and positive for an endothermic reaction.

The final step is to calculate the enthalpy change to bring the reaction system temperature from T_r to $T + dT$ following the same procedure as step one:

$$\Delta H_3 \approx m_t \overline{C}_{pt} (T + dT - T_r)$$

It should be noted that the reaction system composition and temperature change in this step are different from those in the first step so strictly speaking the average heat capacity is also different. Since we are considering an infinitesimal change in both composition and temperature, the average heat capacity can be approximately treated as constant. The reaction system enthalpy changes from time t to $t + dt$ equals the sum of enthalpy changes for the three steps:

$$dH = \Delta H_1 + dH_2 + \Delta H_3 = m_t \overline{C}_{pt} dT + \Delta H_r(-\mathscr{R}_A) V_r dt \qquad (3.80)$$

The heat exchange between the reaction system and environment is:

$$dq = U A_h (T_c - T) dt \qquad (3.81)$$

where U is the overall heat transfer coefficient, A_h is heat transfer area, and T_c the environment temperature.

Substitute Eqs. (3.80) and (3.81) into Eq. (3.79):

$$m_t \overline{C}_{pt} \frac{dT}{dt} = UA_h(T_c - T) - \Delta H_r(-\mathscr{R}_A)V_r \qquad (3.82)$$

Eq. (3.82) represents the correlation between temperature and time for the reaction mixture inside a batch tank reactor. The environment temperature T_c is also the temperature of heat exchange medium. For endothermic reactions heat needs to be provided to the reaction system, so $T_c > T$; for exothermic reactions heat needs to be removed, so $T > T_c$. However, during the start-up, even for exothermic reactions the reactor still needs to be heated to bring the reaction mixture to the desired reaction temperature. The ΔH_r in Eq. (3.82) is the reaction heat at standard temperature, and correspondingly, $\overline{C}_{pt}$ is the average heat capacity of the reaction system, and the temperature range depends on the path selected for ΔH_r calculation.

If reaction temperature is maintained at constant value, i.e., $dT = 0$, then Eq. (3.82) can be simplified as:

$$UA_h(T_c - T) = \Delta H_r(-\mathscr{R}_A)V_r$$

This equation reflects the fact that in order to maintain a constant reaction temperature, the heat released or adsorbed by the reactions must be equal to the heat exchange between the reaction system and the environment. However, it is very hard to meet such a requirement for a batch operation since reaction rate changes with time and so does the heat released or absorbed by the reactions. In order to achieve isothermal operation the heat exchange rate between the reaction system and the environment has to be adjusted according to the heat releasing or adsorbing rate, which is obviously not an easy task. Therefore, in the real world most batch reactors are operated under nonisothermal conditions, especially for reactions with large heat effect. For very exothermic reactions, e.g., nitration of benzene, the reaction temperature has to be controlled very carefully due to safety reasons. If the reaction heat cannot be removed promptly the reaction mixture temperature may increase dramatically and eventually cause an explosion.

For an isothermal batch reactor the reactor design equation can be derived based on the mass balance equation and reaction kinetics. For example, Eq. (3.6) is one of the equations derived before:

$$n_{A0} \frac{dX_A}{dt} = (-\mathscr{R}_A)V_r \qquad (3.83)$$

Since $\mathscr{R}_A$ is a function of conversion and temperature, the relationship between reaction temperature and time is needed to determine $\mathscr{R}_A$. Therefore, for nonisothermal batch reactor designs the mass and energy balance equations, Eq. (3.83) and Eq. (3.82), have to be solved simultaneously.

The relationship between reaction temperature and conversion can be obtained by substituting Eq. (3.83) into Eq. (3.82):

$$m_t \overline{C}_{pt} \frac{dT}{dt} = UA_h(T_c - T) - n_{A0} \Delta H_r \frac{dX_A}{dt} \tag{3.84}$$

From this equation we can see that for a given reaction system the temperature and conversion correlation depends on the heat exchange rate between the reaction system and the environment. For adiabatic operations where there is no heat exchange between the reaction system and the environment, Eq. (3.84) becomes:

$$m_t \overline{C}_{pt} dT = n_{A0}(-\Delta H_r) dX_A \tag{3.85}$$

If $\overline{C}_{pt}$ can be considered a constant, Eq. (3.85) can be integrated as:

$$T - T_0 = \frac{n_{A0}(-\Delta H_r)}{m_t \overline{C}_{pt}} X_A \tag{3.86}$$

T_0 is the initial reaction temperature at zero conversion. Here ΔH_r is the reaction heat calculated at standard temperature T_0, and $\overline{C}_{pt}$ is the average heat capacity between temperatures T_0 and T. For adiabatic reaction processes, integrating a heat balance equation can derive the reaction temperature-conversion correlation, which is a linear algebraic equation. Using this correlation $\mathcal{R}_A$ can be expressed as a function of X_A, and then the reaction time can be calculated by using Eq. (3.83).

For multiple reaction systems the heat balance Eq. (3.82) can be modified to account for heat effects of all the reactions:

$$m_t \overline{C}_{pt} \frac{dT}{dt} = UA_h(T_c - T) - V_r \sum_{j=1}^{M} (\Delta H_r)_j r_j \tag{3.87}$$

Correspondingly Eq. (3.5) needs to be used for mass balance. By solving Eqs. (3.87) and (3.5) simultaneously, the reaction time needed can be calculated for multiple reactions taking place in a vary-temperature batch reactor.

Example 3.10
The n-hexyl maleate can be produced by reacting maleic anhydride (A) with n-hexanol (B):

$$\begin{array}{c} HC-C=O \\ | \qquad \diagdown O \\ HC-C=O \end{array} + C_{16}H_{13}OH \longrightarrow HOOC-CH-CH-COOH(C_{16}H_{13})$$

The reaction rate is first order with respective to both maleic anhydride and n-hexanol with rate constant:

$$k = 1.37 \times 10^{12} \exp(-12,628/T) \ m^3/(kmol \cdot s)$$

(*Continued*)

(Continued)

A batch tank reactor is used, and solid maleic anhydride is first loaded to the reactor and then heated with steam to its melting point of 326K. After all the maleic anhydride is totally melted, n-hexanol is quickly added into the reactor. At this time the concentrations for maleic anhydride and n-hexanol are 4.55 kmol/m³ and 5.34 kmol/m³, respectively.

1. If the reactor is initially operated adiabatically starting at 326K and then switches to isothermal operation when the reactor temperature reaches 373K (since the reaction temperature cannot exceed 373K), calculate the reaction time needed to achieve 98% conversion of A.
2. If the whole reaction process is maintained at 373K, what would be the reaction time needed?
3. Compare the cooling water and steam consumptions for these two operation modes. Assume the temperature increase for the cooling water is 7K, and the saturated steam is at 0.405 MPa. The reaction heat is −33.5 kJ/mol, and average heat capacity of the reaction mixture is 1980 kJ/(m³ · K).

Solution

(1) The conversion and temperature relationship for adiabatic reaction processes is represented in Eq. (3.86), but it needs to be modified based on data given in this problem. If both the numerator and denominator of the right-hand of Eq. (3.86) are divided by reaction volume:

$$T - T_0 = \frac{c_{A0}(-\Delta H_r)}{\rho \overline{C}_{pt}} X_A \qquad (A)$$

where ρ is the density of reaction mixture, $\overline{C}_{pt}$ is the average heat capacity based on unit mass; therefore, $\rho \overline{C}_{pt}$ is the average heat capacity based on unit volume, which is 1980 kJ/m³ · K:

$$T - 326 = \frac{4.55 \times 33.5 \times 10^3}{1980} X_A$$

or:

$$T = 326 + 76.98 X_A \qquad (B)$$

The reaction rate is:

$$-\mathcal{R}_A = r_A = k c_A c_B = k c_{A0}(1 - X_A)(c_{B0} - c_{A0} X_A) \qquad (C)$$

The reaction time can be calculated by substituting Eq. (C) into Eq. (3.8):

$$t = \int_0^{X_{Af}} \frac{dX_A}{k(1 - X_A)(c_{B0} - c_{A0} X_A)} \qquad (D)$$

The reaction takes place under vary-temperature conditions, and k is a function of temperature. In order to integrate Eq. (D) k must be expressed as

(Continued)

(Continued)

a function of conversion X_A by using Eq. (B) in rate constant correlation given in the problem:

$$t = \int_0^{X_{Af}} \frac{\exp[12628/(326 + 76.98X_A)]dX_A}{1.37 \times 10^{12}(1 - X_A)(c_{B0} - c_{A0}X_A)} \tag{E}$$

Since the maximum reaction temperature is 373K and the conversion corresponding to this maximum temperature can be calculated from Eq. (B)

$$373 = 326 + 76.98X_A$$
$$X_A = 0.6105$$

Therefore, the reactor is first operated adiabatically until conversion = 61.05%, and operated isothermally until conversion reaches target value of 98%.

$$t = \int_0^{0.6105} \frac{\exp[12628/(326 + 76.98X_A)]dX_A}{1.37 \times 10^{12}(1 - X_A)(5.34 - 4.55X_A)} \tag{F}$$

This integration has to be calculated numerically:

$$t_1 = 1383 \text{ s} = 23.1 \text{ min}$$

At this moment the temperature reaches 373K. From now on the reactor is operated isothermally until the conversion increases from 61.05% to 98%. The reaction time needed can be calculated using Eq. (D) with integration limit from 0.6105 to 0.98. Eq. (D) can be integrated analytically since k is constant:

$$t = \frac{1}{kc_{A0}(c_{B0}/c_{A0} - 1)} \int_{X_{A0}}^{X_{Af}} \left[\frac{1}{1 - X_A} - \frac{1}{c_{B0}/c_A - X_A}\right] dX_A \tag{G}$$

$$= \frac{1}{kc_{A0}(c_{B0}/c_{A0} - 1)} \ln\frac{(1 - X_{A0})(c_{B0}/c_{A0} - X_{Af})}{(1 - X_{Af})(c_{B0}/c_{A0} - X_{A0})}$$

Rate constant at 373K is:

$$k = 1.37 \times 10^{12}\exp(-12,628/373) = 2.714 \times 10^{-3} \text{ m}^3/(\text{kmol·s})$$

Then using Eq. (G) we can calculate the reaction time for the isothermal period:

$$t = \frac{1}{2.714 \times 10^{-3} \times 4.55(5.34/4.55 - 1)} \ln\frac{(1 - 0.6105)(5.34/4.55 - 0.98)}{(1 - 0.98)(5.34/4.55 - 0.6105)}$$
$$= 886.9 \text{ s} = 14.78 \text{ min}$$

The total reaction time is:

$$t = t_1 + t_2 = 23.1 + 14.78 = 37.88 \text{ min}$$

(*Continued*)

(Continued)

(2) If the whole reaction process takes place at 373K, then the reaction time needed can be calculated using Eq. (G), but now the integration limit has gone from 0 to 0.98:

$$t = \frac{1}{2.714 \times 10^{-3} \times 4.55(5.34/4.55 - 1)} - \ln\frac{(5.34/4.55 - 0.98)}{(1 - 0.98)5.34/4.55} = 984.2 \text{ s}$$
$$= 16.4 \text{ min}$$

We can see that the reaction time needed for an isothermal operation at 373K is less than that needed for an adiabatic-isothermal operation, and this is because the reaction is faster at a higher temperature. During the adiabatic operation period the reaction temperature is lower than 373K.

(3) Since the esterification of maleic anhydride is an exothermic reaction, the reaction heat must be removed to maintain isothermal operation. For the first operation mode the isothermal operation starts at $X_A = 61.05\%$. For 1 m^3 reaction volume the amount of heat released is:

$$q = 1 \times 4.55(0.98 - 0.6105) \times 33.5 \times 10^3 = 56,321 \text{ kJ}$$

This is the amount of heat that has to be removed by cooling water. The temperature increase of cooling water is 7K, so the amount of cooling water needed is:

$$\frac{56,321}{4.186 \times 7} = 1936 \text{ kg} = 1.936 \text{ m}^3$$

For the second operation mode the whole process is operated at 373K, so the amount of heat released from 1 m^3 reaction volume:

$$1 \times 4.55 \times 0.98 \times 33.5 \times 10^3 = 149,380 \text{ kJ}$$

The amount of cooling water needed is:

$$\frac{149,380}{4.186 \times 7} = 5089 \text{ kg} = 5.089 \text{ m}^3$$

Therefore, the consumption of cooling water for the second mode is about 2.5 times of the first operation mode.

The amount of heat needed to melt maleic anhydride and heat n-hexanol to 326K is the same for the two situations, and therefore the steam consumption for this part is also the same. The first operation mode reaction starts adiabatically at 326K. For the second operation mode the reaction mixture needs to be further heated from 326K to 373K, and the heat needed is:

$$1 \times 1980 \times (373 - 326) = 93,060 \text{ kJ}$$

The amount of steam (saturated at 0.405 MPa with latent heat = 2136 kJ/kg) needed is 93,060/2136 = 43.6 kg. This is the additional steam needed for the second operation mode. From these calculations we can see for the second operation mode the reaction time is shorter, but both steam and cooling water consumptions are higher.

3.9 STEADY-STATE OPERATION OF CSTR

Temperature inside the CSTR reactor is uniform and under steady-state operation the reactions take place at constant temperature. If operated at unsteady-state conditions, then the reaction temperature will change with time, but is still uniform inside the reactor. For either steady or unsteady-state operations the reaction temperature is determined by energy and mass balance equations. It is possible that there are multiple solutions from the mass and energy balance equations, which means there are multiple steady-states. If there are multiple steady-states, then we need to know which steady-state(s) are practical in real operations and which are not, and this is the stability issue we will discuss in this section. Stability is a very complicated problem and in this section we will only introduce the most fundamental concepts and knowledge.

3.9.1 Heat Balance for CSTR

CSTR is an open system, and based on the first law of thermodynamics, the heat balance equation is Eq. (3.79). For steady-state operations time is not a variable and when the whole reaction volume is chosen as control volume then Eq. (3.79) can be rewritten as:

$$\Delta H = q \tag{3.88}$$

The enthalpy change ΔH can be calculated following the same procedure as used in deriving the heat balance equation for batch reactor. Here we assume that the reaction feed density ρ is a constant, and choose feed temperature T_0 as the reference temperature, then:

$$\Delta H = Q_0 \rho \overline{C}_{pt}(T - T_0) + (\Delta H_r)_{T_0}(-\mathscr{R}_A)V_r$$

The heat exchange between the reaction system and environment:

$$q = UA_h(T_c - T)$$

Substituting into Eq. (3.88)

$$Q_0 \rho \overline{C}_{pt}(T - T_0) + (\Delta H_r)_{T_0}(-\mathscr{R}_A)V_r = UA_h(T_c - T) \tag{3.89}$$

This is the heat balance equation for a steady-state CSTR. It should be noted that in this equation ΔH_r is the reaction heat at feed temperature and $\overline{C}_{pt}$ is the average heat capacity between T_0 and T at constant pressure. If another temperature is chosen as the reference temperature then Eq. (3.89) needs to be modified correspondingly.

Using Eq. (3.44) in Eq. (3.89) we can obtain the relationship between reaction temperature and conversion:

$$Q_0\left[\rho \overline{C}_{pt}(T - T_0) + c_{A0}X_A(\Delta H_r)_{T_0}\right] = UA_h(T_c - T) \tag{3.90}$$

The amount of heat to be removed from or provided to the reactor to achieve certain conversion can be calculated by using Eq. (3.90), and then the consumption of the heat exchange medium can be determined.

If the reactor is operated under adiabatic conditions, then Eq. (3.90) can be simplified as:

$$T - T_0 = \frac{c_{A0}(-\Delta H_r)_{T_0}}{\rho \overline{C}_{pt}} X_A \qquad (3.91)$$

This equation is the same as Eq. (3.86) derived for an adiabatic batch reactor. The only difference is the variable and notation selected. Therefore, for both batch and continuous tank reactors as long as it is operated under adiabatic conditions the reaction temperature and conversion relationship are the same. For a CSTR under steady-state, although it is operated adiabatically, the reactor is still operated at constant temperature, and the reaction temperature needed to achieve a given conversion can be calculated using Eq. (3.91). The situation is different for a batch reactor. Under adiabatic operation the temperature for a batch reactor will increase with time for exothermic reactions and decrease for endothermic reactions.

If $\overline{C}_{pt}$ is considered as a constant, Eq. (3.91) or Eq. (3.86) can be written as:

$$T - T_0 = \lambda X_A \qquad (3.92)$$

$$\lambda = \frac{c_{A0}(-\Delta H_r)_{T_0}}{\rho \overline{C}_{pt}}$$

λ is called an adiabatic temperature increase, which represents the temperature increase (for exothermic reactions) or decrease (for endothermic reactions) when all the A is converted. When $\overline{C}_{pt}$ is a constant, λ is also a constant, then Eq. (3.92) is a linear equation. Otherwise the relationship between T and X will be nonlinear.

Eq. (3.89) can be generalized for a multiple reaction system by considering reaction heat from all the reactions:

$$Q_0 \rho \overline{C}_{pt}(T - T_0) + V_r \sum_{j=1}^{M} r_j (\Delta H_r)_{j,T_0} = U A_h (T_c - T) \qquad (3.93)$$

Correspondingly, the mass balance equation will be Eq. (3.41). Eqs. (3.93) and (3.41) are the design equations for multiple reactions taking place in a CSTR.

3.9.2 Steady-States of CSTR

For a CSTR under steady-state, the operation temperature and conversion must satisfy both mass and heat balance equations. For a first order, nonreversible exothermic reaction:

$$(-\mathscr{R}_A) = r_A = A\exp[-E/(RT)]c_A = A\exp[-E/(RT)]c_{A0}(1 - X_A)$$

Substituting the above equation into Eqs. (3.44) and (3.89):

$$V_r = \frac{Q_0 X_A}{A\exp[-E/(RT)](1 - X_A)} \tag{3.94}$$

$$Q_0\rho\overline{C}_{pt}(T - T_0) + V_r(\Delta H_r)_{T_0}A\exp[-E/(RT)]c_{A0}(1 - X_A) = UA_h(T_c - T) \tag{3.95}$$

The operation temperature and conversion under steady-state can be obtained by solving Eqs. (3.94) and (3.95) simultaneously. Since $\tau = V_r/Q_0$, from Eq. (3.94):

$$X_A = \frac{A\tau\exp([-E/(RT)]}{1 + A\tau\exp[-E/(RT)]}$$

Substituting into Eq. (3.95)

$$Q_0\rho\overline{C}_{pt}(T - T_0) + UA_h(T - T_c) = \frac{V_r c_{A0}(-\Delta H_r)_{T_0}A\exp[-E/(RT)]}{1 + A\tau\exp[-E/(RT)]} \tag{3.96}$$

The left side of Eq. (3.96) represents the heat removing rate q_r, which includes two parts. The first item represents the heat needed to heat up the feed from T_0 to operation temperature T and the second item represents the heat removed by the cooling medium:

$$q_r = Q_0\rho\overline{C}_{pt}(T - T_0) + UA_h(T - T_c) \tag{3.97}$$

The right side of Eq. (3.96) represents the heat release rate from the reaction:

$$q_g = \frac{V_r c_{A0}(-\Delta H_r)_{T_0}A\exp[-E/(RT)]}{1 + A\tau\exp[-E/(RT)]} \tag{3.98}$$

Eq. (3.97) is a nonlinear algebraic equation, which can be used to calculate steady-state operation temperature. The solution of this equation can be found graphically by plotting q_r and q_g against T, as shown in Fig. 3.13. The heat-removing rate q_r is a straight line, which is clear from Eq. (3.97), and the slope of the line is $Q_0\rho\overline{C}_{pt} + UA_h$. From Eq. (3.98) we know that heat release rate is an S-shaped curve. The intersection points of straight line q_r and curve q_g are the solutions of Eq. (3.96). For example, as shown in

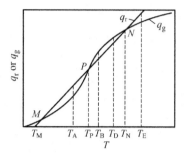

FIGURE 3.13   Steady-state operating temperature of a CSTR.

Fig. 3.13, there are three intersection points, M, P, and N, and corresponding temperatures are steady-state operation temperatures. Therefore, this reaction system has multiple steady-states. T_M, T_P, and T_N all satisfy both mass and heat balance equations. Obviously, the conversion is lowest when operated at T_M and highest at T_N.

In addition to temperature and conversion, there is another important character for each steady-state, which is its stability. The stability of a system reflects its capability of self-adjustment after external disturbance. For example, if there is a fluctuation of feed temperature, it will in turn lead to reaction temperature change from the steady-state. After the feed temperature returns to its normal value, if the reaction temperature can also return to its original steady-state, then the steady-state is stable. Otherwise, the steady-state is not stable.

Next we will analyze steady-state N using Fig. 3.13. The temperature for this steady-state is T_N. Assume that the temperature increases to T_E due to external disturbance. But from the figure we can see that at T_E the heat removal rate is higher than the heat release rate, i.e., $q_r > q_g$. Therefore after the disturbance disappears, the temperature of the reaction system must gradually return to T_N. On the other hand, if temperature decreases to T_D due to external disturbance, the temperature must gradually increase back to T_N after the disturbance disappears since at point T_D $q_g > q_r$. Based on the above analysis we can see that steady-state N is stable. For steady-state M we can do a similar analysis and conclude that it is also stable. However, for point P the situation is different. If the reaction system temperature increases to T_B, then since $q_g > q_r$, even the disturbance disappears, the temperature of the reaction system will keep increasing until it reaches T_N. On the other hand, temperature decreases to T_A, then since $q_g > q_r$ the reaction system temperature will continue decreasing until it reaches to T_M. Therefore steady-state P is not stable. Theoretically, it is a steady-state point but in reality it is almost impossible to operate at this point, since any disturbance, no matter how small, will lead to the reactor state moving to either point M or N. In the real world, it is impossible to eliminate all disturbances. Therefore, when designing a CSTR, the operation conditions should be selected so that the reactor is at a stable steady-state. In most cases the higher steady-state, i.e., point N in Fig. 3.13, should be the choice. Although point M is also stable, the conversion is too low, and in general it is not practical to operate at such low conversion.

From Fig. 3.13 it can be concluded that at steady-state points M and N, the slope of the heat removal curve is greater than that of the heat-releasing curve, i.e.:

$$\frac{dq_r}{dT} > \frac{dq_g}{dT} \tag{3.99}$$

This is the necessary but not the sufficient condition for a steady-state to be stable. In other words, if Eq. (3.99) is satisfied, the steady-state may be stable, but if it not satisfied, then the steady-state is certainly not stable. This condition is also called the slope condition, and it has to be used together with other conditions to form sufficient and necessary conditions for a steady-state to be stable.

Steady-state temperature depends on operation conditions. Fig. 3.14 shows the relationship between feed temperature T_0 and steady-state temperature T. When feed temperature slowly increases from T_G to T_E, the change of steady-state temperature will follow the curve of GAFDE. It should be noted that at point F the curve is not continuous, and the steady-state temperature shows a step increase. This point is called the ignition point. After this point further increasing feed temperature will not lead to a sudden jump of the steady-state temperature. If the feed temperature gradually decreases from T_E to T_G, the steady-state temperature will decrease following the curve EDBAG. There is also a discontinuous point B on this curve. At this point the steady-state temperature shows a steep decrease, and this point is called the extinction point. The ignition and extinction points are very important for reactor operation and control, especially during the startup and shutdown. For example, if the operation temperature is close to ignition point, then a small increase of feed temperature may cause a significant increase in reaction temperature, which may in turn damage the catalyst used or even lead to explosion. If the reactor is operated close to the extinction point, then a small change in feed conditions may lead to a sudden decrease of reaction temperature and thus cease the reaction.

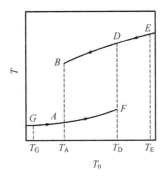

FIGURE 3.14   Ignition and extinction point of a CSTR.

Fig. 3.14 also indicates that in the feed temperature range from T_A to T_D there will be multiple steady-states. Outside this temperature range, there is only one steady-state. When feed temperature is T_A, there are two steady-states, i.e., point A and B on the figure. Similarly, at $T_0 = T_D$ there are also two steady-states. If $T_A < T_0 < T_D$ there will be three steady-states. Curves BD and AF represent the trajectories of upper and lower steady-state temperature, respectively, and BF is the trajectory (not shown in the figure) of the middle steady-state. The number of steady-states depends on the characters of the reactions and reactor operation conditions, such as feed temperature and flow rate, and the heat exchange between the reactor and the environment. Only exothermic reactions can exhibit multiple steady-state behavior. For endothermic reactions there is always only one steady-state.

In order to maintain a stable operation, the stability conditions must be met, which can be achieved by properly selecting reactor design and operation conditions, such as using larger heat transfer areas and maintaining small temperature differences for heat transfer. Most tank reactors are used for liquid phase reactions and due to the large heat capacity of the liquid in the reactor temperature change is usually small. In other words, it is unlikely to see a huge swing of reactor temperature. Therefore, with adequate control and adjustment mechanisms in place it may not be necessary to limit the stable steady-state.

Example 3.11
A mixture of maleic anhydride and n-hexnol is continuously fed to an adiabatic CSTR at 326K and 0.01 m^3/s for esterification reaction. The reaction volume is 2.65 m^3 and feed concentrations of maleic anhydride and n-hexanol are 4.55 and 5.34 kmol/m^3, respectively. Please calculate temperature and conversion at reactor outlet. Use the kinetic and thermodynamic data given in Example 3.10.

Solution
For adiabatic operations Eq. (3.91) can be used to obtain the relationship between reaction temperature and conversion:

$$T - 326 = \frac{4.55 \times 33.5 \times 10^3}{1980} X_A$$

or

$$X_A = 0.01299T - 4.235 \tag{A}$$

Substitute the reaction rate equation into Eq. (3.44):

$$V_r = \frac{Q_0 c_{A0} X_{Af}}{k c_A c_B} = \frac{Q_0 X_{Af}}{1.37 \times 10^{12} \exp(-12,628/T)(1 - X_{Af})(c_{B0} - c_{A0} X_{Af})}$$

Using Eq. (A) and other known data:

$$2.65 = 0.01(0.01299T - 4.235)\exp(12628/T)$$
$$\times \{1.37 \times 10^{12}(1 - 0.01299T + 4.235) \times [5.34 - 4.55(0.01299T - 4.235)]\}^{-1}$$

(Continued)

(Continued)

After simplifying:

$$\frac{(T-326)\exp(12{,}628/T)}{(403-T)(416.4-T)} = 2.146 \times 10^{13}$$

Solving this equation:

$$T = 364 \text{ K}$$

This is the temperature at reactor outlet, which is also the reaction temperature under steady-state. Using this value in Eq. (A):

$$X_{Af} = 0.01299 \times 364 - 4.235 = 0.4986 = 49.86\%$$

SUMMARY

Most tank reactors are used for liquid phase reactions. The selection of operation modes, batch, continuous, or semibatch, depends on production rate, kinetic characters of the reactions, etc. In this chapter it is assumed that both the concentration and temperature of the reaction mixture inside the tank reactor are uniform and generalized design equations can be derived based on mass balance of key components:

$$Q_0 c_{i0} - Q c_i + V_r \mathscr{R}_i = \frac{dn_i}{dt}, \quad i = 1, 2, \ldots, K$$

where

$$\mathscr{R}_i = \sum_{j=1}^{M} \nu_{ij} \bar{r}_j$$

These design equations can be used for any operation modes, single or multiple reaction systems, and can be further simplified based on specific situations.

For batch or semibatch operations, the key task for reactor design is to calculate the reaction time needed to meet a given production rate. For continuous reactors the key task is to calculate the reactor volume. It is important to clearly distinguish between space time and reaction time, which are two totally different concepts. The former is used for continuous reactors and reflects reactor intensity; the latter is usually used for batch reactors.

For single reaction systems, the continuous reactor is usually evaluated by the final conversion at given space time. For multiple reaction systems, it is the final yield and overall selectivity of the desired product that are most critical. For CSTR, the overall selectivity is always equal to instantaneous selectivity, but for batch reactors the instantaneous selectivity will change with time, except for very special cases. Under normal situations, when final conversion is the same, the final yield of a batch tank reactor is higher than that from a CSTR.

With the same space time the performance of multiple CSTR in a series is better than that of a single CSTR, and the higher the numbers of CSTR in series the better the performance will be. But the conclusion is the opposite for abnormal kinetics. There is an optimal reaction volume ratio for multiple CSTR in series. For first order reactions, the optimal solution is that all the reactors have the same volume. For CSTR in parallel, performance is the best when all the reactors have the same space time.

The feed composition and reactor operation modes have to be selected based on the impact of reactant concentration on reaction rate, feed prices, how difficult and expensive to separate the product, etc. The feeding mode to the reactor should be selected based on the requirement for reaction mixture concentrations to achieve optimal yield. For batch reactors, all the feeds are added at the beginning so the concentrations in the reactor depend on only initial feed composition. The same is true for a single CSTR. Only CSTR in a series or semibatch tank reactor can be fed with different modes.

The impact of reaction temperature on reactor operation is more pronounced than concentrations. Under steady-state CSTR is always operated isothermally. For batch reactors, temperature is uniform in the reactor but may change with time. The optimal operating temperature for isothermally operated batch reactors or the optimal temperature progression for nonisothermal reactors need to be determined by pursuing maximized reactor intensity or the final yield of the desired product. The reactor temperature can be decided based upon mass and heat balance and then examined to see whether it meets the required conditions. If not, other parameters have to be adjusted.

Adiabatic CSTR is still operated isothermally, but an adiabatic batch reactor is nonisothermal, and in most situations the conversion—temperature relationship can be approximated by a linear equation.

There might be more than one steady-state for CSTR, i.e., multiple steady-states under certain operation conditions. The steady-state can be either unstable or stable, and in reality it is always desirable to operate the reactor under stable steady-state.

FURTHER READING

Levenspiel O. Chemical reaction engineering. 2nd ed. New York: John Wiley & Son Inc.; 1972.
Denbigh KG, Turner JCR. Chemical reactor theory. 3rd ed. Cambridge University Press; 1985.
Li S. Chemical and catalytic reaction engineering. Beijing: Chemical Industry Press; 1986.
Smith JM. Chemical engineering kinetics. 3rd ed. New York: McGraw-Hill; 1981.

PROBLEMS

3.1 Saponification of ethyl acetate takes place in an isothermal batch reactor:

$$CH_3COOC_2H_5 + NaOH \rightarrow CH_3COONa + C_2H_5OH$$

This reaction is first order with respect to both ethyl acetate and sodium hydroxide. At the beginning the concentration of both ethyl acetate and sodium hydroxide is 0.02 mol/L, and the reaction rate constant is 5.6 L/(mol · min). The required conversion is 95%. Calculate:
1. Reaction time needed when reaction volume is 1 m^3;
2. Reaction time needed when reaction volume is 2 m^3.

3.2 An isothermal batch reactor is used for saponification of ethylene chlorohydrins to make ethylene glycol:

$$\begin{array}{cc} CH_2{-}CH_2 + NaHCO_3 \longrightarrow & CH_2{-}CH_2 + NaCl + CO_2 \\ \quad | \quad\;\; | & \quad | \quad\;\; | \\ \quad Cl \quad OH & \quad OH \quad OH \end{array}$$

The ethylene glycol production rate is 20 kg/h. 15 wt.% $NaHCO_3$ and 30 wt.% ethylene chlorohydrins water solutions are used as feedstock. Before the reaction, the ethylene cholorohydrins to sodium bicarbonate molar ratio is 1:1 in the reactor and the reaction mixture specific gravity is 1.02. This reaction is first order with respect to both ethylene cholorohydrins and sodium bicarbonate. Reaction rate constant at reaction temperature is 5.2 L/(mol · h). The required conversion is 95%.
1. If auxiliary time is 0.5 h, calculate reactor effective volume;
2. Loading factor is 0.75, calculate reactor volume.

3.3 A reaction between sodium propionate and hydrochloric acid is a second order reversible reaction (first order with respect to both sodium propionate and hydrochloric acid). This reaction was studied in the lab using an isothermal batch reactor operated at 50°C.

$$C_2H_5COONa + HCl \leftrightharpoons C_2H_5COOH + NaCl$$

At the beginning of the reaction the molar ratio of the two reactants is 1:1. In order to measure reaction progress, a 10 mL liquid sample of reaction mixture at different reaction time is taken and titrated using 0.515 mol/L NaOH solution to determine the concentration of unreacted hydrochloric acid. The amounts of NaOH solution consumed are summarized in the following table:

Time(min)	0	10	20	30	50	∞
NaOH consumed(mL)	52.5	32.1	23.5	18.9	14.4	10.5

For the commercial production of propanoic acid, a batch reactor operated under the same conditions as the lab study will be used. The propanoic acid production rate is 500 kg/h, and sodium propionate needs to be 90% of equilibrium conversion. Please calculate the reaction volume required. Assumptions: (1) Loading the reactor and heating up the feed to reaction temperature (50°C) will take 20 min, and any reaction during the heating process can be ignored; (2) Unloading and cleaning the reactor will take 10 min; (3) Density of reaction mixture can be treated as a constant during the reaction process.

3.4 Liquid phase adiabatic reaction takes place in a batch reactor:

$$A + B \rightarrow R$$

And its reaction rate equation is:

$$r_A = 1.1 \times 10^{14} \exp\left(-\frac{11,000}{T}\right) c_A c_B, \text{kmol}/(\text{m}^3 \cdot \text{h})$$

Where c_A and c_B are the concentrations of component A and B, respectively, and their unit is kmol/m^3. The unit of temperature is K. Reaction heat is -4000 J/mol. At the beginning there is no R in the solution, and the concentration of both A and B is 0.04 kmol/m^3. The average volumetric specific heat of the reaction mixture is 4.102 kJ/m^{3}K, and initial temperature is 50°C.

1. Calculate reaction time needed to achieve 85% conversion of A and reaction rate at that moment;
2. Is it possible to convert all reactants into product R? Why?

3.5 The following liquid phase reactions take place in a batch reactor:

$$A + B \rightarrow C \quad r_A = k_1 c_A c_B$$
$$C + B \rightarrow D \quad r_D = k_2 c_C c_B$$

Initial concentration of A is 0.1 kmol/m^3, there is no C and D in the feed, and B is excessive. At reaction time t_1, $c_A = 0.055$ kmol/m^3, $c_C = 0.038$ kmol/m^3. At reaction time t_2, $c_A = 0.01$ kmol/m^3, $c_C = 0.042$ kmol/m^3. Please calculate:

1. k_2/k_1;
2. The maximum concentration of C;
3. Conversion of A when concentration of C is at its maximum.

3.6 A liquid phase reaction takes place in an isothermal batch reactor:

$$A_1 \underset{k_2}{\overset{k_1}{\rightleftharpoons}} A_2 \overset{k_3}{\rightarrow} A_3$$

All three reactions are first order reactions. There is no A_2 and A_3 in the feed and initial concentration of A_1 is 2 mol/L. At reaction temperature $k_1 = 4.0$ min^{-1}, $k_2 = 3.6$ min^{-1}, and $k_3 = 1.5$ min^{-1}. Please calculate:

1. Reaction mixture composition when reaction time is 1.0 min;
2. Reaction mixture composition when reaction time approaches infinity;
3. Reaction mixture composition when reaction time approaches infinity if the reactions are:

$$A_1 \xrightarrow{k_1} A_2 \underset{k_3}{\overset{k_2}{\rightleftarrows}} A_3$$

3.7 A reactor system needs to be designed to conduct the following liquid phase reactions isothermally:

$$A + 2B \rightarrow R, \quad r_R = k_1 c_A c_B^2$$
$$2A + B \rightarrow S, \quad r_S = k_2 c_A^2 c_B$$

R is the desired product. B is much more expensive than A and is very difficult to recycle.

1. How would you select the A to B ratio in the feed?
2. If multiple CSTRs in series are used, what feeding mode would you use?
3. If a semibatch reactor is used, what feeding mode would you use?

3.8 Cumene hydroperoxide decomposes in a 300 L reactor to make phenol and acetone.

The reactor is operated isothermally at 86°C, and cumene hydroperoxide concentration in the feed solution is 3.2 kmol/m^3. The reaction is first order and at reaction temperature reaction rate constant is 0.08 s^{-1}. Please calculate production rate of phenol when final conversion is 98.9% for the following situations:

1. A batch reactor is used and the auxiliary time is 15 min;
2. A CSTR reactor is used;
3. Discuss the differences between (1) and (2);
4. If cumene hydroperoxide concentration is doubled and all other conditions are kept the same, what will be the result?

3.9 The following liquid phase reactions take place isothermally in a batch reactor:

$$A + B \rightarrow R, \quad r_R = 1.6c_A$$
$$2A \rightarrow D, \quad r_D = 8.2c_A^2$$

where r_R and r_D is the formation rate of D and R, respectively. Feed is a mixture of A and B, and concentration of A is 2 kmol/m^3:
1. Calculate reaction time needed to achieve 95% conversion of A.
2. When conversion of A is 95%, what is the yield of R?
3. Will it be possible to achieve 70% yield of D without changing reaction temperature?
4. If a CSTR is used, and reaction temperature and feed composition are kept the same, will it be possible to achieve 95% conversion of A when the space time of the CSTR reactor is equal to reaction time of the batch reactor?
5. If a CSTR reactor is used and all the other conditions are kept the same, what is the yield of R if conversion of A is 95%?
6. A semibatch reactor is used, B is loaded into the reactor, and A is gradually added to the reactor at constant flow rate determined based on reaction time calculated in (1). B is 1 m^3, and A is 0.4 m^3. Please calculate the conversion of A and yield of R when all A has been added to the reactor.

3.10 The following liquid phase reactions take place in two isothermal CSTRs in series:

$$A \rightarrow B, \quad r_A = 68c_A^2$$
$$B \rightarrow R, \quad r_R = 14c_B$$

The concentration of A in the feed is 0.2 kmol/m^3, and feed flow rate is 4 m^3/h. The required final conversion of A is 90%.
1. What is the minimum total reaction volume?
2. What is the yield of desired product B?
3. If the objective is to maximize the yield of B, what is final conversion and minimum total reaction volume.

3.11 Hexamethylenetetramine is synthesized in a continuous flow tank reactor with a reaction volume of 490 cm^3 by reaction between ammonia and formaldehyde:

$$4NH_4 + 6HCHO \rightarrow (CH_2)6N_4 + 6H_2O$$
$$\quad\;\;(A) \qquad\;\; (B)$$

The reaction rate equation is:

$$r_A = kc_A c_B^2 \text{ mol/(L·s)}$$
$$k = 1.42 \times 10^3 \exp(-3090/T)$$

The concentrations of ammonia and formaldehyde in the feed are 4.06 mol/L and 6.32 mol/L, respectively, and flow rate for both solutions is 1.50 cm^3/s. Reaction temperature is 36°C, and, and the density of reaction fluid can be considered as a constant. Please calculate ammonia conversion and ammonia and formaldehyde concentrations at reactor outlet (c_A and c_B).

3.12 Multiple tank reactors in series are used for reversible reactions between ethanol and acetic acid. The flow rate for ethanol and acetic acid is 2.2 kg/h and 1.8 kg/l, respectively. The volume of each reactor tank is 0.01 m^3. Reaction temperature is 100°C. The reaction rate for the esterification is 4.76×10^{-4} L/(min · min), and for the backward reaction (ester hydrolysis) is 1.63×10^{-4} L/(min · min). The density of reaction mixture is 864 kg/m^3. Please decide the number of reactors needed to achieve acetic acid conversion of 60%.

3.13 With sulfuric acid as the catalyst, acetic acid reacts with butanol to make butyl acetate. There are two spare reactor tanks with reaction volumes of 3 m^3 and 1 m^3, respectively. It was decided to use these two reactors, operating continuously and isothermally, to make butyl acetate. Acetic acid concentration in the feed is 0.15 kmol/m^3, and butanol is excessive. The reaction is second order with respect to acetic, and at the reaction temperature the rate constant is 1.2 L/(mol · h). If the final acetic acid conversion must be at least 50%, how would you arrange these two reactors to maximize butyl acetate yield? Why? Please calculate butyl acetate yield for your design. If the reaction is first order, how would you arrange these two reactors?

3.14 The following liquid phase reaction takes place isothermally, and the reaction order is 1.5:

$$A \rightarrow B + C$$

The reaction rate constant is 0.158 m$^{1.5}$/(mol$^{0.5}$ · h). The feed solution contains 2 kmol/m^3 of A, and the feed flow rate entering the reactor is 1.5 m^3/h. For the following situations please calculate reaction volume required to achieve 95% conversion of A:

1. One continuous flow reactor;
2. Two continuous flow reactors with equal volume in series;
3. Two continuous flow reactors with the objective to minimize total reaction volume.

3.15 The following liquid phase reactions take place in a continuous flow tank reactor with reaction volume of 20 m^3:

$$A \rightarrow R, \quad r_A = k_1 c_A$$
$$2R \rightarrow D, \quad r_D = k_2 c_R^2$$

where c_A and c_R is the concentration of A and R, respectively, and r_A is the consumption rate of A, and r_D is the formation rate of D. Feed flow rate is $0.5 \text{ m}^3/\text{min}$, and the concentration of A in the feed is 0.1 kmol/m^3. At reaction temperature $k_1 = 0.1 \text{ min}^{-1}$ and $k_2 = 1.25 \text{ L/(mol} \cdot \text{min)}$. Please calculate the conversion of A and the yield of R at the reactor outlet.

3.16 The following liquid phase reactions take place isothermally in a continuous flow tank reactor:

$$2A \underset{k_2}{\overset{k_1}{\rightleftharpoons}} B, \qquad r_B = k_1 c_A^2 - k_2 c_B$$

$$A + C \overset{k_3}{\longrightarrow} D, \qquad r_D = k_3 c_A c_C$$

Feed flow rate is 360 l/h, and the feed contains 25 wt.% A and 5 wt.% C. The density of feed solution is 0.69 kg/m^3. If at the reactor outlet, the conversion of A is 92%, please calculate:

1. Reaction volume needed;

2. Yields of B and D.

At reaction temperature $k_1 = 6.85 \times 10^{-5}$ L/(mol $\cdot$ s), $k_2 = 1.296 \times 10^{-9}$ 1/s, $k_3 = 1.173 \times 10^{-5}$ L/(mol $\cdot$ s). Both B and D have the same relative molecular weight of 140.

3.17 The following reactions take place in a continuous flow tank reactor isothermally at 55°C:

$$C_6H_6 + Cl_2 \overset{k_1}{\longrightarrow} C_6H_5Cl + HCl$$

$$C_6H_5Cl + Cl_2 \overset{k_2}{\longrightarrow} C_6H_4Cl_2 + HCl$$

$$C_6H_4Cl_2 + Cl_2 \overset{k_3}{\longrightarrow} C_6H_3Cl_3 + HCl$$

All three reactions are first order with respect to their reactants. At 55°C, $k_2/k_3 = 30$. Benzene concentration in the feed stream entering the reactor is 10 kmol/m^3. During the reaction process the ratio of chlorine concentration in the liquid phase to that of benzene is equal to 1.4. If $k_1 \tau = 1$ L/mol, please calculate the concentrations of benzene, monochlorobenzene, dichlorobenzene, and trichlorobenzene in the liquid at the reactor outlet.

3.18 Using the conditions and data given in Example 3.11, please discuss whether there will be multiple steady-states. In order to increase the conversion of maleic anhydride, feed is continuously fed into the reactor while all other conditions are kept the same, discuss at what situation there will be three steady-states.

3.19 Use the conditions and data given in Problem 3.3 and change the reactor operation mode from batch to continuous flow:

1. When all the reaction conditions are kept the same, will propanoic acid productivity increase or decrease?
2. If we want to keep both sodium propionate conversion and propanoic acid yield the same as before, calculate the space time needed. Can you use kinetic data given in Problem 3.3 to directly estimate space time needed?
3. If three continuous flow tank reactors are used in series, and the average residence time in each reactor is equal to 1/3 of average residence time of the original reactor, what will be the conversion?

3.20 Based on data given in Problem 3.8, when one continuous flow reactor is used the conversion is 98.9%. Now another reactor of the same size is available and can be used either in parallel or in series with the original one. If conversion is kept the same, what will be the productivity increase for these two situations?

3.21 Hydrolysis of acetic anhydride takes place in a continuous flow tank reactor. The reaction volume is 0.75 m^3. Feed flow rate is 0.05 m^3/min, and acetic anhydride concentration is 0.22 kmol/m^3. Feed temperature is 25°C, and product temperature is 36°C. The reaction is a first order irreversible exothermic reaction, and the heat of reaction is -29 kJ/mol (acetic anhydride). Reaction rate as a function of temperature is:

$$k = 1.8 \times 10^7 \exp\left(\frac{-5526}{T}\right), \quad 1/\min$$

Feed density can be considered as a constant and equals 1050 kg/m^3. Specific heat is 2.94 kJ/(kg · K). No heat exchanger is installed for this reactor, and the only way to remove reaction heat is heat transfer through reactor wall to the surrounding environment. Please calculate:

1. Acetic anhydride concentration at reactor outlet;
2. Heat transfer rate from reactor to surrounding environment.

3.22 In a tank reactor of 1 m^3, propylene oxide dissolved in methanol reacts with water forming 1,2-Propanediol:

$$\text{H}_2\text{C}\!-\!\text{CH}\!-\!\text{CH}_3 + \text{H}_2\text{O} \longrightarrow \text{H}_2\text{C}\!-\!\text{CH}\!-\!\text{CH}_3$$

with OH, OH groups shown below the product.

This reaction is first order with respect to propylene oxide, and at reaction temperature the reaction rate constant is 0.98 l/h. Propylene oxide concentration in the feed is 2.1 kmol/m^3, and the final propylene oxide conversion is 90%.

1. A batch reactor is used, and auxiliary time is 0.65 h. What is the daily production rate of 1,2-propanediol?
2. If a continuous flow reactor is used and all the conditions are the same, what is the daily production rate 1,2-propanediol?
3. Explain why the productivities are different.

3.23 For the reactor system described in Problem 3.11, assume the density of the liquid mixture can be treated as constant and its value is 1.02 g/cm^3, and average specific heat is 4.186 kJ/(kg·K). In addition, the impact of reaction temperature on reaction heat can be ignored, and reaction heat is -2231 kJ/kg hexamethylenetetramine. The temperature at reactor inlet is 25°C:

1. What is the adiabatic temperature increase? Is it possible to achieve 80% conversion under adiabatic condition? If so what will be the reaction temperature?
2. If the reactor is operated isothermally at 100°C, what will be the conversion? What heat removing rate will be required?

3.24 A continuous flow tank reactor is used for condensation polymerization of adipic acid and hexanediol to make alkyd resin. Under normal operating conditions (reaction temperature, feed and product flow rate, etc.) the conversion of adipic acid is 80%. The operator noticed that the conversion decreased to 70%, and then found a leak at a flange at the reactor outlet. After fixing the leak and maintaining normal operation conditions including reaction temperature and flow rate, the conversion still cannot be restored to the original value. Please discuss what the possible causes are and how to restore conversion to 80%.

Chapter 4

Tubular Reactor

Chapter Outline

Any reactor with a large length (height) to diameter ratio can be called a tubular reactor. For tank reactors the height is close to or slightly larger than the diameter. Tubular reactors can be used for both homogeneous and heterogeneous reactions. For example, naphtha crackers used for ethylene production are tubular reactors. Fixed-bed reactors, which are broadly used for gas—solid catalytic reactions, can also be considered tubular reactors. The tubular reactor defined here covers a very broad range of reactors. Most tubular reactors are operated continuously, and there are some tubular reactors that are operated in semibatch mode. However, it is very rare that a tubular reactor is used in batch operation. This chapter will discuss continuous operation with the overall goal of providing a systematic approach for reactor design and analysis.

4.1 PLUG FLOW

Fluid flow in a continuous reactor is a very complex physical phenomenon that has direct impact on chemical reactions taking place inside the reactor, including both reaction rate and the extent of the reactions. Such an impact can be understood by examining the velocity distribution inside the reactor. It is well known that for fluid flow inside a tube, the radial distribution of velocity is nonuniform with the highest velocity at the center and the lowest

Reaction Engineering. DOI: http://dx.doi.org/10.1016/B978-0-12-410416-7.00004-5
© 2017 Chemical Industry Press. Published by Elsevier Inc. All rights reserved.

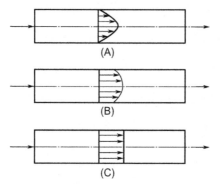

FIGURE 4.1 Radial velocity distributions. (A) Laminar flow; (B) Turbulent flow; (C) Plug flow.

velocity at the wall. For laminar flow, the radial distribution is parabolic, as shown in Fig. 4.1(A). For turbulent flow, the radial velocity distribution becomes flatter when the turbulence increases (see Fig. 4.1(B)). Obviously, residence time of fluid near the center is shorter than that of fluid near the wall. Such a difference in residence time will lead to different extents of reaction.

In addition to velocity distribution, fluid mixing inside the reactor is another important factor. Since the mechanism and degree of fluid mixing will have direct impacts on concentration and temperature, the two factors that will determine reaction rate at various locations in the reactor. Due to the complex nature of fluid flow and the complex relationship between various factors, reasonable simplifications have to be made to establish a realistic physical model, i.e., reactor flow model, to describe fluid flow inside the tube.

Plug flow is the most basic flow model. The fundamental assumption is that the radial velocity distribution is uniform, as shown in Fig. 4.1(C). In other words, all the fluid particles flow from inlet to outlet at exactly the same velocity, just like a piston or a plug moving through the reactor, and therefore it is called plug flow or piston flow. Another assumption is that at any cross-section perpendicular to the flow the concentration and temperature distributions are uniform, i.e., the fluid is completely uniform in the radial direction. In addition, it is assumed that there is no mixing along the axial direction. Mixing in the axial direction is called as axial mixing, or back mixing. Although there is no back mixing, reactant concentration and temperature at various cross-sections can be different due to fluid flow and reaction taking place in the reactor. The so-called back mixing indicates the mixing of particles with different residence times. Since radial velocity distribution is uniform, all fluid particles at a given cross-section must have the same residence time, and therefore there is no back mixing. For the whole reactor, if the flow is plug flow, then the fluid particles entering the reactor at the same time must leave the reaction simultaneously. In other words, all fluid particles must have the same residence time. It is not difficult to understand that reactors following plug flow assumption have the same effect as batch reactors, i.e., if residence times are the same, then final conversions and yields from these two reactors will be the same.

We did not discuss the fluid flow model in the previous chapter when we treated the continuous tank reactor. However, the reactant concentrations inside the reactor are assumed to be uniform, and such an assumption corresponds to another flow model, i.e., complete mixing model. Its basic assumption is that mixing in both the radial and axial directions approach the maximum degree. Maximum radial mixing guarantees that at any given cross-section, the system parameters such as concentrations and temperature are uniform, while maximum mixing in the axial direction eliminates any concentration and temperature differences among difference cross-sections. As a result, both concentration and temperature are uniform throughout the whole reactor. It is clear that the fundamental difference between plug flow and complete mixing model is that for plug flow there is no back mixing while for complete mixing the back mixing approaches its extremity so there is no concentration or temperature difference in the reactor.

Both plug flow and complete mixing model are ideal flow models, and these two flow models are also called ideal flow models. As mentioned above, from a back mixing point of view, these two models represent two ideal situations, one is no back mixing and another is maximum back mixing. We will discuss more on these two ideal models in the next chapter.

Any reactor that can be described by the plug flow model, no matter what its structure, is called a plug flow reactor. Similarly, any reactor that follows the complete mixed assumption is a complete mixed reactor. In reality, there is no reactor that can meet either plug flow or complete mixed assumptions completely. Any model is an approximation of the real world. If the model is exactly the same as the real reactor then there is no need for the model. It is the complexity of the real reactor that makes approximation necessary. The model cannot and should not capture all the details of the real reactor; instead, it needs to reflect the key characters of the real reactor while ignoring aspects that are not critical.

4.2 DESIGN OF ISOTHERMAL TUBULAR REACTOR

Fig. 4.2 is a schematic of a tubular reactor. Feed flows into the reactor at a flow rate of Q_0, and products leave the reactor at the bottom. In this chapter we will only discuss tubular reactors that can be described by the plug flow model and are operated at steady-state. We will examine isothermal reactors in this section and discuss nonisothermal reactors in later sections. The most important task of reactor design is to determine the reactor volume needed in order to achieve a given conversion based on feed flow rate and composition. This is the foundation of any further design calculations. Similar to tank reactors

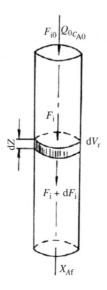

FIGURE 4.2   Tubular reactor.

discussed in the previous chapter, the first step for reactor design is to establish a design equation.

4.2.1   Single Reaction

Design equations under isothermal conditions are based on mass conservation law. Component concentrations inside a tubular reactor are functions of axial locations, so we will select an infinitesimal volume element, dV_r, as the control region (see Fig. 4.2). Based on the plug flow assumption, at steady-state, for any component we can write a mass balance equation for volume dV_r:

$$F_i + dF_i - F_i = \mathscr{R}_i dV_r$$

which can be simplified as:

$$\frac{dF_i}{dV_r} = \mathscr{R}_i \tag{4.1}$$

Eq. (4.1) is the design equation for a tubular reactor based on a plug flow model. It is a very simple but also very important equation that can be used for solving many problems. In this equation component i can be any component in the system, reactants or products. If component i is a reactant, then $\mathscr{R}_i$ is negative, indicating that the mole flow rate of this component will decrease with increasing reactor volume. On the other hand, $\mathscr{R}_i$ is positive if component i is a product. For single reactions, we only need to write the mass balance equation for one component, i.e., the key component. Usually the reactant with limited amount will be selected as the key component. For example, for component A, its mass balance equation is:

$$\frac{dF_A}{dV_r} = \mathscr{R}_A \tag{4.2}$$

To calculate the reactor volume it usually is more convenient to express both F_A and $\mathscr{R}_A$ as functions of conversion:

$$F_A = F_{A0}(1 - X_A)$$

Substitute above equation into Eq. (4.2):

$$F_{A0}\frac{dX_A}{dV_r} = -\mathscr{R}_A(X_A) \tag{4.3}$$

Since $F_{A0} = Q_0 c_{A0}$, Eq. (4.3) can be rewritten as:

$$Q_0 c_{A0}\frac{dX_A}{dV_r} = -\mathscr{R}_A(X_A) \tag{4.4}$$

Eqs. (4.2) to (4.4), although different in format, are all design equations for tubular reactors with a single reaction. Depending on the specific situation under consideration any one of them can be chosen for reactor volume calculation.

By integrating Eq. (4.4), we can obtain reactor volume:

$$V_r = Q_0 c_{A0} \int_0^{X_{Af}} \frac{dX_A}{[-\mathscr{R}_A(X_A)]} \tag{4.5}$$

If reaction rate is constant, then Eq. (4.5) can be simplified as:

$$V_r = \frac{Q_0 c_{A0} X_A}{[-\mathscr{R}_A(X_A)]} \tag{3.43}$$

Eq. (3.43) is the design equation of tank reactors, which can be considered as a special situation of Eq. (4.5).

Based on space time definition, Eq. (4.5) can be converted into Eq. (4.6) by dividing both sides by Q_0:

$$\tau = c_{A0} \int_0^{X_{Af}} \frac{dX_A}{[-\mathscr{R}_A(X_A)]} \tag{4.6}$$

For batch tank reactors discussed in the previous chapter the reaction time is:

$$t = c_{A0} \int_0^{X_{Af}} \frac{dX_A}{[-\mathscr{R}_A(X_A)]} \tag{3.8}$$

Eq. (4.6) looks exactly the same as Eq. (3.8), but it does not necessarily mean that $\tau = t$. In order to discuss this rigorously, we need to examine whether the relationship between $\mathscr{R}_A$ and X_A is the same for these two reactors. Almost all batch tank reactors are operated at constant volume. But this is not necessarily the case for tubular reactors. If both are operated at constant volume, then $\mathscr{R}_A$ for both reactors would be the same, and therefore $\tau = t$. If the tubular reactor is operated under varying-volume conditions, then $\tau \neq t$. This can be illustrated more clearly by examining reaction rate as a function of conversion under these two different scenarios. For the first order reaction under constant volume conditions the reaction rate is:

$$r_A = kc_{A0}(1 - X_A)$$

Under varying-volume conditions:

$$r_A = \frac{kc_{A0}(1 - X_A)}{1 + y_{A0}\delta_A X_A}$$

We can see that these two equations are different. Through the above discussions we can see that under a constant volume situation, the space time needed to achieve a given conversion for a tubular reactor is the same as the reaction time needed to achieve the same conversion using a batch tank reactor operated at the same temperature. If reaction volume changes then this conclusion is not valid anymore.

Although $\tau = t$ is only valid for a constant-volume process, it is still a very important equation, since it shows the relationship between continuous and batch reactors. Under constant volume conditions, all conclusions obtained for batch reactors can be used for plug flow reactors.

Assuming u_0 is the fluid velocity at reactor inlet, Eq. (4.4) can be rewritten to reflect the relationship between conversion and distance from reactor inlet Z:

$$u_0 c_{A0} \frac{dX_A}{dZ} = -\mathscr{R}_A(X_A) \tag{4.7}$$

For constant-volume process Eq. (4.7) can be transformed into an axial concentration distribution equation:

$$u_0 \frac{dc_A}{dZ} = \mathscr{R}_A \tag{4.8}$$

For batch reactor Eq. (3.6) can be rewritten as:

$$\frac{dc_A}{dt} = \mathscr{R}_A \tag{4.9}$$

Eq. (4.8) indicates that for steady-state plug flow reactors, reaction mixture composition will change as a function of axial location, but does not change with time. On the other hand, Eq. (4.9) tells us that for batch tank reactor reactions, mixture composition is a function of time, not location. This is the fundamental difference between these two reactors.

Example 4.1
Using the conditions and data given in Example 3.1, calculate the reactor volume needed if a plug flow tubular reactor is used to produce acetate ester.

Solution
The reaction between ethanol and acetic acid takes place in liquid phase, and therefore can be considered a constant-volume process. Under constant volume conditions, space time needed for a plug flow reactor is the same as that reaction time needed for a batch reactor. In Example 3.1 it was calculated that the reaction time needed is $t = 118.8$ min. If a plug flow reactor is used, to achieve the same conversion the space time should be $\tau = t = 118.8$ min. The feed flow rate is $Q_0 = 4.155$ m³/h, so the reactor volume needed is:

$$V_r = 4.166(118.8/60) = 8.227 \text{ m}^3$$

In Example 3.1, reactor volume needed for a batch reactor is 12.38 m³, which is larger than that of the plug flow reactor. This is because for a batch reactor we need to add the auxiliary times such as feed loading, unloading, reactor cleaning, etc. Excluding auxiliary times the reactor volume will be the same. From this comparison we can see that continuous operation has its advantages.

Example 4.2
Dibutene is produced in a plug flow reactor at 925K by dehydrogenation of butene:

$$C_4H_8 \rightarrow C_4H_6 + H_2$$
$$\text{(A)} \quad \text{(B)} \quad \text{(C)}$$

The reaction rate equation is:

$$r_A = kp_A, \text{ kmol/(m}^3 \cdot \text{h)}$$

Feed is a mixture of butene and steam containing 10 mol% butene. Operating pressure is 10^5 Pa. At 925K $k = 1.079 \times 10^{-4}$ kmol/(h $\cdot$ m³ $\cdot$ Pa). If the required butene conversion is 35%, what will be the space time needed?

Solution
Insert reaction rate equation for $\mathcal{R}_A$ in Eq. (4.6):

$$\tau = c_{A0} \int_0^{0.35} \frac{dX_A}{kp_A} \tag{A}$$

In order to integrate Eq. (A), p_A needs to be expressed as a function of X_A. Based on ideal gas law:

$$p_A = c_A RT \tag{B}$$

(Continued)

(Continued)

For a butene dehydrogenation reaction the reactor is a varying-volume system, and based on stoichiometry $\delta_A = (1 + 1 - 1)/1 = 1$, therefore:

$$c_A = \frac{c_{A0}(1 - X_A)}{1 + y_{A0}\delta_A X_A} = \frac{c_{A0}(1 - X_A)}{1 + 0.1 X_A} \tag{C}$$

Combine Eqs. (B) and (C) and then insert into Eq. (A):

$$\tau = \int_0^{0.35} \frac{(1 + 0.1 X_A)}{kRT(1 - X_A)} dX_A$$

$$= \frac{1}{kRT} \int_0^{0.35} \left[\frac{1.1}{(1 - X_A)} - 0.1 \right] dX_A = \frac{1}{kRT} \left[1.1 \ln \frac{1}{1 - X_A} - 0.1 X_A \right]_0^{0.35}$$

Then space time can be calculated as:

$$\tau = \frac{1}{1.079 \times 10^{-4} \times 8.314 \times 10^3 \times 923} \left[1.1 \times \ln \frac{1}{1 - 0.35} - 0.1 \times 0.35 \right]$$

$$= 5.37 \times 10^{-4} \text{ h} = 1.933 \text{ s}$$

This space time is calculated based on feed volumetric flow rate at reactor inlet temperature and pressure. Since it is a varying-volume process, space time not equal to average residence time in the reactor. The reactive gas volume increases during the course of the reaction, as a result gas velocity inside the reactor also increases, leads to an average residence time inside the reactor less than 1.933 s.

For constant-volume process,

$$\tau = \frac{1}{kRT} \int_0^{0.35} \frac{dX_A}{(1 - X_A)} = \frac{1}{kRT} \ln \frac{1}{1 - X_A}$$

Based on this equation it is calculated that $\tau = 1.898$ s, which is shorter than the value for the varying-volume process, suggesting a smaller reactor volume will be needed. If we treat this process as constant-volume, the reactor volume will be underdesigned, i.e., 35% conversion cannot be achieved. The reason for this is that for design calculation based on constant-volume assumption, the residence time changes caused by reaction volume expansion were not considered.

If the reaction volume decreased with conversion, and you designed the reactor assuming constant volume, what would be the result? Can you explain why?

4.2.2 Multiple Reactions

When multiple reactions take place in a reactor, the reaction process cannot be described by a single reaction variable. Similar to tank reactor design, in order to obtain reactor design equations we need to set up mass balance equation for each key component. The procedure is the same as what we just discussed in the previous section for a single reaction, the only difference is that we need multiple equations, i.e., applying Eq. (4.1) for each key component:

$$\frac{dF_i}{dV_r} = \mathscr{R}_i \quad i = 1,2,\ldots,K \tag{4.1}$$

K represents the numbers of key components. This equation can also be written as:

$$\frac{dF_i}{dV_r} = \sum_{j=1}^{M} \nu_{ij}\bar{r}_j \quad i = 1,2,\ldots,K \tag{4.10}$$

where M is the total numbers of reactions taking place in the reactor.

For a single reaction inside an isothermal tubular reactor, Eq. (4.1) can be integrated as long as we can express F_i and $\mathscr{R}_i$ as functions of conversion. For multiple reactions, a group of ordinary differential equations need to be solved. This will be an initial value problem, and initial values are:

$$V_r = 0, F_i = F_{i0} \quad i = 1,2,\ldots,K$$

In order to solve these equations, we first need to select reaction variables, which can be the conversions or yields of key components. Another option is to choose the reaction extent of key components, ξ_i, as the reaction variable. The next step is to express F_i and r_i as functions of reaction variables, then Eq. (4.10) can be solved. In most situations a numerical method has to be used.

It is also possible to directly use F_i as the reaction variables. In this case we need to express F_i as a function of $\bar{r}_j$.

For constant-volume systems it doesn't matter which one is selected as the reaction variable. However, for varying-volume processes it is usually more convenient to choose F_i as the reaction variable. In many situations, the reaction rates are expressed as functions of concentrations, i.e., $\bar{r}_j = f(c)$, therefore if F_i is chosen as the reaction variables, we need to express c_i as functions of F_i $(i = 1, 2, \ldots N)$, where N is the number of components in the reaction system.

Based on ideal gas law:

$$c_i = \frac{p_i}{RT} = \frac{p y_i}{RT} = \frac{F_i p}{RT \sum_{i=1}^{N} F_i} \tag{4.11}$$

Using Eq. (4.11), concentrations can be expressed as functions of mole flow rate. Since not all the components are key components, we need to express mole flow rates of nonkey components as functions of those of key components, which are based on reaction stoichiometry. Example 4.3 shows how to perform such a conversion.

If the total numbers of moles do not change during the reaction, then

$$\sum_{i=1}^{n} F_i = F_{t0},$$

that is, at any time the mole flow rate of the mixture is a constant and equal to the mole flow rate at the reactor inlet. If the reactions take place under isothermal and constant-pressure conditions, it must be a constant-volume process.

If using concentrations as reaction variables, substitute $F_i = Qc_i$ into Eq. (4.10);

$$\frac{d(Qc_i)}{dV_r} = \sum_{j=1}^{M} \nu_{ij}\bar{r}_j \quad i = 1,2,\ldots,K \tag{4.12}$$

Since $dV_r = A_r dZ$, the design equations can be rewritten in the format of axial concentration distribution of reaction components, i.e.:

$$\frac{d(uc_i)}{dZ} = \sum_{j=1}^{M} \nu_{ij}\bar{r}_j \quad i = 1,2,\ldots,K \tag{4.13}$$

For a constant-volume process, $Q = Q_0 = $ constant, $u = u_0 = $ constant, Eqs. (4.12) and (4.13) can be simplified as:

$$Q\frac{dc_i}{dV_r} = \sum_{j=1}^{M} \nu_{ij}\bar{r}_j \quad i = 1,2,\ldots,K \tag{4.14}$$

$$u\frac{dc_i}{dZ} = \sum_{j=1}^{M} \nu_{ij}\bar{r}_j \quad i = 1,2,\ldots,K \tag{4.15}$$

Eqs. (4.14) and (4.15) are more convenient to use for constant-volume processes.

Example 4.3

Gases A_1 and A_2 are fed into an isothermal and constant-pressure plug flow reactor for gas phase reactions:

$$2A_1 + A_2 = A_3 \quad \bar{r}_1 = k_1 c_1^2 c_2$$

$$A_3 + A_2 = A_4 \quad \bar{r}_2 = k_2 c_3 c_2$$

$$2A_4 = A_5 \quad \bar{r}_3 = k_3 c_4^2$$

c_i is the concentration of A_i $(i = 1,2,3,4)$, k_j is reaction rate constant of reaction j $(j = 1,2,3)$. Please write reactor design equations by using F_i and V_r as reaction variables and assuming no reaction product in the feed to the reactor.

Solution

1. Select key components. There are three reactions, and all are independent, so the number of key components is also three. When selecting the key components we need to make sure that there is at least one key component in

(Continued)

(Continued)

each reaction. For example, selecting A_1, A_2, and A_3 as key components will not meet this requirement since none of them participates in the third reaction. Here we select A_1, A_2, and A_5 as key components, and assume mole flow rate is F_1, F_2, and F_5, respectively. Mole flow rate of A_1 and A_2 at reactor inlet is F_{10} and F_{20}, respectively. Since there is no product in the feed, $F_{30} = F_{40} = F_{50} = 0$.

2. Express mole flow rates of nonkey components as functions of those of key components. Eq. (2.56) can be used to express mole flow rate of A_3 and A_4, F_3 and F_4, respectively, as functions of mole flow rates of key components:

$$F_i = F_{i0} + \sum_{j=1}^{M} \nu_{ij}\xi_j \tag{A}$$

Here ξ_j is the reaction extent per unit time for reaction j. Eq. (A) is valid for any component so

$$F_1 = F_{10} - 2\xi_1 \tag{B}$$

$$F_2 = F_{20} - \xi_1 - \xi_2 \tag{C}$$

$$F_3 = \xi_1 - \xi_2 \tag{D}$$

$$F_4 = \xi_2 - 2\xi_3 \tag{E}$$

$$F_5 = \xi_3 \tag{F}$$

(D) − (C) + (B) gives:

$$F_3 = F_{10} - F_{20} - F_1 + F_2 \tag{G}$$

(C) + (E) + (F) − 0.5 (B):

$$F_4 = F_{20} - 0.5F_{10} + 0.5F_1 - F_2 - 2F_5 \tag{H}$$

3. Use Eq. (4.11) to express component concentrations as functions of mole flow rates of key components.

4. List the design equations. From Eq. (4.1):

$$\frac{dF_1}{dV_r} = -2\bar{r}_1 = -2k_1 c_1^2 c_2 \tag{I}$$

$$\frac{dF_2}{dV_r} = -\bar{r}_1 - \bar{r}_2 = -(k_1 c_1^2 + k_2 c_3)c_2 \tag{J}$$

$$\frac{dF_5}{dV_r} = \bar{r}_3 = k_3 c_4^2 \tag{K}$$

Use Eq. (4.11) for c_1, c_2, c_3, and c_4, and Eqs. (G) and (H) for F_3 and F_4 in Eqs. (I)−(K):

$$\frac{dF_1}{dV_r} = -2\psi^3 k_1 F_1^2 F_2 \tag{L}$$

(Continued)

(Continued)

$$\frac{dF_2}{dV_r} = \psi^2 F_2 \left[\psi k_1 F_1^2 + k_2 (F_{10} - F_{20} - F_1 + F_2) \right] \quad \text{(M)}$$

$$\frac{dF_5}{dV_r} = \psi^2 k_3 (F_{20} - 0.5F_{10} + 0.5F_1 - F_2 - 2F_5)^2 \quad \text{(N)}$$

where $\psi = p/(RT \sum\limits_1^5 F_i)$

and $\sum\limits_1^5 F_i = F_1 + F_2 + F_3 + F_4 + F_5 = 0.5F_{10} + 0.5F_1 + F_2 - F_5$

Eqs. (L)–(N) are the design equations.

From this example we can see that when designing a tubular reactor with multiple reactions taking place inside, the first step is to decide the number of independent reactions and select key components, and then express the concentrations of reaction components as functions of mole flow rates of key components, and finally list the design equations.

Example 4.4
Trimethyl naphthalene dealkylation reactions take place in a tubular reactor operated at 4.05 MPa and 936K:

The reaction rates of these three reactions are:

$$r_1 = k_1 c_T c_H^{0.5}$$

$$r_2 = k_2 c_D c_H^{0.5}$$

$$r_3 = \vec{k}_3 (c_M c_H^{0.5} - c_N c_G / c_H^{0.5} K)$$

(Continued)

(Continued)

Subscripts T, D, M, H, G, and N represent trimethyl naphthalene, dimethyl naphthalene, methyl naphthalene, hydrogen, methane, and naphthalene, respectively. The unit for all reaction rate constants is $kmol/(m^3 \cdot s)$, and K is the equilibrium coefficient of the third reaction. Under reaction temperature:

$$K = 5$$

$$k_1 = 5.66 \times 10^{-6} \ m^{1.5}/(mol^{0.5} \cdot s)$$

$$k_2 = 5.866 \times 10^{-6} \ m^{1.5}/(mol^{0.5} \cdot s)$$

$$\vec{k}_3 = 2.052 \times 10^{-6} \ m^{1.5}/(mol^{0.5} \cdot s)$$

Feed gas composition is 25 mol% trimethyl naphthalene and 75 mol% hydrogen. The superficial velocity at reactor inlet is 0.1 m/s (under reaction temperature and pressure). If the required trimethyl naphthalene conversion is 80%, calculate:
1. The reactor length needed;
2. Yields of dimethyl naphthalene, methyl naphthalene, and naphthalene.

Solution
All three reactions are independent reactions, and there are three key components. We can select methyl naphthalene, dimethyl naphthalene, and methyl naphthalene as the key components. Since it is a constant-volume process, we can use Eq. (4.14) to describe mass balance for these three components:

$$-Q_0 \frac{dc_T}{dV_r} = k_1 c_T c_H^{0.5} \tag{A}$$

$$Q_0 \frac{dc_D}{dV_r} = k_1 c_T c_H^{0.5} - k_2 c_D c_H^{0.5} \tag{B}$$

$$Q_0 \frac{dc_M}{dV_r} = k_2 c_D c_H^{0.5} - \vec{k}_3 \left[c_M c_H^{0.5} - c_N c_G / K c_H^{0.5} \right] \tag{C}$$

In order to solve these three equations, we need to express all the concentrations as functions of three reaction variables (conversion of trimethyl naphthalene X_T, dimethyl naphthalene yields Y_D, and methyl naphthalene yield Y_M):

$$c_T = c_{T0}(1 - X_T)$$

$$c_D = c_{T0} Y_D$$

$$c_M = c_{T0} Y_M$$

$$c_H = c_{H0} - [c_{T0} X_T + c_{T0}(X_T - Y_D) + c_{T0}(X_T - Y_D - Y_M)] = c_{H0} - c_{T0}(3X_T - 2Y_D - Y_M)$$

Yield of naphthalene is equal to $(X_T - Y_D - Y_M)$, so $c_N = c_{T0}(X_T - Y_D - Y_M)$. The amount of methane formed should equal the amount of hydrogen consumed, so

$$c_G = c_{T0}(3X_T - 2Y_D - Y_M)$$

(Continued)

(Continued)

Since $\tau = V_r/Q_0$, using equations derived above, Eqs. (A)–(C) can be rewritten as:

$$\frac{dX_T}{d\tau} = k_1(1 - X_T)H \tag{D}$$

$$\frac{dY_D}{d\tau} = [k_1(1 - X_T) - k_2 Y_D]H \tag{E}$$

$$\frac{dY_M}{d\tau} = k_2 Y_D H - \vec{k}_3\left[Y_M H - \frac{c_{T0}(X_T - Y_D - Y_M)(3X_T - 2Y_D - Y_M)}{KH}\right] \tag{F}$$

$$H = [c_{H0} - c_{T0}(3X_T - 2Y_D - Y_M)]^{1/2} \tag{G}$$

Based on feed gas composition, temperature and pressure we can calculate initial concentrations of trimethyl naphthalene and hydrogen:

$$c_{T0} = \frac{pT}{RT} = \frac{4{,}050{,}000 \times 0.25}{8.314 \times 10^3 \times 936} = 0.1303 \text{ kmol/m}^3$$

$$c_{H0} = \frac{4{,}050{,}000 \times 0.75}{8.314 \times 10^3 \times 936} = 0.3909 \text{ kmol/m}^3$$

Substitute these values into Eqs. (D) to (G):

$$\frac{dX_T}{d\tau} = 1.79 \times 10^{-4}(1 - X_T)H \tag{H}$$

$$\frac{dY_D}{d\tau} = \left[1.79 \times 10^{-4}(1 - X_T) - 1.855 \times 10^{-4} Y_D\right]H \tag{I}$$

$$\frac{dY_M}{d\tau} = 1.855 \times 10^{-4} Y_D H - 6.49 \times 10^{-4}$$

$$\left[Y_M H - \frac{0.1303(X_T - Y_D - Y_M)(3X_T - 2Y_D - Y_M)}{5H}\right] \tag{J}$$

$$H = [0.3909 - 0.1303(3X_T - 2Y_D - Y_M)]^{1/2} \tag{K}$$

Initial values of above equations are: $\tau = 0$, $X_T = 0$, $Y_D = 0$, and $Y_M = 0$. These equations have to be solved numerically and Runge–Kutta–Gill methods can be used. The results are listed in Table 4A, where the yield of naphthalene $Y_N = X_T - Y_D - Y_M$.

Using data from Table 4A, we can plot X_T, Y_D, Y_M, and Y_N as functions of τ, as shown in Fig. 4A. We can see for intermediate product dimethyl naphthalene there are maximum yields. Similarly methyl naphthalene also has a maximum yield but in the range of space time we examined Y_M is increasing. At higher space time Y_M will decline.

Using the data from Table 4A, the space time needed to achieve 80% conversion is 17,042 s, and the reactor length needed is:

$$17{,}042 \times 0.1 = 1704.2 \text{ m}$$

(Continued)

(Continued)

TABLE 4A Product Distribution for Trimethyl Naphthalene Dealkylation

$\tau \times 10^{-3}/s$	X_T	Y_D	Y_M	Y_N
0	0	0	0	0
1	0.1049	0.0990	0.0052	0.0007
2	0.1972	0.1756	0.0167	0.0049
3	0.2783	0.2340	0.0304	0.0139
4	0.3498	0.2777	0.0443	0.0278
5	0.4129	0.3097	0.0573	0.0460
6	0.4686	0.3322	0.0687	0.0677
7	0.5179	0.3471	0.0788	0.0920
8	0.5617	0.3562	0.0876	0.1179
9	0.6005	0.3605	0.0953	0.1447
10	0.6351	0.3612	0.1022	0.1717
11	0.6660	0.3591	0.1086	0.1983
12	0.6936	0.3548	0.1146	0.2242
13	0.7184	0.3488	0.1205	0.2491
14	0.7407	0.3416	0.1263	0.2728
15	0.7607	0.3335	0.1322	0.2950
16	0.7788	0.3247	0.1383	0.3158
17	0.7952	0.3156	0.1446	0.3350
18	0.8101	0.3062	0.1511	0.3528
19	0.8236	0.2966	0.1579	0.3691
20	0.8359	0.2870	0.1649	0.3840

Obviously, it is not practical to build such a long reactor. In addition, the pressure drop for such a long reactor would be prohibitive. The above calculation is based on the assumption that the reactor is under constant-pressure conditions. More accurately we need to treat this reactor as a varying-volume process. Then Eq. (4.14) is not valid, and Eq. (4.10) or Eq. (4.12) has to be used. In addition, a momentum balance equation has to be used to describe pressure distribution along the reactor. Such a momentum equation has to be solved simultaneously with mass balance equations. The pressure change will affect concentrations and the volumetric flow rate of the gas mixture.

(Continued)

(Continued)

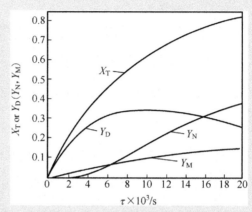

FIGURE 4A Product distribution for trimethyl naphthalene dealkylation.

As mentioned above, it is not practical to build a very long reactor. In reality, multitube reactors will be used so that the pressure drop will be much lower. For example, 480 tubes with length of 3.55 m will give a total reactor length of 1074 m, which is approximately equal to the reactor length needed as calculated above. If the gas flow inside each tube still can be described as plug flow then the product distribution should be the same as the calculation results listed in Table 4A. It should be noted that since multiple tubes are used, gas velocity in each tube will be lower. When using 480 tubes in parallel, velocity in each tube will be $0.1/480 = 2.08 \times 10^{-4}$ m/s. At such low velocity it is questionable whether the flow inside the reactor tube can still be treated as plug flow. We will discuss this in the next chapter.

The yields of methyl naphetene, dimethyl naphetene, and naphetene at $\tau = 17{,}042$ s can also be obtained by using data shown in Table 4A, and the results are: $Y_D = 0.3152$, $Y_M = 0.1449$, and $Y_N = 0.3358$.

4.2.3 Pseudo Homogeneous Model

For heterogeneous catalytic systems, the chemical reactions take place on the catalyst surface. The reactants in the fluid need to transfer to a solid catalyst surface. On the other hand, products need to transfer back into fluid phase. For most chemical reactions there will be reaction heat, and as result there will be also heat transfer between fluid phase and solid catalysts. Due to these mass and heat transfer processes, concentrations of reacting components on the catalyst surface will be different from those in the fluid phase, and catalyst temperature will also be different from the value of fluid phase. If mass and heat transfer are very fast, then the concentration and temperature differences between catalyst surface and fluid phase will be small. In this case, although rigorously speaking it is a heterogeneous catalytic reaction system, it is possible to ignore such small differences and treat the reactor system as pseudo homogeneous. Reactor models based on such an assumption is called a pseudo homogeneous model.

If the tubular reactor where heterogeneous catalytic reactions take place can be treated using the pseudo homogeneous assumption, then the reactor design equation discussed in previous sections can be used. However, it should be noted that for homogeneous reactions, the reactor volume is the whole space where reactions take place. For heterogeneous reactions, the reactor volume is the packed volume of the catalyst, which is the volume of catalyst particles plus the void space among those particles.

If the reaction rates are based on catalyst weight, then Eq. (4.10) can be rewritten as:

$$\frac{dF_i}{dV_r} = \rho_b \sum \nu_{ij}\bar{r}_j \quad i = 1,2,\ldots,k \tag{4.16}$$

where ρ_b is bulk density of the catalysts. This equation can also be written as:

$$\frac{dF_i}{dW} = \sum \nu_{ij}\bar{r}_j \quad i = 1,2,\ldots,k \tag{4.17}$$

Example 4.5
Ethyl benzene dehydrogenation reaction takes place at 0.12 MPa and 898K, and the reaction rate equation is:

$$r_A = k[p_A - p_S p_H / K_P] \quad \text{mol/(kg·s)} \tag{A}$$

where p is partial pressure, subscripts A, S, and H represents ethyl benzene, styrene, and hydrogen, respectively. At reaction temperature $k = 1.68 \times 10^{-10}$ kmol/(kg·s·Pa), and equilibrium coefficient $K_P = 3.727 \times 10^4$ Pa. If the reaction takes place in a plug flow reactor and feed is a mixture of ethyl benzene and steam with ethyl benzene to steam mole ratio $= 1:20$, please calculate the amount of catalyst needed to achieve 60% conversion when feed ethyl benzene flow rate is 1.7×10^3 mol/s.

Solution
Assuming the pseudo homogeneous model applies and there is no side reaction, this is a varying-mole reaction:

$$\delta_A = (1 + 1 - 1)/1 = 1$$

Next we need to express partial pressure of each components as functions of conversion:

$$p_A = p_{A0}(1 - X_A)/(1 + y_{A0}\delta_A X_A) \tag{B}$$

$$p_S = p_{A0}X_A/(1 + y_{A0}\delta_A X_A) \tag{C}$$

$$p_H = p_{A0}X_A/(1 + y_{A0}\delta_A X_A)$$

(Continued)

(Continued)

Ethyl benzene mole fraction in feed: $y_{A0} = 1(1 + 20) = 1/21$

$$p_{A0} = py_{A0} = 1.2 \times 10^5 \times (1/21))$$

Substitute y_{A0} and p_{A0} into Eqs. (B) and (C):

$$p_A = 1.2 \times 10^5 (1/21)(1 - X_A) - (1 + X_A/21) = 1.2 \times 10^5 (1 - X_A)/(21 + X_A) \quad \text{(D)}$$

$$p_S = p_H = 1.2 \times 10^5 (1/21)X_A/(1 + X_A/21) = 1.2 \times 10^5 X_A/(21 + X_A) \quad \text{(E)}$$

Use Eqs. (D) and (E) in Eq. (A) and substitute into the value of k:

$$r_A = 1.684 \times 10^{-10} \left[\frac{1.2 \times 10^5 (1 - X_A)}{(21 + X_A)} - \frac{(1.2 \times 10^5 X_A)^2}{3.727 \times 10^4 (21 + X_A)^2} \right]$$

$$= 2.02 \times 10^{-5} \frac{21 - 20X_A - 4.22X_A^2}{441 + 42X_A + X_A^2} \quad \text{(F)}$$

$Q_0 c_{A0} = 1.7 \times 10^{-3}$ kmol/s, then substituting this value and Eq. (F) into Eq. (4.5) we can obtain the catalyst weight needed:

$$W = \frac{1.7 \times 10^{-3}}{2.021 \times 10^{-5}} \int_0^{0.6} \frac{441 + 42X_A + X_A^2}{21 - 20X_A - 4.22X_A^2} dX_A$$

The integration can be calculated either analytically or numerically. Here we use the Simpson method to calculate the integration numerically, so

$$W = \frac{1.7 \times 10^{-3} \times 20.5}{2.021 \times 10^{-5}} = 1725 \text{ kg}$$

Since the reaction rate is based on catalyst weight, what we have calculated is catalyst weight instead of volume.

4.3 COMPARISON OF REACTOR VOLUMES OF TUBULAR AND TANK REACTORS

In this section we will compare reactor volumes needed for tank and tubular reactors when feed flow rate and composition, reaction conditions (temperature and pressure) and final conversion are the same. In the previous chapter and sections we performed reactor design calculations for acetate ester production (see Examples 3.4, 3.6, and 4.1), and reactor volumes needed for those reactors were summarized in Table 4.1. From this table it is clear that the reactor volume for a tubular reactor is the smallest, the volume for a single tank reactor is the largest, and the volume for multiple tank reactors in series is in between. For tank reactors in series, the higher the number of reactor in series, the smaller the total reactor volume needed. This conclusion was obtained for a specific reaction, and then the question is whether such a conclusion can be extended to other reactions. In Section 3.5 we already discussed reactor volume

TABLE 4.1 Reactor Volume Comparison for Various Type Reactors

Reactor Type	Tubular Reactor	Tank Reactor		
		Single	Two in Series	Three in Series
Reactor volume/m³	8.227	14.68	10.88	9.897

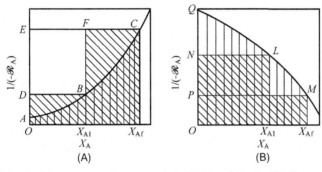

FIGURE 4.3   Continuous reactor volume comparison. (A) Normal kinetics; (B) Abnormal kinetics.

comparison for single and multiple tank reactors, here we will focus on the comparison between tank and tubular reactors.

We can plot $[1/(-\mathscr{R}_A)]$ against X_A, as shown in Fig. 4.3. Fig. 4.3(A) shows the situation for normal kinetics, where reaction rate decreases with increasing X_A. From Eq. (4.5) we know the area below curve ABC (shadow with vertical lines) is the integration in the equation. Rectangular area $OECX_{aAf}$ is $X_{Af}/(-\mathscr{R}_{Af})$. Divide Eq. (4.5) by Eq. (3.43):

$$\frac{V_{r_p}}{V_{r_M}} = \frac{\int_0^{X_{Af}} \frac{dX_A}{(-\mathscr{R}_A)}}{X_{Af}/(-\mathscr{R}_{Af})} < 1 \qquad (4.18)$$

It is obvious that area $OABCX_{Af}$ is smaller than the area of rectangular $OECX_{Af}$, and therefore V_{r_p}/V_{r_M} is less than 1, i.e., volume for tank reactor V_{r_M} is larger than that of tubular reactor V_{r_p}. When two reactors are in series the total volume $V_{r_{M-2}}$ is proportional to area $ODBFCX_{Af}$ (shadow area with diagnostic lines in Fig. 4.3(A)). Therefore:

$$V_{r_M} > V_{r_{M-2}} > V_{r_p}$$

If more than two tanks are used in series then the total volume will be even smaller. This can be easily confirmed by comparing areas in Fig. 4.3. We can reach the conclusion that when an infinite number of tank reactors are used in series then the total reactor volume would equal to that of the tubular reactor. We will next examine this conclusion using a first order irreversible reaction as an example.

For a first order reaction taking place in a tubular reactor, from Eq. (4.5) we can easily obtain that reactor volume needed:

$$V_{r_p} = \frac{Q_0}{k} \ln \frac{1}{1 - X_{Af}} \tag{4.19}$$

If multiple tank series reactors are used, then based on Eq. (3.50) the total volume is:

$$V_{r_{M-N}} = \frac{Q_0 N}{k} \left[\left(\frac{1}{1 - X_{Af}} \right)^{1/N} - 1 \right] \tag{4.20}$$

Dividing Eq. (4.19) by Eq. (4.20):

$$\frac{V_{r_p}}{V_{r_{M-N}}} = \frac{1}{N} \left[\left(\frac{1}{1 - X_{Af}} \right)^{1/N} - 1 \right]^{-1} \ln \frac{1}{1 - X_{Af}} \tag{4.21}$$

When $N \to \infty$, then the limit of the right hand of Eq. (4.21) is 1, then $V_{r_{M-N}} = V_{r_p}$.

From the above discussion we can see that for normal kinetics, tubular reactors are better than tank reactors in the sense that smaller reactors will be needed to achieve the same production rate. We can also say that tubular reactors are more efficient and have a higher production intensity. However, for abnormal kinetics the conclusion will be the opposite. Using Fig. 4.3(B) and following the same logic we can obtain the following conclusion:

$$V_{r_p} > V_{r_{M-3}} > V_{r_{M-2}} > V_{r_M}$$

In reality, this rarely happens, and that is why it is called abnormal kinetics.

Fig. 4.4 shows another special situation of $X_A \sim [1/(-\mathscr{R}_A)]$ curve, i.e., there is a maximum reaction rate at certain conversion. In this

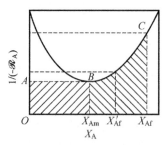

FIGURE 4.4 $X_A - [1/(-\mathscr{R}_A)]$ curve which has a minimum.

situation, the relative size of the tank and tubular reactor depends on final conversion. If the final conversion is smaller than the conversion at maximum reaction rate X_{Am}, then the $1/(-\mathscr{R}_A) \sim X_A$ curve will be similar to Fig. 4.3(B), so $V_{r_p} > V_{r_M}$. If final conversion is larger than X_{Am}, then V_{r_p} could be either larger or smaller than V_{r_M}. For final conversion X_{Af} shown in Fig. 4.4, it obvious that $V_{r_p} < V_{r_M}$. If the final conversion is X'_{Af}, then similarly from Fig. 4.4 we can see $V_{r_p} > V_{r_M}$. For this special situation, the optimal solution is to use two reactors in series. The first reactor uses a tank reactor to reach conversion of X_{Am}, and then sends to a tubular reactor to further convert to final conversion X_{Af}. Such a combination will lead to smallest reactor volume.

For a single reaction we just discussed the relative reactor size needed to achieve a given conversion when using a tank or a tubular reactor. Following the same logic we can discuss when to use the same size reactor and which one will lead to higher conversion.

For multiple reactions, in Chapter 3, Tank Reactor, we compared continuous and batch operations for tank reactors. The performance of a batch tank reactor is the same as that of a tubular reactor with the same volume, so the conclusions of comparison between continuous and batch tank reactors are also valid for comparisons between continuous tank and tubular reactors. For comparisons between tubular and tank reactors, it is more important to examine at the same final conversion which one will lead to higher yield of target products. The relationship between instantaneous selectivity and conversion is shown in Fig. 3.10, and there are two possible situations. One is S decreases with increasing X_A (Fig. 3.10(A)), and another is S increases with increasing X_A (Fig. 3.10(B)). For tubular reactors conversion increases with axial distance, while for tank reactors composition inside the reactor always corresponds to those at final conversion. From Eq. (3.55) we know that the rectangular area in Fig. 3.10(A) and (B) represents final yield of tank reactor, and the area below the curve is the final yield for tubular reactor. From Fig. 3.10(A) we can see that final yield in a tubular reactor is higher than that of a tank reactor, while the situation in Fig. 3.10(B) is the opposite, i.e., a tank reactor leads to higher yield. Therefore it is the instantaneous selectivity $\sim X_A$ relationship that will decide which reactor will lead to a higher yield.

From Fig. 3.10(A) and (B) we can also see that the yield of multiple tank reactors in series will be between those of single tank and tubular reactors.

The relationship between instantaneous selectivity and conversion actually reflects the dependence of reaction selectivity on reactant concentration. If the selectivity is favored by lower reactant concentration, a tank reactor will obviously be better than a tubular reactor. On the other hand, if higher reactant will lead to higher selectivity, then a tubular reactor will be the better option.

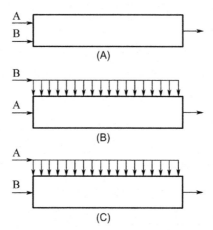

FIGURE 4.5 Different feed schemes for a continuous reactor. (A) Requires high concentrations of both A and B; (B) Requires high concentration of A but low concentration of B; (C) Requires high concentration of B but low concentration of A.

Based on the impact of reactant concentration on selectivity, different feeding schemes can be used to achieve better selectivity. If both A and B are reactants, and it is desirable to have high concentrations of both A and B, then A and B should be added together at the reactor inlet, as shown in Fig. 4.5(A). If a high concentration of A and a low concentration of B is desired, we can either bring excessive A into the reactor or use the feeding scheme shown in Fig. 4.5(B), i.e., feed all A into the reactor inlet while adding B along the reactor. Compared with the approach of using excessive A the latter approach has the advantage of lower product separation cost. Based on the same logic, we can use the scheme shown in Fig. 4.5(C) if a higher concentration of B is desired.

Example 4.6
Hydrochloride acid reacts with mixed octanol and dodecanol under isothermal conditions:

$$HCl + CH_3(CH_2)_6CH_2OH \rightarrow CH_3(CH_2)_6CH_2Cl + H_2O$$

$$HCl + CH_3(CH_2)_{10}CH_2OH \rightarrow CH_3(CH_2)_{10}CH_2Cl + H_2O$$

Using A, B, and C to represent hydrochloride, octanol, and dodecanol, and the reaction rates for these two reactions are:

$$r_B = k_1 c_A c_B$$

$$r_C = k_2 c_A c_C$$

(Continued)

(Continued)

At reaction temperature $k_1 = 1.6 \times 10^{-6}$ m³/(mol·min), $k_2 = 1.92 \times 10^{-6}$ m³/(mol·min). In the feed, the concentration of hydrochloride, octanol, and dodecanol are 2.3 kmol/m³, 2.2 kmol/m³, and 2 kmol/m³, respectively. Feed flow rate is 2 m³/h, and the required octanol conversion is 30%. Please calculate the reactor volume needed for (1) tubular reactor; (2) tank reactor.

Solution

1. For tubular we can calculate reactor volume by substituting the octanol reaction rate equation into Eq. (4.5):

$$V_r = Q_0 c_{B0} \int_0^{0.3} \frac{dX_B}{k_1 c_A c_B} \tag{A}$$

To integrate Eq. (A) we need to express c_A and c_B as functions of conversion X_B:

$$c_B = c_{B0}(1 - X_B) \tag{B}$$

Hydrochloride acid participates in both reactions, so based on stoichiometry:

$$c_{A0} - c_A = (c_{B0} - c_B) + (c_{C0} - c_C) \tag{C}$$

i.e., the amount of HCl converted equal to the sum of those of octanol and dodecanol. The correlation between c_B and c_C can be obtained by using two rate equations:

$$\frac{dc_B/d\tau}{dc_C/d\tau} = \frac{k_1 c_A c_B}{k_2 c_A c_C}$$

or:

$$\frac{dc_B}{dc_C} = \frac{k_1 c_B}{k_2 c_C}$$

Integrating the above equation:

$$c_C = c_{C0}(c_B/c_{B0})^{k_2/k_1} \tag{D}$$

Substitute the above equation into Eq. (C)

$$c_A = c_{A0} - (c_{B0} - c_{C0}) + c_B + c_{C0}(c_B/c_{B0})^{k_2/k_1}$$

Then substitute Eq. (B) into above equation:

$$c_A = c_{A0} - c_{C0} - c_{B0}X_B + c_{C0}(1 - X_B)^{k_2/k_1} \tag{E}$$

Insert Eqs. (B) and (E) into Eq. (A):

$$V_r = Q_0 c_{B0} \int_0^{0.3} \frac{dX_B}{k_1 c_{B0}(1 - X_B)[c_{A0} - c_{C0} - c_{B0}X_B + c_{C0}(1 - X_B)^{k_2/k_1}]}$$

We know:
$Q_0 = 2$ m³/h = 0.033 m³/min,
$c_{B0} = 2.2$ kmol/m³,
$c_{C0} = 2$ kmol/m³
$c_{A0} = 2.3$ kmol/m³,

(Continued)

(Continued)

$k_1 = 1.6 \times 10^{-6} \, \text{m}^3/(\text{mol} \cdot \text{min})$,
$k_2 = 1.92 \times 10^{-6} \, \text{m}^3/(\text{mol} \cdot \text{min})$.

Then:

$$V_r = 20.83 \int_0^{0.3} \frac{dX_B}{(1 - X_B)[0.3 - 2.2 X_B + 2(1 - X_B)^{1.2}]} = 5.024 \, \text{m}^3$$

2. For the CSTR reactor, the volume needed can be calculated using Eq. (3.43), and using octanol consumption rate we can obtain:

$$V_r = \frac{Q_0 c_{B0} X_B}{k_1 c_A c_B} \tag{F}$$

Based on mass balance of component B and C;

$$\frac{c_{B0} - c_B}{k_1 c_A c_B} = \frac{c_{C0} - c_C}{k_2 c_A c_C}$$

Substitute Eq. (B) into above equation:

$$c_C = \frac{k_1 c_{C0}(1 - X_B)}{k_1(1 - X_B) + k_2 X_B}$$

Then substitute it into Eq. (C):

$$c_A = (c_{A0} - c_{B0} - c_{C0}) + c_{B0}(1 - X_B) + \frac{k_1 c_{C0}(1 - X_B)}{k_1(1 - X_B) + k_2 X_B} \tag{G}$$

Insert Eqs. (B) and (G) into (F):

$$V_r = \frac{Q_0 c_{B0} X_B}{k_1 c_{B0}(1 - X_B)\left[c_{A0} - c_{C0} - c_{B0} X_B + \frac{k_1 c_{C0}(1 - X_B)}{k_1(1 - X_B) + k_2 X_B} \right]}$$

$$= (2/60) \times 0.3 \left\{ 1.6 \times 10^{-3}(1 - 0.3) \left[2.3 - 2 - 2.2 \times 0.3 + \frac{2 \times (1 - 0.3)}{(1 - 0.3) + 1.92 \times 0.3/1.6} \right] \right\}^{-1}$$

$$= 9.293 \, \text{m}^3$$

From the above calculation it is clear that for these reactions, the volume of CSTR is larger than that of a tubular reactor. Although both reactors can achieve an octanol conversion of 30%, dodecanol conversions from these two reactors are different. Please discuss why this is the case.

4.4 RECYCLE REACTOR

For some commercial reaction processes, such as ammonium synthesis, methanol synthesis, and ethanol production through hydration of ethylene, the single pass conversion is limited due to the equilibrium constraints. One common practice to increase feed utilization is to separate unreacted feed from the reactor outlet stream and return it to the entrance of the reactor and combine it with fresh feeds. This type of reactor is called a recycle reactor.

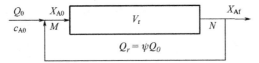

FIGURE 4.6 Recycle reactor.

Fig. 4.6 is the schematics of a recycle reactor. Assuming that this reactor can be treated as a plug flow reactor, the reactor volume can be calculated using Eq. (4.5). However, we need to first decide feed flow rate and conversion at reactor entrance. For reactors without recycle, the actual flow rate into the reactor is simply the feed flow rate. The reactor inlet conversion can be easily calculated based on feed composition. For recycle reactors, the actual flow rate into the reactor depends on both the feed flow rate and the recycle rate. If the ratio of recycle flow rate Q_r to fresh feed flow rate Q_0 is ψ, then $Q_r = \psi Q_0$. Therefore the flow rate at reactor entrance is:

$$Q_0 + Q_r = (1 + \psi)Q_0 \tag{4.22}$$

For component A we can do a mass balance at point M:

$$Q_0 c_{A0} + \psi Q_0 c_{A0}(1 - X_{Af}) = (1 + \psi)Q_0 c_{A0}(1 - X_{A0})$$

Simplifying the above equation:

$$X_{A0} = \frac{\psi X_{Af}}{1 + \psi} \tag{4.23}$$

In Eq. (4.5), Q_0 is the flow rate entering the reactor. For recycle reactors we need to use Eq. (4.22) to calculate the actual flow rate at the reactor entrance. Using both Eqs. (4.22) and (4.23) we can obtain the reactor volume calculation equation for the recycle reactor:

$$V_r = (1 + \psi)Q_0 c_{A0} \int_{\frac{\psi X_{Af}}{1+\psi}}^{X_{Af}} \frac{dX_A}{(-\mathscr{R}_A)} \tag{4.24}$$

When $\psi \to 0$, from Eq. (4.23) we know $X_{A0} = 0$, then Eq. (4.24) is simplified as Eq. (4.5); on the other hand, if $\psi \to \infty$, from Eq. (4.23) $X_{A0} = X_{Af}$, under this condition the reactor is equivalent to CSTR. In fact, if ψ is sufficiently large, such as $\psi = 25$, the reactor can be considered as being operated at constant concentration. In lab research it is very common to use a reactor with large recycle ratio since this will make the reaction kinetic analysis as well as reactor temperature control much easier.

4.5 NONISOTHERMAL TUBULAR REACTOR

Most chemical reaction processes encountered in commercial production take place under nonisothermal conditions. One reason for this is that most chemical reactions have a heat effect and some of them have a very large heat effect. For commercial reactors, although various heat transfer schemes are used to remove (for exothermic reactions) or provide (for endothermic reactions) heat, it is very hard to achieve homogeneous temperature distribution throughout the reactor. For gas−solid fixed bed catalytic reactors it is especially difficult to achieve isothermal operations. Another reason for nonisothermal operation is that for many reactions isothermal operation is not the optimal operating mode. Instead, certain temperature distributions will lead to optimal reactor performance. For example, for reversible exothermic reactions such as ammonium synthesis and methanol synthesis, there is an optimal temperature distribution that will lead to the highest conversion. For multiple reactions, due to the difference of activation energies of main and side reactions, temperature impacts on main and side reactions are also different. Therefore, product distribution can be adjusted by changing reaction temperature to achieve the maximum yield of the desired product. In summary, it is very rare that a commercial reactor is operated isothermally. Instead, most commercial reactors are operated under nonisothermal conditions.

In this section we will discuss the fundamental principles of designing nonisothermal tubular reactors based on the heat balance equation. More detailed discussions on design and analysis of nonisothermal tubular reactors will be presented in Chapter 7, Analysis and Design of Heterogeneous Catalytic Reactors.

4.5.1 Heat Balance Equation for Tubular Reactor

For the tubular reactor under consideration we assume that the flow pattern is plug flow, at any cross-section that is perpendicular to the flow direction the temperature distribution is uniform and temperature only changes along the axial direction. Choosing an infinitesimal volume element dV_r as the control volume to perform heat balance, we can obtain the heat balance equation for a tubular reactor. If the changes of kinetic and potential energies can be ignored and there is no work input, for constant-pressure process based on First Law in thermodynamics:

$$dH = dq \qquad (4.25)$$

If the mass flow rate of the fluid is G, diameter of the tubular reactor is d_t, then $dV_r = (\pi/4)d_t^2 dZ$. If the fluid temperature change in the differential volume is dT, the enthalpy change under steady-state is:

$$dH = [(-\mathscr{R}_A)(\Delta H_r)_{T_r} dZ + GC_{pt} dT](\pi/4)d_t^2$$

where T_r is a reference temperature. The heat exchange between the differential volume and the surrounding is:

$$dq = U(T_C - T)\pi d_t dZ$$

where T_C is the temperature of heat exchange medium. Substitute dH and dq into Eq. (4.25) we can obtain:

$$GC_{pt}\frac{dT}{dZ} = (-\mathscr{R}_A)(-\Delta H_r)_{T_r} - 4U(T_C - T)/d_t \qquad (4.26)$$

Eq. (4.26) is the axial temperature distribution equation for a tubular reactor. The format of this equation is very similar to the heat balance equation for the batch tank reactor, Eq. (3.82). The difference is the independent variable. For batch reactors, the independent variable is time while for a tubular reactor it is the distance along the axial direction. Another difference is that for batch tank reactors the heat balance is for the whole reactor and for tubular reactors the heat balance is for differential reaction volume.

Since

$$Q_0 c_{A0} = \frac{G(\pi d_t^2/4)w_{A0}}{M_A}$$

And:

$$dV_r = \frac{\pi}{4}d_t^2 dZ$$

where w_{A0} is initial mass fraction of component A, M_A is the molecular weight of A, Eq. (4.4) can be rewritten as:

$$\frac{Gw_{A0}}{M_A}\frac{dX_A}{dZ} = -\mathscr{R}_A(X_A) \qquad (4.27)$$

Substituting the above equation into Eq. (4.26), we can obtain the equation describing the relationship between reaction temperature and conversion:

$$GC_{pt}\frac{dT}{dZ} = \frac{Gw_{A0}(-\Delta H_r)_{T_r}}{M_A}\frac{dX_A}{dZ} - \frac{4U}{d_t}(T - T_C) \qquad (4.28)$$

The heat balance equation derived above is for a single reaction and it can be easily extended to describe the heat balance for a multiple reaction system. The first term on the right-hand side of Eq. (4.26) is the reaction heat. For a multiple reaction system we need to account for reaction heats of all reactions:

$$GC_{pt}\frac{dT}{dZ} = \sum_{j=1}^{M}(-\Delta H_r)_j|\nu_{ij}r_j| - \frac{4U}{d_t}(T - T_C) \qquad (4.29)$$

In Eq. (4.29) the reaction heat is also based on the reference temperature, although the subscript T_r is omitted. M is the number of reactions.

For a single reaction, Eqs. (4.26) and (4.27) need to be solved simultaneously to design nonisothermal tubular reactors. For a multiple reaction system, Eqs. (4.13) and (4.29) need to be solved.

4.5.2 Adiabatic Tubular Reactor

If the reactions take place under adiabatic conditions, then Eq. (4.28) can be simplified as:

$$dT = \frac{w_{A0}(-\Delta H_r)_{T_r}}{M_A C_{pt}} dX_A$$

At the reactor entrance $T = T_0$ and $X_A = 0$. If we ignore the possible change of heat capacity due to the changes of composition and temperature and use average value of heat capacity at T_0 and T, by choosing $T_r = T_0$ the above equation can be integrated:

$$T - T_0 = \lambda X_A \qquad (4.30)$$

$$\lambda = w_{A0}(-\Delta H_r)_{T_r} / \overline{C}_{pt} M_A$$

Eq. (4.30) is exactly the same as Eq. (3.91) which is derived for a CSTR. The reason for this is that both equations reflect the relationship between reaction temperature and conversion under adiabatic conditions. This equation is also the same as that for batch tank reactor Eq. (3.86). Therefore Eq. (4.30) is an adiabatic equation that can be used for different types of reactors.

If we plot conversion against temperature T we will get a straight line with slope of $1/\lambda$. For exothermal reactions $\lambda > 0$ and the inclination is less than 90°. For endothermic reactions $\lambda < 0$ and the inclination is greater than 90°. For isothermal reaction, $\lambda = 0$, and the inclination is 90°.

It should be noted that although Eq. (4.30) can describe the relationship between reaction temperature and conversion for all three types of reactors operated under adiabatic conditions, there are still fundamental differences. For tubular reactors, it reflects the relationship between temperature and conversion at different axial locations. For batch tank reactors it reflects the relationship between reaction temperature and conversion at different time. For CSTR, no matter whether there is heat exchange with the surrounding environment or not, the reactor itself is isothermal, i.e., temperature is uniform throughout the reactor, so adiabatic equation Eq. (4.30) reflects the relationship between reaction temperature and conversion when the reactor is operated under adiabatic conditions.

From Fig. 4.7 we can see that for endothermic reactions under adiabatic conditions with the increases of conversion the reaction temperature will decrease. The situation for exothermic irreversible reactions is the opposite, i.e., temperature increases with conversion. Therefore, for adiabatic tubular reactors, higher feed temperature will normally lead to higher reaction rate. However, for reversible exothermic reactions the situation is more complicated. Fig. 4.8 shows the conversion–temperature relationship for a reversible exothermic reaction. AD, BE, and CF are adiabatic operating lines for a adiabatic tubular reactor with inlet temperature at T_A, T_B, and T_C, respectively.

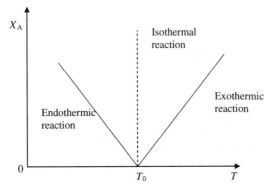

FIGURE 4.7 Conversion−temperature relationship for an adiabatic reaction process.

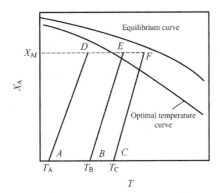

FIGURE 4.8 Conversion−temperature relationship for reversible exothermic reaction.

For a given conversion, when reaction temperature is below optimal temperature, reaction rate increases with temperature. If reaction temperature is higher than optimal temperature then the reaction rate will decrease with temperature due to the reversibility of the reaction. If the reactor is operated following line BE, then the average reaction rate is higher than that following the line AD. This means higher feed temperature will lead to higher reaction rate and hence smaller reactor volume. If the feed temperature is further increased to T_C, then the average reaction rate from the CF line operation may not be higher than that from the BE line operation, since at the final stage the composition is too close to equilibrium, leading to a lower reaction rate. Therefore, for reversible exothermic reactions there is an optimal feed temperature that will lead to the smallest reactor volume. Fig. 4.9 schematically shows the reactor volume as a function of feed temperature and each line on the figure corresponds to a given final conversion. The higher the final conversion, the lower optimal feed temperature, i.e., if $X''_{Af} > X'_{Af} > X_{Af}$, then the optimal feed temperature $T''_0 < T'_0 < T_0$.

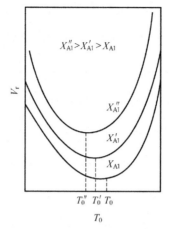

FIGURE 4.9   Optimal feed temperature for an adiabatic tubular reactor.

If multiple reactions take place in an adiabatic reactor, then the heat balance equation Eq. (4.29) can be simplified as:

$$GC_{pt}\frac{dT}{dZ} = \sum_{j=1}^{M}(-\Delta H_r)_j|\nu_{ij}\bar{r}_j| \qquad (4.31)$$

Example 4.7
Ethyl benzene dehydrogenation reaction takes place in an adiabatic tubular reactor with ID = 1.22 m. Feed temperature is 898K and all other conditions are the same as those in Example 4.5. The reaction rate constant as a function of temperature is:

$$k = 3.452 \times 10^{-5}\exp(-10{,}983/T), \text{ kmol/(s·kg·Pa)}$$

Reaction equilibrium coefficients at different temperatures can be calculated by:

$$K = 3.96 \times 10^{11}\exp(-14{,}520/T), \text{ 1/Pa}$$

The average heat capacity of reaction mixture is 2.177 kJ/(kg·K), and reaction heat is 1.39×10^5 J/mol. The bulk density of the catalyst bed is 1440 kg/m^3. Please calculate:
1. Amount of catalyst needed;
2. Axial temperature and conversion distribution in the reactor.

Solution
In Example 4.5 we already obtained ethyl benzene reaction rate as a function of conversion:

$$r_A = k\left[\frac{1.2 \times 10^5(1 - X_A)}{21 + X_A} - \frac{(1.2 \times 10^5 X_A)^2}{K(21 + X_A)^2}\right] \qquad (A)$$

(Continued)

(Continued)

In Example 4.5 the reactor is operated isothermally. For an adiabatic operation we need to consider the impact of temperature changes on k and K by using correlation given above in Eq. (A):

$$r_A = 3.452 \times 10^{-5} \exp(-10{,}983/T) \times 1440 \times 1.2 \times 10^5$$

$$\times \left[\frac{(1 - X_A)}{21 + X_A} - \frac{1.2 \times 10^5 X_A{}^2}{3.96 \times 10^{11} \exp(-14{,}520/T)(21 + X_A)^2} \right] \qquad (B)$$

The rate equation given above is based on catalyst weight, Eq. (B) needs to be multiplied by bulk density of catalyst bed ($1440\ kg/m^3$) to convert it to volume based.

In the feed, the ethyl benzene to steam mole ratio is 1:20, so:

$$y_{A0} = 1/(20 + 1) = 1/21$$

and adiabatic temperature increases is (Please note that here the total mole change during the reaction is neglected. What would the results be if the total mole change is not neglected?):

$$\lambda = \frac{y_{A0}(-\Delta H_r)}{\overline{C}_{pt}} = \frac{-1.39 \times 10^5}{21 \times 2.177 \times 22.19} = -137\ K$$

In this equation the 22.19 is the average molecular weight of the reaction mixture. Substitute value of λ into Eq. (4.30) and we can obtain the correlation between reaction temperature and conversion:

$$T = 898 - 137 X_A \qquad (C)$$

In Eq. (4.27) Gw_{A0}/M_A on the left is the ethyl benzene mole flow rate per cross-section area of the reactor, so

$$Gw_{A0}/M_A = 1.7 \times 10^{-3}/(\pi \times 1.22^2/4) = 1.454 \times 10^{-3}\ kmol/(s \cdot m^2)$$

Substituting this value as well as Eqs. (B) and (C) into Eq. (4.27), we can get the axial conversion distribution equation:

$$\frac{dX_A}{dz} = \frac{4.104 \times 10^6}{21 + X_A} \exp\left(-\frac{10{,}983}{898 - 137 X_A} \right)$$

$$\times \left[1 - X_A - \frac{3.03 \times 10^{-7} X_A^2}{21 + X_A} \exp\left(\frac{14{,}520}{898 - 137 X_A} \right) \right] \qquad (D)$$

The initial value of Eq. (D) is $Z = 0$, $X_A = 0$. Integrating Eq. (D) we can obtain axial conversion distribution, i.e., conversion at various reactor height, and hence reactor volume. Eq. (D) can be integrated either analytically by using the variable-separating method or numerically. The calculation results using the Runge–Kutta–Gill method are given in Table 4B. For an adiabatic reaction process the temperature is a linear function of conversion and using Eq. (C) we can calculate the temperature at each conversion, which is also listed in Table 4B.

(Continued)

(Continued)

TABLE 4B Axial Distributions of Conversion and Temperature
for Ethyl Benzene Dehydrogenation

Z/m	T/K	X_A
0.0	898	0
0.2	877.3	0.1510
0.4	863.5	0.2517
0.6	853.5	0.3250
0.8	845.6	0.3811
1.0	839.7	0.4235
1.2	834.8	0.4610
1.4	830.9	0.4900
1.6	827.6	0.5140
1.8	824.8	0.5340
2.0	822.6	0.5507
2.2	820.6	0.5647
2.4	819.0	0.5766
2.6	817.6	0.5866
2.8	816.5	0.5951
3.0	815.5	0.6023
3.2	814.7	0.6084

Using the data in Table 4B, we can plot X_A and T as functions of Z, as shown in Fig. 4B, and these two curves are temperature and conversion distribution, respectively.

From Table 4B and Fig. 4B it is clear that the reaction rate is very fast at the beginning. After the reaction mixture flows through the first meter of the reactor the conversion is about 42.35%. However, for the final meter of the reactor, the net conversion is only $0.6084 - 0.5647 = 0.0437$, or 4.37%, indicating the reaction rate is much slower. One reason for this slower reaction rate is the lower reactant concentrations. Another reason is that the temperature also becomes lower. The combination of these two factors leads to a significant decrease in reaction rate. Since ethyl benzene dehydrogenation is an endothermic reaction, under adiabatic conditions the reaction temperature will gradually decrease, which will in turn lead to a lower reaction rate. From this perspective, the
(Continued)

(Continued)

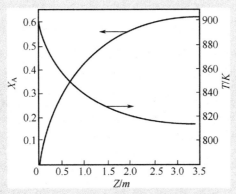

FIGURE 4B Axial distributions of conversion and temperature for ethyl benzene dehydrogenation.

adiabatic reactor is not an ideal choice. It is worth pointing out that many other factors need to be taken into consideration when selecting reaction conditions.

In order to calculate catalyst weight needed to achieve 60% conversion, we first need to know the bed height. Based on the data presented in Table 4B we can estimate that the bed height needed to achieve 60% conversion is:

$$L_r = 2.8 + \frac{3.0 - 2.8}{0.6032 - 0.5951} \times (0.6 - 0.5951) = 2.94 \text{ m}$$

Then the amount of catalyst needed is:

$$V_r = \frac{\pi}{4} \times 1.22^2 \times 2.94 = 3.437 \text{ m}^3$$

$$W = 3.437 \times 1440 = 4949 \text{ kg}$$

From Example 4.5 we know that under isothermal conditions, the amount of catalyst needed to achieve the same conversion is 1725 kg, which is about one-third of the value under adiabatic conditions. The reason for this big difference is obvious. For an adiabatic operation, except at the reactor entrance, the reaction temperature is lower than 898K, so more catalyst is needed.

4.5.3 Nonadiabatic Nonisothermal Tubular Reactor

If reaction heat effect is high, regardless of whether the reaction is exothermic or endothermic, adiabatic operations will lead to large temperature changes in the reactor. For exothermic reactions, the reaction temperature will increase along the axial direction. For irreversible reactions this is okay from a reaction rate perspective. However, in many situations, due to various reasons, reaction temperature has to be controlled in a certain range. Then, adiabatic reactors may not be a viable choice. If the reaction is reversible, higher temperature will lead to lower equilibrium conversion and, therefore, it is impossible to achieve high conversion for reversible exothermic reactions taking place in an adiabatic reactor. For endothermic reactions, no matter reversible or irreversible, reaction temperature always decreases along the axial direction, leading to slower reaction rate. For reversible reactions the equilibrium conversion will also decrease, therefore it is impossible to achieve high conversion if an adiabatic reactor is used. The above discussions indicate that in many situations, in order to achieve reasonable conversion, there must be heat exchange between the reactor and surrounding environment. The reactor needs to be cooled for exothermic reactions and heated for endothermic reactions in order to control the reaction temperature in a given range, which is usually required to achieve safe operation and high conversion. Reaction temperature control is especially important for reactions that may become explosive at high temperatures. Reaction temperature control is also critical to avoid catalyst or equipment damage.

The selection of heat exchange media usually depends on reaction temperature. For high temperature reactions, flue gas from combustions of liquid or gas fuels, melted salts, and high-pressure steam are often used. For low temperature processes, water and air are commonly used. To improve energy efficiency, it is very common, if not a certainty, that some form of heat integration will be implemented. For example, feed stream can be used to cool the reactor while being heated up itself.

For nonadiabatic tubular reactors, reactions and heat transfer take place simultaneously. In order to match the heat transfer rate with the heat-generating rate from the reactions, it is very important to provide sufficient heat transfer area when designing the reactor. Multitube reactors are widely used for reactions with large heat effect, which uses multiple tubes with small diameters in parallel to provide the heat transfer area needed. In addition, a small diameter also avoids large cross-section area so that the radial temperature difference will be small.

For multitube reactors, the operation conditions and status of these tubes are usually considered to be the same, and therefore we only need to examine one tube in order to design or analyze the reactor. The fundamental design equations for multitube reactors are also mass and heat balance equations. The only difference, compared with the design of adiabatic reactors, is the heat balance equation, i.e., heat exchange between the reactor and external environment has to be considered.

For tubular reactors, heat needs to be transferred through the reactor wall, and heat transfer area per unit reactor volume can be increased by reducing tube diameter. In addition to the heat exchange area, another factor that will influence the heat transfer rate is the temperature of the heat exchange medium. In order to control the reaction temperature at a certain level, both heat transfer area and heat exchange medium temperature need to be carefully selected. For example, benzene oxidation is an exothermic reaction with a strong heat effect, and a melted salt bath with temperature of $\sim 300°C$ is used to cool the reactor. The reason to select such a high cooling medium temperature is to maintain adequate temperature differential across the reactor wall to avoid dramatic temperature decrease caused by high heat transfer rate.

Example 4.8

Instead of using an adiabatic reactor with ID = 1.22 m for ethyl benzene dehydrogenation as discussed in Example 4.7, a multitube reactor is used. There are a total of 144 tubes with ID = 0.101 m in parallel. Flue gas at constant temperature of 1100K is used outside the tube and the overall heat transfer coefficient between gas inside the reactor and flue gas outside is 2.85 kW/(m$^2 \cdot$ K). All the other conditions and requirements are the same as those in Example 4.7, please calculate the axial distribution of conversion and temperature, as well as the amount of catalyst needed to achieve 60% conversion.

Solution

As mentioned above, for multitubes in parallel, we only need to examine one tube. The total cross-section area of the 144 tubes is equal to that of the ID = 1.22 m reactor discussed in Example 4.7, so feed composition and the mass flow rate for these two reactors are the same. As a result, conversion distribution (D) derived in Example 4.7 can be used here. The only difference is the correlation between reaction temperature T and conversion X_A. For nonadiabatic reactors this correlation is not linear. Eq. (D) from Example 4.7 needs to be modified:

$$\frac{dX_A}{dZ} = \frac{4.104 \times 10^6}{21 + X_A} \exp\left(\frac{-10,983}{893 - 137X_A}\right) \times \left[1 - X_A - \frac{3.03 \times 10^{-7} X_A^2}{21 + X_A} \exp\left(\frac{-14,520}{T}\right)\right]$$

(A)

Ethyl benzene feed flow rate is 1.7×10^3 kmol/s, and the ethyl benzene to steam mole ratio is 1:20, so steam flow rate is 3.4×10^2 kmol/s, and total feed mass flow rate is:

$$G = \frac{1.7 \times 10^{-3} \times 106 + 3.4 \times 10^{-2} \times 18}{1.44 \times \pi(0.1016)^2/4} = 0.678 \text{ kg(s} \cdot \text{m}^2)$$

From Example 4.7 we know $Gw_{A0}/M_A = 1.454 \times 10^{-3}$ kmol/sm^2, dividing Eq. (4.28) by GC_{pt}, and substituting all the values needed:

$$\frac{dT}{dZ} = \frac{-1.39 \times 10^5 \times 1.454 \times 10^{-3}}{0.678 \times 2.177} \frac{dX_A}{dZ} + \frac{4 \times 2.85 \times 10^3(1100 - T)}{0.678 \times 2.177 \times 0.1016}$$

$$= 0.07605(1100 - T) - 137\frac{dX_A}{dZ}$$

(B)

Eq. (B) is the temperature distribution equation. Initial values for Eqs. (A) and (B) are: $Z = 0$, $X_A = 0$, $T = 898$K. These two equations can only be solved numerically, and the results obtained from using the Rung–Kutta–Gill method are given in Table 4C. Based on the data presented in Table 4C we can plot the conversion and temperature distribution curves, as shown in Fig. 4C. For comparison, the conversion and temperature distribution for adiabatic tubular reactor (obtained in Example 4.7) are also shown in the figure by the dotted line.

(Continued)

(Continued)

TABLE 4C Axial Distributions of Conversion and Temperature for Catalytic Ethyl Benzene Dehydrogenation

Z/m	X_A	T/K
0	0.0	898
0.2	0.1533	880.2
0.4	0.2593	869.1
0.6	0.3391	861.8
0.8	0.4022	856.8
1.0	0.4537	853.5
1.2	0.4967	851.3
1.4	0.5333	850.1
1.6	0.5647	849.6
1.8	0.5922	849.7
2.0	0.6164	850.2
2.2	0.6379	851.0
2.4	0.6571	852.1
2.6	0.6746	853.5
2.8	0.6904	855.1
3.0	0.7050	856.8
3.2	0.7184	858.6
3.4	0.7308	860.6
3.6	0.7424	862.7
3.8	0.7533	864.7

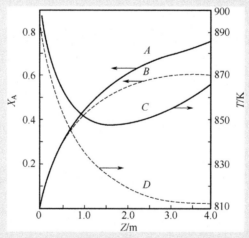

FIGURE 4C Conversion and temperature comparison for ethyl benzene dehydrogenation under adiabatic and nonadiabatic conditions.

(Continued)

(Continued)

From Fig. 4C, we can see that at any location, in the reactor conversion under nonadiabatic operation is higher than that under adiabatic conditions. The reason for this is that at the same axial location, the temperature under nonadiabatic conditions is higher, thanks to the continuous heat supply from the environment. For adiabatic operations, no heat is supplied to the reactor, so the reaction temperature monotonically decreases due to the endothermic nature of the reaction. For nonadiabatic operations, the reaction temperature also decreases at an early stage of the reaction, but after passing a minimum the reaction temperature gradually increases. At early stages the reaction rate is fast, the heat transfer rate is lower than the heat consumption rate of the reaction, so the reaction temperature decreases. Later on, the reaction rate becomes slower and less heat is consumed by the reaction, and when heat transfer from the environment is higher than heat consumption rate of the reaction, reaction temperature will gradually increase. It should be noted that the minimum on the temperature distribution curve only exists under certain circumstances. If the heat supply rate is always higher than the heat consumption rate, then the reaction temperature will increase monotonically. If less heat is supplied, then the reaction temperature will decrease monotonically.

Similar to the procedure used in Example 4.7, the amount of catalyst needed to achieve 60% conversion can be calculated using data in Table 4C. The reactor bed height needed is:

$$L_r = 1.8 + \frac{2.0 - 2.8}{0.6164 - 0.5922} \times (0.6 - 0.5922) = 1.865 \text{ m}$$

$$V_r = 144 + \frac{\pi}{4} \times (0.1016)^2 = 1.865 = 2.177 \text{ m}^3$$

$$W = 2.177 \times 1440 = 3135 \text{ kg}$$

Compared with the results from Example 4.7, to achieve the same 60% conversion, the catalyst weight needed for a nonadiabatic reactor is less than that needed for an adiabatic reactor. This is expected, because heat supplied from the environment makes the reaction temperature higher, which is obvious when comparing curve C and D in Fig. 4C. If feed composition, temperature, and mass flow rate are the same, this conclusion is always valid, unless the correlation between reaction rate and temperature is abnormal.

4.6 OPTIMAL TEMPERATURE SEQUENCE FOR TUBULAR REACTORS

Selecting the proper operating temperature is an important task for reactor design. For a single reaction, it is very common to use production intensity as the optimization target, i.e., selecting an operating temperature to maximize production intensity. The production intensity is defined as the productivity per unit time and unit reactor volume. For multiple reactions, the yield of desired product is usually the optimization target. Next we will discuss these two situations.

4.6.1 Single Reaction

First we will discuss isothermal reactors by exploring at what temperature we can achieve maximized production intensity. For irreversible or reversible endothermic reactions, reaction rate always increases with temperature, therefore the higher the reaction temperature the higher the reactor production intensity. In this case, higher operating temperature is preferred. In reality, the actual operating temperature is limited by many practical reasons, such as reactor construction materials, energy consumption, thermal stability of the catalyst used, etc. As a result, reaction temperature still needs to be carefully selected by considering all these factors that will influence reactor operation.

For reversible exothermic reactions, the situation is more complicated since reactor production intensity does not necessarily increase with temperature. Assuming $A \rightleftharpoons P$ is a first order reversible reaction, and the rate equation is:

$$r_A = \vec{k} \, c_{A0}[(1 - X_A) - X_A/K]$$

where K is equilibrium coefficient. Substituting the above equation into Eq. (4.5):

$$V_r = \frac{Q_0}{\vec{k}} \int_0^{X_{Af}} \frac{dX_A}{1 - (1 + 1/K)X_A} = \frac{Q_0}{\vec{k}(1 + 1/K)} \ln \frac{1}{1 - (1 + 1/K)X_{Af}} \quad (4.32)$$

Both reaction rate constant k and equilibrium coefficient K are functions of temperature. For a given feed flow rate Q_0 and final conversion X_A, higher temperature will lead to an increase of k but a decrease of K. As a result, the logarithmic term on the right-hand side of Eq. (4.32) will increase, but the value of the term in front of the logarithmic term will decrease. Therefore, the reactor volume needed, which is the product of the two terms we just discussed, could either increase or decrease with the increase in reaction temperature. Therefore, there must be an optimal reaction temperature that will lead to minimum reactor volume. Or we can say that for a given reactor volume and feed flow rate maximum final conversion can be achieved at an optimal reaction temperature.

We can discuss this from another perspective. At an early stage of the reaction, a higher temperature will lead to a higher reaction rate since it is still far away from equilibrium. However, at the final stage, a higher temperature will make the reaction closer to equilibrium, leading to a lower net reaction rate. Therefore, there is an optimal temperature to achieve maximum average reaction rate. For a given reaction rate equation, if we know the temperature dependences of both the reaction rate constant and equilibrium constant, it is not difficult to obtain the relationship between reactor volume V_r and temperature, such as Eq. (4.32). Then the optimal operating temperature can be calculated by differentiating Eq. (4.32) with respect to T and setting the derivative $dV_r/dT = 0$.

The above discussion is for isothermal operations. In reality, most tubular reactors are operated under nonisothermal conditions. So we need to select an optimal temperature progression, i.e., the optimal temperature profile along the axial direction. For irreversible and reversible endothermic reactions, the optimal temperature progression should be from low to high. In other words, for tubular reactors the temperature should gradually increase from the reactor inlet to outlet. This is because with the progress of the reaction, reactant concentrations gradually decrease, which will in turn lead to lower reactor rate. Gradually increasing reaction temperature will compensate for the reaction rate decline caused by the decrease of reactant concentrations, increase the average reaction rate, and therefore improve production intensity of the reactor. In addition, for reversible endothermic reactions, high conversion can only be achieved by maintaining high reactor exit temperature. Otherwise the final conversion will be limited by the reaction equilibrium.

If a reversible exothermic reaction takes place in a tubular reactor, then the optimal temperature progression is from high to low. This is the opposite to the situation for reversible endothermic reactions. Based on the discussions in Chapter 2, Fundamentals of Reaction Kinetics, we know that for reversible exothermic reactions there is an optimal reaction temperature that decreases with the increase of conversion. So we need to control the reactor temperature following this pattern to maximize reaction rates at each location inside the reactor, and therefore maximize reaction production intensity. Of course, it would be very difficult to operate the reactor completely following the optimal temperature profile, but we should try to get as close as possible. We will further discuss this issue in Chapter 7, Analysis and Design of Heterogeneous Catalytic Reactors.

For adiabatic reactors the reaction temperature always changes monotonically, an increase for exothermic and a decrease for endothermic reactions, respectively. Obviously, the temperature profile of an adiabatic reactor is just the opposite of the optimal temperature progression for all reactions except irreversible exothermic reactions. From this perspective, the adiabatic tubular reactor is not a good choice. However, it should be noted that we cannot simply abandon adiabatic reactors. In the real world different reactor design proposals including reactor type and operating conditions need to be evaluated carefully through calculations and comparisons. On the other hand, temperature progression in adiabatic tubular reactors depends on feed temperature, so selecting the proper reactor inlet temperature is critical to maintaining an adequate reaction temperature. For reversible exothermal reactions detailed analyses have already been presented in the previous section.

4.6.2　Multiple Reactions

The optimal temperature progression for the single reaction discussed above is based on maximum production intensity. As mentioned earlier, for multiple reactions we can also use the maximum yield of desired product as the target for optimization of reaction temperature. Different optimization targets will lead to different optimal temperature distributions. We will use the parallel reactions:

$$A + B \xrightarrow{\ \ 1\ \ } P$$

$$A + B \xrightarrow{\ \ 2\ \ } Q$$

as an example to discuss this. Assume P is the desired product and the activation energy of the first reaction E_1 is smaller than that of the second reaction E_2. If the target is to maximize production intensity, the temperature progression should be from low to high. This is because low temperature favors the yield of P. Low temperature at the early stages will lead to a higher yield of P. At later stages the reaction rate will decrease due to the lower reactant concentrations so high temperatures will help compensate so that reaction rate won't be too slow. Of course, the amount of by-product Q will also increase. The strategy just described is for maximized productivity of P per unit time and unit reactor volume. If we want to maximize the yield of P then we should maintain low temperature for the whole reaction process. Obviously this will reduce production intensity and lead to larger reactor volume, but it will increase selectivity to product P. Which strategy is a better choice depends on the prices of feed products as well as reactor construction cost. In other words, process economics will determine which one is better.

In the previous chapter we discussed consecutive reactions taking place in an isothermal batch reactor:

$$A \xrightarrow{\ \ k_1\ \ } P \xrightarrow{\ \ k_2\ \ } Q$$

If P is the desired product, then there is an optimal reaction time to achieve maximum yield of P. Such a conclusion is also valid for a plug flow reactor, i.e., the space time for a plug flow reactor should be equal to the optimal reaction time of a batch reactor to achieve maximum yield of P.

For isothermal reactors there is no optimal temperature if the maximum yield of P is the target. If P is the desired product and $E_1 < E_2$, then the lower the temperature the higher the yield of P; if $E_1 > E_2$, then the higher the temperature the better. However, for nonisothermal operations, temperature progression should be from high to low if $E_1 < E_2$. The reason to use high temperatures initially is to enhance the first reaction and therefore promote the formation of P. After a certain amount of P being already accumulated, temperature should be lowered to reduce the formation of by-product Q.

Now let's examine a more complex example:

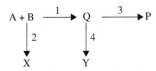

Assuming P is the desired product and $E_2 < E_1 < E_3$. If only reaction 1 and 2 are considered, higher temperature favors the formation of P. However, if reaction 1 and 3 are considered, then a lower temperature should be selected to favor the formation of P. Therefore, there must be an optimal temperature that is neither too high nor too low to give a maximized yield of P.

The next figure is a famous optimization problem in reaction engineering:

$$A + B \xrightarrow{\ 1\ } Q \xrightarrow{\ 3\ } P$$

with $\downarrow 2$ below A+B to X, and $\downarrow 4$ below Q to Y.

Assuming P is the desired product, and $E_1 < E_2$, $E_3 > E_4$. In order to produce more P, more Q has to be produced first. Since $E_1 < E_2$, lower temperature will favor Q formation over X. On the other hand, if $E_3 > E_4$, lower temperature also favors conversion of Q to by-product Y over P. Therefore, operating at low temperature cannot enhance the yield of P. High operating temperature will favor high yield of X but not Q, and hence will not lead to high yield of P either. From this discussion we see that it is impossible to achieve a high yield of P under isothermal operation no matter what operating temperature is selected. For tubular reactors the only way to achieve high yield of desired product P is to choose a temperature progression from low to high. Initial low temperature will favor the formation of Q, and high temperatures at later stage favors the conversion of Q to P. If the relative order of activation energy changes, the optimal temperature progression will also change. Following the same thought we can conclude that if $E_1 > E_2$ and $E_3 > E_4$ then we should keep high temperatures for the whole reaction process; if $E_1 < E_2$ and $E_3 < E_4$, then we should keep the reaction temperature low; if $E_1 > E_2$ and $E_3 < E_4$, then the optimal temperature progression is from high to low.

The above discussion on optimal temperature profile is qualitative and focuses on relatively simple situations. Quantitative analysis will require the use of suitable optimization procedures. For some multiple reaction systems even qualitative analysis can be complicated enough that only through sophisticated simulation can a conclusion of optimal temperature progression be reached.

Example 4.9
Parallel reactions as shown in Example 3.8 take place in an isothermal tubular reactor. Feed is pure A, and space time is 1 h. Please calculate the maximum yield of P and the operating temperature to achieve maximum yield. Compare the results with those from Example 3.8. The reaction rates are the same as given in Example 3.8, and assume the fluid flow in the tubular reactor is plug flow.

Solution
Both main and side reaction are first order irreversible reactions, so:

$$-\mathscr{R}_A = (k_1 + k_2)c_A = (k_1 + k_2)c_{A0}(1 - X_A)$$

Substitute into Eq. (4.6)

$$\tau = \int_0^{X_{Af}} \frac{dX_A}{(k_1 + k_2)(1 - X_A)} = \frac{1}{k_1 + k_2}\ln\left(\frac{1}{1 - X_{Af}}\right)$$

Or:

$$X_{Af} = 1 - \exp[-(k_1 + k_2)\tau] \tag{A}$$

Based on the definition of instantaneous selectivity:

$$S = \frac{k_1}{k_1 + k_2} \tag{B}$$

Since instantaneous selectivity is independent of the concentrations and the reactions take place under isothermal conditions, the overall selectivity of S is equal to instantaneous selectivity. Therefore, the final yield of desired product P is:

$$Y_{Pf} = S_0 X_{Af} = \frac{k_1 X_{Af}}{k_1 + k_2}$$

Substitute Eq. (A) into above equation:

$$Y_{Pf} = \frac{k_1(1 - \exp[-(k_1 + k_2)]\tau)}{k_1 + k_2} \tag{C}$$

Using Arrhenius correlations:

$$k_1 = A_1 \exp[-E_1/(RT)], \; k_2 = A_2 \exp[-E_2/(RT)] \tag{D}$$

Combining Eqs. (C) and (D), then differentiating Y with respect to T, and setting $dY_{Pf}/dT = 0$, we can obtain:

$$E_2 - E_1 = \left[\tau(k_1 E_1 + k_2 E_2)\left(1 + \frac{k_1}{k_2}\right) + E_2 - E_1\right]\exp[-(k_1 + k_2)\tau] \tag{E}$$

Again substitute Eq. (D) into Eq. (E), and

$$A_1 = 4.368 \times 10^5 \; \text{h}^{-1}, \; E_1 = 41,800 \; \text{J/(mol·K)}, \; A_2 = 3.533 \times 10^{18} \; \text{h}^{-1},$$
$$E_2 = 141,000 \; \text{J/(mol·K)}$$

(Continued)

(Continued)

Then we can obtain the optimal operation temperature is 390.1K. At this temperature:

$$k_1 = 4.368 \times 10^5 \exp[-41,800/(8.314 \times 390.1)] = 1.102 \text{ h}^{-1}$$
$$k_2 = 3.533 \times 10^{18} \exp[-141,000/(8.314 \times 390.1)] = 0.4626 \text{ h}^{-1}$$

using these k_1 and k_2 values in Eq. (A):

$$X_{Af} = 1 - \exp[-(1.102 + 0.4626) \times 1] = 0.7908$$

And using Eq. (C) we can obtain yield of P:

$$Y_{Pf1} = \frac{1.102 \times 0.7908}{1.102 \times 0.4626} = 0.557$$

In Example 3.8 we obtained an optimal operating temperature for CSTR of 389.3K, based on this we can obtain $k_1 = 1.074 \text{ h}^{-1}$, $k_2 = 0.4231 \text{ h}^{-1}$, using these values in Eq. (F) in Example 3.8 we can obtain the final yield of P:

$$Y_{Pf} = \frac{1.074 \times 1}{1 + (1.0740.4231) \times 1} = 0.4301$$

Using Eq. (E) in Example 3.8 we can calculate corresponding conversion:

$$X_{Af} = \frac{(1.074 + 0.4231) \times 1}{1 + (1.0740.4231) \times 1} = 0.5995$$

Comparing the above results we can see that at the same space time the optimal operating temperature, conversion, and yield of the desired product for a tubular reactor are higher than those of CSTR. However, the final selectivity of the tubular reaction, $S_0 = 0.557/0.7908 = 0.7044$, is lower than the value of CSTR, which is $(0.4301/0.5995) = 0.7174$. This is due to the higher operation temperature of the tubular reactor. We must point out that these conclusions are specific for these two examples, not a general conclusion.

FURTHER READING

Froment GF, Bischoff KB. Chemical reactor analysis and design. New York: John Wiley & Son Inc.; 1979.

Denbigh KG, Turner JCR. Chemical reactor theory. 3rd ed. Cambridge University Press; 1985.

Nauman EB. Chemical reactor design. New York: John Wiley & Son Inc.; 1987.

PROBLEMS

4.1 The following homogeneous gas phase reaction takes place in a plug flow reactor at constant pressure and 800°C:

$$C_6H_5CH_3 + H_2 \rightarrow C_6H_6 + CH_4$$

And reaction rate equation under such conditions can be expressed as:

$$r = 1.5c_T c_H^{0.5}, \text{mol}/(\text{L} \cdot s)$$

where c_T and c_H represent the concentrations of toluene and methane, respectively. Feed flow rate is 2 kmol/h, and toluene to hydrogen molar ratio is 1:1. If the reactor diameter is 50 mm, please calculate the reactor length needed to achieve toluene conversion of 95%.

4.2 Use the data and conditions given in Problem 3.2, calculate reactor volume needed if a plug flow tubular is used to produce ethyl glycol, and compare the value with that of the batch tank reactor.

4.3 Nitrogen monoxide oxidation reaction takes place in a plug flow reactor at 1.013×10^5 Pa and 20°C, and reactor volume is 0.5 m³:

$$2NO + O_2 \rightarrow 2NO_2$$

$$r_{NO} = 1.4 \times 10^4 c_{NO}^2 c_{O_2}, \text{kmol}/(\text{m}^3 \cdot s)$$

The unit for all the concentrations in the rate equation is kmol/m³. The feed contains 10% NO, 1% NO_2, 9% O_2, and 80% N_2, and feed flow rate under standard conditions is 0.6 m³/h. Please calculate the gas composition at reactor outlet.

4.4 A plug flow reactor with ID = 76.2 mm is used for ethane cracking to make ethylene;

$$C_2H_6 \leftrightarrows C_2H_4 + H_2$$

The reaction pressure is 2.026×10^5 Pa and temperature is 815°C. The feed contains 50 mol% ethane, and the rest is steam, and the flow rate is 0.178 kg/s. The rate equation is

$$\frac{dp_A}{dt} = kp_A$$

where p_A is ethane partial pressure. At 815°C the rate constant $k = 1.0$ 1/s, and reaction equilibrium constant $K_P = 7.49 \times 10^4$ Pa. Assume all side reactions can be ignored, please calculate:

1. Equilibrium conversion under these conditions;

2. Reactor length needed to achieve 50% of ethane equilibrium conversion.

4.5 The following gas–solid catalytic reaction takes place in a plug flow reactor at 277°C and 1.013×10^5 Pa:

$$\begin{array}{cccc} C_2H_5OH + CH_3COOH \rightarrow CH_3COOC_2H_5 + H_2O \\ \text{(A)} \quad\quad \text{(B)} \quad\quad\quad \text{(P)} \quad\quad\quad \text{(Q)} \end{array}$$

The packing density of the catalyst bed is $700 \, \text{kg/m}^3$. At $277°C$, the consumption rate for reactant B is:

$$r_B = \frac{4.096 \times 10^{-7}(0.3 + 8.885 \times 10^{-6} p_Q)\left(p_B - \frac{p_P p_Q}{9.8 p_A}\right)}{3600(1 + 1.515 \times 10^{-4} p_P)}, \, \text{kmol/(kg·s)}$$

The unit for all partial pressures in the above equation is Pa. The transport resistance between gas and solid phase can be ignored. The feed contains 23 wt.% B, 46 wt.% A, and 31 wt.% Q. Calculate the weight of catalyst needed and the reactor volume to achieve $X_B = 35\%$ when the production rate of ethyl acetate is 2083 kg/h.

4.6 Chlorodifluoromethane decomposition is a first order reaction:

$$2CHFCl_2(g) \rightarrow C_2F_4(g) + 2HCl(g)$$

2 kmol/h pure $CHFCl_2$ is heated to $700°C$ in a preheater and fed to a plug flow reactor operated isothermally at $700°C$. $CHFCl_2$ is already partially decomposed in the preheater, and the conversion in the preheater is 20%. If the superfacial velocity at the reactor inlet is 20 m/s, what is the superfacial velocity at the reactor outlet if the total conversion of $CHFCl_2$ is 40.8%? What is the reactor length needed to achieve this conversion? The whole system is operated at 1.013×10^5 Pa. At $700°C$ reaction constant is 0.97 1/s. If the feed flow rate is doubled, what is the reactor length needed to achieve the same conversion?

4.7 Please design an isothermal reactor for the following liquid phase reactions:

$$A + B \rightarrow R, \, r_R = k_1 c_A c_B$$

$$2A \rightarrow S, \, r_S = k_2 c_A^2$$

The desired product is R, and separating R from B is extremely difficult.

1. How would you select the feed composition?
2. If a plug flow reactor is used, what feeding mode should be selected?
3. If a semibatch reactor is used, what feeding mode should be used?

4.8 An endothermic irreversible gas phase reaction takes place isothermally at $400°C$ in a tubular reactor. The activation energy is 39.77 kJ/mol. How can you increase production rate by 35% using the same reactor and feed and achieve the same conversion? Assume the reactor is still operated under isothermal conditions and maintains plug flow regime.

4.9 Using the conditions and data given in Problem 3.8, calculate phenol production rate when a plug flow reactor is used, and compare the results from different types of reactors.

4.10 Using the conditions and data of Problem 3.9, estimate whether 95% conversion can be achieved when a plug flow reactor is used. The reaction temperature and feed composition are the same, and the space time of the plug reactor equals the reaction time in the batch reactor. Please also estimate the yield of R.

4.11 Using the conditions and data from Problem 3.14, calculate the following for a plug flow reactor:
1. Reaction volume;
2. Total reaction volume if two plug flow reactors in series are used.

4.12 Gas phase elementary reaction takes place in a tubular reactor:

$$A + B \rightarrow C$$

Feed A is a gas, B is a liquid, and product C is also a gas. B accumulates at the reactor bottom (as shown in the figure), and the volume it occupies can be ignored. The gas phase is saturated with B and the reaction takes place in the gas phase.

The operation pressure is 1.013×10^5 Pa, the saturated vapor pressure of B is 2.532×10^4 Pa, reaction temperature is 340°C, and reaction rate constant is $100 \, \text{m}^3/(\text{mol} \cdot \text{min})$. Please calculate the consumption rate of A when its conversion is 50%, and the reaction volume needed if the flow rate of A is $0.1 \, \text{m}^3/\text{min}$.

4.13 The following reaction takes place in a plug flow reactor:

$$A \xrightarrow{k_1} P \xrightarrow{k_2} Q$$

Both reactions are first order and at the reaction temperature the rate constant $k_1 = 0.30 \, \text{min}^{-1}$ and $k_2 = 0.10 \, \text{min}^{-1}$. Feed A flow rate is $3 \, \text{m}^3/\text{h}$, and there is no P or Q in the feed stream. Please calculate the maximum yield of P, the overall selectivity and the reaction volume needed to achieve the maximum yield.

4.14 The following parallel reactions take place in liquid phase:

$$A + B \rightarrow P, \ r_P = c_A c_B^{0.3}, \ \text{kmol}/(\text{m}^3 \cdot \text{min})$$

$$a(A + B) \rightarrow Q, \ r_Q = c_A^{0.5} c_B^{1.3}, \ \text{kmol}/(\text{m}^3 \cdot \text{min})$$

where a is the stoichiometric coefficient. The desired product is P.
1. Please derive the equation for calculating instantaneous selectivity;
2. If $a = 1$, calculate the overall selectivity under the following conditions:
   a. Plug flow reactor, $c_{A0} = c_{B0} = 10 \, \text{kmol/m}^3$, $c_{Af} = c_{Bf} = 1 \, \text{kmol/m}^3$

b. Continuous flow tank reactor, the concentrations are the same as in (a)

c. Plug flow reactor, and reactant A and B are fed into the reactor as shown in figure below:

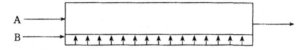

   A is fed continuously into the reactor from one end and B is added to the reactor continuously at different locations to maintain a uniform concentration of B at 1 kmol/m^3 throughout the reactor volume. The concentration of A at reactor inlet and outlet is 19 kmol/m^3 and 1 kmol/m^3, respectively.

4.15 The following gas phase reactions take place in a plug flow reactor isothermally and at constant pressure:

$$A \rightarrow P, \; r_P = 5.923 \times 10^{-6} p_A, \; \text{kmol}/(\text{m}^3 \cdot \text{mol})$$

$$A \rightarrow 2Q, \; r_Q = 1.777 \times 10^{-5} p_A, \; \text{kmol}/(\text{m}^3 \cdot \text{mol})$$

$$A \rightarrow 3R, \; r_R = 2.961 \times 10^{-6} p_A, \; \text{kmol}/(\text{m}^3 \cdot \text{mol})$$

where p_A is the partial pressure of A in Pa. The feed contains 10 mol% A, and the rest is inert gas. If the feed flow rate is 1800 Standard m^3/h, please calculate the reaction volume needed and the yield of Q when the conversion of A is 90%.

4.16 A plug flow reactor packed with vanadium catalyst is used to produce maleic anhydride (MA):

$$B \xrightarrow{k_1} MA \xrightarrow{k_2} CO, CO_2, H_2O$$
$$\text{(with } k_3 \text{ path from B)}$$

All three reactions are first order, and the activation energies are:

$$E_1 = 70{,}800 \text{ kJ/kmol}; \quad E_2 = 193{,}000 \text{ kJ/kmol};$$
$$E_3 = 124{,}800 \text{ kJ/kmol};$$

The preexponential factors [kmol/(kg · h · Pa)] are:

$$A_1 = 0.2171; \quad A_2 = 1.372 \times 10^8; \quad A_3 = 470.8$$

The reactions take place isothermally at 1.013×10^5 Pa and 704K. The feed is a mixture of benzene vapor and air with benzene molar fraction = 0.018. The production rate of MA is 1000 kg/h, and the desired yield is 42%. Assume the reactions can be considered a constant-volume process and the reactor can be described by pseudo homogeneous model, please calculate:

1. The final conversion of benzene;
2. The feed flow rate needed;
3. The amount of catalyst needed.

4.17 (1) Derive the differential equation to describe correlation between reactant temperature and conversion in a adiabatic tubular reactor; (2) Under what conditions can this equation be simplified into a linear algebraic equation;

(3) Calculate adiabatic temperature increase for toluene hydrogenation reaction $C_6H_5CH_3 + H_2 \rightarrow C_6H_6 + CH_4$. The feed temperature is 873K, and hydrogen to benzene ratio in the feed is 5. The reaction heat $\Delta H_{298} = -49,974$ J/mol, and the specific heats are:

$$H_2 : C_v = 20.786, \text{ J/(mol·K)}$$

$$CH_4 : C_v = 0.04414T + 27.87, \text{ J/(mol·K)}$$

$$C_6H_6 : C_v = 0.1067T + 103.187, \text{ J/(mol·K)}$$

$$C_5H_6CH_3 : C_v = 0.03535T + 124.85, \text{ J/(mol·K)}$$

(4) Under the condition specified in (3), calculate the reactor outlet temperature when the final conversion of toluene is 70%.

4.18 Ammonia (A) reacts with ethylene oxide (B) can form monoethanolamine (M), diethanolamine (D), and triethanolamine (T):

$$NH_3 + C_2H_4O \xrightarrow{k_1} H_2NCH_2CH_2OH$$

$$H_2NCH_2CH_2OH + C_2H_4O \xrightarrow{k_2} HN(CH_2CH_2OH)_2$$

$$HN(CH_2CH_2OH)_2 + C_2H_4O \xrightarrow{k_3} N(CH_2CH_2OH)_3$$

The rate equations for these three reactions are:

$$r_1 = k_1 c_A c_B$$

$$r_2 = k_2 c_M c_B$$

$$r_3 = k_3 c_D c_B$$

The reaction system is maintained at constant temperature and the desired product is monoethanolamine.

1. Please propose how to select feed composition and why;
2. Select more suitable reactor type and operation mode;
3. Based on (2), select feeding mode;
4. How the reaction time should be controlled and why?

4.19 Two plug flow reactors with reaction volume of 1 m^3 are used to decompose a cumene hydroperoxide solution to produce phenol and aceton. The concentration of cumene hydroperoxide in the feed solution is 3.2 kmol/m^3. The decomposition reaction is a first order irreversible reaction, and takes place at 86°C. At this temperature, the rate constant is 0.08 1/s. The flow rate of cumene hydroperoxide is 2.4 m^3/min. Please calculate the cumene hydroperxide conversion under the following conditions:

1. The two reactors are in series;
2. The two reactors are in parallel, and the feed is equally distributed to the two reactors, i.e., flow rate to each reactor is 1.2 m^3/min;
3. The two reactors are in parallel and the flow rate ratio is 1:2, i.e., flow rate for one reactor is 0.8 m^3/min and for another is 1.6 m^3/min;
4. Use one plug flow reactor with reaction volume of 2 m^3;
5. If concentration of cumene hydroperoxide is increased to 4 kmol/m^3 while all the other conditions are kept the same, will productivity of phenol change?
6. Compare and discuss the results from above calculations to see what conclusions can be reached.

4.20 Butadiene and ethylene react in a adiabatic plug flow reactor to make cyclohexene:

$$C_4H_6 + C_2H_4 \rightarrow C_6H_{10}$$
$$\text{(A)} \qquad \text{(B)} \qquad \text{(R)}$$

This is a gas phase reaction and the reaction rate is:

$$r_A = kc_A C_B$$

$$k = 3.16 \times 10^7 \exp\left(-\frac{13,840}{T}\right) L/(mol \cdot s)$$

The feed is an equal molar mixture of butadiene and ethylene at 440°C, and the reactor pressure is 1.013 × 10^5 Pa. The reaction heat is −1256 × 10^5 J/mol. Molar specific heats for all components can be considered constants, and the values are:

$$c_{pA} = 154 \text{ J}/(mol \cdot K)$$

$$c_{pB} = 85.6 \text{ J}/(mol \cdot K)$$

$$c_{pR} = 249 \text{ J}/(mol \cdot K)$$

The required conversion of butadiene is 12%, please calculate:

1. Space time, average residence time, and temperature at reactor outlet;
2. If the reaction takes place isothermally at 440°C, redo the calculations in (1);
3. The amount of heat that needs to be removed if the reaction takes place at 440°C isothermally.

4.21 Ethylene oxide reacts with water can make ethylene glycol with diethylene glycol as by-product:

$$C_2H_4O + H_2O \xrightarrow{k_1} CH_2OHCH_2OH$$

$$C_2H_4O + CH_2OHCH_2OH \xrightarrow{k_2} (CH_2CH_2OH)_2O$$

Both reactions are first order with respect to their reactants, and the reaction rate constant ratio $k_2/k_1 = 2$. The molar ratio of water to ethylene oxide in the feed is 20 and there is no product in the feed stream.

1. What type of reactor should be selected?
2. What the conversion will be if the yield of ethylene glycol is at its maximum?
3. Someone suggested that using plug flow reactor would be beneficial for achieving high ethylene glycol yield, but ethylene oxide conversion would be low. Therefore a recycle reactor should be used to increase the overall conversion. Do you agree with such suggestion? If the recycle ratio $\psi = 25$, and the space time is kept the same as in that in (2), compare the overall conversion and the ethylene glycol yields with that in (2).

4.22 The reaction rate equation for autocatalytic $A \rightarrow P$ is:

$$r_A = kc_Ac_P \ \text{kmol}/(\text{m}^3 \cdot \text{min})$$

At reaction temperature $k = 1 \ \text{m}^3(\text{kmol} \cdot \text{min})$, and $c_{A0} = 2 \ \text{kmol/m}^3$. The feed contains 99% (mol) A and the rest is P. The feed flow rate is 1000 mol/h, and the required final conversion is 90%.

1. To minimize reactor volume, what type of reactor should be used? Calculate the reaction volume based on your selection;
2. If a recycle reactor is used, please determine the optimal recycle ratio and calculate the reaction volume;
3. What will be the reaction volume if the recycle ratio $\psi = \infty$?
4. What will be the reaction volume if the recycle ratio $\psi = 0$?

4.23 Methane pyrolysis takes place in a tubular reactor at high temperature and ambient pressure:

$$CH_4 \xrightarrow{k_1} C_2H_4 \xrightarrow{k_2} C_2H_2 \xrightarrow{k_3} C$$
$$\text{(A)} \qquad\quad \text{(B)} \qquad\quad \text{(D)}$$

The feed contains 90% methane.

Rate equations are:

$$-\frac{dc_A}{d\tau} = k_1c_A$$

$$\frac{dc_B}{d\tau} = \frac{1}{2}k_1 c_A - k_2 c_B$$

$$\frac{dc_D}{d\tau} = k_2 c_B - k_3 c_D$$

And the rate constants:

$$k_1 = 4.5 \times 10^{13} \exp\left(-\frac{45{,}800}{T}\right), 1/s$$

$$k_2 = 2.6 \times 10^{8} \exp\left(-\frac{20{,}100}{T}\right), 1/s$$

$$k_3 = 1.7 \times 10^{5} \exp\left(-\frac{15{,}100}{T}\right), 1/s$$

1. If C_2H_4 is the desired product, and the third step can be ignored, what is the maximum yield of C_2H_4?
2. If the third step has to be considered, with the maximum yield of C_2H_4 change?
3. Illustrate concentrations of each component as a function of space time (using plots or tables)
4. Will product distribution change if methane concentration in the feed changes?
5. Will product distribution change if reaction temperature changes? If temperature increases will C_2H_4 yield increase or decreases? How about yield of acetylene?

4.24 Pure acetone vapor enters a plug flow reaction with diameter of 26 mm at 1000K and at flow rate of 8 kg/s. In the reactor acetone decomposes into ethenone and methane:

$$CH_3COCH_3 \rightarrow CH_2CO + CH_4$$

This is a first order reaction, and reaction rate constant as a function of temperature can be expressed as:

$$\ln k = 34.34 - 34{,}222/T$$

The unit of k is 1/s. The reactor is operated at 162 kPa, and is heated by a heat source at constant 1300K. The overall heat transfer coefficient between the heat source and the reactant gas is 110 W/(m$^2 \cdot$ K). The acetone conversion is 20%. The molar specific heats as functions of temperature are:

$$CH_3COCH_3 : C_p = 26.63 + 0.183T - 45.86 \times 10^{-6}T^2$$

$$CH_2CO:C_p = 20.04 + 0.0945T - 30.95 \times 10^{-6}T^2$$

$$CH_4:C_p = 13.39 + 0.077 - 18.71 \times 10^{-6}T^2$$

At 298K the reaction heat is 80.77 kJ/mol.
1. Calculate the reaction volume required;
2. Plot axial temperature profile and acetone concentration profile.

4.25 A first order reversible reaction takes place in an isotheral plug flow reactor. The forward and backward reaction rate constants as functions of temperature are:

$$\overrightarrow{k} = 2 \times 10^6 \exp\left(-\frac{5000}{T}\right)$$

$$\overleftarrow{k} = 3.5 \times 10^9 \exp\left(-\frac{9000}{T}\right)$$

The final conversion is 90%. Please determine the reaction temperature to minimize reaction volume.

Chapter 5

Residence Time Distribution and Flow Models for Reactors

Chapter Outline

In Chapter 3, Tank Reactor and Chapter 4, Tubular Reactor, we discussed two different types of flow reactors, the continuous tank reactor and the tubular reactor. Under the same operating conditions, the performances of these two types of reactors are significantly different because the flow patterns of the reacting materials vary in these two types of reactors, i.e., the residence time distributions (RTDs) are different. More details will be discussed in this chapter to elaborate the quantitative delineation and experimental determination of the RTD in different flow systems.

In preceding discussions, the design of the continuous tank reactor was based on the assumption that the concentrations of the reacting materials are uniform in the reactor. Plug flow assumption was made to deal with the tubular reactor. If these two assumptions are not satisfied, new flow models are needed for the design and analysis of reactors. Flow model development depends on the RTD, which is one of the primary subjects of this chapter. Additionally, design characteristics and calculations of different reactors will be discussed on the basis of flow models developed in this chapter.

Reaction Engineering. DOI: http://dx.doi.org/10.1016/B978-0-12-410416-7.00005-7

Mixing of fluid in reactors has direct impact on the progress of chemical reactions. Therefore, the mixing of fluids in flow reactors will be briefly discussed to clarify several basic concepts at the end of this chapter.

5.1   RESIDENCE TIME DISTRIBUTION

5.1.1   Overview

Completeness of a chemical reaction is related to the residence time of the reacting material in the reactor. The longer the residence time, the more complete is the reaction. Therefore, it is crucial to study the residence time of the reacting material in the reactor. For a batch reactor, this problem is quite simple because all reacting materials enter the reactor at the same time and leave the reactor simultaneously. Consequently, the residence times of all materials in the reactor are the same at any instance, and there is no RTD. Thus, the measurement and control of the residence time of reacting material in a batch reactor is very straightforward.

It is a different story for flow systems. Since fluid continuously enters and leaves the system, the residence time of the fluid is complicated. Generally, residence time refers to the total time a fluid spent in the system from the moment it entered the system to the moment it left the system. A question is inevitably raised of whether the fluids entering the system at the same time would leave the system simultaneously. Fluid moves continuously, fluid molecules, however, move disorderly. It is thus impossible for all fluid molecules to move forward in the system following the same pattern since molecular movement is a complete random process. Consequently, it is not feasible to study the residence time of a single molecule. Instead, residence time of an aggregate of fluid molecules is usually studied. The fluid composed by an aggregate of fluid molecules is called a fluid particle or fluid element. The volume of the fluid particle is negligible compared to the volume of the system; however, it consists of sufficient fluid molecules to exhibit the nature of statistical average. Would fluid particles that enter a system at the same time leave the system simultaneously? In other words, are their residence times the same? It is very hard to find such a system in the real world, but it cannot exclude the possibility that they are approximately the same. Plug flow assumption for tubular reactors as stated in Chapter 4, Tubular Reactor, is one of these cases.

Due to the nonuniform distribution of flow velocity in the system, molecular diffusion and turbulent diffusion, forced convection by agitation, dead zones (stagnant regions) caused by inappropriate installation of equipments, and channeling and short circuiting, residence times of fluid particles are either short or long: some leave the system very quickly, others may stay for a very long time, consequently forming a distribution of residence times.

There are two different types of RTDs, one is life distribution and another is age distribution. The former refers to the time elapsed from the instant the fluid particles entered the system to the moment they left the system as

Fluid inlet
System
Outlet

FIGURE 5.1 Schematic diagram of closed system.

defined previously. The latter definition is used to represent the residence time of the fluid particles inside the system since they entered the system. Life and age are two different concepts. The former refers to the residence time of the fluid particles at the outlet of the system, and the latter addresses the residence time for fluid particles inside the system. The distribution that is experimentally measured and has significance in applications is life distribution. Normally, the RTD refers to life distribution, which is widely accepted. It is worthwhile to note that in some literature, life distribution is denoted as "age distribution," and the age distribution is denoted as "internal age distribution." These different denotations should not be confused.

The RTD discussed in this book is limited to closed systems with only one inlet and one outlet as illustrated in Fig. 5.1. Fluid continuously enters through an inlet tube on one side of the system and exits on the other side. A primary assumption for a closed system is that fluid particles would never go back to the inlet tube once they have entered the system, and the fluid particles eluted from the system would never reenter the system. Simply speaking, there is only entrance and no exit for fluid particles at the inlet and only exit and no entrance for fluid particles at the outlet. The assumption for closed systems is appropriate for most cases in reality.

The theory of RTD is not merely an important component of chemical reaction engineering; it is extensively applied in the design and modeling of equipments in various separation processes such as adsorption, extraction, distillation, and crystallization, as well as some other fields with fluid flow. Applications of RTD are mainly in two aspects: one is to analyze the characteristics of the existing equipment and provide useful information to improve its performance through measurement of RTD. This application is diagnostic; it may provide some insights on how to improve the performance of the equipments that are not efficiently operated. For example, measurement of RTD might be used to check the presence of dead zones or short circuiting in packed towers and fixed-bed reactors. Another application is in the analysis and design of reactors and other equipment. Appropriate flow models may be developed based on the RTD and can be used as a basis for material, energy, and momentum balance calculations.

5.1.2 Quantitative Delineation of RTD

For simplicity, a closed system with constant flow as illustrated in Fig. 5.1 is discussed here. It is assumed that the density of the fluid is constant and no chemical reactions occur in the system. The fluid entering the system is assumed to be colorless. When flow reaches steady-state, 100 red particles

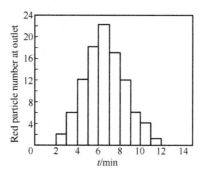

FIGURE 5.2 Bar diagram of residence time distribution.

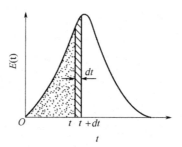

FIGURE 5.3 Density function of residence time distribution.

are suddenly injected to the system at some point ($t = 0$) in as short a time as possible. The number of red particles at the outlet is then measured at different times. Plotting the number of red particles in the outlet as a function of time, a histogram of RTD as illustrated in Fig. 5.2. The figure shows that 18 red particles are found in the outlet between 5 and 6 min since they are injected to the system. That is to say the residence times of 18% of red particles are between 5 and 6 min. Assuming the only difference between the red particles and the main fluid is the color and other characteristics are the same, it thus can be believed that the percentage of main fluid with residence times between 5 and 6 min is 18%. This is the so-called tracer response technique, and is the primary basis for the experimental measurement of RTD that will be discussed in the next section. The red particles are called tracers.

RTD is derived by observing the tracer. Here the red particles in the outlet as described above is a discrete distribution. If a red fluid is used as a tracer and the concentration of tracer in the outlet is monitored continuously, and if the time interval between two measurements is small enough, a continuous curve of RTD as shown in Fig. 5.3 will be obtained. The slash covered area, $E(t)dt$, represents the fraction of fluid particles entered at time zero and exited between time t and $t + dt$. According to probability theory, $E(t)dt$ is the probability that the residence time of fluid particles in the system is between time t and $t + dt$. Hence, $E(t)$ is a function of RTD and is independent of the characteristics of the system. $E(t)$ is called the RTD density function, its dimension is $[\text{time}]^{-1}$.

In practical applications, treating $E(t)dt$ as probability has a more direct significance than taking $E(t)$ as probability density.

Based on the properties of $E(t)$, the following equations are obviously satisfied.

$$E(t) = 0 \quad (t < 0) \tag{5.1}$$

$$E(t) \geq 0 \quad (t \geq 0) \tag{5.2}$$

$$\int_0^\infty E(t)dt = 1 \tag{5.3}$$

Eq. (5.3) is called normalization condition. Since $E(t)dt$ is the fraction of fluid particles with residence time between t and $t + dt$, summation of all $E(t)dt$ equals one.

Integrating $E(t)$ with respect to t from 0 to t, we obtain

$$\int_0^t E(t)dt = F(t) \tag{5.4}$$

The black dot occupied area in Fig. 5.3 equals the integral value of $F(t)$. Since this value includes contributions from all fluid particles with residence times less than t, it is not hard to understand that $F(t)$ is the fraction of fluid particles with residence times less than t, and it is a dimensionless value. $F(t)$ is called the RTD function. From the perspective of probability theory, $F(t)$ is the probability of fluid particles whose residence times are less than t. Fig. 5.4 is a typical $F(t)$ curve. $F(t)$ curves are different from $E(t)$ curves and are monotonically increasing curves. Its maximum value is 1, or it can be written as $F(\infty) = 1$, while its minimum value is 0 or $F(t) = 0$ (at $t \leq 0$). Anyway, $F(t)$ is always positive. Since $F(t)$ represents the fraction of fluid particles with residence time less than t, $1 - F(t)$ is the fraction of fluid particles whose residence times are greater than t.

Eq. (5.4) can be rewritten as

$$E(t) = \frac{dF(t)}{dt} \tag{5.5}$$

Hence, when the $F(t)$ curve is known, draw a tangent at one point on the $F(t)$ curve like line AP shown in Fig. 5.4. The slope of this line is the corresponding $E(t)$ value. Instead, when the $E(t)$ curve is known, integrating $E(t)$ gives the corresponding $F(t)$ value. Therefore, one of the two RTD forms can be derived given another one is known.

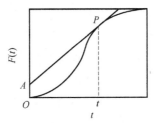

FIGURE 5.4   Function of residence time distribution.

For simplicity, a dimensionless residence time θ is usually used as defined below.

$$\theta = \frac{t}{\bar{t}} \qquad (5.6)$$

where $\bar{t}$ is the mean residence time. For fluid in a closed system, if the density of fluid is constant, the mean residence time equals V_t/Q

$$\bar{t} = \frac{V_t}{Q} \qquad (5.7)$$

If the residence time of a fluid particle falls in the interval $(t, t + dt)$, its dimensionless residence time must fall in the interval $(\theta, \theta + d\theta)$. Since they refer to the same event, the probability that t and θ fall in this interval must be same. Consequently,

$$E(t)dt = E(\theta)d\theta$$

Substituting Eq. (5.6) into the above equation, the relationship between $E(\theta)$ and $E(t)$ is obtained after some simplification.

$$E(\theta) = \bar{t}E(t) \qquad (5.8)$$

$F(t)$ is a cumulative probability and θ is deterministic function of t. Based on the principle that the probability of the deterministic function of a random variable should equal the probability of that random variable, we have

$$F(\theta) = F(t) \qquad (5.9)$$

Clearly, Eqs. (5.3) to (5.5) can be described with dimensionless residence time as the following.

$$\int_0^\infty E(\theta)d\theta = 1 \qquad (5.10)$$

$$\int_0^\theta E(\theta)d\theta = F(\theta) \qquad (5.11)$$

$$E(\theta) = \frac{dF(\theta)}{d\theta} \qquad (5.12)$$

A quantitative description of RTD was discussed in the above section. Due to the fact that it is a random process and residence time is a random variable, we used probability distribution density or distribution function to describe RTD. Even if one is not quite familiar with probability theory, it is not hard to understand this representation.

5.2   EXPERIMENTAL DETERMINATION OF RTD

As mentioned previously, the most frequently used RTD measurement method is the trace response technique, which uses the tracer to track the residence time of fluid particles in a system. Depending on the injection method of the tracer, this technique is categorized as three methods, i.e., pulse input, step input, and cycle input. Only the first two methods are going to be discussed here.

5.2.1 Pulse Experiments

In a pulse experiment, a certain amount of tracer is suddenly injected into the fluid entering the system in as short a time as possible. Fig. 5.5 is the schematic diagram of the pulse input method for measuring RTD. The emphasis on the fast injection of the tracer is to make sure that all tracers are injected into the system at the same time, which usually takes at time 0 ($t = 0$). Only this way can the RTD be accurately measured. This pulse input is called an ideal pulse, which is shown as $c_0(t) - t$ graph in the lower left of Fig. 5.5. Mathematically, this input response can be described using the δ function, which is a common mathematical tool in dealing with the physics problems that concentrate on one point.

As soon as the tracer is injected, the outlet concentration of the tracer $c(t)$ is immediately measured as a function of time. Its systematic diagram is shown as the lower right graph in Fig. 5.5. This measurement is continuous; otherwise, a discrete result is received. For gas systems, thermal analysis is frequently used and conductivity analysis is often used for electrolyte solutions. It definitely can be measured using other methods such as radiation rate analysis for radioactive materials. Concentration is determined by changes of some physical properties.

The curve of the effluent tracer concentration $c(t)$ as a function of t is called a response curve, through which the RTD curve can be obtained. From the definition of $E(t)$, we obtain

$$Qc(t)dt = mE(t)dt$$

So

$$E(t) = \frac{Qc(t)}{m} \tag{5.13}$$

where m is the amount of the tracer injected. RTD density function can be derived from the response curve based on Eq. (5.13). So, the pulse input method gives $E(t)$ directly. If $F(t)$ is desired, it can be derived from $E(t)$ from Eq. (5.4).

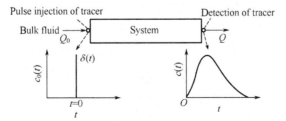

FIGURE 5.5   Measurement of residence time distribution with pulse method.

Sometimes, the amount of tracer, m, cannot be measured accurately, but it can be calculated from the following equation

$$m = \int_0^\infty Qc(t)dt \qquad (5.14)$$

If Q is a constant, the area under the response curve times the volumetric flow rate of the main fluid should equal the amount of tracer. Substituting Eq. (5.14) into (5.13), we have

$$E(t) = \frac{c(t)}{\displaystyle\int_0^\infty c(t)dt} \qquad (5.15)$$

If some other parameters instead of the tracer concentration is measured at the outlet, as long as this parameter is linear with concentration, the response value of this parameter can be directly applied in Eq. (5.15) to get $E(t)$, and it is not necessary to change back to concentration. Also please note that if the tail of the curve is too long or residence time of a small fraction of fluid is too long, the integral in the denominator of right side of Eq. (5.15) is not easily calculated correctly. In this situation, it is desirable to know the amount of tracer injected to avoid the error caused by integration.

Example 5.1
The regenerator in the fluidized catalytic cracking unit combusts coke on silica-alumina catalyst with air to regenerate the catalysts. The flow rate of air entering the regenerator is 0.84 kmol/s. RTD of air in the regenerator is determined by pulse experiments with helium as a tracer. The amount of helium injected is 8.84×10^{-3} kmol. Helium concentration c (as indicated by the molar ratio of helium to other gases) in the effluent of the regenerator is measured as a function of time as shown in the following table:

t/s	0	9.6	15.1	20.6	25.3	30.7	41.8	46.8	51.8
$c \times 10^6$	0	0	143	378	286	202	116	73.5	57.7

Determine the RTD density function and RTD function at time $t = 35$ s.

Solution
$E(t)$ is derived from Eq. (5.13). The given flow rate is for air at the inlet of the regenerator, while Q in Eq. (5.13) is the flow rate at outlet. However, since 1 kmol of carbon dioxide is produced per 1 kmol of oxygen consumption in the carbon burning process, the molar flow rate of the gas is not changed, therefore, the flow rate at outlet is still 0.84 kmol/s. Substituting $t = 15.1$ s and $c = 1.43 \times 10^{-4}$ into Eq. (5.13), $E(t)$ is obtained:

$$E(t) = \frac{0.84 \times 1.43 \times 10^{-4}}{8.84 \times 10^{-3}} = 0.0136s^{-1}$$

(Continued)

(Continued)

Similarly, $E(t)$ at other times can be calculated and results are listed in Table 5A.

Plot $E(t)$ as a function of time using data from Table 5A as shown in Fig. 5A. The figure shows that at $t = 35$ s, $E(t)$ equals 15.5×10^{-3} 1/s. It can be solved in an alternative way, i.e., plotting $c(t)$ over t, and reading c value at $t = 35$ s from the plot, substituting c into Eq. (5.13) to get corresponding $E(t)$. This solution saves more effort. However, since $F(t)$ at $t = 35$ s is also required, and it can be obtained from integration of $E(t)$ as shown by Eq. (5.4), $E(t)$ at different times needs to be determined. From Eq. (5.4), we have:

$$F(35) = \int_0^{35} E(t)dt$$

The integral of the right side should be equal to the area shaded by the slash under the curve, and the value is 0.523, which is the value of the RTD function at time $t = 35$ s.

TABLE 5A Relation of $E(t)$ versus t

t/s	0	9.6	15.1	20.6	25.2	30.7	41.8	46.8	51.8
$E(t) \times 10^3/s^{-1}$	0	0	13.6	35.9	27.2	19.2	11.0	6.98	5.48

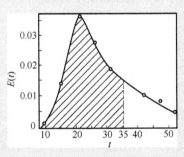

FIGURE 5A $E(t)$ curve.

5.2.2 Step Experiments

The essence of step experiments is that fluid with constant flow in the system is switched with fluid that contains a tracer at the same flow rate, or vice versa. The former input is called the increasing step (or positive step) and the latter is called the decreasing step (or negative step). The difference between step experiments and pulse experiments is that the tracer is continuously injected in the former method and all tracers are injected all at once in a very short time in the latter method.

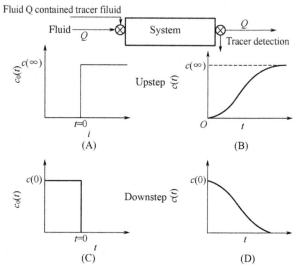

FIGURE 5.6 Measurement of residence time distribution with step method. (A) Step decrease input; (B) Output response; (C) Step increase input; (D) Output response.

Fig. 5.6 is the systematic diagram of the RTD measured by step input, where (A) and (B) represent the input signal and output response curve of an increasing step, respectively. Assuming $c(\infty)$ is the tracer concentration in the fluid, which is constant during the input process, and taking the time when the fluid with tracer is switched as time zero, the step-input function can be written as

$$
\begin{aligned}
c_0(t) &= 0 & t < 0 \\
c_0(t) &= c(\infty) = constant & t \geq 0
\end{aligned}
\tag{5.16}
$$

No matter whether an increasing step or a decreasing step is used, the flow rate of the fluid before and after switch must be equal. For the step increase method the tracer concentration of the effluent fluid increases monotonically with time (see Fig. 5.6(B)), from zero to the same concentration at the inlet, $c(\infty)$. In the time interval of $t - dt$ to t, the amount of tracer exited from the system is $Qc(t)dt$. The residence time of this fraction of tracer must be smaller than or equal to t, while the amount of tracer injected in the corresponding time period equals $Qc(\infty)dt$. Consequently, from the definition of $F(t)$, we get,

$$
F(t) = \frac{Qc(t)dt}{Qc(\infty)dt} = \frac{c(t)}{c(\infty)}
\tag{5.17}
$$

Hence, the step-input response curve gives RTD function, while the pulse input response curve gives RTD density.

Decreasing step input uses fluid without a tracer to replace the fluid with tracer, its input function is

$$c_0(t) = c(0) = contsant, \quad t < 0$$
$$c_0(t) = 0 \quad t \geq 0 \tag{5.18}$$

Its geometric diagram is shown in Fig. 5.6(C) and the corresponding output response curve is shown in Fig. 5.6(D). The tracer concentration $c(t)$ monotonically decreases from $c(0)$ to zero. Since fluid without tracer is used to replace the fluid with a tracer, the residence time of the tracer that was detected between time t and $t + dt$ must be greater than or equal to t. Thus, the ratio of $c(t)/c(0)$ should be the fraction of material whose residence time is greater than t. Consequently,

$$1 - F(t) = c(t)/c(0) \tag{5.19}$$

From Eq. (5.19), the RTD function can be obtained from the decreasing step response curve.

In the above discussion, RTD was measured through the switch between the fluid with and without tracers. It can also be measured by continuously injecting tracer to the main fluid. As long as the flow of tracer is continuous and stable, and it is much smaller than the flow rate of the main fluid, the results should be identical to those produced using the switching fluids method, and they all belong to the step-input method.

For either pulse or step-input experiments, a tracer that usually does not react with the main fluid is inevitably needed. Besides this specific requirement, the selection of tracer should follow several rules: (1) Tracer should be easily soluble (mixable) with the main fluid; they should have as many as similar physical properties except the one that is distinguishable from main fluid and is detectable; (2) It should be detectable even at very low concentrations so that the tracer concentration can be reduced to avoid any impact on the main fluid; (3) The concentration of tracer should have broader linearity range with the physical property that is going to be measured so that the experimental data can be used directly without any correction or transformation; (4) Tracers used in multiphase systems should not transfer from one phase to another, e.g., the gas tracer should not be absorbed by liquid, while liquid tracer should not evaporate to the gas phase; (5) The tracer itself should have or should be easily changed to an electrical signal or optical signal so that modern instruments and computers can be used to collect data and do real-time analysis of the data to improve the efficiency of the experiment and the precision of data.

The advantage of pulse input is that $E(t)$ can be derived directly from experimental data, which then has more practical applications than $F(t)$. Additionally, the pulse input method is simple with a small amount of tracer. The greatest difficulty of pulse input is how to make sure that the injection period is shortest, which is even more difficult for systems whose mean residence times are very short. Step input is relatively easy, but it needs a large amount of tracer. The increasing step and decreasing step can be employed alternately, but it may be better to deal with the decreasing step data. The directly derived function from the step-input method is $F(t)$, which is different from the pulse method. In practical applications, no matter which method is employed, one must make sure that there is no back mixing between the injection point and the cross-section of the system entrance. That is to say the system must be a closed system to make it possible to obtain accurate RTD data.

5.3 STATISTICAL EIGENVALUES OF RTD

To compare different RTDs, their statistical eigenvalues are usually compared, just like other statistical distributions. There are two statistical characteristics that are commonly used: one is *mathematical expectation* and the other is *variance*.

Mathematical expectation is actually the mean. For RTD, it is the mean residence time $(\bar{t})$. The mean measured is of the first moment to the origin. Hence, from the definition of the first moment, mean residence time is

$$\bar{t} = \mu_1 = \frac{\displaystyle\int_0^\infty tE(t)dt}{\displaystyle\int_0^\infty E(t)dt} = \int_0^\infty tE(t)dt \tag{5.20}$$

Variance is measured using the second moment to the mean. From the definition of the moment, variance of RTD is

$$\sigma_t^2 = \mu_2' = \int_0^\infty (t-\bar{t})^2 E(t)dt = \int_0^\infty t^2 E(t)dt - \bar{t}^2 \tag{5.21}$$

Variance is an indication of dispersion from the mean; the greater the variance, the broader the distribution, i.e., the greater is the degree of uneven residence times for RTD. Therefore, comparing only the mean residence time will not provide enough information for the two RTDs; their variance must be compared when more exact conclusions are desired.

If dimensionless time is used, we obtain dimensionless average residence time $(\bar{\theta})$ and dimensionless variance δ_θ^2 by substituting its definition from Eq. (5.6) into Eq. (5.20) $\sim$ Eq. (5.21).

$$\bar{\theta} = \int_0^\infty \theta E(\theta)d\theta \tag{5.22}$$

$$\sigma_\theta^2 = \frac{\sigma_t^2}{\bar{t}^2} = \int_0^\infty \theta^2 E(\theta)d\theta - 1 \tag{5.23}$$

Example 5.2

The response curve of a flow system was measured by three methods: (1) pulse input; (2) upstep; (3) downstep, respectively. Derive the relationship of mean residence time $\bar{t}$, and variance δ_θ^2 with $c(t)$.

Solution

1. Pulse input.

   Substituting Eq. (5.20) into Eq. (5.21) gives the required relationships

$$\bar{t} = \frac{\int_0^\infty tc(t)dt}{\int_0^\infty c(t)dt}$$

$$\sigma_t^2 = \frac{\int_0^\infty t^2 c(t)dt}{\int_0^\infty c(t)dt} - (\bar{t})^2$$

   Hence, the mean residence time and variance can be calculated from the response curve.

2. Increasing step input.

   Substituting Eq. (5.5) into Eq. (5.20) gives

$$\bar{t} = \int_0^\infty tE(t)dt = \int_0^1 tdF(t) \tag{A}$$

   Assuming the injected tracer concentration is $c(\infty)$, from Eq. (5.17), we have $dF(t) = dc(t)/c(\infty)$. Substituting it into Eq. (A) yields

$$\bar{t} = \frac{1}{c(\infty)} \int_0^{C(\infty)} tdc(t) \tag{B}$$

   Fig. 5B is the tracer response curve of increasing step, where the slash-covered area represents the integral value on the right side of Eq. (B). From this figure we can see that the slash-covered area equals to the area of rectangle OABE minus the area of OAB, hence:

$$\int_0^{C(\infty)} tdc(t) = \int_0^T c(\infty)dt - \int_0^T c(t)dt \tag{C}$$

   where T is the time when the effluent tracer concentration equals $c(\infty)$.

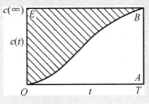

FIGURE 5B   Response curve for step method.

(Continued)

(Continued)

Substituting Eq. (C) into Eq. (B) gives,

$$\bar{t} = \int_0^T \left[1 - c(t)/c(\infty)\right] dt \tag{D}$$

Following the same way as we derived Eq. D, after substituting Eq. (5.6) into Eq. (5.21), substituting Eq. (5.17) yields

$$\sigma_t^2 = \frac{1}{c(\infty)} \int_0^{c(\infty)} t^2 dc(t) - (\bar{t})^2 \tag{E}$$

The above equation can be rewritten by integration

$$\sigma_t^2 = \frac{2}{c(\infty)} \left[\left[\int_0^T tc(\infty)dt - \int_0^T tc(t)dt - (\bar{t})^2 = 2 \int_0^T t\left[1 - \frac{c(t)}{c(\infty)}\right]\right]dt - (\bar{t})^2 \tag{F}$$

Consequently, variance δ_θ^2 can be calculated from the response curve based on Eq. (E) or Eq. (F).

3. Decreasing step input.

Following a similar procedure as used in the increasing step method, we derived

$$\bar{t} = \int_0^T \frac{c(t)}{c(0)} dt$$

$$\sigma_t^2 = 2 \int_0^T \frac{tc(t)}{c(0)} dt - (\bar{t})^2$$

where T is the time when effluent tracer concentration is zero

---

Example 5.3
RTD in a constant flow reactor was measured using the pulse input method. The relationship between effluent tracer concentration $c(t)$ and t is listed in the following table. Calculate the mean residence time and variance

t/min	0	2	4	6	8	10	12	14	16	18	20	22	24
$c(t)$/(g/min)	0	1	4	7	9	8	5	2	1.5	1	0.6	0.2	0

Solution
From the given data we obtain $E(t)$. Then, mean residence time and variance can be calculated using Eqs. (5.20) and (5.21), respectively. However, Example 5.2 showed that mean residence time and variance can be directly derived from the response curve $c(t) - t$, i.e., calculated based on the following two equations.

(Continued)

(Continued)

$$\bar{t} = \frac{\int_0^\infty tc(t)dt}{\int_0^\infty c(t)dt} \qquad (A)$$

$$\sigma_t^2 = \frac{\int_0^\infty t^2 c(t)dt}{\int_0^\infty c(t)dt} - (\bar{t})^2 \qquad (B)$$

To obtain $\bar{t}$ and δ_θ^2, the values of three definite integrals need to be determined, then the integrand at different times can be calculated and are listed in Table 5B

TABLE 5B $tc(t)$ and $t^2c(t)$ at Different Times

t/min	c(t)/ (g/m³)	tc(t)/ (min(g/m³))	t²c(t)/ (min²(g/m³))	t/min	c(t)/ (g/m³)	tc(t)/ (min(g/m³))	t²c(t)/ (min²(g/m³))
0	0	0	0	14	2	28	392
2	1	2	4	16	1.5	24	384
4	4	16	64	18	1	18	324
6	7	42	252	20	0.6	12	240
8	9	72	576	22	0.2	4.4	96.8
10	8	80	800	24	0	0	0
12	5	60	720				

From the data in Table 5B, the three integrals in Eq. (A) and Eq. (B) can be calculated using the graphical method and some other approximate formulas can be used too. The following formula is used here

$$\int_{x_0}^{x_n} f(x)dx = \frac{h}{3}\left[f_0 + f_n + 2\sum_{i=1}^{\frac{n}{2}-1} f_{2i} + 4\sum_{i=1}^{\frac{n}{2}} f_{2i-1} \right] \qquad (C)$$

Where $h = (x_n - x_0)/n$, n is even. Hence, Eq. (C) can only be used when data points are odd and are equidistance. The data in Table 5B completely satisfy these requirements.

$$h = (24 - 00)/12 = 2 \text{ min}$$

$$\int_0^\infty c(t)dt = \int_0^{24} c(t)dt$$

$$= \frac{2}{3}[0 + 4(1) + 2(4) + 4(7) + 2(9) + 4(8) + 2(5) + 4(2)$$

$$+ 2(1.5) + 4(1) + 2(0.6) + 4(0.2) + 0]$$

$$= 78 \text{ min·g/m}^3$$

(Continued)

(Continued)

$$\int_0^\infty tc(t)dt = \int_0^{24} tc(t)dt$$

$$= \frac{2}{3}[0 + 4(2) + 2(16) + 4(42) + 2(72) + 4(80) + 2(60)$$

$$+ 4(28) + 2(24) + 4(18) + 2(12) + 4(4.4) + 0]$$

$$= 710.4 \ min^2 \cdot g/m^3$$

$$\int_0^\infty t^2 c(t)dt = \int_0^{24} t^2 c(t)dt$$

$$= \frac{2}{3}[0 + 4(4) + 2(64) + 4(432) + 2(576) + 4(800) + 2(720) + 4(392)$$

$$+ 2(384) + 4(324) + 2(240) + 4(96.8) + 0]$$

$$= 8108.8 \ min^3 \cdot g/m^3$$

Substituting these integrals into Eq. (A) and Eq. (B) yields mean residence time and variance.

$$\bar{t} = \frac{710.4}{78} = 9.11 \ min$$

$$\sigma_t^2 = \frac{8108.8}{78} - (9.11)^2 = 20.97 \ min^2$$

5.4 RTD OF IDEAL REACTORS

From the viewpoint of RTD, plug flow and completely-mixed flow, the two assumptions made in describing tubular reactors and continuous tank reactors in the previous two chapters, are two extreme conditions, or two ideal conditions. Therefore, reactors whose flow patterns can be described with plug flow or completely-mixed flow are all called ideal reactors. In this section, these two ideal flow patterns will be further discussed; the mathematical description of their RTDs will be clarified too. Due to the fact that flow patterns of real reactors all are somewhere between these two extreme situations and ideal flow models are the foundations for building nonideal flow models, it is necessary to clarify these two ideal flow models.

5.4.1 Plug-Flow Model

The physical essence of the plug flow model has been clarified in Chapter 4, Tubular Reactor. If analyzing from the concept of RTD, plug flow means that all fluid particles on the cross-section perpendicular to flow have the same ages. Hence, there is no mixing between fluid particles with different ages, in other words, there is no mixing of fluid particles with different residence times. This mixing is on the macro scale, therefore, it is called macromixing. The extensity of mixing can be denoted by RTD. Although the particles at the same cross-section have the same ages, ages of particles from different cross-sections are different. Therefore, plug flow is normally thought to have no axial mixing or have zero back-mixing. Obviously, plug flow represents an extreme situation.

Overall, the characteristic of residence time of a plug flow is that fluid particles that entered the system at the same time will exit the system simultaneously, i.e., fluid particles at the outlet of system have the same life. Apparently, if tracer is injected to the inlet of a plug-flow reactor at $t = 0$ as a pulse with a form of δ function, tracer concentration in the effluent will exhibit as δ function too as illustrated in Fig. 5.7. Although the response curve to a pulse input represents tracer concentration distribution curve, $E(t)$ curve exhibits the same shape as the response curve as indicated by Eq. (5.13) or Eq. (5.15). Thus, instead of taking $c(t)$ as ordinate, Fig. 5.7 is presented as $E(t)$ curve directly. Then, RTD density function for plug flow model is

$$E(t) = \delta(t - \bar{t}) \tag{5.24}$$

Using dimensionless time

$$E(\theta) = \bar{t}E(t) = \delta(\theta - 1) \tag{5.25}$$

Based on the properties of δ function, dimensionless mean residence time $\bar{\theta}$ and variance δ_θ^2 are derived from Eqs. (5.22) and (5.23),

$$\bar{\theta} = \int_0^\infty \theta\delta(\theta - 1)d\theta = \theta|_1 = 1 \tag{5.26}$$

$$\sigma_\theta^2 = \int_0^\infty \theta^2\delta(\theta - 1)d\theta - 1 = \theta^2|_1 - 1 = 0 \tag{5.27}$$

From Eq. (5.27), one can see that dimensionless variance δ_θ^2 of RTD for plug-flow reactors is zero indicating that all fluid particles have the same residence times in the reactor. The smaller the variance, the more concentrated the distribution and the narrower the distribution curve. The feature that

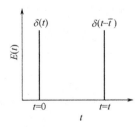

FIGURE 5.7 $E(t)$ plot of PFR.

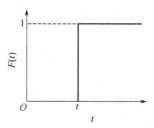

FIGURE 5.8 $F(t)$ plot of PFR.

variance of RTD equals zero indicates that there is no backmixing in the system. Hence, reaction efficiency of a plug flow reactor may be the same as that of a batch reactor when its residence time is the same as that of the batch reactor.

Fig. 5.8 is the $F(t)$ curve of RTD for plug-flow reactors. $F(t)$ is a step function, and its mathematical expression is

$$F(t) = \begin{cases} 0 & t < \bar{t} \\ 1 & t \geq \bar{t} \end{cases} \tag{5.28a}$$

Substituting with dimensionless time gives

$$F(\theta) = \begin{cases} 0 & \theta < 1 \\ 1 & \theta \geq 1 \end{cases} \tag{5.28b}$$

It is not hard to understand this figure. Because residence times of plug-flow reactors are uniform and they all equal to the mean residence time $\bar{t}$, it is not possible to have fluid particles with residence times less than $\bar{t}$. Consequently, $F(t)$ equals zero when $t < \bar{t}$ and $F(t)$ equals one when $t = \bar{t}$ since all fluid particles have the same residence time t. This result can be interpreted from the viewpoint of the measurement method. Since $F(t)$ is obtained by the step input of tracer, and there is no backmixing in the system, a step input must result in a step output.

5.4.2 Perfectly-Mixed Flow Model

When dealing with the design of continuous tank reactors previously, we assumed that concentration and temperature of reaction materials in the reactor are uniform, which is essentially the intuitive result of a perfectly-mixed flow model. This uniformity is due to vigorous agitation. This flow model will be analyzed from the viewpoint of RTD theory.

Under the force of agitation, some fluid particles that entered the reactor may escape through the exit immediately, which results in very short residence times; while some other particles may be agitated back from the exit when they just intend to leave, which makes their residence times extremely long. Hence, residence times of fluid particles in continuous tank reactors are very uneven: some are short and some are long, which causes different levels of backmixing, i.e., mixing among fluid particles with different residence times. The extent of unevenness is related to the force of agitation: the stronger the agitation, the more significant the backmixing. When back-mixing reaches the maximum level, concentrations of the reaction materials are uniform everywhere inside the reactor. In other words, mixing among fluid particles with different residence times reaches the maximum level, or perfectly mixed. This is the so-called perfectly-mixed flow model.

Assuming the tracer concentration in the fluid that continuously entered the perfectly-mixed flow reactor is c_0, then the amount of tracer entered into the reactor and exited from the reactor in a unit time are Qc_0 and Qc, respectively. Since tracer concentrations in the reactor are uniform and all equal to the tracer concentration in the effluent, the accumulation of tracer in the reactor within a unit of time is $V_t dc/dt$. Taking the mass balance of the tracer in the reactor gives:

$$V_r \frac{dc}{dt} = Qc_0 - Qc$$

Recalling $V_r/Q = \tau$, the above equation can be rewritten as

$$\frac{dc}{dt} + \frac{1}{\tau}c = \frac{1}{\tau}c_0 \tag{5.29}$$

This is the mathematical description of perfectly-mixed flow model. The initial condition is

$$t = 0 \quad c = 0$$

Integrating Eq. (5.29) yields

$$\int_0^c \frac{dc}{c_0 - c} = \frac{1}{\tau}\int_0^t dt$$

Rewritten as

$$\ln \frac{c_0 - c}{c_0} = -\frac{t}{\tau} \tag{5.30}$$

Or

$$1 - \frac{c}{c_0} = e^{-t/\tau}$$

From the definition of $F(t)$, the above equation becomes

$$F(t) = \frac{c(t)}{c(\infty)} = \frac{c}{c_0} = 1 - e^{-t/\tau} \tag{5.31}$$

Differentiating Eq. (5.31) with respect to t gives

$$E(t) = \frac{1}{\tau}e^{-t/\tau} \tag{5.32}$$

FIGURE 5.9 $E(t)$ plot of CSTR.

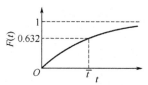

FIGURE 5.10 $F(t)$ plot of CSTR.

Eqs. (5.31) and (5.32) are the mathematical expressions of the RTD function and distribution density for perfectly-mixed flow reactors. Its dimensionless form is

$$F(\theta) = 1 - e^{-\theta} \tag{5.33}$$

$$E(\theta) = e^{-\theta} \tag{5.34}$$

Figs. 5.9 and 5.10 are the $E(t)$ and $F(t)$ curves for perfectly-mixed flow reactors, which are plotted based on Eqs. (5.32) and (5.31), respectively. RTD density $E(t)$ curve for a general flow system usually exhibits the form as illustrated by Fig. 5.3: a form like mountain peaks, in other words a form that is high in the middle and low at both ends. However, $E(t)$ curve for perfectly-mixed flow reactors monotonically decreases as time increases, and $E(t) \to 0$ when $t \to \infty$. This indicates that residence times of fluid particles in perfectly-mixed flow reactors are extremely uneven, ranging from zero to infinity. The maximum level of backmixing is one of the extreme situations of the macromixing. Fig. 5.10 is the $F(t)$ curve for perfectly-mixed flow reactors, which exhibits the same form as that of a general flow system and both increase with time. The difference is that the slope of the former curve increases as time decreases, while the latter holds a maximum slope.

Substituting Eq. (5.34) into Eqs. (5.22) and (5.23) respectively, the mean residence time and variance of RTD in perfectly-mixed flow reactors can be obtained as follows:

$$\bar{\theta} = \int_0^\infty \theta e^{-\theta} d\theta = -\int_0^\infty \theta de^{-\theta} = -\left[(\theta e^{-\theta}) \Big|_0^\infty - \int_0^\infty e^{-\theta} d\theta \right] = 1 \tag{5.35}$$

$$\sigma_\theta^2 = \int_0^\infty \theta^2 e^{-\theta} d\theta - 1 = 1 \qquad (5.36)$$

It shows that the dimensionless variance δ_θ^2 of residence distribution equals 1 when backmixing reaches the maximum level compared to the zero variance when there is no backmixing. Hence, variance usually falls between 0 and 1, and the bigger the variance, the more dispersive is the RTD.

Plug flow reactors and perfectly-mixed flow reactors are compared in the former chapter, and the conclusion is that plug-flow reactors are usually advantageous over perfectly-mixed flow reactors for normal reaction kinetics. Reactant concentration changes in these two reactors are also compared based on their design equations to verify this conclusion. Further details will be described here based on the difference between the RTDs. It is usually assumed that the reactions carried out in these two reactors are the same and their corresponding mean residence times are equal. For plug flow reactors, all fluid particles have the same residence time that is also equal to the mean residence time. However, perfectly-mixed flow reactors behave differently. From Eq. (5.31), one can see that the fraction of fluid particles with residence time shorter than t over the total fluid is $F(t) = 1 - e^{-1} = 0.632$. There is no doubt that the conversion of this fraction of fluid is lower than that in the plug-flow reactor. On the other hand conversion of the other 36.8% of fluid whose residence time is greater than the mean residence time is greater than that in a plug flow reactor, which, however, cannot compensate for the loss of conversion due to the short residence time. Consequently, conversion in a plug flow reactor is greater than that of a perfectly-mixed flow reactor. Therefore, concentrating RTD can improve the production intensity of the reactors. Of course, this is only an analysis from the point of residence time of the fluid; conversion also depends on the mixing among fluid molecules, which is the so-called micromixing. This will be discussed at the end of this chapter.

Example 5.4
The gradientless internal circulating reactor frequently used in measuring kinetics is actually a perfectly-mixed flow reactor, which needs to be tested before use to make sure perfectly-mixed flow is satisfied. Nitrogen now is selected as main fluid and hydrogen is used as tracer. The tracer is injected as a upstep with concentration c_0. Outlet hydrogen concentration c is monitored by conductivity analyzer. Results are listed in the following table.

z/cm	0	4	9	14	24	34	44
c/c_0	0	0.333	0.597	0.757	0.908	0.963	0.986

(*Continued*)

(Continued)

where z is the moving distance of the recording paper. Is the flow pattern in this reactor a completely-mixed flow?

Solution

Since input is a positive step, if it is perfectly-mixed flow, relationship between response time and output tracer concentration should follow Eq. (5.37)

$$-\ln\left[1 - \frac{c}{c_0}\right] = \frac{t}{\tau}$$ (5.37)

Since outlet hydrogen concentration was monitored continuously, and the recording paper moved forward at a fixed speed, the corresponding time then can be determined from the distance z that the recording paper moved and its moving velocity u,

$$t = z/u$$ (A)

Substituting into Eq. (5.37) yields

$$-\ln\left[1 - \frac{c}{c_0}\right] = \frac{z}{ut}$$ (B)

When flow rate is fixed, τ should be a constant as well as u. Therefore, if it is perfectly-mixed flow, $\ln[1 - c/c_0]$ should exhibit linear relationship with z. Plot with the given data, as shown in Fig. 5C. The plot shows good linearity indicating that perfectly-mixed flow was satisfied. It is worthy of note that this test was done at specific operating conditions and flow pattern may change when conditions change.

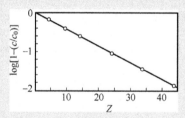

FIGURE 5C $Z \sim \log(1 - c/c_0)$.

5.5 NONIDEAL FLOW PHENOMENON

Two ideal flow situations, plug flow and perfectly-mixed flow, were discussed above. However, flow patterns for real reactors are usually somewhere between these two ideal conditions: some are close to them, and some deviate significantly from them. All flows that do not satisfy ideal flow conditions are called nonideal flow. The reasons that real reactors' flow deviate from ideal flow can be categorized into the following situations.

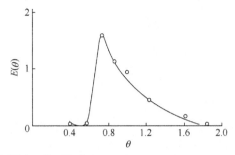

FIGURE 5.11   Real $E(t)$ in a fixed bed reactor.

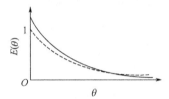

FIGURE 5.12 $E(\theta)$ of tank reactor with dead zone.

1. Existence of the stagnant zone.

The so-called stagnant zone refers to the section in a reactor where fluid flows very slowly or does not flow at all. Hence, the stagnant zone is also called the dead zone. The presence of a stagnant zone makes some fluid have extremely long residence time and their corresponding RTD density function $E(t)$ have long tails. Fig. 5.11 is the measured RTD curve for a fixed-bed reactor. In general, flow patterns in fixed-bed reactors approximate that of plug-flow. However, this measured result has long tails—some fluid's residence time is 1.8 times of the mean residence time—which is due to the presence of the stagnant zone.

The flow pattern of continuous-tank reactors is close to perfectly-mixed flow. However, when a stagnant zone is present, its RTD exhibits the form as illustrated in Fig. 5.12, where the dashed line represents the $E(\theta)$ curve for a perfectly-mixed reactor, and the solid line is the $E(\theta)$ curve for a continuous-tank reactor with stagnant zone, respectively. Since the RTD density for perfectly-mixed reactor $E(\theta) = e^{-\theta}$, one can see that $E(\theta) = 1$ when $\theta = 0$. However, $E(\theta) > 1$ when a stagnant zone is present.

Stagnant zones mainly arise at the dead regions in the equipment, e.g., the two ends of the equipment, the connection between the baffle and the equipment wall. Additionally, stagnant zones are easily formed when the equipment has other obstacles. Reducing the stagnant zone mainly depends on the design. For the existing equipments, one can check if a stagnant zone is present by measuring RTD.

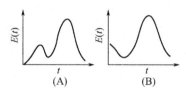

FIGURE 5.13 Bypass and shortcut. (A) Bypass; (B) Shortcut.

FIGURE 5.14 Cyclic flow.

2. Existence of channeling and short circuit.

   In fixed-bed reactors, packed towers and trickle-bed reactors, uneven loading of the catalyst particles or packing results in channels with low resistance, which caused part of the fluid flowing through the channel very quickly and formed channeling. Fig. 5.13(A) shows the RTD curve when channeling is present in the flow system. The feature of this figure is that $E(t)$ curve has double peaks. If the channeling is not significant, then the first peak may not be apparent. Bad equipment design may cause short circuit of flow, i.e., very short residence time for the fluid in the equipment. For example, when the outlet and inlet of the equipment are too close, short circuit will occur. Fig. 5.13(B) is the RTD curve in the presence of a short circuit. If channeling or a short circuit exist in the system, the mean residence time calculated from the measured RTD will be smaller than V_r/Q, When stagnant zone is present instead of channeling and short circuit, the situation is just opposite, i.e., $\bar{t} > V_r/Q$

3. Cyclic flow

   Cyclic movement of fluid always occurs in real tank reactors, such as bubble tower and fluidized-bed reactors. In recent years, people intentionally install draft tubes in the reactor to strengthen or control cyclic flow with some driving mode like stripping or jet. Fig. 5.14 is the RTD curve when cyclic flow is present, and multipeaks is one of its features.

4. Uneven flow distribution

   Due to the uneven flow distribution in the radial direction of the reactor, residence times of fluid in the reactor can be either shorter or longer, which was discussed in Chapter 4, Tubular Reactor. The plug flow model assumes that flow is uniform in a radial direction. If flow in the reactor is

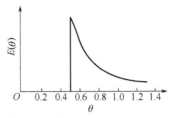

FIGURE 5.15   Residence time distribution of laminar flow reactor.

laminar flow then deviation from plug flow is significant and the flow in radial direction displays a parabolic form. If, disregarding molecular diffusion, and assuming no stagnant zone and no channeling and short circuit in the system as well, then RTD density function for laminar flow reactors can be derived from radial flow distribution,

$$E(\theta) = 0 \quad \theta < 0.5 \tag{5.38}$$

$$E(\theta) = \frac{1}{2\theta^3} \quad \theta \geq 0.5$$

Fig. 5.15 is the $E(\theta)$ curve for laminar flow reactors that is derived from Eq. (5.38). The characteristic of this distribution is that fluid particles whose residence times are shorter than half of the mean residence time are zero. When fluid is in turbulent flow in the reactor, radial flow distribution is pretty flat, which makes the RTD closer to that of a plug flow; however, it cannot satisfy the assumptions for plug flow.

5. Diffusion

Due to molecular diffusion and eddy diffusion, mixing among fluid particles occurs and causes RTD to deviate from that for ideal situations. Deviation from plug flow is more significant.

It is possible that the reasons that cause nonideal flow as discussed above all exist in a flow system or only some of them are present. Taking the measured RTD for fixed-bed reactors as illustrated in Fig. 5.11 as an example, the long tail of the $E(\theta)$ curve indicates that stagnant zone may exist. However, the shape of curve is very close to that of the $E(\theta)$ curve for laminar-flow reactor as shown in Fig. 5.15. Therefore, this result could be due to the laminar flow caused by extremely low flow rate. On the other side, the mean residence time $\bar{\theta}$ should be equal to one, however, the value calculated from the measured curve is only 0.7, which thus cannot exclude the possibility that channeling and short circuit may exist. Therefore, RTD is affected by many factors, and it is difficult to identify the reasons that RTD deviated from that for ideal flow.

5.6 NONIDEAL FLOW MODELS

From the above discussions, one can see that not all continuous tank reactors have the features of perfectly-mixed flow, and not all tubular reactors satisfy the assumptions for plug flow. In order to measure or calculate the conversion and yield in a nonideal reactor, appropriate flow models need to be developed based on its flow pattern. The basis for developing flow models is the RTD in the reactor, and the most commonly used technique is to make corrections on ideal flow models, or combine the ideal flow model with stagnant zone, channeling, and/or short circuit etc. The mathematical model developed should be easy for mathematical treatment, and model parameters should not exceed two. Additionally, the model should correctly reflect the actual physical phenomena. Three nonideal flow models will be introduced in the following section.

5.6.1 Segregation Model

If there is no material interchange between the fluid particles in the reactor, or in other words there is no micromixing between the fluid particles, fluid particles will move from the inlet to outlet of the reactor like individuals with boundaries. This kind of flow is called segregation flow. Since each fluid particle has no relation with its surroundings, it reacts like a batch reactor and the extent of reaction depends on the residence time of this particle in the reactor. Assuming the concentration of reactant A in the fluid at the inlet of reactor is c_{A0}, and concentration is $c_A(t)$ at reaction time t, from the RTD of the reactor, one can see that the fraction of fluid particles with residence times between t and $t + dt$ is $E(t)$ dt. Consequently, contribution from this part of fluids on c_A, the concentration of A in effluent of the reactor, should be $c_A(t) E(t) dt$. Summing up all these contributions yields, the average concentration A at the exit of reactor $\overline{c_A}$,

$$\overline{c}_A = \int_0^\infty c_A(t)E(t)dt \tag{5.39}$$

The reason that we use the average concentration is that c_A is different for fluid particles with different residence times, while the concentration of A at the outlet of the reactor is actually an average result. $c_A(t)$ can be obtained by integrating the reaction rate equation. Thus, as long as the RTD for the reactor and the reaction rate equation are known, conversion of the reaction can be predicted. Of course, the premise is that the assumption for segregation flow is satisfied. Eq. (5.39) is the model equation for segregation flow. Since it directly uses RTD to design the reactor, some refer to the segregation model as the RTD model.

From the definition of conversion, Eq. (5.39) can be rewritten as

$$1 - \overline{X}_A = \int_0^\infty [1 - X_A(t)]E(t)dt$$

$$= \int_0^\infty E(t)dt - \int_0^\infty X_A(t)E(t)dt$$

$$= 1 - \int_0^\infty X_A(t)E(t)dt$$

So

$$\overline{X}_A = \int_0^\infty X_A(t)E(t)dt \tag{5.40}$$

Since the segregation model introduces the RTD density function directly to the mathematical model equation, it does not have model parameters. However, if it is set up by mathematical simulation on the known RTD, it then has model parameters.

When using Eq. (5.39) to calculate irreversible reactions, please pay attention to the upper limit of the integral. The integral upper limit in this equation is the complete reaction time t^*, that is the time to reach $c_A = 0$. For example, the reaction rate equation for half-order reactions is

$$-\frac{dc_A}{dt} = kc_A^{0.5}$$

Integrating it yields

$$\sqrt{c_A} - \sqrt{c_{A0}} = -kt/2$$

At complete conversion, $c_A = 0$, thus, the complete reaction time is

$$t^* = 2\sqrt{c_{A0}}/k$$

When residence times of fluid particles are greater than t^*, c_A is still zero, so it does not make any contribution. Integration therefore should stop at complete reaction time t^*. Of course for some reactions, such as first order reactions, c_A equals zero only when residence time is infinity. Consequently, the integral upper limit would be infinity. Finally, it needs to be pointed out that the above discussions are all for isothermal situations.

Example 5.5
Liquid phase reaction $2A \rightarrow R + P$ is carried out in a flow reactor isothermally. The volume of reactor is $4.55 \, m^3$. The reaction is second order; the reaction rate constant at the reacting temperature is $2.4 \times 10^{-3} \, m^3/(mol \cdot min)$; the feed rate is $0.5 \, m^3/min$, and concentration of A is $1.6 \, kmol/m^3$. The RTD in this reactor is the same as that in Example 5.3. Calculate the conversion of A at the outlet of reactor by (1) segregation model, (2) plug flow model.

Solution
(1) Using segregation model.
Eq. (5.39) can be used to calculate the concentration of A at the outlet of reactor when segregation model is used. Thus, the relationship between c_A and t needs to be determined first, which can be obtained by integrating the second order reaction rate equation.

$$-\frac{dc_A}{dt} = kc_A^2$$

Integrating yields

$$-\int_{c_{A0}}^{c_A} \frac{dc_A}{c_A^2} = \int_0^t k \, dt$$

Or

$$c_A = \frac{c_{A0}}{1 + kc_{A0}t} \tag{A}$$

From Eq. (A), one can see that c_A equals zero when $t \rightarrow \infty$. Consequently, the reaction can only be completed at infinite long time, thus, the integral upper limit therefore should be ∞. $E(t)$ is also needed when using Eq. (5.39), which can be determined from the measured data in Example 5.3 based on the following equation.

$$E(t) = \frac{c(t)}{\int_0^\infty c(t) \, dt} \tag{B}$$

The integral in the denominator of Eq. (B) was obtained from Example 5.3, and it is $78 \, min/m^3$. Results for $E(t)$ are listed in Table 5C, where the data in the first and second columns are the results of tracer response measurement given by Example 5.3, and $c(t)$ is the concentration of tracer. The data in the fourth column of Table 5C is calculated by using Eq. (A). Substitute Eq. (A) and Eq. (B) into Eq. (5.39), we obtain:

$$\bar{c}_A = \int_0^\infty c_A E(t) \, dt = \int_0^\infty \frac{c_{A0}}{1 + kc_{A0}t} E(t) \, dt$$

Using the data in the fifth column of Table 5C, the integral can be calculated using the Simpson method, and the value is $0.05447 \, kmol/m^3$, which is the concentration of A at the exit of the reactor. So, conversion is:

$$\bar{X}_A = \frac{1.6 - 0.05447}{1.6} = 0.966$$

(Continued)

(Continued)

TABLE 5C $E(t)$ and c_A at Different Time

t/min	$c(t)/(g/m^3)$	$E(t) \times 10^3/min^{-1}$	$c_A \times 10^2/$ (kmol/m^3)	$c_A E(t) \times 10^5/$ [kmol/(m$^3 \cdot$ min)]
0	0	0	160	0
2	1	12.82	18.43	236.3
4	4	51.28	9.78	501.5
6	7	89.74	6.656	597.3
8	9	115.4	5.044	582.1
10	8	102.6	4.061	416.7
12	5	64.1	3.398	217.8
14	2	25.64	2.922	74.92
16	1.5	19.23	2.562	49.27
18	1	12.82	2.281	29.24
20	0.6	7.692	2.057	15.82
22	0.2	2.564	1.872	4.8
24	0	0	0	0

Of course, the exit conversion of reactor can be calculated directly from Eq. (5.40), but c_A in Eq. (A) needs to be changed as a function of conversion.
(2) Using plug-flow model
Space time can be obtained from the given data

$$\tau = \frac{V_r}{Q_0} = \frac{4.55}{0.5} = 9.11 \text{ min}$$

For plug flow reactors, the performance and reaction time are the same as that of a batch reactor with equivalent space time. Consequently, substituting τ into Eq. (A) gives the concentration of A at the outlet of the reactor.

$$\overline{c}_A = c_A = \frac{c_{A0}}{1 + kc_{A0}\tau} = \frac{1.6}{1 + 2.4 \times 1.6 \times 9.11}$$

$$= 0.04447 \text{ kmol/m}^3$$

So, exit conversion is

$$\overline{X}_A = \frac{1.6 - 0.04447}{1.6} = 0.9722$$

(Continued)

(Continued)

The concentration of A at reactor outlet can also be obtained by substituting RTD density for plug flow reactor $E(t) = \delta(t-9.11)$ and Eq. (A) into Eq. (5.39). The result is exactly the same as the above calculation.

Thus, for this problem, results calculated using the segregation model and plug flow model are very close. The reason is that RTD of this reactor does not deviate significantly from that of plug flow, otherwise, the difference between these two results will be much bigger.

5.6.2   Tanks-in-Series Model

Single CSTR, plug flow reactors and multiple CSTR in series were compared in Section 3.4. It was found that the properties of multiple CSTR in series are somewhere between those of the first two reactors. In addition, the more tanks, the more closely its behavior approaches that of a plug flow reactor. When the tank number approaches infinity, its behavior is identical to that for plug flow. Therefore, we can use N CSTR in series to model a real reactor. N is a model parameter: when $N = 1$, it is a CSTR, and when $N = \infty$, it becomes plug flow. The value of N reflects the different degree of mixing, which needs to be determined by RTD.

Therefore, RTD for tanks in series must be determined first. Assuming N CSTR with volume V_r are operated in series, there is no backmixing between each tank, and the time that fluid flow through the pipeline connecting each tank can be ignored, Fig. 5.16 shows the schematic diagram of tanks in series, where Q is the flow rate, and c is the concentration of tracer. Assuming the temperature of each tank is identical, mass balance of tracer on the pth tank is,

$$Qc_{p-1}(t) - Qc_p(t) = V_r \frac{dc_p(t)}{dt} \tag{5.41}$$

Or

$$\frac{dc_p(t)}{dt} = \frac{1}{\tau}\left[c_{p-1}(t) - c_p(t)\right] \tag{5.42}$$

Where τ is the mean residence time of fluid in one tank, it equals V_r/Q.

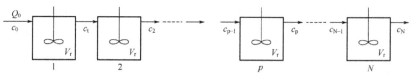

FIGURE 5.16   Tanks in series.

If tracer is injected as a step increase with concentration c_0, the initial condition of Eq. (5.42) is

$$t = 0, \quad c_p(0) = 0, \quad p = 1, 2, \ldots \ldots, N$$

When $p = 1$, Eq. (5.42) can be written as

$$\frac{dc_1(t)}{dt} = \frac{1}{\tau}[c_0(t) - c_1(t)]$$

This is actually the material balance equation for the first tank, which is Eq. (5.29), the solution has been derived as

$$c_1(t) = c_0(1 - e^{-t/\tau}) \tag{5.43}$$

For the second tank, from Eq. (5.42), we get

$$\frac{dc_2(t)}{dt} = \frac{1}{\tau}\left[c_1(t) - c_2(t)\right]$$

Substituting Eq. (5.43) gives

$$\frac{dc_2(t)}{dt} + \frac{c_2(t)}{\tau} = \frac{c_0}{\tau}(1 - e^{-t/\tau}) \tag{5.44}$$

Solving this first order linear differential equation yields

$$\frac{c_2(t)}{c_0} = 1 - \left(1 + \frac{t}{\tau}\right)e^{-t/\tau} \tag{5.45}$$

The same procedure can be used to solve the other tanks in series, and the results for the Nth tank can be generalized by the mathematical induction method, which is

$$F(t) = \frac{c_N(t)}{c_0} = 1 - e^{-t/\tau} \sum_{p=1}^{N} \frac{(t/\tau)^{p-1}}{(p-1)!} \tag{5.46}$$

This is the RTD function for a tanks-in-series system. Substituting the total mean residence time of the system $\tau_t = N\tau$ into Eq. (5.46) yields

$$F(t) = 1 - e^{-Nt/\tau_t} \sum_{p=1}^{N} \frac{(Nt/\tau_t)^{p-1}}{(p-1)!} \tag{5.47}$$

It can be written as the dimensionless form

$$F(\theta) = 1 - e^{-N\theta} \sum_{p=1}^{N} \frac{(N\theta)^{p-1}}{(p-1)!} \tag{5.48}$$

Notice that here $\theta = t/\tau_t$, which is defined based on the total mean residence time of the system τ_t, not the mean residence time τ for each tank.

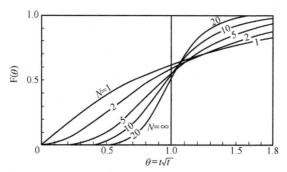

FIGURE 5.17 $F(\theta)$ plot of tanks in series.

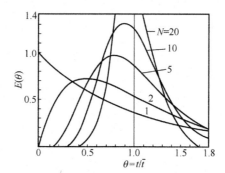

FIGURE 5.18 $E(\theta)$ plot of tanks in series.

RTD function for different numbers of tanks operating in series was calculated from Eq. (5.48) as illustrated by Fig. 5.17, which demonstrates that as the number of the tanks increases, its behavior approaches that of a plug flow reactor.

Differentiating Eq. (5.48) with respective to θ gives the RTD density for the tanks-in-series model

$$E(\theta) = \frac{dF(\theta)}{d\theta} = \frac{N^N}{(N-1)!}\theta^{N-1}e^{-N\theta} \tag{5.49}$$

Fig. 5.18 is the result from Eq. (5.49) at different N values. The figure indicates that different N values can model different RTDs, and when N increases, RTD becomes narrower. Substituting Eq. (5.49) into Eq. (5.22) yields the mean residence time for the tanks-in-series model.

$$\overline{\theta} = \int_0^\infty \frac{N^N\theta^N e^{-N\theta}}{(N-1)!}d\theta = 1 \tag{5.50}$$

Substituting Eq. (5.49) into Eq. (5.23), we get variance

$$
\sigma_\theta^2 = \int_0^\infty \frac{N^N \theta^{N+1} e^{-N\theta}}{(N-1)!} d\theta - 1
$$
$$
= \frac{N+1}{N} - 1 = \frac{1}{N}
$$

(5.51)

Clearly, from Eq. (5.51), we can see that when $N = 1$, $\delta_\theta^2 = 1$, which is consistent with CSTR model; while when $N \to \infty$, $\delta_\theta^2 = 0$, which is identical to plug-flow model. Hence, as long as N is positive, its variance is somewhere between 0 and 1. Consequently, different N values can be selected to model different RTDs.

When using the tanks-in-series model to model the flow in a real reactor, we need to measure the RTD in the reactor first, and then calculate variance and substitute it into Eq. (5.51) to obtain model parameter N. That is to say that the RTD of this reactor is equivalent to that of a series of N CSTR with equal volumes: their mean residence times are equal as well as their variances. But we cannot say that the two distributions are identical. When using the above method to estimate model parameter N, it is possible that N is a noninteger. Rounding to the nearest integer is a rough approximation method for this problem, and the more accurate method is to treat the fractional part as a tank with small volume.

Example 5.6
Benzoic acid is used as a tracer and injected as a pulse to a liquid phase reactor with a volume of 1735 cm³ to measure the RTD. The volumetric flow rate of liquid is 40.2 cm³/min, and the tracer amount is 4.95 g. Tracer concentration in the effluent liquid as a function of time $c(t)$ is listed in the following table. If we use the tanks-in-series model to model this reactor, determine the model parameter N.

t/min	$c(t) \times 10^3$/ (g/cm³)	t/min	$c(t) \times 10^3$/ (g/cm³)	t/min	$c(t) \times 10^3$/ (g/cm³)	t/min	$c(t) \times 10^3$/ (g/cm³)
10	0	40	3.520	65	0.910	90	0.131
15	0.113	45	2.840	70	0.619	95	0.094
20	0.863	50	2.270	75	0.413	100	0.075
25	2.210	55	1.735	80	0.300	105	0.001
30	3.340	60	1.276	85	0.207	110	0
35	3.720						

Solution
Substituting Eq. (5.13) into Eqs. (5.20) and (5.21) gives the mean residence time and variance.

$$
\bar{t} = \int_0^\infty t E(t) dt = \frac{Q}{m} \int_0^\infty t c(t) dt
$$

(A)

$$
\sigma_t^2 = \int_0^\infty t^2 E(t) dt - \bar{t}^2 = \frac{Q}{m} \int_0^\infty t^2 c(t) dt - \bar{t}^2
$$

(B)

(Continued)

(Continued)

TABLE 5D Relation of t versus $tc(t)$ and $t^2c(t)$

t/min	$tc(t)$ $\times 10^3$	$t^2c(t)$ $\times 10^3$	t/min	$tc(t)$ $\times 10^3$	$t^2c(t)$ $\times 10^3$	t/min	$tc(t) \times 10^3$	$t^2c(t)$ $\times 10^3$
10	0	0	45	127.8	5751	80	24.00	1920
15	1.695	25.43	50	113.5	5675	85	17.60	1496
20	17.26	345.2	55	95.43	5248	90	11.79	1061
25	55.25	1381	60	76.56	4594	95	8.93	848.4
30	100.2	3006	65	59.15	3845	100	7.50	750
35	130.2	4557	70	43.33	3033	105	0.11	11.03
40	140.8	5632	75	30.98	2323	110	0	0

In order to obtain the integral in Eqs. (A) and (B), $tc(t)$ and $t^2c(t)$ are calculated from the given relationship between $c(t)$ and t, and results are listed in Table 5D. Using the results from Table 5D and the Simpson method, we have

$$\int_0^\infty tc(t)dt = \frac{5}{3}\Big[0 + 4(1.695 + 55.26 + 130.2 + \cdots + 0.11)$$

$$+ 2(17.26 + 100.2 + 140.8 + \cdots + 7.5) + 0\Big] \times 10^{-3} = 5.297$$

Substituting the relevant data into Eq. (A) yields

$$\bar{t} = \frac{40.2}{4.95} \times 5.297 = 43.02 \text{ min}$$

Which is particularly close to the value calculated from $t = V/Q$. Similarly, variance is obtained as

$$\sigma_t^2 = \frac{40.2}{4.95} \times \frac{5}{3}\Big[0 + 4(25.43 + 1381 + 4557 + \cdots + 750)$$

$$+ 2(345.2 + 3006 + 5632 + \cdots + 11.03) + 0\Big] \times 10^{-3} - 43.02^2$$

$$= 233.4 \text{ min}^2$$

$$\sigma_\theta^2 = \frac{\sigma_t^2}{\bar{t}^2} = 233.4/43.02^2 = 0.1261$$

Model parameter is calculated from Eq. (5.51)

$$N = \frac{1}{0.1261} = 7.93 \approx 8$$

Thus, the RTD in this reactor can be approximated with eight equal-sized CSTR in series.

5.6.3   Axial Dispersion Model

When real flow deviates from ideal flow due to molecular diffusion, eddy diffusion, and nonuniform distribution of flow, it can be described by the axial dispersion model. The dispersion model is particularly suitable for tubular reactors. The model assumes: (1) fluid moves through the system at a constant flow rate u; (2) concentration distribution is uniform in the radial direction on the cross-section perpendicular to the flow, i.e., complete mixing along radial direction; and (3) the diffusion caused by different transfer mechanisms, like turbulent mixing, molecular diffusion and velocity distribution, only occurs in the axial direction. The combination of these effects is denoted by the axial dispersion coefficient D_a and can be described by Fick's law.

Since backmixing is opposite to the flow direction, then

$$J = -D_a \frac{\partial c}{\partial Z} \tag{5.52}$$

In addition, it is assumed that the axial dispersion coefficient in the same reactor does not change with time and location; it, however, depends on the structure of the reactor, operating conditions, and properties of fluid.

Based on the above assumptions, mathematical model equations for axial dispersion model can be developed. Because this kind of system is a distribution parameter system, the infinitesimal volume dV_r is used as control volume. Reactors are usually cylindrical, if the cross-sectional area is A_r, then $dV_r = A_r dZ$ (see Fig. 5.19). Mass balance of tracer on this infinitesimal volume yields the model equation. Input includes two items: one is from convection and the other is through diffusion. Hence, input is

$$uA_r c - D_a A_r \left(\frac{\partial c}{\partial Z}\right)_Z$$

Where the first part represents contribution by convection and the second part represents contribution from diffusion. Similarly, output includes two parts too, i.e.,

$$uA_r \left[c + \left(\frac{\partial c}{\partial Z}\right)_Z dZ\right] - D_a A_r \left[\left(\frac{\partial c}{\partial Z}\right)_Z + \frac{\partial}{\partial Z}\left(\frac{\partial c}{\partial Z}\right)_Z dZ\right]$$

The accumulation form is

$$\frac{\partial c}{\partial t} A_r dZ$$

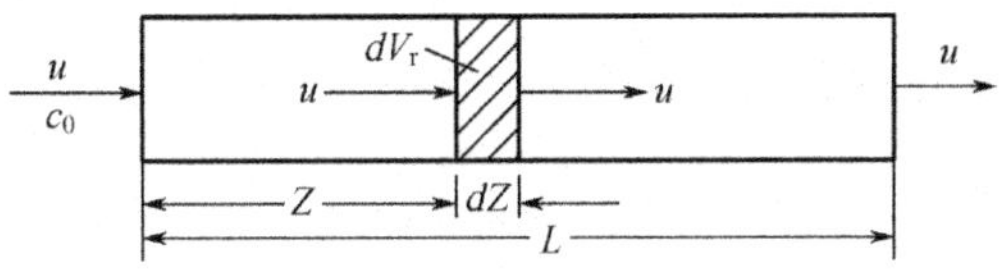

FIGURE 5.19   Axial dispersion model.

If no reaction occurs in the reactor, substituting these parts into the relationship input = output + accumulation and rearranging gives

$$\frac{\partial c}{\partial t} = D_a \frac{\partial^2 c}{\partial Z^2} - u \frac{\partial c}{\partial Z} \qquad (5.53)$$

This is the axial dispersion model equation. There are two independent variables, one is time t and another is space, i.e., the axial distance X. Therefore, the model equation is a partial differential equation. Eq. (5.53) indicates that the axial dispersion model is actually a plug-flow model superimposed with a diffusion term, that is the first term on the right-hand of Eq. (5.53). This term reflects the degree of mixing in the system. If $D_a = 0$, Eq. (5.53) becomes a plug flow model

$$\frac{\partial c}{\partial t} = -u \frac{\partial c}{\partial Z} \qquad (5.54)$$

With different D_a values, the axial dispersion model can describe any nonideal flow that is anywhere between plug flow and perfectly-mixed flow. However, practical experience has shown that it is applicable only when the backmixing is not too serious.

Due to its convenience, the dimensionless form of Eq. (5.53) is usually used. Hence, several dimensionless variables are introduced here,

$$\theta = \frac{tu}{L_r}, \quad \psi = \frac{c}{c_0}, \quad \zeta = \frac{Z}{L_r}, \quad P_e = \frac{uL_r}{D_a}$$

Substituting these variables into Eq. (5.53) yields the dimensionless equation for axial dispersion model

$$\frac{\partial \psi}{\partial \theta} = \frac{1}{P_e} \frac{\partial^2 \psi}{\partial \zeta^2} - \frac{\partial \psi}{\partial \zeta} \qquad (5.55)$$

Where P_e is the Peclet number, its physical meaning can be inferred from its definition,

$$P_e = \frac{uL_r}{D_a} = \frac{\text{transport rate by convection}}{\text{transport rate by diffusion}}$$

P_e represents the relative significance of convection to diffusion and reflects the extent of backmixing. Please be aware that there are different definitions for Peclet number in literatures, the difference is caused by using different characteristic length. For example, in a fixed-bed reactor, the characteristic length is usually taken as the diameter of the filling solid particles d_p, then $P_e = ud_p/D_a$. The reactor diameter d_t can also be used as a characteristic length to define the Peclet number, that is $P_e = ud_t/D_a$. Hence, we need to be cautious when using the Peclet number. However, no matter what the definition is, their meanings are identical and they all reflect the degree of backmixng. The reciprocal of the Peclet number, D_a/uL, sometimes is called dispersity.

When $P_e \to 0$, the transport rate by convection is much slower compared to the rate by diffusion, this is the case for perfectly-mixed flow. Instead, when $P_e \to \infty$, or $D_a = 0$, the case changes to plug flow when transport rate by diffusion can be neglected compared to the convection rate. Consequently, the bigger the Peclet number, the smaller the extent of the backmixing. Peclet number P_e is the model parameter in the axial dispersion model. Like the tanks-in-series model, the axial dispersion model belongs to the one-parameter model too.

The initial and boundary conditions in Eq. (5.55) vary with the input method of the tracer. For a closed system and step input (decreasing step), the initial and boundary conditions are

$$\psi(0, \zeta) = 1, \quad 0 < \zeta < 1 \tag{5.56}$$

$$0 = \psi(\theta, 0^+) - \frac{1}{P_e}\left(\frac{\partial \psi}{\partial \zeta}\right)_{0^+} \tag{5.57}$$

$$\left(\frac{\partial \psi}{\partial \zeta}\right)_{1^-} = 0 \tag{5.58}$$

Eqs. (5.55)–(5.58) can be solved using the separation of variables method, that is substituting $\psi(\theta, \zeta) = f(\theta)g(\zeta)\exp\left(\frac{\zeta P_e}{2}\right)$ into Eq. (5.55), rearranging it into a couple of ordinary differential equations (ODEs), and then from the definition of RTD function, we get

$$F(\theta) = 1 - e^{P_e/2} \sum_{n=1}^{\infty} \frac{8 w_n \sin w_n \exp\left[-\left(P_e^2 + 4 w_n\right)\theta/(4P_e)\right]}{P_e^2 + 4P_e + 4 w_n^2} \tag{5.59}$$

Where w_n is the positive roots of the following equation

$$\tan w_n = \frac{4 w_n P_e}{4 w_n^2 - P_e^2} \tag{5.60}$$

Differentiating Eq. (5.59) with respect to θ gives RTD density

$$E(\theta) = \frac{dF(\theta)}{d\theta} = e^{P_e/2} \sum_{n=1}^{\infty} \frac{(-1)^{n+1} 8 w_n^2 \exp\left[-\left(P_e^2 + 4 w_n\right)\theta/4P_e\right]}{P_e^2 + 4P_e + 4 w_n^2} \tag{5.61}$$

Plotting $F(\theta)$ and $E(\theta)$ versus θ based on Eqs. (5.59) and (5.61), $F(\theta)$ and $E(\theta)$ curves are obtained as illustrated in Figs. 5.20 and 5.21, respectively. It can be found from the figure that RTD becomes narrower as the reciprocal of the Peclet number decreases. The mean residence time and variance are

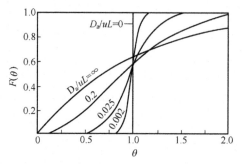

FIGURE 5.20 $F(\theta)$ of axial dispersion model.

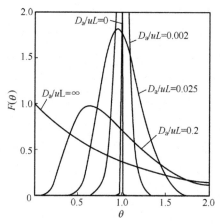

FIGURE 5.21 $E(\theta)$ of axial dispersion model.

$$\bar{\theta} = 1$$

$$\sigma_\theta^2 = \frac{2}{P_e} - \frac{2}{P_e^2}(1 - e^{-P_e}) \tag{5.62}$$

If the RTD of a real system is known, the variance of the distribution can be obtained and substituted into Eq. (5.62), then the model parameter Peclet number P_e can be calculated by using the trial and error method. Curve fitting is another option, then the least−square method, which minimizes the summation of the squares of the difference between $E(t)$ calculated from Eq. (5.61) and the measured $E(t)$, is used to determine the model parameter Peclet number P_e, i.e.,

$$\int_0^\infty \left[\hat{E}(t) - E(t)\right]^2 dt = \min$$

The Peclet number P_e determined this way is more accurate, but of course more calculation work is involved too.

When designing a reactor, the RTD is usually unknown, the Peclet number P_e then can be estimated from some correlations. For example, for empty tubular reactors, we have

$$\frac{1}{P_e} = \frac{1}{S_c R_e} + \frac{R_e S_c}{192} \tag{5.63}$$

The Peclet number P_e in the above equation is defined based on the tube diameter. Schmidt number $S_c = \mu/\rho\mathscr{D}$. Eq. (5.63) is applicable when $1 < R_e < 2000$, and $0.23 < S_c < 1000$. For turbulent flow, the following equation should be used

$$P_e = R_e^{0.125} \tag{5.64}$$

Finally, it needs to be pointed out that the solution for the axial dispersion model varies with the initial and boundary conditions. However, when backmixing is not significant, the results are not significantly different.

Example 5.7
Model the RTD discussed in Example 5.6 with the axial-dispersion model and calculate the model parameter.

Solution
The variance of RTD in Example 5.6 was calculated and equals 0.1261. Substituting it into Eq. (5.62) gives

$$0.1261 = \frac{2}{P_e} - \frac{2}{P_e^2}\left(1 - e^{-P_e}\right) \tag{A}$$

Solving the equation by trial and error method

$$P_e = 14.79$$

This is the required model parameter. If backmixing is not significant, Eq. (A) can be approximately written as

$$\sigma_\theta^2 \approx \frac{2}{P_e} \tag{B}$$

If calculated from Eq. (B), then

$$P_e = 2/\sigma_\theta^2 = 2/0.1261 = 15.86$$

Deviation from the exact value is only 3%, which is acceptable. If the variance $\delta_\theta^2 < 0.2$, using Eq. (B) to calculate model parameter, the error will not exceed 6%.

5.7 DESIGN OF NONIDEAL REACTORS

Designs on nonideal reactors are based on the flow models. In the previous section, we discussed how to use the segregation model to calculate conversion. For the tanks-in-series model, when the number of tanks is determined, the method described in Section 3.5.2 is used for calculations. If N is a noninteger, the last tank can be replaced with a smaller tank. Therefore, only the calculations of the axial dispersion model will be discussed in this section.

When using the axial dispersion model to model the reactors operated at steady-state, the mass balance for the key component A is its model equation, which is developed similarly as Eq. (5.53). Since the operation is at steady-state, $\partial c_A/\partial t = 0$, there is no time variable in the model. In addition, due to the presence of a chemical reaction, the consumption of A by chemical reaction should be added to the model equation. After these two amendments, the model equation is obtained as

$$D_a\frac{d^2 c_A}{dZ^2} - u\frac{dc}{dZ} + \mathscr{R}_A = 0 \tag{5.65}$$

Boundary conditions are as follows

$$Z = 0, \quad uc_{A0} = uc_A - D_a\frac{dc_A}{dZ}\bigg|_{0^+} \tag{5.66}$$

$$Z = L_r, \quad \frac{dc_A}{dZ}\bigg|_{L_r^-} = 0 \tag{5.67}$$

If a first order irreversible reaction is carried out isothermally in the reactor, then $r_A = kc_A$. Substituting into Eq. (5.65), we have

$$D_a\frac{d^2 c_A}{dZ^2} - u\frac{dc_A}{dZ} - kc_A = 0 \tag{5.68}$$

This is a second order linear ODE, which can be solved analytically. Combining with the boundary conditions in Eqs. (5.66) and (5.67), the solution is

$$\frac{c_A}{c_{A0}} = \frac{4\alpha}{(1+\alpha)^2 \exp\left[-\frac{P_e}{2}(1-\alpha)\right] - (1-\alpha)^2 \exp\left[-\frac{P_e}{2}(1+\alpha)\right]} \tag{5.69}$$

where $\alpha = (1 + 4k\tau/P_e)^{1/2}$

As $P_e \to \infty$, $\alpha \to 1$, α can be expanded as

$$\alpha = 1 + \frac{1}{2}\left(\frac{4k\tau}{P_e}\right) - \frac{1}{8}\left(\frac{4k\tau}{P_e}\right)^2 + \cdots \tag{5.70}$$

Substituting Eq. (5.70) into Eq. (5.69), after rearranging, we have

$$\frac{c_A}{c_{A0}} = \exp(-k\tau) \tag{5.71}$$

Clearly, Eq. (5.71) is the same result when using the plug flow model to calculate a first order reaction, which further demonstrates that the axial dispersion model is actually a plug flow model superimposed with an axial dispersion term.

As $P_e \to 0$, after carrying out series expansion on $\exp[-P_e(1 - \alpha)/2]$ and disregarding the high-order items, substituting into Eq. (5.69) yields

$$\frac{c_A}{c_{A0}} = \frac{4\alpha}{(1+\alpha)^2\left(1 - \dfrac{P_e}{2} + \alpha\dfrac{P_e}{2}\right) - (1-\alpha)^2\left(1 - \dfrac{P_e}{2} - \alpha\dfrac{P_e}{2}\right)}$$

$$= \frac{4\alpha}{4\alpha - \alpha P_e + \alpha^3 P_e} = \frac{1}{1 + k\tau}$$

(5.72)

This equation is the same as the formula for first order reactions in continuous tank reactors as discussed in Chapter 3, Tank Reactor.

In summary, the axial dispersion model with closed boundary conditions can be used to describe any backmixing between plug flow and perfectly-mixed flow when the model parameter P_e takes different values. Now, taking $\frac{D_a}{uL}$ as model parameter, c_A/c_{A0} versus $k\tau$ based on Eq. (5.69) is plotted as illustrated in Fig. 5.22. The figure shows that conversion in a real reactor increases as the reciprocal of P_e decreases. The bigger the space time, the more significant the effect by deviation from ideal flow will be.

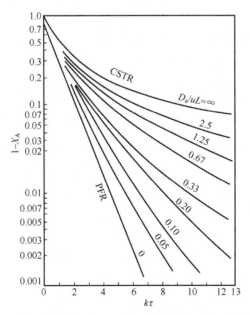

FIGURE 5.22   Conversion of a first order reaction calculated by the axial dispersion model.

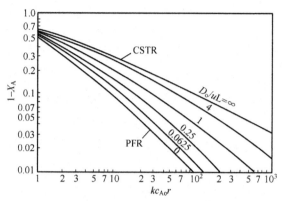

FIGURE 5.23 Conversion of a second order reaction calculated by the axial dispersion model.

For nonfirst order reactions, Eq. (5.65) is a nonlinear second order ODE, which is difficult if not impossible to solve analytically, but it can be solved numerically. Fig. 5.23 shows the results for second-order irreversible reaction.

Comparing Figs. 5.22 and 5.23 we can see that when all the other conditions are identical, the conversion of a second order reaction is more significantly influenced by backmixing compared to a first order reaction. Generally speaking, the effect of backmixing on the reaction results increases as the reaction orders increase.

Example 5.8
Liquid phase reaction A→P is carried out isothermally in a perfectly-mixed reactor operated in a laboratory. When space time equals 43.02 min, the conversion of A is 82%. The reactor is scaled up for pilot test. The RTD of the reactor, a tubular reactor, in a pilot study is measured and shown in Example 5.6. If the operating temperature and space time in the pilot study are the same as those in the lab test, determine the conversion of A at the exit of reactor using (1) the tanks-in-series model and (2) the axial dispersion model.

Solution
First, the reaction rate constant at the operating temperature is obtained from the given lab data. Because

$$-\mathscr{R}_A = r_A = kc_A = kc_{A0}(1 - X_A)$$

Substituting into Eq. (3.44), we have

$$V_r = \frac{Q_0 c_{A0} X_A}{kc_{A0}(1 - X_A)}$$

Or

$$\tau = \frac{V_r}{Q_0} = \frac{X_A}{k(1 - X_A)}$$

(Continued)

(Continued)

Substituting the given data into this equation yields

$$43.02 = \frac{0.82}{k(1 - 0.82)}$$

Solving it gives

$$k = 0.1059 \text{ min}^{-1}$$

1. Using tanks-in-series model

    It had been shown in Example 5.6 that the model parameter N is 8 when using the tanks-in-series model to model the RTD of this reactor. Consequently, the conversion can be determined based on 8 equal-volume tanks-in-series. From Eq. (3.51) we get

$$\tau = \frac{1}{k}\left[\left(\frac{1}{1 - X_{AN}}\right)^{1/N} - 1\right] \tag{3.51}$$

   Substituting k, N, and $\tau = 43.02/8$, we have

$$\frac{43.02}{8} = \frac{1}{0.1059}\left[\left(\frac{1}{1 - X_{AN}}\right)^{1/8} - 1\right]$$

   Then, exit conversion is

$$X_{AN} = 0.9728$$

2. Using axial dispersion model

    Model parameter P_e was calculated as 15.35 in Example 5.7. Consequently

$$a = (1 + 4k\tau/P_e)^{1/2}$$
$$= 91 + 4 \times 0.1059 \times 43.02/15.32^{1/2} = 1.479$$

   Substituting into Eq. (5.70) gives

$$\frac{C_A}{C_{A0}} = 4 \times 1.479\left\{(1 + 1.479)^2 \exp[-15.35(1 - 1.479)/2]\right.$$

$$\left. -(1 - 1.479)^2 \exp[-15.35(1 + 1.479)/2]\right\}^{-1}$$

$$= 0.02437$$

   Therefore, the exit conversion is

$$X_A = 1 - \frac{C_A}{C_{A0}} = 1 - 0.02437 = 0.9756$$

Hence, the results calculated from these two models are very consistent, but deviate significantly from the lab results although the temperature and space time in the lab test and pilot study are identical. This is because the flow patterns of these two operations are different. The lab test is carried out at perfectly-mixed flow, while the pilot study is operated when the extent of

(Continued)

(Continued)

backmixing is small, which definitely will make its corresponding conversion higher compared to that of the lab test. In general, when the reactor is scaled up, its conversion is always reduced due to many reasons. This example is, however, an exception.

5.8 MIXING OF FLUIDS IN FLOW REACTORS

The segregation model was introduced in Section 5.6. Its primary assumption is that there is no material interchange from the time the fluid particles enter the reactor until they leave the reactor. In other words, there is no mixing between fluid particles. This is called complete segregation, which means that particles are isolated and independent from each other. If the mixing between fluid particles is in the molecular scale, it is called micromixing. When there are no more segregated fluid particles in the reactor, the micromixing reaches maximum, which is called complete micromixing or maximum micromixing. This explains two extremes of mixing states: one is no micromixing, i.e., complete segregation, this fluid is called macrofluid; another case is, on the other hand, no segregation, i.e., complete micromixing, the corresponding fluid is called microfluid. Mixing between these two extremes is called partial segregation or partial micromixing, which means that both circumstances exist in the system.

Different mixing states exhibit different effects on chemical reactions. Assume that two fluid particles with equal volume and different concentrations c_{A1} and c_{A2} react as a α-order irreversible reaction. If the two particles are completely segregated, the reaction rates of these two particles are $r_{A1} = kc_{A1}^{\alpha}$ and $r_{A2} = kc_{A2}^{\alpha}$, respectively. Their average reaction rate is

$$\langle r_A \rangle = \frac{1}{2}\left(r_{A1} + r_{A2}\right) = \frac{k}{2}\left(c_{A1}^{\alpha} + c_{A2}^{\alpha}\right)$$

If micromixing occurs between these two particles and the level of mixing is maximum, the concentration of A after mixing equals $(c_{A1} + c_{A2})/2$. The reaction is carried out automatically at this concentration. Consequently, the average reaction rate in this situation should be

$$\langle r_A' \rangle = k[(c_{A1} + c_{A2})/2]^{\alpha}$$

This indicates that the different extent of micromixing will influence the reaction rates.

When micromixing is zero or at complete segregation, the average reaction rate is $\langle r_A \rangle$. On the other hand, when micromixing is maximum, reaction rate is $\langle r_A' \rangle$. Which rate is bigger depends on the value of α. For first order

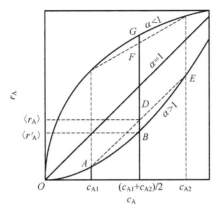

FIGURE 5.24   Influence of fluid mixing on the reaction rate.

reactions, as we can see from the above two equations, $\langle r_A \rangle = \langle r'_A \rangle$. Since the reaction rate exhibits a linear relation with concentration, their average results should be same. If $\alpha > 1$, the curve for r_A and c_A is concave; on the other hand, if $\alpha < 1$, the curve is convex as shown in Fig. 5.24. It can be seen from the figure that when $\alpha > 1$, the average reaction rate for complete micromixing $\langle r'_A \rangle$ corresponds to the ordinate of point B, while the average reaction rate for complete segregation corresponds to the ordinate of point D. This can be easily verified by plane geometry. Therefore, when reaction orders are greater than 1, $\langle r_A \rangle > \langle r'_A \rangle$, which means that micromixing reduces average reaction rate. Similarly, when $\alpha < 1$, $\langle r_A \rangle < \langle r'_A \rangle$, micromixing increases the average reaction rate.

The previous discussion focused on the effect of fluid mixing on reaction rate, at the core of this discussion is that fluid particle concentrations are different. Now, the effect of fluid mixing on reactor performance will be discussed. Let's look at a batch reactor first. At any time, all fluid particles in batch reactors have the same residence time, their compositions are thus the same too. Consequently, the extent of micromixing will not affect the performance of batch reactors. Then, let's look at plug flow reactors. In plug flow reactors, all fluid particles in the same cross-sectional area have the same residence time, their compositions are correspondingly the same. Consequently, the extent of micromixing has no effect on the performance of plug flow reactors. However, the situation is different for perfectly-mixed flow reactors. Since the residence times of fluid particles at the same cross-sectional area are different in perfectly-mixed flow reactors, their corresponding compositions are different. Then, the extent of micromixing will affect the performance of the rectors except for first-order reactions. This effect varies with the RTD: the more severe the backmixing, the more significant the effect it will have.

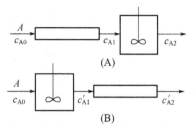

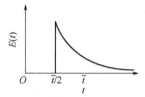

FIGURE 5.25 CSTR and PFR in series. (A) PFR in front of a CSTR; (B) CSTR in front of a PFR.

FIGURE 5.26 Residence time distribution of CSTR and PFR in series.

The calculation result of a single reaction in a perfectly-mixed flow reactor verified that conversion by complete segregation is not significantly different from that of the maximum micromixing, the difference is only several percentages at most cases. In addition, this difference decreases as the RTD becomes narrower. Consequently, the effect of micromixing can be neglected for this case. Of course, if multiple reactions are carried out simultaneously, the effect of micromixing on product distribution will be significant. For some fast reactions and multiphase reactions, the effect of micromixing usually cannot be ignored.

It is noteworthy to ask that if two reactors having the same RTD and same level of micromixing would have the same performance. To answer this question, the two reaction systems as illustrated by Fig. 5.25 will be studied. Both diagrams (A) and (B) in Fig. 5.25 are the cases where a plug flow reactor and perfectly-mixed reactor are connected in series; the difference is the sequence of these two reactors. Clearly, the RTDs of these two cases should be same, as illustrated in Fig. 5.26, where $\bar{t}$ is the mean residence time that fluid flows through the system. If the reaction volumes of the two reactors are identical, the residence time in the plug flow reactor would be $\bar{t}/2$. Consequently, the distribution curve jumps at $\bar{t}/2$. If the extent of micromixing in these two systems are the same, e.g., they all reach complete micromixing, then, if the same chemical reactions occur under the same space time and same temperature, do they yield same conversions? The calculation results (see Example 5.9) demonstrate that the conversions in these two systems are not same except for first order reactions. The reason is the sequence of mixing. Mixing in case (A) in Fig. 5.25 occurs in late-stage, while mixing in case (B) occurs in early stage. The former mixing occurs at lower concentrations, while the latter mixing occurs at higher concentrations. Hence, although their extent of mixing is the same, due to the different concentrations after mixing, the change of reaction rate is inevitably different. Consequently, the corresponding ultimate conversions of these two systems deviate from each other.

In summary, the performance of flow reactors is related to the reaction kinetics and RTDs, as well as the mixing of fluid, which includes the extent of micromixing (or extent of segregation) and the sequence of mixing. The sequence of mixing actually reflects the concentration levels at mixing.

Example 5.9
The same reactions are carried out under the same space time and same temperature in the two reactors connected in series as illustrated in Fig. 5.25. If the space times for the plug-flow reactor and perfectly-mixed flow reactor connected in series are 1 min, and the inlet concentration $c_{A0} = 1$ kmol/m^3, calculate the conversions of these two situations. Assume that the reaction is (1) first order and (2) second order, and the reaction rates at reaction temperature are 1 min^{-1} and 1×10^{-3} m^3/(mol $\cdot$ min), respectively.

Solution
1. First-order reaction
   Calculation formula for plug flow reactors is

   $$c_A = c_{A0} e^{-k\tau} \tag{A}$$

   Substituting relative data gives component A concentration in the effluent of the plug flow reactor as illustrated in Fig. 5.25(A).

   $$c_{A1} = 1 \times e^{1 \times 1} = 0.368 \text{ kmol/m}^3$$

   Calculation formula for perfectly-mixed reactors is

   $$c_{A2} = \frac{c_{A1}}{1 + k\tau} \tag{B}$$

   Hence, the effluent concentration of A in case (a) is

   $$c_{A2} = \frac{0.368}{1 + 1 \times 1} = 0.184 \text{ kmol/m}^3$$

   The conversion calculated from this equation is $1 - 0.184/1 = 0.816$, i.e., 81.6%
   Following a similar procedure, case (b) can be calculated using Eqs. (A) and (B)

   $$c'_{A1} = \frac{1}{1 + 1 \times 1} = 0.5 \text{ kmol/m}^3$$

   (Continued)

(Continued)

$$c'_{A2} = 0.5 \times e^{-1 \times 1} = 0.184 \ \text{kmol/m}^3$$

The effluent concentration is the same as that of case (a). Consequently, the conversion is 81.6% too. Therefore, the sequence of mixing has no effect on the first order reaction.

2. Second order reactions

Calculation formula for plug flow reactors is

$$c_{A1} = \frac{c_{A0}}{1 + k\tau c_{A0}} \tag{C}$$

Substituting relative data gives component A concentration in the effluent of the plug flow reactor

$$c_{A1} = \frac{1}{1 + 1 \times 1 \times 1} = 0.5 \ \text{kmol/m}^3$$

Calculation formula for perfectly-mixed flow reactors is

$$c_{A2} = \frac{1}{2k\tau} \left[-1 + \sqrt{1 + 4k\tau c_{A1}} \right] \tag{D}$$

Consequently, effluent A concentration in case (a) is

$$c_{A2} = \frac{1}{2 \times 1 \times 1} \left[-1 + \sqrt{1 + 4 \times 1 \times 1 \times 1} \right] = 0.366 \ \text{kmol/m}^3$$

Hence, the final conversion is $1 - 0.366/1 = 0.634$, i.e., 63.4%
In a similar way, case (b) can be calculated. From Eq. (D), we have

$$c'_{A1} = 0.618 \ \text{kmol/m}^3$$

From Eq. (C), we obtain

$$c'_{A2} = 0.382 \ \text{kmol/m}^3$$

Consequently, the final conversion is $1 - 0.382/1 = 0.618$ or 61.8%. This shows that the sequence of mixing has an impact on second order reactions, and later mixing is beneficial to second order reactions. Apparently this effect is not significant.

FURTHER READING

Li S. Chemical and catalytic reaction engineering. Beijing: Chemical Industry Press; 1986.

Levenspel O. Chemical reaction engineering. 2nd ed. New York: John Wiley & Sons; 1972.

Nauman EB, Buffham BA. Mixing in continuous flow systems. New York: John Wiley & Sons; 1983.

Himmelblau DM, Bischoff KB. Process analysis and simulation. New York: John Wiley & Sons; 1968.

PROBLEMS

5.1 $F(\theta)$ and $E(\theta)$ are RTD function and RTD density in a closed flow system, respectively, and θ is dimensionless time.
  1. If the reactor is a plug flow reactor, determine
    a. $F(1)$;
    b. $E(1)$;
    c. $F(0.8)$;
    d. $E(0.8)$;
    e. $E(1.2)$.
  2. If the reactor is a perfectly-mixed flow reactor, determine
    a. $F(1)$;
    b. $E(1)$;
    c. $F(0.8)$;
    d. $E(0.8)$;
    e. $E(1.2)$.
  3. If the reactor is a nonideal flow reactor, determine
    a. $F(\infty)$;
    b. $F(0)$;
    c. $E(\infty)$;
    d. $E(0)$;
    e. $\int_0^\infty E(\theta)d\theta$;
    f. $\int_0^\infty \theta E(\theta)d\theta$.

5.2 The step experiment is used to measure the RTD in a closed flow system, and the exiting tracer concentration is obtained as a function of time as shown below

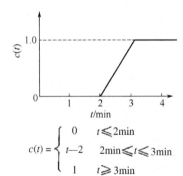

$$c(t) = \begin{cases} 0 & t \leqslant 2\text{min} \\ t-2 & 2\text{min} \leqslant t \leqslant 3\text{min} \\ 1 & t \geqslant 3\text{min} \end{cases}$$

Calculate
  1. RTD function $F(\theta)$ and RTD density for this reactor;
  2. Mathematical expectation $\bar{\theta}$ and variance δ_θ^2;
  3. If the reactor is modeled with the tanks-in-series model, determine the model parameter;

4. If the reactor is modeled with the axial dispersion model, determine the model parameter;

5. If a first order reaction is carried out in the reactor, the reaction rate constant $k = 1$ min^{-1} and there is no side reaction, calculate the exit conversion.

5.3 The step experiment is used to measure the RTD in a close flow system, and the exiting tracer concentration is obtained as a function of time as shown below

t/s	0	15	25	35	45	55	65	75	90	100
$c(t)$/(g/cm^3)	0	0.5	1.0	2.0	4.0	5.5	6.5	7.0	7.7	7.7

1. Calculate the RTD of this reactor and the mean residence time;

2. If the fluid in this reactor is microfluid, the reaction is first-order irreversible reaction, and reaction rate constant $k = 0.05$ 1/s, estimate the exit conversion of this reactor;

3. If the fluid in this reactor is macrofluid, and all the conditions did not change, then what the exit conversion would be?

5.4 In order to measure the RTD, pulse experiment method is used and tracer concentration in the effluent is obtained as shown in the table.

t/min	0	1	2	3	4	5	6	7	8	9	10
$c(t)$/(g/L)	0	0	3	5	6	6	4.5	3	2	1	0

Calculate

1. The mean residence time $\bar{t}$ and variance δ_θ^2 of the reaction material;

2. The fraction of material whose residence time is less than 4.0 min.

5.5 The RTD density function for an isothermal, closed liquid-phase reactor is $E(t) = 16t \exp(-4t)$, min^{-1}, calculate

1. Mean residence time;

2. Space time;

3. Space velocity;

4. Fraction of material with residence time less than 1 min;

5. Fraction of material with residence time greater than 1 min;

6. If this reactor is modeled by the tanks-in-series model, this reactor is equivalent to how many equal-volume perfectly-mixed tanks operated in series;

7. If modeled by the axial dispersion model, what is the model parameter P_e?

8. If the reacting material is microfluid, the reaction is a first order irreversible reaction, and the reaction rate constant $k = 6$ min^{-1}, $c_{A0} = 1$ mol/L, calculate the exit conversion using the axial dispersion model and the tanks-in-series model, separately, and compare the results.

9. If the reacting material is macrofluid and all other conditions are the same as above, estimate the exit conversion and compare with the results of microfluid.

5.6 A second order irreversible liquid phase reaction of the microfluid is carried out in an isothermal, nonideal tubular reactor that is 10 m long. The reaction rate constant k is 0.266 L/(mol · s), feed concentration c_{A0} is 1.6 mol/L, superficial velocity inside the reactor is 0.25 m/s. The exit conversion determined by experiment is 80%. In order to reduce the backmixing, the length of reactor is extended to 40 m, and all the other conditions are kept the same, estimate the exit conversion.

5.7 Zero order reaction $A \rightarrow B$ is carried out isothermally in a perfectly-mixed reactor. The reaction rate $r_A = 9$ mol/(min · L), feed concentration $c_{A0} = 10$ mol/L, and the mean residence time of fluid in the reactor $\bar{t}$ is 1 min, calculate the exit conversion at the following conditions,

1. Reacting material is microfluid;
2. Reacting material is macrofluid.

Compare the results and discuss along with Problem 5.5.

5.8 A first order irreversible reaction $A \rightarrow P$ is carried out isothermally in a reactor with the following RTD. The reaction rate constant is 2 (1/min).

$$E(t) = \begin{cases} 0 & t < 1 \text{ min} \\ \exp(1 - t) & t \geq 1 \text{ min} \end{cases}$$

Calculate the exit conversion using the axial dispersion model and the segregation model separately, and compare the results.

Chapter 6

Chemical Reaction and Transport Phenomena in Heterogeneous System

Chapter Outline

Many chemical reactions occur in heterogeneous systems, such as ammonia synthesis, ethylene oxidization to produce ethylene oxide, the synthesis of trichloroethylene with acetylene and hydrogen chloride, etc. Since there are two or more phases involved in heterogeneous reactions, *transport processes*, mostly mass and heat transfer, are necessary for chemical reactions to take place.

As described in Chapter 1, Introduction, the heterogeneous chemical reaction systems can be classified as gas−solid, gas−liquid, liquid−solid,

Reaction Engineering. DOI: http://dx.doi.org/10.1016/B978-0-12-410416-7.00006-9

solid—solid, and gas—liquid—solid systems. Meanwhile, they can be classified as catalytic or noncatalytic reaction systems according to whether a catalyst is present. Despite the large variety of heterogeneous reactions, there are only three basic situations defined by the location of the reaction: (1) The reactions take place at the interface of phases: all gas—solid reactions, catalytic and noncatalytic, belong to this situation; (2) The reactions take place within a certain phase: most gas—liquid reactions take place in liquid phase. The phase where reactions take place is called the reactive phase; (3) The reactions take place in two phases: some liquid—liquid reactions belong to this situation. Some reaction systems may belong to more than one situation. For example, the reaction occurs at the interface and in one of the phases. In this book we will only discuss the first two situations which are relatively simple but typical for many applications.

As an example of the first situation, the gas—solid catalyzed reactions will be discussed more comprehensively in this chapter, since they are widely applied in commercial applications and technically more matured. The most important application of the second situation is the gas—liquid reactions which will be examined in Chapter 9, Multiple-Phase Reactors. The focus of these two chapters are the effects of transport processes on chemical reactions, and we will discuss methods and procedures for quantitative analysis.

6.1 STEPS IN HETEROGENEOUS REACTIONS

Since heterogeneous catalytic reactions occur on the surface of solid catalysts, reactants have to move from bulk fluid to the surface in order to participate in the reaction. On the other hand, products also need to be transferred from the catalyst surface back to the bulk fluid constantly as well. In order to understand the reaction and transport processes the macroscopic structures and properties of solid catalysts have to be examined first.

6.1.1 Macroscopic Structures and Properties of Solid Catalyst Particles

Most solid catalysts are porous media. Specifically, there are pore networks formed by many irregularly shaped and interconnected pores inside the particles. The pore-network structure leads to a large surface area where chemical reactions occur. The surface area of a catalyst is characterized by the surface area per unit mass, i.e., specific surface area in unit m^2/g. Usually, the BET method is used to measure this property experimentally; its value can reach $200-300$ m^2/g in some catalysts.

Obviously, the specific surface area is related to the size of these pores. Smaller pores generally lead to greater surface area. Since the pore size inside catalyst particles is not uniform, pore size distribution, which is usually calculated based on pore volume distribution, is used to characterize the catalyst. The pore volume is the total pore volume per unit mass, cm^3/g. If the pore size is not too small, e.g., greater than 100×10^{-10} m, a mercury intrusion porosimetry can be used to measure pore volume distribution since different pressures applied correspond to different pore sizes. For catalysts with pore size smaller than 100×10^{-10} m, other methods have to be used. If the information of pore volume distribution is not needed, the total pore volume can be measured by using the CCl_4 method, which gives the total volume of all pores, i.e., the total pore capacity. If not specified, the pore volume is usually referred to the total pore capacity.

For the convenience of comparison and calculations, the pore size is usually characterized by the mean pore radius. Given the distribution function of pore radius r_a, the mean pore radius is

$$\langle r_a \rangle = \frac{1}{V_g} \int_0^{V_g} r_a dV, \tag{6.1}$$

where V is the pore volume of pores with radius r_a, normalized with unit mass; V_g is the total pore volume. If the pore volume distribution is unavailable, the mean pore radius can be estimated with the specific surface area S_g and total pore capacity V_g.

Assuming that there are n parallel cylindrical pores inside the particle, all of them have a mean pore radius $\langle r_a \rangle$ and length L, so:

$$S_g = n(2\pi \langle r_a \rangle \overline{L})$$
$$V_g = n(\pi \langle r_a \rangle^2 \overline{L})$$

Combining these two equations, we obtain

$$\langle r_a \rangle = 2 \frac{V_g}{S_g} \tag{6.2}$$

The values of mean pore radius calculated with Eqs. (6.1) and (6.2) may be quite different. Results by Eq. (6.1) are more accurate. Eq. (6.2) only gives an approximate estimation, since the shape and intersection of pores inside the catalyst are quite complex.

In addition to the pore volume, porosity ε_P is another important parameter, and is defined as the ratio between pore volume and particle volume, which is always less than unity. The difference between porosity and pore volume is that the former is normalized by the particle volume and the latter by particle mass. They can be converted to each with

$$\varepsilon_P = V_g \rho_P, \tag{6.3}$$

where ρ_P is particle density. In addition, true density ρ_t and bulk density are also used, and their definitions are:

$$\rho_P = \frac{\text{solid mass}}{\text{particle volume}}, \quad \rho_t = \frac{\text{solid mass}}{\text{solid volume}}, \quad \rho_b = \frac{\text{solid mass}}{\text{bed volume}}.$$

Density is defined as mass per unit volume, and the difference in these three definitions lies in the volume used. ρ_P is based on the volume of the particle (including solid volume and pore volume), while ρ_t only includes the solid volume. So $\rho_t > \rho_P$. The bulk density is based on the volume of the packed bed, which includes the volume of the particles and the volume of the void space between particles. Among these, the true density is the highest and the bulk density is the lowest.

Readers should pay attention to the difference between bed porosity ε and particle porosity ε_P. The bed porosity is volume ratio of the void space between catalyst particles to the whole packed bed; the particle porosity is the volume ratio of the pores to the whole particle.

For solid particles, especially the very fine and irregularly shaped ones, their size can be described by sieving. For example, if solid particles can pass through a 40 mesh sieve but cannot pass through a 60 mesh sieve, then they are called particles of 40–60 mesh, and the size of the particle can be expressed as the arithmetic average of the net width of pores of the two sieves.

The size of the particle can also be characterized by the diameter of a sphere equivalent. There are three kinds of equivalencies: (1) sphere with the same volume; (2) sphere with the same external surface area; (3) sphere with the same specific external surface area. All three of these definitions have been seen in the literature, so readers should be aware of the differences. This is especially important when using correlations involving particle diameter, since for the same particle the diameter values calculated based on these three definitions can be very different.

Another commonly used parameter of solid particles is the shape parameter, ψ_S. It is the ratio of external surface area of a spherical particle with the same volume a_S to the actual external surface area of the particle a_P, i.e.,

$$\psi_S = \frac{a_S}{a_P} \tag{6.4}$$

Since the surface area of a sphere is the smallest of all shapes with the same volume, $a_S < a_P$, hence $\psi_S < 1$. ψ_S describes how the particle shape approximates a sphere, and thus is also called sphericity.

Example 6.1
A catalyst particle weights 1.083 g. Its volume, pore capacity, and specific surface area are 1.033 cm^3, 0.255 cm^3/g, and 100 m^2/g. Calculate the ε_P, ρ_P, and $\langle r_a \rangle$.

Solution
According to the definitions,

$$\rho_P = 1.083/1.033 = 1.048 \text{ g/cm}^3$$
$$\varepsilon_P = 0.255/(1/1.048) = 0.267$$

According to Eq. (6.2)

$$\langle r_a \rangle = 2\, V_g/S_g = 2 \times 0.255/100 \times 10^4 = 50.1 \times 10^{-10} \text{ m}$$

6.1.2   Steps in a Catalytic Reaction

After discussing macroscopic properties and key parameters of porous solid particles, we now examine the steps involved in an irreversible reaction A (g) → B (g) taking place on a porous catalyst particle, as depicted in Fig. 6.1. There are many interconnected pores inside the catalyst particle. The particle is surrounded by a laminar boundary layer of gas which represents transport resistance between the bulk gas phase and catalyst external surface. Since the reaction takes place on the catalyst surface, reactant A has to move from bulk gas phase to the surface of catalyst. On the other hand, product B formed on the catalyst surface must leave the catalyst surface to move back to the bulk gas phase. The whole process includes the following steps (See Fig. 6.1):

(1) Reactant A diffuses from bulk gas phase to the external surface of the catalyst.
(2) Reactant A diffuses from external surface into pores inside catalyst particle to reach the active sites for absorption/reaction.
(3), (4), and (5) is absorption of A, reaction of A to form product B, and desorption of B from the surface, respectively. These three steps constitute the surface reaction process, and determine the intrinsic kinetics of this catalytic reaction.
(6) Product B diffuses from interior of the particle to external surface.
(7) Product B diffuses from external surface to bulk gas phase.

Steps (1) (7) are external diffusion; Steps (2) (6) are internal diffusion; Steps (3)−(5) are surface reaction steps. Among them, internal diffusion and surface reactions occur inside the catalyst particle simultaneously, and they are parallel processes. The surface reaction Steps (3)−(5) are sequential. The external diffusion between bulk fluid and the external surface of the catalyst is

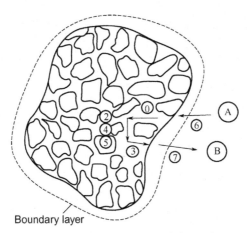

Boundary layer

FIGURE 6.1   Schematic of steps for a heterogeneous catalytic reaction.

the mass transfer process between two phases. The external and internal diffusion steps and the surface reaction take place consecutively, and therefore there will be a rate-determining step, as discussed in Chapter 2, Fundamentals of Reaction Kinetics. At steady-state all steps proceed at the same rate, which equals that of the rate-determining step. For parallel processes, there is no rate-determining step.

Due to the influence of diffusion, the reactant concentrations in the bulk fluid, on the external surface and at the center of the catalyst particles, c_{AG}, c_{AS}, and c_{AC}, are different, and $c_{AG} > c_{AS} > c_{AC} > c_{Ae}$ (c_{Ae} is the equilibrium concentration). For the product, this order is reversed.

6.2 HEAT AND MASS TRANSFER BETWEEN BULK FLUID AND THE CATALYST EXTERNAL SURFACE

The first step of a heterogeneous catalyzed reaction is the reactant transfer to the external surface of catalyst, its rate can be expressed as

$$N_A = k_G a_m (c_{AG} - c_{AS}) \tag{6.5}$$

where a_m is the external surface area per unit catalyst mass and k_G is the mass transfer coefficient. The concentration difference $(c_{AG} - c_{AS})$ is the driving force for mass transfer. At steady-state, the mass transfer rate is equal to the chemical reaction rate, i.e., $N_a = (-\mathscr{R}_a)$.

Chemical reactions always have heat effects, either endothermic or exothermic. Therefore, when a reactant moves to the external surface, there must be a heat transfer between the bulk fluid and external surface of the catalyst. Heat will be transferred from the external surface to bulk fluid for exothermic reactions, or in the opposite direction for endothermic reactions. Heat transfer rate can be expressed as:

$$q = h_s a_m (T_S - T_G) \tag{6.6}$$

where h_S is the heat transfer coefficient; T_S and T_G are the temperatures at bulk fluid and the external surface of the catalyst. Under steady-state the heat transfer rate is equal to the heat generated (or absorbed) by the reaction, i.e.,

$$q = (-\mathscr{R}_A)(-\Delta H_r) \tag{6.7}$$

Eqs. (6.5)−(6.7) are the fundamental equations describing interphase transport processes.

6.2.1 Transport Coefficient

Eqs. (6.6) and (6.7) include mass and heat transport coefficients, k_G and h_S, respectively. These transport coefficients reflect the magnitude of film resistance, controlled by the thickness of the laminar boundary layer surrounding the catalyst. There are differences in both concentration and temperature across the laminar boundary layer. It is usually assumed that the temperature and concentration are uniform on the external surface of the catalyst. Similarly temperature and concentration are assumed uniform in the bulk fluid. These assumptions imply that the thickness of the laminar layer is uniform. With these assumptions, the analyses of interphase transport processes can be greatly simplified with acceptable accuracy, since the process can be treated as a one-dimensional problem.

The mass and heat transfer rates depend on the resistance and more resistance leads to smaller transport coefficients. The mass transfer coefficients between bulk fluid and catalyst particles are affected by the shape and size of the catalyst, the flow condition, and the fluid properties. The heat transfer coefficients are affected by the same factors. According to the similarity theory of mass and heat transfer, j factors can be used to correlate experimental data for gas–solid mass and heat transfer. The j factor of mass j_D and heat transfer j_H are defined as

$$j_D = \frac{k_G \rho}{G} Sc^{2/3} \tag{6.8}$$

and

$$j_H = \frac{h_S}{G C_P} Pr^{2/3} \tag{6.9}$$

where Sc and Pr are the Shmidt number and the Prandtl number, $\left(Sc = \frac{\mu}{\rho D}, \ Pr = \frac{C_P \mu}{\lambda_P} \right)$, respectively.

Both j_H and j_D are functions of the Reynolds number, the correlations depend on the structure of the catalyst beds. For example, the correlation for a packed bed can be expressed by

$$\varepsilon j_D = 0.357/Re^{0.359}, \tag{6.10}$$

within the ranges $3 < Re < 1000$, $0.6 < Sc < 5.4$; and

$$\varepsilon j_H = 0.395/Re^{0.36}, \tag{6.11}$$

within the ranges $0.6 \leq Pr \leq 3000$, $30 \leq Re \leq 10^5$. The Reynolds number in Eqs. (6.10) and (6.11) is defined based on the particle diameter:

$$Re = d_P G/\mu$$

According to the similarity theory,

$$j_D = j_H \tag{6.12}$$

Generally, Eq. (6.12) is a good approximation for packed bed reactors. We can take advantage of this similarity to estimate the mass transfer coefficient based on the heat transfer coefficient, or vice versa. Estimating mass transfer coefficients based on heat transfer coefficient is more important since heat transfer coefficients can be measured more accurately, and often more easily, than mass transfer coefficients. However, some literature does show large discrepancies between j_H and j_D in packed bed reactors.

The j_D versus Re correlation Eq. (6.10) indicates mass transfer coefficient k_G increases with mass flow rate G, and so does the external mass transfer rate. On the other hand, low mass flow rate will lead to large mass transfer resistance, making the external mass transfer the rate-determining step.

In commercial applications it is uncommon that a gas–solid catalytic reaction is controlled by external diffusion, since it is always desirable to operate at high mass flow rate to enhance productivity. One exception is the production of nitric acid by oxidizing ammonia on platinum gauze, which is controlled by external diffusion. There are two reasons for this phenomenon: (1) the intrinsic kinetic rate is very fast due to high reaction temperature of 800–900°C; (2) mass flow rate of ammonia–air mixture cannot be too high in order to prevent platinum gauze damage caused by high mechanical friction. For a gas–solid noncatalytic reaction, such as coal combustion, the external diffusion is usually the rate-determining step because of the very fast combustion rate at high temperatures.

6.2.2 Concentration and Temperature Difference Between the External Surface of Catalyst and Bulk Fluid

The combination of Eqs. (6.5)–(6.7) gives:

$$k_G a_m (c_{AG} - c_{AS})(-\Delta H_r) = h_S a_m (T_S - T_G)$$

Substituting Eqs. (6.8) and (6.9) into this equation, we obtain

$$T_S - T_G = (c_{AG} - c_{AS}) \frac{(-\Delta H_r)}{\rho C_P} \left(\frac{Pr}{Sc}\right)^{2/3} \left(\frac{j_D}{j_H}\right) \tag{6.13}$$

Since $Pr/Sc \approx 1$ for most gases, j_D approximately equals j_H in packed bed reactors, Eq. (6.13) can be simplified as:

$$T_S - T_G = \frac{(-\Delta H_r)}{\rho C_P}(c_{AG} - c_{AS}), \tag{6.14}$$

which implies that the temperature difference, $\Delta T = T_S - T_G$ and the concentration difference, $\Delta c = c_{AG} - c_{AS}$, are proportional to each other. For reactions without very strong heat effects ΔT will be significant only when Δc is large. On the other hand, for reactions with strong heat effect ΔT can reach extreme values even when Δc is small or moderate. Although high ΔT can occur for both exothermic and endothermic reactions, it should be adequately addressed especially for exothermic reactions. Usually it is the bulk fluid temperature that is measured, and for exothermic reactions very high ΔT means $T_S \gg T_G$, which may lead to catalyst damage.

For a reaction under adiabatic conditions, when the concentration in the fluid decreases from c_{AG} to c_{AS}, the fluid temperature changes can be calculated based on heat balance:

$$(\Delta T)_{ad} = \frac{(-\Delta H_r)}{\rho C_P}(c_{AG} - c_{AS}) \tag{6.15}$$

Comparing Eq. (6.14) with Eq. (6.15) gives $\Delta T = (\Delta T)_{ad}$, assuming the concentration difference between bulk fluid and catalyst external surface is the same as the fluid concentration changes in the adiabatic reaction system.

Example 6.2

To remove O_2 in H_2, a deoxygenator with Pt/Al_2O_3 catalyst is used to oxidize hydrogen:

$$2H_2 + O_2 = 2H_2O$$

The rate equation is

$$r_A = 3.09 \times 10^{-5} \exp[-2.19 \times 10^4/(RT)]p_A^{0.804} \text{ mol}/(g \cdot s)$$

where p_A is O_2 partial pressure (Pa). The porosity of the packed bed is $\varepsilon = 0.35$, gas mass flow rate is $G = 1250$ kg $(m^2 \cdot h)$. The diameter of catalyst particle is $d_P = 1.86$ cm. The external surface area $a_m = 0.5434$ m^2/g.

If pressure $p = 0.1135$ MPa, temperature $T = 373$K, the volume fraction of H_2 and O_2 are 96% and 4%, respectively, evaluate whether or not the external mass and heat transfer resistance can be neglected (no internal diffusion resistance).

Data: the diffusion coefficient of O_2 is 0.414 m^2/h. The gas viscosity is 1.03×10^{-5} Pa $\cdot$ s, density is 0.117 kg/m^3. The heat of reaction is 2.424×10^5 J/mol, heat transfer coefficient is 2.424×10^6 J/(m$^3 \cdot$ h $\cdot$ K).

Solution

First, we evaluate

$$Re = d_P G/\mu = 1.86 \times 10^{-2} \times 1250/0.03708 = 627$$
$$Sc = \mu/\rho D = 0.03708/0.117/0.414 = 0.7685$$
$$Sc^{2/3} = 0.8368$$

Eq. (6.10) gives

$$j_D = 0.357/(0.35)(627)^{0.359} = 0.101$$

The reaction occurs on the surface of the catalyst where the temperature and concentration is unknown, but can be calculated using the trial and error method.

Step 1. Assume

$$c_{AS} = c_{AG} = p_{AG}/RT$$
$$= 0.01135 \times 0.04/0.008314 \times 373 = 1.464 \times 10^{-3} \text{ mol/L}$$
$$T_S = T_G = 373 \text{ K}$$

$$p_{AS} = p_{AG} = 0.1135 \times 0.04 = 0.00454 \text{ MPa}$$

then the reaction rate can be calculated

(Continued)

(Continued)

$$(-\mathscr{R}_A) = 3.09 \times 10^{-3} \exp[-2.19 \times 10^4/(8.314 \times 373)](4540)^{0.804}$$
$$= 2.308 \times 10^{-5} \, mol(g \cdot s)$$

Then Eqs. (A) and (B) can be derived from Eqs. (6.5), (6.8) and Eqs. (6.6), (6.7) and Eq.(6.9), respectively:

$$c_{AG} - c_{AS} = \frac{-\mathscr{R}_A \rho}{j_D G a_m}(Sc)^{2/3} \tag{A}$$

$$T_S - T_G = \frac{(-\mathscr{R}_A)(-\Delta_r H)}{h_S a_m} \tag{B}$$

According to the reaction rate calculated, the concentration difference is

$$c_{AG} - c_{AS} = \frac{2.308 \times 10^{-3} \times 0.117 \times 3600}{0.101 \times 1250 \times 0.5434} \times 0.8368$$
$$= 1.1858 \times 10^{-4} \, mol/L$$

Then we obtain the first corrections

$$(c_{AS})_1 = c_{AG} - 1.1858 \times 10^{-4} = 1.3454 \times 10^{-3} \, mol/L$$
$$(T_S)_1 = T_G + (-\mathscr{R}_A)(-\Delta_r H)/(h_S a_m)$$
$$= 373 + 2.038 \times 10^{-5} \times 3600 \times 2.424 \times 10^5/$$
$$(2.424 \times 10^6 \times 0.5434 \times 10^{-3})$$
$$= 373 + 15.29 = 388.29K$$

Step 2. Calculate the reaction rate with corrected $(c_{AS})_1$ and $(T_S)_1$

$$(p_{AS})_1 = (c_{AS})_1 R(T_S)_1$$
$$(-\mathscr{R}_A) = 3.09 \times 10^{-3} \exp[-2.19 \times 10^4/(8.314 \times 388.29)](4343)^{0.804}$$
$$= 2.938 \times 10^{-5} \, mol/(g \cdot s)$$

Then we use the updated $(-\mathscr{R}_A)$ in Eqs. (A) and (B) to get new corrections.

$$(c_{AS})_2 = 1.3130 \times 10^{-3} \, mol/L, \quad (T_S)_2 = 392.46 \, K$$

Step 3. Repeat the previous steps:

$$(c_{AS})_3 = 1.3034 \times 10^{-3} \, mol/L, \quad (T_S)_3 = 393.71K$$
$$(c_{AS})_4 = 1.3005 \times 10^{-3} mol/L, \quad (T_S)_4 = 394.09K$$
$$(c_{AS})_5 = 1.2996 \times 10^{-3} \, mol/L, \quad (T_S)_5 = 394.20K$$
$$(c_{AS})_6 = 1.2993 \times 10^{-3} \, mol/L, \quad (T_S)_6 = 394.24K$$

Now the $(c_{AS})_5$ and $(c_{AS})_6$, $(T_S)_5$ and $(T_S)_6$ are quite close. So the final results are

$$c_{AG} - c_{AS} = 1.464 \times 10^{-3} - 1.2993 \times 10^{-3} = 0.165 \times 10^{-3} \, mol/L,$$
$$(c_{AG} - c_{AS})/c_{AG} = 0.165/1.464 = 11.2\%,$$
$$T_S - T_G = 394.24 - 373 = 21.24 \, K.$$

As such, the external mass and heat transfer resistances cannot be neglected in this example.

6.2.3 Effect of External Diffusion on Heterogeneous Catalytic Reactions

6.2.3.1 Single Reaction

The effect of external diffusion on a heterogeneous catalytic reaction can be described by the external effectiveness factor, which is defined as:

$$\eta_X = \frac{\text{Reaction rate under the influence of external diffusion}}{\text{Reaction rate without the influence of external diffusion}}. \qquad (6.16)$$

The concentration of reactant on the external surface of a catalyst is always lower than that in the bulk fluid, so $\eta_X \leq 1$ for reactions with positive order, and $\eta_X \geq 1$ for reactions with negative order.

For simplicity, in the following discussion we will only consider interphase mass transfer resistance, i.e., assuming that there is no temperature difference between particle surface and bulk gas phase and internal diffusion resistance can be neglected. For a first order irreversible reaction, the reaction rate is $k_W c_{AG}$ without external diffusion resistance and $k_W c_{AS}$ with external diffusion resistance. According to Eq. (6.16), we obtain

$$\eta_X = \frac{k_W c_{AS}}{k_W c_{AG}} = \frac{c_{AS}}{c_{AG}} \qquad (6.17)$$

At steady-state,

so

$$k_G a_m (c_{AG} - c_{AS}) = k_W c_{AS}, \qquad (6.18)$$

$$c_{AS} = \frac{c_{AG}}{(1 + D_a)} \qquad (6.19)$$

$$D_a = \frac{k_W}{k_G a_m} \qquad (6.20)$$

According to Eq. (6.17), we can obtain the external effectiveness factor for first order irreversible reactions

$$\eta_X = \frac{1}{(1 + D_a)} \qquad (6.21)$$

where D_a is Damkohler Number, defined as

$$D_a = \frac{k_W c_{AG}^{\alpha-1}}{k_G a_m} \qquad (6.22)$$

Similarly, we obtain the η_X for other irreversible reactions of different orders:

$$\alpha = 2 \quad \eta_X = \frac{1}{4 D_a^2} (\sqrt{1 + 4 D_a} - 1)^2 \qquad (6.23)$$

$$\alpha = \frac{1}{2} \quad \eta_X = \left[\frac{2 + D_a^2}{2} \left(1 - \sqrt{1 - \frac{4}{(2 + D_a^2)}} \right) \right]^{1/2} \qquad (6.24)$$

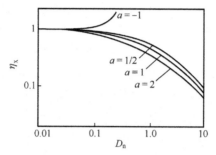

FIGURE 6.2   External effectiveness factor at constant temperature.

$$\alpha = -1 \quad \eta_X = \frac{2}{1 + \sqrt{1 - 4D_a}} \tag{6.25}$$

Fig. 6.2 shows the dependence of η_X on D_a, the external effectiveness factor decreases with increasing D_a, except for reactions with negative order. This trend is more obvious for higher order reactions. η_X is approaching unity while D_a approaches zero for reactions with any order. For higher order reactions it is more critical to reduce external diffusion resistance and increase external effectiveness factor.

6.2.3.2   Multiple Reactions

Assume that the temperature at the catalyst surface is the same as that of the bulk gas phase. Consider the parallel reactions

$$A \rightarrow B \text{ (desired reaction)},$$
$$A \rightarrow D \text{ (unwanted reaction)},$$

The reaction order for forming desired product B is α and for forming D is β. The rate equations for these two reactions:

$$r_B = k_1 c_{AS}^\alpha$$
$$r_D = k_2 c_{AS}^\beta$$

The instantaneous selectivity of product B is:

$$S = \frac{r_B}{r_B + r_D} = \frac{1}{1 + k_2 c_{AS}^{\beta - \alpha} / k_1} \tag{6.26}$$

If the impact of external diffusion negligible, then $c_{AS} = c_{AG}$, and the instantaneous selectivity is

$$S' = \frac{1}{1 + k_2 c_{AG}^{\beta - \alpha} / k_1} \tag{6.27}$$

Eqs. (6.26) and (6.27) show that the impact of external diffusion on the selectivity of parallel reactions depends on the sign of $(\beta - \alpha)$. If $\alpha > \beta$, $S < S'$, indicating that external diffusion will reduce the selectivity of B. This is because the external diffusion reduces the reactant concentration at the surface, i.e., $c_{AS} < c_{AG}$. The concentration effect on reaction rate is stronger for reactions of higher order. Therefore, when the order for the desired reaction α is higher than the order of side reaction β, external diffusion will lower selectivity of B. On the other hand, if $\alpha < \beta$, $S > S'$

For first order irreversible consecutive reactions

$$A \xrightarrow{k_1} B \xrightarrow{k_2} D$$

where B is the desired product, if the mass transfer coefficient for A, B, and D are equal, then

$$k_G a_m (c_{AG} - c_{AS}) = k_1 c_{AS} \tag{6.28}$$

$$k_G a_m (c_{BS} - c_{BG}) = k_1 c_{AS} - k_2 c_{BS} \tag{6.29}$$

$$k_G a_m (c_{DS} - c_{DG}) = k_2 c_{BS} \tag{6.30}$$

According to Eqs. (6.28)−(6.30), we obtain

$$c_{AS} = \frac{c_{AG}}{(1 + D_{a1})}, \tag{6.31}$$

$$c_{BS} = \frac{D_{a1} c_{AG}}{(1 + D_{a1})(1 + D_{a2})} + \left(\frac{c_{BG}}{1 + D_{a2}} \right), \tag{6.32}$$

where $D_{a1} = k_1/k_G a_m$, $D_{a2} = k_2/k_G a_m$.

The instantaneous selectivity is

$$S = \frac{k_1 c_{AS} - k_2 c_{BS}}{k_1 c_{AS}} = 1 - \frac{k_2 D_{a1}}{k_1(1 + D_{a2})} - \frac{k_2 c_{BG}(1 + D_{a1})}{k_1 c_{AG}(1 + D_{a2})} \tag{6.33}$$

Since $D_{a2} = k_2 D_{a1}/k_1$, we obtain

$$S = \frac{1}{1 + D_{a2}} - \frac{k_2 c_{BG}(1 + D_{a1})}{k_1 c_{AG}(1 + D_{a2})} \tag{6.34}$$

If the external diffusion has no influence,(i.e., $D_{a1} = D_{a2} = 0$), Eq. (6.34) reduces to

$$S' = 1 - \frac{k_2 c_{BG}}{k_1 c_{AG}} \tag{6.35}$$

Eqs. (6.34) and (6.35) show that external diffusion reduces the selectivity of consecutive reactions, even if the main and side reactions have the same reaction order. This is different from a parallel reaction system. Therefore, for consecutive reactions external diffusion resistance should be minimized to increase the selectivity. For example, for naphthalene oxidation to produce phthalic anhydride, or benzene oxidation to produce maleic anhydride, external diffusion can cause over-oxidization and reduce the yields of desired products.

If there is a temperature difference between bulk gas phase and catalyst surface due to heat transfer resistance, then $T_S > T_G$ for exothermic reactions. The impact of external heat transfer resistance on selectivity of multiple reactions depends on activation energies of these reactions, as illustrated by Example 6.3.

Example 6.3

$A \xrightarrow{k_1} B \xrightarrow{k_2} D$ are consecutive first order irreversible exothermic reactions. If B is the desired product, estimate the selectivity.

Data

$T_G = 450K$ $c_{BG}/c_{AG} = 0.5$, $(T_S - T_G) = 10K$, $(k_G a_m)_A = (k_G a_m)_B = 40$ cm^3/(g•s)

$k_1 = 6.0 \times 10^8 \exp[-E_1/(RT)]$ cm^3/(g·s)

$k_2 = 1.2 \times 10^6 \exp[-E_1/(RT)]$ cm^3/(g·s)

$E_1 = 80.0$ KJ/mol, $E_2 = 60.0$ KJ/mol

Solution

Case 1. Only consider external mass transfer resistance but ignore heat transfer resistance, i.e., assume $T_S = T_G = 450K$. We can calculate $k_1 = 0.310$ mol/(g·s), $k_2 = 0.130$ mol/(g·s) and

$$D_{a1} = 0.310/40 = 7.75 \times 10^{-3}, \quad D_{a2} = 0.130/40 = 3.25 \times 10^{-3}.$$

The selectivity under the influence of external diffusion can be calculated using Eq. (6.34) as

$$S = \frac{1}{1 + 3.25 \times 10^{-3}} - \frac{0.130(1 + 7.75 \times 10^{-3})0.5}{0.310(1 + 3.25 \times 10^{-3})} = 0.786$$

If external diffusion resistance can be ignored, the selectivity can be calculated using Eq. (6.35)

$$S' = 1 - 0.130 \times 0.5/0.310 = 0.790$$

Case 2. Consider both external mass and heat transfer resistance. At $T_S = 460K$, $k_1 = 0.494$ mol/(g·s), $k_2 = 0.184$ mol/(g·s), $D_{a1} = 0.01235$, $D_{a2} = 0.0046$. According to Eq. (6.34), the selectivity is

$$S = \frac{1}{1 + 0.0046} - \frac{0.184(1 + 0.1235)0.5}{0.494(1 + 0.0046)} = 0.8077$$

Case 3. Only consider the film resistance of heat transfer. The selectivity calculated using Eq. (6.35) is

$$S'' = 1 - 0.84 \times 0.5/0.494 = 0.8138$$

According to these results, when $T_S = T_G$ the external mass transfer resistance always reduces the selectivity of consecutive reactions. For exothermic reactions, the external heat transfer resistance results in $T_S > T_G$. If the activation energy of the main reaction is greater than that of the side reaction, S' is higher than S but lower than S''. Interested readers can consider what the conclusion will be if the activation energy of the main reaction is lower than that of the side reaction.

6.3 GAS DIFFUSION IN POROUS MEDIA

For heterogeneous catalytic reactions, the chemical reactions mainly occur at the internal surface inside catalyst particles. Therefore, after moving from the bulk gas phase to the external surface of the catalyst particles, the reactant molecules have to continue diffusing into the particles. Next, we will first discuss diffusion through a single pore and then move on to discuss diffusions inside porous catalyst particles.

6.3.1 Diffusion in Pores

Given components in the bulk fluid have to diffuse into pores if there is no pressure driven laminar flow because of the absence of a pressure difference inside and outside of the pores. There are two types of diffusion within pores, depending on the relative magnitude between pore radius r_a and the mean free path of the moving molecule λ.

When $\lambda/2r_a \leq 0.01$, diffusion inside the pore is ordinary diffusion, identical to diffusion in bulk fluid. The diffusion rate is controlled by the collision frequency between molecules and independent of pore size. The ordinary diffusion coefficients in a binary gas system should be obtained from handbooks, though it can be measured by conducting experiments or estimated with empirical correlations.

While $\lambda/2r_a \geq 10$, diffusion inside the pore is called the Knudsen diffusion. Most collisions occur between molecules and pore wall, and the collisions between molecules have little impact on the transport process. Therefore the Knudsen diffusion coefficient D_K is determined by the pore radius, and not influenced by other gas components in the system. Its value can be calculated by:

$$D_K = 9.7 \times 10^3 r_a \sqrt{T/M} \tag{6.36}$$

in which the unit of r_a is cm, the unit of D_K is cm^2/s.

The mean free path λ can be estimated with

$$\lambda = 1.013/p \text{ cm} \tag{6.37}$$

The unit of p in Eq. (6.37) is Pa.

if the relative magnitude of λ and r_a lies between these two extreme cases, both ordinary and Knudsen diffusions should be taken into account, and the transition diffusion coefficient for a binary system is:

$$D_A = \frac{1}{1/(D_K)_A + (1 - by_A)/D_{AB}} \tag{6.38}$$

$$b = 1 + N_B/N_A \tag{6.39}$$

where N_A, N_B is the diffusion flux of gas component A and B, y_A is the mole fraction of component A, $(D_K)_A$ is the Knudsen diffusion coefficient of A, and D_A is the transition diffusion coefficient. Implementation of Eq. (6.38) is not straightforward, since it contains the parameter b, which is related to N_A and N_B. For equimolar counterdiffusion $N_A = -N_B$, Eq. (6.38) reduces to

$$D_A = \frac{1}{1/(D_K)_A + 1/D_{AB}} \qquad (6.40)$$

6.3.2 Diffusion in Porous Particles

Diffusion processes in a single pore have been discussed in the previous section. The diffusion flux of component i inside a porous particle is

$$N_i = -\frac{P}{RT} D_{ei} \frac{dy_i}{dZ} = -D_{ei} \frac{dc_i}{dZ} \qquad (6.41)$$

where D_{ei} is the effective diffusion coefficient of component i, Z is the diffusion distance.

If the pores in the catalyst with porosity ε_P are oriented randomly, the ratio of pore open area to the external surface area is ε_P as well. Pores are interconnected and their shape and size are not uniform. These characters make the diffusion distance different from the diffusion in idealized cylindrical pores. Tortuosity τ_m is introduced to correct the diffusion distance, and the modified diffusion distance is $\tau_m Z$. Therefore, the effective diffusion coefficient becomes

$$D_{ei} = \frac{\varepsilon_P D_i}{\tau_m}. \qquad (6.42)$$

The value of τ_m usually lies between 3 and 5, depending on the structure of pores in the catalysts, and should be measured experimentally.

In addition to the Knudsen diffusion, transport of gas components inside catalysts may also include surface diffusion, i.e., the absorbed gas component can transfer along the pore walls. The direction of surface is the same as the direction of concentration gradient in the absorbed layer, which in turn is the same direction as the concentration gradient of this component in the pore.

Example 6.4
Thiophene (C_4H_4S) (Component A) diffuses through hydrogen (Component B) in a catalyst particle at 600K and 3.04 MPa. The specific surface area measured by the BET method is 180 m^2/g. The porosity is 0.4. The particle density is 1.4 g/cm^3. It is known that the pore size distribution is very narrow and the tortuosity is 3.0. The binary diffusivity of thiophene and hydrogen is 0.0457 cm^2/s. Estimate the effective diffusion coefficient.

(Continued)

(Continued)

Solution

1. Calculate $(D_K)_A$. According to Eq. (6.3), $V_g = \varepsilon_P/\rho_P$. By substituting it into Eq. (6.2), we obtain

$$\langle r_a \rangle = 2\varepsilon_P/S_g\rho_P = 2 \times 0.4/(180 \times 1.4 \times 10^4) = 31.7 \times 10^{-8} \text{ cm}$$

Then D_K can be calculated using Eq. (6.36)

$$(D_K)_A = 9.7 \times 10^3 \times 31.7 \times 10^{-8} \times (600/84)^{1/2} = 8.22 \times 10^{-3} \text{ cm}^2/s$$

2. Calculate the combined diffusion coefficient according to Eq. (6.40)

$$D_A = \frac{1}{1/(D_K)_A + 1/D_{AB}} = \frac{1}{1000/8.22 + 1/0.0454} = 6.97 \times 10^{-3} \text{ cm}^2/s$$

3. Calculate the effective diffusion coefficient using Eq. (6.42)

$$D_{eA} = D_A \, \varepsilon_P/\tau_m = 6.97 \times 10^{-3} \times 0.4/3 = 9.29 \times 10^{-4} \text{ cm}^2/s$$

6.4 DIFFUSION AND REACTION IN POROUS CATALYSTS

During heterogeneous catalytic reactions, the reactant molecules will first diffuse through the laminar boundary layer to reach the external surface of the catalyst. The reaction will take place once the reactant molecules reach the external surface. The internal surface formed by the pore walls is much larger than the external surface. For example, for a spherical catalyst with a diameter of 0.3 cm, a specific surface area of 150 m^2/g, and density of 1.2 g/cm^3, its external surface area is 16.7 cm^2/g, which only represents a very small part of the total surface area of 150 m^2/g. Thus, most reactant molecules have to diffuse into the catalyst through pores. This process is called internal diffusion. The difference between internal and external diffusion lies in that the reactant has to reach the external surface before a reaction occurs, while reaction and diffusion happen simultaneously during the internal diffusion process, i.e., the reaction happens on the pore walls while the reactants diffuse inside the pores. Due to the diffusion resistance and reaction consumption, the reactant concentration decreases gradually during the diffusion process. Correspondingly, the reaction rate decreases as well. At the center of the catalyst, both the reactant concentration and reaction rate drop to their minimal value or possibly reach zero. If a reaction rate is relatively slow and the effective diffusion coefficient of reactant is large, a very small concentration gradient can provide enough driving force to provide enough reactants for reaction on the pore walls. As a result, the reactant concentration at any location on the pore wall is almost equal to the concentration at the pore entrance, and the internal surface of the particle is considered to be fully utilized. On the other hand, if the reaction rate is high, it only takes a short distance for all the reactant molecules to be consumed (for irreversible reactions). In order to provide enough reactants for reactions on internal surface a much higher concentration gradient is required. In this situation, most reactions occur within a thin layer next to the external surface, and the rest of the internal surface is not utilized since the reactant concentration is so low that the reaction can be ignored.

6.4.1 Reactant Concentration Profile in Porous Catalysts

Reactant concentration is not uniform inside a catalyst particle. The reactant concentration is highest at the external surface and lowest at the center, forming a decaying profile from the outside to the inside. Product concentration has a reversed distribution. At a given temperature, the reaction rate is solely determined by the concentration, so it is important to determine the concentration distribution inside the catalyst particle.

We consider a first order irreversible reaction taking place isothermally on a slab catalyst with the thickness of $2L$, and assume the pore structure is uniform and isotropic (i.e., statistically homogeneous). A reaction-diffusion differential equation is needed to determine the concentration distribution of reactant A, and can be established by the mass balance of reactant A on an infinitesimal with thickness of dZ inside the catalyst (shown in Fig. 6.3). Assuming the slab thickness is much smaller than its length and width, the diffusion can be considered as a one-dimensional process, (i.e., we only consider diffusion from the left to right surface; diffusion through the other four surfaces is neglected). At steady-state, the mass conservation gives

$$
\begin{pmatrix} \text{Reactant A diffuses} \\ \text{into the elementary} \\ \text{section} \end{pmatrix} - \begin{pmatrix} \text{Reactant A diffuses} \\ \text{out of the elementary} \\ \text{section} \end{pmatrix} = \begin{pmatrix} \text{Reactant A} \\ \text{disappearance} \\ \text{by reaction} \end{pmatrix},
$$

i.e.,

$$
D_e a \left(\frac{dc_A}{dZ} \right)_{Z+dZ} - D_e a \left(\frac{dc_A}{dZ} \right)_Z = k_P c_A a dZ \tag{6.43}
$$

where D_e and a are the effective diffusion coefficient and diffusion area, k_P is the reaction rate constant based on unit catalyst volume. Because

$$
\left(\frac{dc_A}{dZ} \right)_{Z+dZ} = \left(\frac{dc_A}{dZ} \right)_Z + \frac{d}{dZ} \left(\frac{dc_A}{dZ} \right) dZ
$$

Eq. (6.43) is converted to

$$
\frac{d^2 c_A}{dZ^2} = \frac{k_P}{D_e} c_A \tag{6.44}
$$

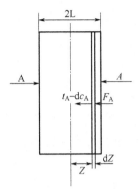

FIGURE 6.3 The catalyst plate.

Eq. (6.44) is the reaction-diffusion differential equation for first order irreversible reactions inside a slab catalyst. Its boundary conditions (BCs) are:

$$Z = L, \quad c_A = c_{AS} \qquad (6.45)$$

$$Z = 0, \quad \frac{dc_A}{dZ} = 0 \qquad (6.46)$$

The solutions to Eqs. (6.44)−(6.46) give the reactant concentration distribution. By introducing the following dimensionless variables:

$$\xi = c_A/c_{AS}, \quad \zeta = Z/L, \quad \phi^2 = L^2 \frac{k_P}{D_e}$$

We rearrange Eqs. (6.44)−(6.46) to

$$\frac{d^2 \xi}{d\zeta^2} = \phi^2 \xi \qquad (6.47)$$

$$\zeta = 1, \quad \xi = 1 \qquad (6.48)$$

$$\zeta = 0, \quad \frac{d\xi}{d\zeta} = 0 \qquad (6.49)$$

Eq. (6.47) is a second order homogenous ordinary differential equation with constant coefficients. Its general solution is

$$\xi = Ae^{\phi\zeta} + Be^{-\phi\zeta} \qquad (6.50)$$

Combining with BCs, Eqs. (6.48) and (6.49), the parameter A, B can be obtained as

$$A = B = 1/(e^{\phi} + e^{-\phi})$$

Then Eq. (6.50) gives

$$\xi = \frac{e^{\phi\zeta} + e^{-\phi\zeta}}{e^{\phi} + e^{-\phi}} = \frac{\cosh(\phi\zeta)}{\cosh(\phi)}, \qquad (6.51)$$

i.e.,

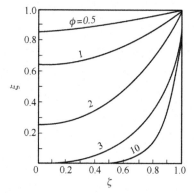

FIGURE 6.4   Reactant concentration distribution in a slab catalyst.

$$\frac{c_A}{c_{AS}} = \frac{\cosh(\phi\zeta)}{\cosh(\phi)} \tag{6.52}$$

Eq. (6.52) describes the concentration profile in a slab catalyst. Fig. 6.4 shows the dependence of concentration distribution on ϕ. The dimensionless concentration ξ monotonically decreases while the dimensionless distance ζ decreases (i.e., the concentration of reactant A decreases from the external surface to the center of the catalyst). The steepness of the decline depends on the value of ϕ; the greater ϕ, the faster the reactant concentration decreases. For example, the concentration decreases very quickly when $\phi = 10$, and is almost zero at $\zeta = 0.5$. With smaller ϕ, the concentration distribution is more uniform. For example, reactant concentration inside the catalyst is almost equal to that on the external surface when $\phi = 0.5$. The concentration distribution is affected by both diffusion and reaction processes. The dimensionless group ϕ determines the character of concentration distributions, whose value illustrates the influence of pore diffusion resistance on reaction processes. This dimensionless group ϕ, Thiele modulus, is an essential parameter in modeling diffusion and reaction processes.

The physical meaning of Thiele modulus can be understood according to its definition which can be expressed as

$$\phi^2 = L^2 \frac{k_P}{D_e} = \frac{aLk_Pc_{AS}}{D_ea(c_{AS} - 0)/L} = \frac{\text{Surface reaction rate}}{\text{Internal diffusion rate}},$$

which shows that the Thiele modulus characterizes the relative magnitude of surface reactions and internal diffusion rates. Faster surface reactions and slower internal diffusion lead to greater ϕ, meaning internal diffusion resistance has a stronger influence on reactions.

6.4.2   Internal Effectiveness Factor

The internal effectiveness factor is introduced to calculate the reaction rates in the catalyst, and its definition is:

$$\eta = \frac{\text{Reaction rate under the influence of internal diffusion}}{\text{Reaction rate without the influence of internal diffusion}}. \tag{6.53}$$

As discussed above, the internal diffusion leads to nonuniform concentration distribution inside the catalyst particles, so the average value of actual reaction rate:

$$\langle r_A \rangle = \frac{1}{L} \int_0^L k_P c_A dZ$$

should be used in Eq. (6.53). The average reaction rate of the example shown in Section 6.4.1 is

$$\langle r_A \rangle = \frac{k_P c_{AS}}{L \cosh \phi} \int_0^L \cosh\left(\frac{\phi h}{L}\right) dZ = \frac{k_P c_{AS} \tanh(\phi)}{\phi}.$$

The reactant concentration inside the catalyst should be equal to the concentration c_{AS} on the external surface of the catalyst if there is internal diffusion resistance. The corresponding reaction rate is $k_P c_{AS}$. Applying these rate equations in Eq. (6.53), the internal effectiveness factor of a first order irreversible reaction in a slab catalyst is:

$$\eta = \frac{\tanh(\phi)}{\phi} \tag{6.54}$$

which describes the dependence of η on ϕ, as shown in Fig. 6.5. η decreases monotonically while ϕ increases, i.e., the larger ϕ is, the stronger the effects of internal diffusion. Increasing the internal effectiveness factor can improve reaction rates and therefore enhance the reactor productivity in chemical plants. According to the definition of ϕ, reducing particle size of the catalyst can lead to smaller ϕ and hence greater η. Moreover, increasing the porosity

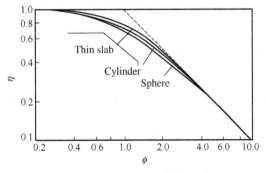

FIGURE 6.5   The effectiveness factors of catalysts in different shapes.

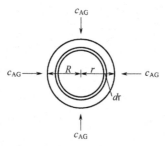

FIGURE 6.6   The spherical catalyst.

and pore size can enhance effective diffusion coefficient D_e, which in turn will lead to smaller ϕ and greater η.

The estimation of internal effectiveness factor can be summarized as the following steps: (1) establish the differential equation that describes reactant concentration distribution inside the catalyst particles, i.e., the diffusion-reaction equation, and determine the BCs; solve the differential equation to obtain the concentration distributions; (2) calculate the average reaction rate based on the concentration distribution; (3) calculate the internal effectiveness factor based on its definition. The internal effectiveness factors of catalysts with other shapes can be estimated using the following steps.

Consider a first order irreversible reaction in a spherical catalyst. As shown in Fig. 6.6, the mass balance on an infinitesimal with thickness dr at radius r gives

$$\frac{d^2 c_A}{dr^2} + \frac{2}{r}\frac{dc_A}{dr} = \frac{k_P}{D_e}c_A \tag{6.55}$$

Similarly, the diffusion-reaction equation for a cylindrical catalyst with either infinite length or nonpermeable ends on both sides is

$$\frac{d^2 c_A}{dr^2} + \frac{1}{r}\frac{dc_A}{dr} = \frac{k_P}{D_e}c_A. \tag{6.56}$$

The BCs

$$r = R_P, \quad c_A = c_{AS}$$
$$r = 0, \quad dc_A/dr = 0$$

are applicable for both Eqs. (6.55) and (6.56).

Both of these two equations are ordinary differential equations with non-constant coefficients. Eq. (6.55) can be converted to homogeneous ODE with constant coefficients by a change of variables $c_A r = u$; Eq. (6.56) is a zero order Bessel equation and can be solved using the standard method. Solving these two equations with the corresponding BCs, we obtain

$$\text{Spherical:} \quad \frac{c_A}{c_{AS}} = \frac{R_P \sinh(3\phi r/R_P)}{r \sinh(3\phi)} \tag{6.57}$$

Cylindrical: $\quad \dfrac{c_A}{c_{AS}} = \dfrac{I_0(2\phi r/R_P)}{I_0(2\phi)}$ (6.58)

where I_0 is zeroth order modified Bessel function.
ϕ can be defined as:

$$\phi = \frac{V_P}{a_P}\sqrt{\frac{k_P}{D_e}}$$ (6.59)

which can be used for catalysts in different shapes. V_P and a_P are a particle's volume and external surface area, respectively. Eq. (6.59) is consistent with the Thiele modulus definition discussed in the previous section for a slab catalyst.

Based on concentration distribution described by Eqs. (6.57) and (6.58), the internal effectiveness factors can be derived as:

Spherical: $\quad \eta = \dfrac{1}{\phi}\left[\dfrac{1}{\tanh(3\phi)} - \dfrac{1}{3\phi}\right]$ (6.60)

Cylindrical: $\quad \eta = \dfrac{I_1(2\phi)}{\phi I_0(2\phi)}$ (6.61)

where I_1 is first order modified Bessel function, and its value can be easily found in mathematical handbooks.

Using Eqs. (6.54), (6.60), and (6.61). the effectiveness factor η as a function of ϕ for catalysts with different shapes is shown in Fig. 6.5. The three curves in Fig. 6.5 almost overlap to each other, especially for very small or large value of ϕ. The difference is more pronounced within the region $0.4 < \phi < 3$. And even in this region the difference remains within $10-20\%$. Thus, as long as ϕ is calculated using Eq. (6.59), the internal effectiveness factor for catalysts with any shapes can be estimated with acceptable accuracy using any one of Eqs. (6.54), (6.60), or (6.61).

Meanwhile, Fig. 6.5 shows that if $\phi < 0.4$ the effects of internal diffusion can be ignored $\eta \approx 1$. On the other hand, if $\phi > 3$, the internal pore diffusion will have very strong effects, and the three curves and their asymptotes (Eq. (6.62)) will overlap each other:

$$\eta = 1/\phi$$ (6.62)

Under such conditions, Eq. (6.62) can be used to estimate the internal effectiveness factors of the catalysts in any shape.

These discussions are valid only for isothermal first order irreversible reactions. In the next section we will consider reactions of other orders.

Example 6.5

Butane is dehydrogenated on a Cr-Al catalyst. The rate equation is

$$r_A = k_W c_A \text{ mol}/(g \cdot s)$$

$k_W = 0.92 \text{ cm}^3/(g \cdot s)$ at pressure of 0.1013 MPa and temperature of 773K. The catalyst is a plate with a thickness of 8 mm. Its average pore radius is 4.8 nm, porosity is 0.35 cm^3/g, and tortuosity is 2.5. Estimate the internal effectiveness factor.

Solution

According to Eq. (6.37), the mean free length is 10^{-5} cm

$$\frac{\lambda}{2r_a} = \frac{10^{-5}}{2 \times 48 \times 10^{-8}} = 10.4 > 10$$

so the gas diffusion is Knudsen diffusion. The diffusion coefficient can be calculated using Eq. (6.36)

$$D_K = 9700 \times (4.8 \times 10^{-8}) \times (773/58)^{1/2} = 1.70 \times 10^{-2} \text{ cm}^2/s$$

The effective diffusion coefficient can be calculated by substituting Eq. (6.3) into Eq. (6.42)

$$D_K = V_g \rho_P D_K / \tau_m = 0.35 \times 1.70 \times 10^{-2} \rho_P / 2.5 = 2.38 \times 10^{-3} \rho_P \text{ cm}^2/s$$

The given rate constant is based on unit mass of a catalyst, and needs to be converted to a rate constant based on unit volume.

$$k_P = k_W \rho_P = 0.92 \rho_P \text{ s}^{-1}$$

According to Eq. (6.54), the Thiele modulus is

$$\phi = \frac{0.8}{2} \left(\frac{0.92 \rho_P}{2.38 \times 10^{-3} \rho_P} \right) = 7.86$$

The effectiveness factor can be calculated using Eq. (6.54)

$$\eta = \tanh(7.86)/7.86 = 0.127$$

Since $\phi = 7.86 > 3$, the influence of internal diffusion is very strong and the internal effectiveness factor can be estimated using Eq. (6.62), which gives the same result.

Example 6.6

Derive the equation to calculate the internal effectiveness factor of a first order reversible reaction A⇌B taking place isothermally on a spherical catalyst.

Solution

The rate equation of first order reversible reaction is

$$r_A = k_A c_A - k_B c_B$$

(Continued)

(Continued)

Assuming c_{A0} is the initial concentration of A, then $c_B = c_{A0} - c_A$, and the rate equation is

$$r_A = (k_A + k_B)c_A - k_B c_{A0} \tag{A}$$

$r_A = 0$ when the reaction reaches equilibrium,

$$(k_A + k_B)c_{Ae} - k_B c_{A0} = 0$$

where c_{Ae} is the equilibrium concentration. We rearrange Eq. (A)

$$r_A = (k_A + k_B)(c_A - c_{Ae}) \tag{B}$$

To calculate the internal effectiveness factor, the diffusion-reaction equation should be developed first, which is identical to Eq. (6.55) except for the reaction term, i.e.,

$$\frac{d^2 c_A}{dr^2} + \frac{2}{r}\frac{dc_A}{dr} = \frac{k_A + k_B}{D_e}(c_A - c_{Ae}) \tag{C}$$

The BCs are also identical to those for Eq. (6.55). We assign

$$u = c_A - c_{Ae},$$

and

$$\phi = \frac{R_P}{3}\left(\frac{k_A + k_B}{D_e}\right)^{\frac{1}{2}} \tag{D}$$

Eq. (C) can be rearranged as:

$$\frac{d^2 u}{dr^2} + \frac{2}{r}\frac{du}{dr} = \frac{9\phi^2}{R_P^2}u$$

And the BCs are

$$r = R_P, u = u_S = c_{AS} - c_{Ae}$$

$$r = 0, du/dr = 0$$

Now, the equation and its BCs are identical to those for Eq. (6.55). So its solution should be the same as Eq. (6.57)

$$\frac{u}{u_S} = \frac{c_A - c_{Ae}}{c_{AS} - c_{Ae}} = \frac{R_P \sinh(3\phi r/R_P)}{r \sinh(3\phi)} \tag{E}$$

The equation to calculate the internal effectiveness factor should be the same as well.

$$\eta = \frac{1}{\phi}\left[\frac{1}{\tanh(3\phi)} - \frac{1}{3\phi}\right] \tag{F}$$

It should be noted that the definition of ϕ is different, and Eq. (6.59) is replaced by Eq. (D), i.e., k_P is replaced by $(k_A + k_B)$. Obviously k_A and k_B. Obviously $\bar{k}$ and $\bar{k}$ are rate constants based on catalyst particle volume.

6.4.3 Internal Effectiveness Factor for Non-first Order Reactions

In the previous section, we discussed the diffusion-reaction issues for first order reactions. In reality, most reactions are not first order, and their rate equations can be very complicated. Generally, the methodology described in Section 6.4.2 is applicable for an estimation of the internal effectiveness factors of these reactions, but the mathematical manipulations become much more difficult.

Consider a reaction taking place isothermally on a slab catalyst, and the reaction rate equation is:

$$r_A = k_P f(c_A),$$

Assume a constant effective diffusion coefficient D_e, using the same methods to derive Eq. (6.44), we obtain the diffusion-reaction equation:

$$\frac{d^2 c_A}{dZ^2} = \frac{k_P}{D_e} f(c_A) \tag{6.63}$$

Since $\frac{d^2 c_A}{dZ^2} = \frac{dc_A}{dZ}\left[\frac{d}{dc_A}\frac{dc_A}{dZ}\right]$

And assign $p = dc_A/dZ$, and Eq. (6.63) and its BCs can be expressed as

$$p\frac{dp}{dc_A} = \frac{k_P}{D_e} f(c_A)$$

$$Z = L, c_A = c_{AS}, p = p_S = \frac{dc_A}{dZ}\Big|_S \tag{6.64}$$

$$Z = 0, c_A = c_{AC}, p = \frac{dc_A}{dZ} = 0$$

where c_{AC} is the concentration of A at the center of the catalyst, $(dc_A/dZ)_S$ is the concentration gradient at the external surface. The integration of Eq. (6.64) leads to

$$p_S = \frac{dc_A}{dZ}\Big|_S = \left[\frac{2k_P}{D_e}\int_{c_{AC}}^{c_{AS}} f(c_A)dc_A\right]^{1/2}.$$

The flux of A diffusing into the catalyst is

$$D_e a_P = \frac{dc_A}{dZ}\Big|_S = D_e a_P \left[\frac{2k_P}{D_e}\int_{c_{AC}}^{c_{AS}} f(c_A)dc_A\right]^{1/2},$$

and its value equals the total reaction rate of A in the catalyst under a steady-state conditions. If there is no internal diffusion resistance, the concentration of A inside the catalyst is equal to its concentration at the external surface c_{AS}, and the corresponding amount of component A reacted is

$La_P k_P f(c_{AS})$. Based on Eq. (6.53), the internal effectiveness factor can be calculated based on reaction rate with and without internal diffusion resistance:

$$\eta = \frac{\sqrt{2D_e}}{L\sqrt{k_P f(c_{AS})}} = \left[\int_{c_{AC}}^{c_{AS}} f(c_A) dc_A \right]^{1/2}. \tag{6.65}$$

Since the concentration at the center of catalyst c_{AC} is unknown and cannot be measured experimentally, Eq. (6.65) is not useful to estimate η. If the effects of internal diffusion resistance is significant, the concentration at the center of the catalyst $c_{AC} = 0$ for irreversible reactions and $c_{Ac} = c_{Ae}$ for reversible reactions, where c_{Ae} is the equilibrium concentration, which can be determined using thermodynamic analysis. Using $c_{AC} = 0$ for irreversible reactions or $c_{AS} = c_{Ae}$ for reversible reactions Eq. (6.65) can provide an accurate prediction of η if the effects of internal diffusion resistance are pronounced, but would only be an approximation if the internal diffusion resistance is not very large. Eq. (6.65) can be readily reduced to Eq. (6.54) for first order irreversible reactions. It should be noted that the internal effectiveness factor for non first order reactions depends on the concentration at the external surface, while for first order reactions the internal effectiveness factor is independent on the concentration at the external surface.

Since $L = V_P/a_P$, Eq. (6.65) can be rewritten as

$$\eta = \frac{a_P \sqrt{2D_e}}{V_P \sqrt{k_P f(c_{AS})}} = \left[\int_{c_{AC}}^{c_{AS}} f(c_A) dc_A \right]^{1/2} \tag{6.66}$$

With the plate thickness L being replaced with the characteristic length V_P/a_P, Eq. (6.66) is a more general form of Eq. (6.65), and can be used for catalysts with different shapes.

Example 6.7

The hydrogenolysis reaction of toluene occurs isothermally on a spherical catalyst with a 8 mm diameter.

$$C_6H_5CH_3 + H_2 \rightarrow C_6H_6 + CH_4$$
$$\text{(A)} \qquad \text{(B)}$$

The rate equation at the reaction temperature is

$$r_A = 0.32 c_A c_B^{0.5} \text{ kmol}/(\text{s} \cdot \text{m}^3 \text{ particle})$$

The initial concentration of toluene and hydrogen are 0.1 and 0.48 kmol/m³, respectively. The external diffusion resistance can be neglected. The effective diffusion coefficient of toluene in the catalyst is 8.42×10^{-8} m²/s. Estimate the internal effectiveness factor when toluene conversion is 10%.

(Continued)

(Continued)

Solution

Since it is not a first order reaction, the internal effectiveness factor should be calculated using Eq. (6.66). When toluene concentration is c_A, the hydrogen concentration is $c_{B0} - (c_{A0} - c_A) = (c_{B0} - c_{A0}) + c_A$. According to the data provided, we obtain

$$f(c_A) = c_A c_B^{0.5} = c_A[(c_{B0} - c_{A0}) + c_A]^{0.5} = c_A(0.38 + c_A)^{0.5},$$

and its integration:

$$\int_0^{c_{AS}} c_A(0.38 + c_A)^{0.5} dc_A = \frac{4 \times 0.38^{2.5}}{15} - \frac{2(2 \times 0.38 - 3c_{AS})(0.38 + c_{AS})^{3/2}}{15}.$$

Since the external diffusion resistance can be neglected, the concentration at the catalyst external surface is equal to that of bulk gas phase. When toluene conversion is 10%, toluene concentration is

$$c_{AS} = 0.09 \, \text{kmol/m}^3$$

Then

$$\int_0^{c_{AS}} c_A(0.38 + c_A)^{0.5} dc_A = 2.686 \times 10^{-3}$$

According to Eq. (6.66), the internal effectiveness factor is

$$\eta = \frac{4\pi(0.004)^2 \sqrt{2 \times 8.42 \times 10^{-8}} \times (2.686 \times 10^{-3})^{1/2}}{\frac{4}{3}\pi(0.004)^3 \times \sqrt{0.32} \times 0.09 \times (0.38 + 0.09)^{1/2}} = 0.4624$$

6.4.4 Effectiveness Factor Under the Influences of Both Internal and External Diffusions

The external effectiveness factor, η_X, and internal effectiveness factor, η, have been discussed in the previous sections. If both of them have to be considered, the overall effectiveness factor is defined as

$$\eta_O = \frac{\text{Reaction rate under the influence of both internal and external diffusion}}{\text{Reaction rate without diffusion influence}}$$

Based on what had been discussed in the previous sections, for first order reactions,

$$(-\mathscr{R}_A) = k_G a_m(c_{AG} - c_{AS}) = \eta k_W c_{AS} = \eta_O k_W c_{AG} \tag{6.67}$$

Their equalities are equivalent at steady-state: the first expression indicates that the reaction rate is equal to the rate of external diffusion; the second expression gives the reaction rate based on the internal effectiveness factor, in which the c_{AS} implies the influence of external diffusion; the third expression contains the overall effectiveness factor. Using these equations we obtain

$$c_{AS} = c_{AG} \Big/ \left(1 + \frac{k_W}{k_G a_m} \eta \right) \tag{6.68}$$

$$(-\mathscr{R}_A) = \eta k_W c_{AG} \Big/ \left(1 + \frac{k_W}{k_G a_m} \eta \right) = \left(\frac{\eta}{1 + \eta D_a} \right) k_W c_{AG} \tag{6.69}$$

Comparing Eq. (6.67) with Eq. (6.69), we obtain

$$\eta_O = \left(\frac{\eta}{1 + \eta D_a} \right). \tag{6.70}$$

If the internal diffusion can be ignored, i.e., $\eta = 1$, and only the external diffusion has to be consider, Eq. (6.70) reduces to

$$\eta_O = \left(\frac{1}{1 + D_a} \right), \tag{6.71}$$

Comparing to Eq. (6.21), we can see that the overall effectiveness factor η_O equals the external effectiveness factor η_X. This obviously makes sense when the internal diffusion can be ignored, the overall effectiveness factor is caused by the external diffusion.

If the external diffusion can be neglected and only the internal diffusion will have impact, then $c_{AG} = c_{AS}$, and Eq. (6.70) reduces to

$$\eta_O = \eta. \tag{6.72}$$

Substituting Eq. (6.54), which is the equation for the internal effectiveness factor of first order irreversible reactions in a slab catalyst, and Eq. (6.20), the definition of the Damkohler number, into Eq. (6.70), we obtain

$$\eta_O = \frac{\tanh(\phi)}{\phi \left(1 + \frac{k_W}{k_G a_m} \tanh(\phi) \right)}. \tag{6.73}$$

Since

$$\frac{k_W}{k_G a_m \phi} = \frac{k_W \phi}{k_G a_m \phi^2} = \frac{k_W \phi D_e}{k_G a_m L^2 k_P},$$

and $k_W = a_m L k_P$, therefore:

$$\frac{k_W}{k_G a_m \phi} = \frac{\phi D_e}{k_G L} = \frac{\phi}{Bi_m}.$$

Substituting these equations into Eq. (6.73), we obtain

$$\eta_O = \frac{\tanh(\phi)}{\phi\left(1 + \dfrac{\phi \tanh(\phi)}{Bi_m}\right)}, \tag{6.74}$$

where $Bi_m = k_G L/D_e$ is the Biot number of mass transfer, which characterizes the relative magnitude of external and internal diffusion resistance. When $Bi_m \to \infty$, the external diffusion resistance can be neglected. Eq. (6.74) reduces to

$$\eta_O = \frac{\tanh(\phi)}{\phi} = \eta$$

When the internal diffusion resistance can be neglected (i.e., $\tanh(\phi)/\phi = 1$), Eq. (6.73) gives

$$\eta_O = \frac{1}{\left(1 + \dfrac{k_W}{k_G a_m}\right)} = \frac{1}{1 + D_a} = \eta_X$$

In this section, we discuss diffusion and reaction processes involved in a heterogeneous catalytic reaction systems, and introduce the important concept of effectiveness factor, which can simplify the descriptions of such complex systems. The internal effectiveness factor is more useful. It characterizes the utilization of the internal surface of a catalyst particle and can provide guidance for catalyst manufacture and use. Using the concept of the internal effectiveness factor also makes the design calculation of heterogeneous catalytic reactors easier.

These discussions and results are valid only for isothermal particles, i.e., the temperature is uniform in the catalyst. If not, the differential equations for temperature distribution have to be established, and solved together with the differential equations describing the concentration distributions to obtain the internal effectiveness factors. Obviously, the processes are much more difficult. Fortunately, for most gas−solid catalytic reactions, the heat transfer resistance mostly exists in the laminar boundary layer surrounding the catalyst particle and the mass transfer resistance mainly exists inside the catalyst particles. Therefore, except for a few exceptions, using the isothermal assumption when analyzing the diffusion-reaction problem inside catalyst particles will not cause significant errors.

6.5   EFFECT OF INTERNAL DIFFUSION ON SELECTIVITY OF MULTIPLE REACTIONS

The diffusion-reaction processes for a single reaction inside a catalyst particle have been examined in the previous section. Generally, internal diffusion resistance lowers reactant concentration inside the catalyst, and hence reduces the reaction rate. If there are multiple reactions taking place inside the catalyst, what is the influence of internal diffusion resistance on selectivity? We first consider the following parallel reactions:

$$A \xrightarrow{k_1} B, \quad r_B = k_1 c_A^\alpha, \ \alpha > 0$$
$$A \xrightarrow{k_{21}} D, \quad r_D = k_2 c_A^\beta, \ \beta > 0$$

The concentration of reactant in the catalyst and at the external surface should be equal if there is no internal diffusion resistance. If B is the desired product, its instantaneous selectivity is

$$S' = \frac{\mathscr{R}_B}{-\mathscr{R}_A} = \frac{r_B}{r_B + r_D} = \frac{k_1 c_{AS}^\alpha}{k_1 c_{AS}^\alpha + k_2 c_{AS}^\beta} = \frac{1}{1 + \dfrac{k_2}{k_1} c_{AS}^{\beta-\alpha}} \tag{6.75}$$

If the internal diffusion has to be considered, the instantaneous selectivity is

$$S = \frac{1}{1 + \dfrac{k_2}{k_1} \langle c_A \rangle^{\beta-\alpha}} \tag{6.76}$$

where $\langle c_A \rangle$ is the average concentration of reactant A in the catalyst. Since $c_{AS} > \langle c_A \rangle$, the relative magnitude of S and S' is dependent on the difference between the orders of the two reactions. According to Eqs. (6.75) and (6.76), we know

if $\alpha = \beta$, $S' = S$
if $\alpha > \beta$, $S' < S$
if $\alpha < \beta$, $S' > S$

We can see when the reaction orders of the two reactions are equal, the internal diffusion has no impact on selectivity. If the reaction order of the main reaction is higher, the internal diffusion resistance will lower the selectivity. On the other hand, if the reaction order of the main reaction is lower, the internal diffusion will increase the selectivity.

Next we consider the influence of internal diffusion resistance on the selectivity of consecutive first order reactions:

$$A \xrightarrow{k_1} B \xrightarrow{k_2} D$$

At steady-state, the reaction rate of A is equal to the diffusion rate of A from the external surface into the interior of the catalyst; the reaction rate of B is equal to the diffusion rate of B from interior of the catalyst to the external surface. Using the methodology we applied to study single reactions, we can build the diffusion-reaction equations and solve them to obtain the concentration distributions and concentration gradients for both components A and B. If the pore internal resistance is significant and the effective diffusion coefficients of A and B are equal, $D_{eA} = D_{eB}$, we obtain the instantaneous selectivity as

$$S = \frac{\mathscr{R}_B^*}{-\mathscr{R}_A^*} = \frac{1}{1 + \sqrt{k_2/k_1}} - \sqrt{\frac{k_2}{k_1}} \frac{c_{BS}}{c_{AS}} \tag{6.77}$$

If the internal diffusion resistance is absent, the instantaneous selectivity is

$$S' = \frac{\mathscr{R}_B}{-\mathscr{R}_A} = \frac{k_1 c_{AS} - k_2 c_{BS}}{k_1 c_{AS}} = 1 - \frac{k_2}{k_1} \frac{c_{BS}}{c_{AS}} \tag{6.78}$$

Eqs. (6.77) and (6.78) show that the internal diffusion resistance reduces the instantaneous selectivity.

After discussing the effect of internal diffusion on selectivities of parallel and consecutive reactions, next we consider the effect of internal diffusion resistance on the yield of desired product. We will use the first order consecutive reaction system discussed above as an example, and assume the desired product is B. In Chapter 3, Tank Reactor, the correlation between yield and conversion of first order irreversible homogeneous consecutive reactions was derived, i.e., Eq. (3.68), which is valid for heterogeneous reactions if the internal resistance can be ignored. The upper curve in Fig. 6.7 is obtained using Eq. (3.68) when $k_1/k_2 = 4$. It shows the existence of a maximal yield. For heterogeneous consecutive reactions under the influence of internal diffusion, the correlation between yield and conversion (the details of how this correlation is derived is beyond the scope of this chapter and hence is not presented here) shown as the lower curve in Fig. 6.7, in which $k_1/k_2 = 4$, $D_{eA} = D_{eB} = 1$. The shapes of these two curves are similar, and the yield of desired product B also has a maximal value. The internal diffusion resistance reduces the yield, and the more severe the internal diffusion impact the lower the yield will be. For the example shown in Fig. 6.7, the yield is reduced by more than half, which indicates that the internal diffusion resistance has to be reduced to maintain the yield. This example illustrates the importance of understanding the interactions between reaction and transport processes.

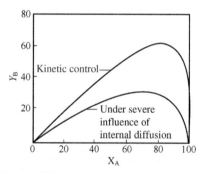

FIGURE 6.7   Effect of internal diffusion resistance on yield of consecutive reactions.

In this section we discussed the impact of internal diffusion on selectivity of multiple reactions using specific examples, but the conclusions are qualitatively applicable to most reaction systems. Of course, in order to obtain a quantitative conclusion for an specific reaction system, more detailed analysis is required.

6.6 DETERMINATION OF DIFFUSION IMPACT ON HETEROGENEOUS REACTIONS

The steps involved in heterogeneous catalytic reactions have been discussed in Section 6.1. According to the concentration distribution from the center of catalysts to bulk fluid, several scenarios can be distinguished based on rate-determining steps. For a reaction kinetic study, we have to first understand what is the rate-determining step so that we can properly select the experimental conditions. If the objective is to obtain the intrinsic kinetics, the experimental conditions selected should ensure that the reaction rate is controlled by the surface reaction steps and the effects of both external and internal diffusion are eliminated. When designing a commercial reactor, we need not only the intrinsic kinetic rate equations, but also the internal and external effectiveness factors under the operation conditions for the catalyst used.

Therefore, for both lab studies of heterogeneous catalytic reaction kinetics and engineering designs of industrial reactors it is necessary to determine the magnitude of the effects of internal and external diffusion on overall reaction behavior.

6.6.1 Determination of the Effects of External Diffusion

As discussed previously, the mass flow rate G through the catalyst bed has great impact on external diffusion resistance. When G increases, the mass transfer coefficient k_G increases as does the external diffusion rate. But G does not affect the internal diffusion process. According to this feature, the effect of external diffusion can be investigated by conducting experiments with various values for G while keeping other conditions (e.g., temperature, pressure, and space time) the same. For example, if we conduct an experiment on a tubular packed bed reactor with catalyst volume V_1 and reactant volume flow rate Q_1, the mass flow rate G_1 and space time Q_1/V_1 can be calculated. Under such conditions the conversion is measured as X_1. Then a series of experiments are conducted using the same reactor with various catalyst volumes V_2, V_3, ..., the reactant volume flux Q_2, Q_3, ..., and the corresponding mass flux G_2, G_3, ..., but the space time is maintained as Q_1/V_1 in all experiments. The dependence of conversion at the reactor outlet on the mass flow rate G is plotted as shown in Fig. 6.8.

Fig. 6.8 shows that there is a critical value of G_0, beyond which the conversion remains constant although G keeps increasing. It implies that the

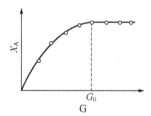

FIGURE 6.8 The influence of mass flux on the fractional conversion.

influence of external diffusion is eliminated if $G > G_0$. When $G < G_0$, conversion increases with mass flow rate indicating that the external diffusion has an impact, and can even be the rate-determining step.

It should be noted that the flow pattern inside the reactor in all experiments should be maintained at the same value. For example, plug flow should be maintained in packed bed reactors. Otherwise, different flow patterns lead to different degrees of backmixing in the reactor which will also influence conversion. As a result, it is difficult to distinguish the conversion changes caused by a different degree of backmixing associated with different flow patterns and/or external diffusion.

For the same catalyst, the higher the reaction temperature, the higher the mass flow rate required to eliminate the external diffusion impact. For the same reaction, the higher activity the catalyst has, the higher the mass flow rate required to eliminate the external diffusion impact.

This experimental method, i.e., investigating the influence of external diffusion by conducting experiments with a series of mass flow rates, is commonly applied in lab studies. However, it is very difficult to use in industrial reactors for commercial production. Certain criteria can be applied to determine whether the effects of external diffusion can be neglected. For mass transfer, the criteria is

$$\frac{\mathscr{R}_A^* L}{c_{AG} k_G} < \frac{0.15}{\alpha} \qquad (6.79)$$

where $\mathscr{R}$, α are the apparent reaction rate and reaction order, respectively. Inequality (6.79) may be used to estimate whether the concentration difference between the catalyst external surface and the bulk fluid can be neglected. Similarly, whether the temperature difference can be neglected may be determined by:

$$\frac{\mathscr{R}_A^* L(-\Delta H_r)}{T_G h_S} < 0.15 \frac{R T_G}{E}. \qquad (6.80)$$

If inequality (6.80) is valid, the error caused by neglecting the temperature difference is less than 5%. Comparing inequalities (6.79) and (6.80), the former is easier to be satisfied, which implies it is easier to eliminate the influence of external mass transfer, and much more difficult, though not impossible, to eliminate the influence of external heat transfer.

6.6.2 Determining the Effects of Internal Diffusion

When investigating the effects of internal diffusion resistance, it is definitely desirable to conduct the experiments without the influence of external diffusion. Under such conditions, $c_{AS} = c_{AG}$, $T_S = T_G$. Since the heat transfer resistance mainly exists in the gas film, the temperature can be assumed as a constant if $T_S = T_G$ is maintained.

The effects of internal diffusion can be characterized with an internal effectiveness factor, which is a function of Thiele modulus ϕ. For the same reactions occurring in the catalyst with the same temperature and compositions, ϕ only depends on the size of catalyst particles. So the effects of internal diffusion can be examined by conducting experiments using different particle sizes.

Consider an irreversible reaction with reaction order α, the apparent rate equation without the influence of external diffusion is

$$\mathscr{R}_A^* = -\frac{1}{W}\frac{dN_A}{dt} = k_W\eta c_{AG}^\alpha$$

If the experiments are conducted with two particle sizes, R_1 and R_2, while other conditions remain unchanged, the ratio of the measured reaction rates is

$$\frac{\mathscr{R}_{A1}^*}{\mathscr{R}_{A2}^*} = \frac{k_W\eta_1 c_{AG}^\alpha}{k_W\eta_2 c_{AG}^\alpha} = \frac{\eta_1}{\eta_2}$$

Without the influence of internal diffusion, $\eta = 1$, and hence $\mathscr{R}_{A1} = \mathscr{R}_{A2}$, i.e., the reaction rates are independent of particle size. If the influence of internal diffusion is strong,

$$\frac{\mathscr{R}_{A1}^*}{\mathscr{R}_{A2}^*} = \frac{\eta_1}{\eta_2} = \frac{\phi_2}{\phi_1} = \frac{R_2}{R_1}$$

i.e., the reaction rate is inversely proportional to the radius of particles. Generally, reaction rate decreases with particle size. Experiments conducted on catalysts with the same shape but different sizes usually give results similar to those shown in Fig. 6.9. Due to the influence of internal diffusion, the measured reaction rates increase when the particle sizes decrease until a critical size R_c is reached. After that the reaction rate becomes independent of particle size R, which implies that the internal diffusion does not have impact on the reaction when $R < R_c$. The reaction rate measured under such conditions is the intrinsic kinetic rate if the influence of external diffusion impact is absent as well. The internal effectiveness

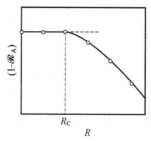

FIGURE 6.9   The influence of catalyst size on the reaction rate.

factors for larger catalyst particles can be calculated using this intrinsic kinetic rate.

The value of R_c depends on temperature and concentration (except for first order reactions). The higher the reaction temperature, the smaller the R_c. For reactions of order α, $f(c_A) = c_A^\alpha$. According to Eq. (6.65), we obtain

$$\eta = \frac{\sqrt{2D_e}}{L\sqrt{k_P c_{AS}^\alpha}} \left[\int_0^{c_{AS}} c_A^\alpha dc_A \right]^{1/2} = \frac{1}{L} \sqrt{\frac{2D_e}{(\alpha + 1)k_P c_{AS}^{\alpha-1}}}. \tag{6.81}$$

So, when the concentration increases, the catalyst of size L has to decrease to maintain the same η.

If it is determined that the influence of external and internal diffusion have been eliminated under high temperature and high concentration conditions, then the influence is absent under lower temperature and concentration conditions as well, except for reactions of negative orders.

If experiments are conducted only on one particle size, the extent of the internal diffusion can still be estimated. For example, the reaction rate of a first order irreversible reaction with no external diffusion influence is measured as $\mathscr{R}_A^*$

$$\mathscr{R}_A^* = \eta k_P c_{AG}.$$

If k_P is known, η can be calculated to evaluate the extent of internal diffusion. Unfortunately, k_P is usually unknown. According to the definition of the Thiele modulus, we obtain

$$k_P = D_e \phi^2 / L^2,$$

and

$$\mathscr{R}_A^* = D_e \phi^2 \eta c_{AG} / L^2$$

Assigning $\phi_S = \frac{\mathscr{R}_A^* L^2}{D_e c_{AC}}$, we obtain

$$\phi_S = \phi^2 \eta, \tag{6.82}$$

where ϕ_S is a dimensionless variable, can be determined experimentally. We already know $\eta = 1$ if $\phi \ll 1$. So if

$$\phi_S \ll 1, \tag{6.83}$$

the influence of internal diffusion can be neglected.

Eq. (6.83) is derived for first reactions. It can be applied to the reaction of other orders, though the definition of ϕ needs to be modified.

$$\mathscr{R}_A^* = \eta k_P f(c_A) \tag{6.84}$$

According to Eq. (6.65), the generalized Thiele modulus can be defined as

$$\phi^2 = \frac{L^2 k_P [f(c_{AG})]^2}{2D_e \int_{c_{AC}}^{c_{AS}} f(c_A) dc_A} \tag{6.85}$$

Comparing Eq. (6.84) with (6.85), we obtain

$$\phi_S^2 = \frac{\mathscr{R}_A^* L^2 f(c_{AG})}{2D_e \int_{c_{AC}}^{c_{AS}} f(c_A) dc_A} = \eta \phi^2 \ll 1 \tag{6.86}$$

if inequality (6.86) is valid, the influence of internal diffusion is eliminated.

6.7 EFFECTS OF DIFFUSION ON EXPERIMENTAL MEASUREMENT OF REACTION RATE

For the experimental study of heterogeneous reaction kinetics, both external and internal diffusion resistance have to be eliminated to obtain a true description of intrinsic kinetics such as reaction order, activation energy, etc. The intrinsic kinetic parameters cannot be obtained if the experiments were conducted under the influence of diffusion resistance.

If the reaction rate is measured when the process is controlled by the external diffusion, experimental results will show linear dependence of reaction rate on the reactant concentration no matter what the intrinsic kinetics of the reaction are. All reactions seem to be first order, as shown by Eq. (6.5), when the external diffusion dominates the overall kinetics. Therefore, if the experimental conditions are not properly selected to eliminate the impact of external diffusion, a wrong conclusion of first order will be reached. This is especially the case for lab scale experiments since at lab scale, the mass flow is usually low and the effects of interphase mass transfer phases may be significant.

Without the influence of external diffusion, the rate equation of reaction of order α is

$$\mathscr{R}_A^* = \eta k_P c_{AG}^\alpha \tag{6.87}$$

or in the logarithmic form

$$\ln \mathscr{R}_A^* = \ln \eta + \ln k_P + \alpha \ln c_{AG} \qquad (6.88)$$

Taking the derivative with respect to $\ln c_{AG}$, we obtain

$$\frac{d \ln \mathscr{R}_A^*}{d \ln c_{AG}} = \alpha + \frac{d \ln \eta}{d \ln c_{AG}} \qquad (6.89)$$

Assuming the rate equation was obtained under the influence of internal diffusion by no external diffusion:

$$\mathscr{R}_A^* = k_a c_{AG}^{\alpha_a} \qquad (6.90)$$

Theoretically, Eqs. (6.87) and (6.90) should be equivalent. However the order of reaction is different. The α in Eq. (6.87) is the intrinsic reaction order; α_a in Eq. (6.90) is the apparent reaction order. α is the property of the reaction, and is a constant; α_a varies with the extent of internal diffusion influence.

According to Eq. (6.90), we obtain

$$\frac{d \ln \mathscr{R}_A^*}{d \ln c_{AG}} = \alpha_a$$

Combining it with Eq. (6.89), we obtain

$$\alpha_a = \alpha + \frac{d \ln \eta}{d \ln c_{AG}} = \alpha + \frac{d \ln \eta}{d \ln \phi} \frac{d \ln \phi}{d \ln c_{AG}} \qquad (6.91)$$

From Eq. (6.85) the Thiele Modular for a slab catalyst and a αth order reaction is

$$\phi = L \sqrt{\frac{k_p}{2 D_e}} \frac{c_{AG}^\alpha}{\left[\int_0^{c_{AS}} c_A^\alpha dc_A \right]^{1/2}} = L \left[\frac{(\alpha+1) k_P}{2 D_e} \right]^{1/2} c_{AG}^{(\alpha-1)/2} \qquad (6.92)$$

Taking the logarithm of both sides of Eq. (6.92) and then taking the derivative with respect to $\ln c_{AG}$ we obtain

$$\frac{d \ln \phi}{d \ln c_{AG}} = \frac{\alpha - 1}{2}$$

By substituting it into Eq. (6.91), we obtain

$$\alpha_a = \alpha + \frac{\alpha - 1}{2} \frac{d \ln \eta}{d \ln \phi} \qquad (6.93)$$

Eq. (6.93) shows that the intrinsic and apparent reaction orders are equal if the internal diffusion resistance is absent (i.e., $\eta = 1$). If the influence of internal diffusion is severe, $\eta = 1/\phi$, then $d \ln \eta / d \ln \phi = -1$. Therefore, according to Eq. (6.93), the apparent reaction order is

$$\alpha_a = \frac{\alpha + 1}{2} \tag{6.94}$$

when the intrinsic reaction order is 0, 1, and 2, the apparent reaction order is 0.5, 1, and 1.5, respectively. $\alpha = \alpha_a$ is valid only for first order reactions, since the effect of internal diffusion on first order reactions is not related to the concentration. The apparent reaction orders of other reactions varies between $(\alpha + 1)/2$ and α, depending on the extent of internal diffusion impact.

Next we will examine the impact of internal diffusion on activation energy. Taking the logarithm of both sides of Eq. (6.90) and then taking derivative with respect to $(1/T)$ we obtain

$$\frac{d \ln \mathscr{R}_A^*}{d(1/T)} = \frac{d \ln k_a}{d(1/T)} = -\frac{E_a}{R} \tag{6.95}$$

Assuming that the dependence of the intrinsic rate constant k_P on temperature can be described by the Arrhenius equation, taking derivative of Eq. (6.88) with respect $1/T$ yields

$$\frac{d \ln \mathscr{R}_A^*}{d(1/T)} = \frac{d \ln \eta}{d(1/T)} + \frac{d \ln k_P}{d(1/T)} = -\frac{E}{R} + \frac{d \ln \eta}{d(1/T)} \tag{6.96}$$

where E is the intrinsic activation energy. Combining Eqs. (6.95) and (6.96) and rearranging we obtain

$$E_a = E - R \frac{d \ln \eta}{d \ln \phi} \frac{d \ln \phi}{d(1/T)} \tag{6.97}$$

Taking the logarithm of both sides of Eq. (6.92) and then taking the derivative with respect to $1/T$ gives:

$$\frac{d \ln \phi}{d(1/T)} = \frac{d \ln k_P}{2d(1/T)} = -\frac{E}{2R}$$

Substituting it into Eq. (6.97), we obtain

$$E_a = E + \frac{E}{2} \frac{d \ln \eta}{d \ln \phi} \tag{6.98}$$

Eq. (6.98) shows that the apparent activation energy depends on the severity of the pore diffusion impact. Fig. 6.10 is obtained from experimental measurements on the reaction rate of cumene cracking on Si-Al catalyst. At a given temperature the reaction rates were measured for different particle size. The severity of internal diffusion impact depends on catalyst particle sizes. For a given particle size experimental data can regress into a straight line, and the activation energy can by calculated by the slope. Fig. 6.10

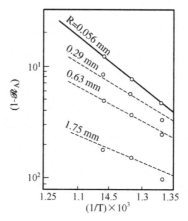

FIGURE 6.10   The influence of internal diffusion on the activation energy.

shows that the smaller the particle size, the steeper the slope becomes. So the apparent activation energy decreases with the increase of severity of internal diffusion influence.

When ϕ value is very small, $\eta \approx 1$, the apparent and intrinsic activation energy are equal; when ϕ value is large, $d \ln \eta / d \ln \phi = -1$, Eq. (6.98) becomes

$$E_a = \frac{E}{2} \tag{6.99}$$

The apparent activation energy is merely half of the intrinsic value.

Fig. 6.11 shows the relationship between $\ln(\mathscr{R}_A^*)$ and $1/T$. According to the temperature range, the effect of internal diffusion on activation energy can be classified into five regions:

I. High temperature region: External diffusion is the rate-determining step. The relationship between $\ln(\mathscr{R}_A^*)$ and $1/T$ is characterized by a straight line of slope E_D/R. $E_D = 4-12$ kJ/mol in this region, which is the lowest among the five regions. The lower panel of Fig. 6.11 shows the concentration distribution of reactant A for the five regions individually, in which the X-axis is the distance, the Y-axis is the concentration. The left side of mark R is the catalyst; the right side is the fluid, mark O represents the catalyst center. The zone between the solid and dashed line is the lamina boundary layer.

II. Transition region from external diffusion control to internal diffusion control. In this region both the external and internal diffusion resistance cannot be neglected. The apparent activation energy varies with temperature, and its correlation with the intrinsic activation energy is given by Eq. (6.98). Region IV is also a transition region from the internal diffusion control to kinetic control, in which the dependence of the apparent activation energy on temperature is even pronounced.

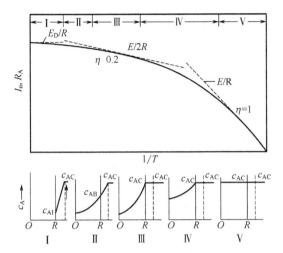

FIGURE 6.11   The activation energy within different temperature range.

III. Internal diffusion control. The external diffusion can be neglected in this region. The apparent activation energy is almost constant, which is close to half of the intrinsic one. The reactant concentration at the center of the catalyst is close to equilibrium, which is lower than the value for Region IV. The effectiveness factor is very low, and usually $\eta < 0.2$.

IV. Kinetic control. Reaction temperature is relatively low, and the effectiveness factor is close to one. Within this region, both external and internal diffusion resistance are eliminated. The apparent and intrinsic activation energy is equal, reflecting the intrinsic characters of the chemical reaction, and its value does not depend on temperature.

Generally, the apparent activation energy of heterogeneous reactions is not constant, but rather varies with temperature. It can be considered as constant only within certain temperature ranges, i.e., when the overall process is controlled by external diffusion, internal diffusion, or intrinsic kinetics. The measured activation energy becomes higher while the controlling step switches from external diffusion to internal diffusion to intrinsic kinetics. This phenomenon is caused by the different sensitivity of reaction and diffusion processes to temperature. The reaction process is very sensitive to temperature while the diffusion process is less sensitive. Fig. 6.11 also shows that temperature change can alter the controlling steps. The influence of diffusion is less pronounced at lower temperature.

FURTHER READING

Satterfield CN. Mass transfer in heterogeneous catalysis. London: M.I.T. Press; 1970.

Pertersen EE. Chemical reaction analysis. Englewood Cliffs NJ: Prentice-Hall; 1965.

Jackson R. Transport in porous catalysts. New York: Amsterdam Elsevier Illus, Elsevier Scientific; 1977.

Butt JR. Reaction kinetics and reactor design. Englewood Cliffs NJ: Prentice-Hall; 1980.

PROBLEMS

6.1 The gas phase reaction $A \rightleftharpoons B$ takes place in an isothermal spherical catalyst with radius R. Draw the schematic figure to show the concentration distribution of product B. The X-axis should be the distance r, the Y-axis should be c_B, referring to three conditions

1. Kinetic control

2. Film resistance control

3. Both film and pore resistance cannot be neglected

Mark the positions of concentrations at the bulk fluid, external surface, center of the catalyst, and the equilibrium concentration.

6.2 The heat transfer coefficient of the gas film surrounding the catalyst is 421 kJ/(m$^2 \cdot$ h $\cdot$ k). The density and heat capacity of the gas phase are 0.8 kg/m^3 and 2.4 kJ/(kg $\cdot$ K), respectively. Estimate the mass transfer coefficient of the gas film.

6.3 The true density, particle density, and specific surface area of a catalyst are measured as 3.60 g/cm^3, 1.65 g/cm^3, and 100 m^2/g, respectively. Estimate the pore volume, porosity, and average pore radius.

6.4 The bulk density and particle density of a Fe catalyst are 2.7 g/cm^3 and 3.8 g/cm^3, respectively, and the specific surface area is 16 m^2/g. Calculate the number of particles per cm^3 of bed and the surface area per cm^3 of bed.

6.5 Derive Eqs. (6.23) and (6.24).

6.6 Reaction $2C_2H_2 + 3H_2O \rightarrow CH_3COCH_3 + CO_2 + 2H_2$ takes place in a packed bed reactor filled with $ZnO\text{-}Fe_2O_3$ catalyst. At a certain position in the reactor, the pressure and temperature are 0.101 MPa and 400°C, respectively. The mole fraction of C_2H_2 is 3%. The rate equation is

$$r = kc_A$$

where c_A is the concentration of C_2H_2, $k = 7.06 \times 10^7 \exp$ $[-61,570/(RT)]$ s^{-1}. Calculate the external effectiveness factor.

Data: the particle diameter is 0.5 cm, the particle density is 1.6 g/cm^3, the diffusion coefficient of C_2H_2 is 7.3×10^{-5} m^2/s, the gas phase viscosity is 2.35×10^{-5} Pa $\cdot$ s, the mass flow rate in the bed is 0.24 kg/(m$^2 \cdot$ s)

6.7 A lab experiment is conducted to study the oxidization of ethylene in a tubular reactor packed with silver catalyst of diameter of 6.35 mm. The bed length is 80 cm and internal diameter is 2.1 cm. The mole fraction of ethylene in the feed gas is 2.25%, and the rest is air. At a location inside the reactor, the pressure and temperature are measured as 1.06×10^5 Pa, $T_G = 470K$, respectively, the conversion is 35.7%, the yield of ethylene oxide is 23.2%. The heat of reactions is

$$C_2H_4 + \frac{1}{2}O_2 \rightarrow C_2H_4O \quad \Delta H_1 = -9.61 \times 10^4 \text{ J/mol } C_2H_4$$

$$C_2H_4 + \frac{1}{2}O_2 \rightarrow 2CO_2 + 2H_2O \quad \Delta H_2 = -1.25 \times 10^6 \text{ J/mol } C_2H_4$$

The heat transfer coefficient is 210 kJ/($m^2 \cdot h \cdot K$). The particle density is 1.89 g/cm^3. If the total reaction rate of ethylene is 1.02×10^{-2} kmol(kg $\cdot$ h), calculate the temperature difference between the external surface of the catalyst and the bulk fluid.

6.8 The first order consecutive reactions

$$A \xrightarrow{k_1} B \xrightarrow{k_2} D$$

takes place at 0.101 MPa and 350°C.
$k_1 = 4.368$ s^{-1}, $k_2 = 0.4173$ s^{-1}. The particle density is 1.3 g/cm^3. Both $(k_G a_m)_A$ and $(k_G a_m)_B$ are equal to 20 cm^3/(g $\cdot$ s). Calculate the instantaneous selectivity of the desired product B when $c_{BG}/c_{AG} = 0.4$ and when the external diffusion impact can be neglected.

6.9 The oxidization reaction of dilute CO and air takes place on a Pt/Al$_2$O$_3$ catalyst. The pore volume of the catalyst is 0.3 cm^3/g, the specific surface area is 200 m^2/g, the particle density is 1.2 g/cm^3, and the tortuosity is 3.7. The ordinary diffusion coefficient of CO is 0.192 cm^2/s. Calculate the effective diffusion coefficient of CO.

6.10 Derive Eq. (6.60).

6.11 The decomposition reaction of gas A takes place on a spherical catalyst. The reaction is first order, irreversible, and exothermic. The diameter of the catalyst is 0.3 cm, the effective diffusion coefficient of the gas inside the particle is 4.5×10^{-5} m^2/h. The mass and heat transfer coefficient is 310 m/h and 161 kJ/($m^2 \cdot h \cdot K$), respectively. The heat of reaction is -162 kJ/mol. The concentration of A in bulk fluid is 0.20 mol/L. The measured apparent reaction rate is 1.67 mol/(L $\cdot$ min). Estimate

1. The influence of external diffusion
2. The influence of internal diffusion
3. The temperature difference between catalyst external surface and bulk fluid

6.12 The first order irreversible reaction

$$A \rightarrow B \tag{A}$$

takes place on a solid catalyst. The reaction rate constant is k. The mass transfer coefficient is $k_G a_m$, the internal effectiveness factor is η. C_{AG} is the concentration of A in bulk fluid

1. Derive

$$(-\mathscr{R}_A) = \cfrac{c_{AG}}{\cfrac{1}{k\eta} + \cfrac{1}{k_G a_m}} \tag{B}$$

2. Derive reaction rate equation similar to Eq. (B) when reaction (A) is reversible.

6.13 The gas phase hydrogenation of benzene takes place on a Ni catalyst of particle diameter 100 μm at a temperature of 150°C. Since the hydrogen is excessive in the feed, the reaction can be treated as first order (with respect to benzene). The intrinsic rate constant is measured and when both external and internal diffusion impact are eliminated $k_P = 5$ min^{-1}. The effective diffusion coefficient of benzene in the catalyst is 0.2 cm^2/s.

1. At operating pressure of 0.101 MPa, what is the maximum particle diameter to maintain $\eta = 0.80$.
2. If the operating pressure is 2.02 MPa, and the effective diffusion coefficient of benzene is inversely proportional to pressure, recalculate the maximum particle diameter to maintain $\eta = 0.80$.
3. If the reaction takes place in a liquid phase, where the effective diffusion coefficient of benzene in the catalyst is 10^{-5} cm^2/s but the reaction rate constant is the same, calculate the maximum particle diameter to maintain $\eta = 0.80$.

6.14 The first order irreversible reaction A→B takes place in a differential packed bed reactor filled with spherical catalyst particles. At 400°C, and reactant concentration is 0.05 kmol/m^3, the reaction rate is 2.5 kmol/(m^3 bed · min). The intrinsic kinetic rate per unit bed volume is $k_V = 50$ (1/s). The bed porosity is 0.3. The effective diffusion coefficient of A is 0.03 cm^2/s. Neglecting the external diffusion resistance, estimate

1. The internal effectiveness factor under reaction conditions.
2. The radius of the catalyst in the bed.

6.15 Butane dehydrogenation reaction takes place at 0.101 MPa and 530°C. A Al-Cr catalyst with diameter of 5 mm is used. The specific surface area, pore capacity, particle density, and tortuosity of the catalyst particles are 120 m^2/s, 0.35 cm^3/g, 1.2 g/cm^3, and 3.4, respectively. At reaction conditions the reaction can be treated as a first order irreversible reaction and the intrinsic kinetic rate constant is 0.94 cm^3/ (g · s). The external diffusion resistance can be neglected. Calculate the internal effectiveness factor.

6.16 The first order irreversible reaction takes place in a isothermal fixed bed reactor packed with spherical catalyst particles with diameter of

6 mm. At the operating temperature, the effective diffusion coefficient of reacting components is $0.2 \, \text{cm}^2/\text{s}$ and the intrinsic kinetic rate constant is $0.1 \, \text{min}^{-1}$. It is suggested to decrease the catalyst diameter to 3 mm to enhance productivity. Please evaluate whether using a smaller catalyst would increase the production and determine the magnitude of the production increase. Assume that both physical and chemical properties of the catalyst do not change with particle size and the reactor temperature is maintained at the same value

6.17 The oxidization of naphthalene to produce phthalic anhydride takes place on V_2O_5/SiO_2 catalyst at $1.013 \times 10^5 \, \text{Pa}$ and 350°C. The mole fraction of naphthalene in the naphthalene-air mixture is 0.10%. The rate equation is

$$r_A = 3.821 \times 10^5 \, p_A^{0.38} \exp\left(-\frac{135,360}{RT}\right), \quad \text{kmol}/(\text{kg} \times \text{h})$$

where p_A is the partial pressure of naphthalene. The diameter and density of the catalyst are 0.5 cm and $1.3 \, \text{g/cm}^3$, respectively. The effective diffusion coefficient is $3 \times 10^{-3} \, \text{cm}^2/\text{s}$. Assuming the external diffusion resistance can be neglected, calculate the reaction rate when naphthalene conversion is 80%.

6.18 Ethylbenzene dehydrogenation takes place on a spherical catalyst with diameter of 0.4 cm at 0.101 MPa and 600°C. The feed is a mixture of ethylbenzene and water steam with molar ratio 1:9. The rate equation is

$$r = k_P p_{EB}$$

where p_{EB} is the partial pressure (Pa) of ethylbenzene. $k = 0.1244 \exp[-9.13 \times 10^4/(RT)]$, kmol ethylbenzene/(kg·h·Pa). The density, porosity, and tortuosity of the catalyst are $1.45 \, \text{g/cm}^3$, 0.35, and 0.3, respectively. Calculate

1. The internal effectiveness factor if pore size is big enough so that the diffusion inside the pores is ordinary diffusion and the diffusion coefficient $1.5 \times 10^{-5} \, \text{m}^2/\text{s}$.

2. The internal effectiveness factor when average pore radius is $100 \times 10^{-10} \, \text{m}$.

6.19 The oxidization of benzene (B) on V catalyst to make maleic anhydride (MA):

All three reactions are first order. At a given location in the reactor the mole fraction of B and MA are measured as 1.27% and 0.55%, respectively. The temperature at the external surface of catalyst is 623K. At this temperature $k_1 = 0.0196\,\mathrm{s}^{-1}$, $k_2 = 0.0158\,\mathrm{s}^{-1}$, $k_3 = 0.00198\,\mathrm{s}^{-1}$, and $k_G a_m$ of both benzene and maleic anhydride is $1.0 \times 10^{-4}\,\mathrm{m^3/(s \cdot kg)}$. The particle density is $1500\,\mathrm{kg/m^3}$. Calculate the instantaneous selectivity and compare the calculated values with the instantaneous selectivity without the impact of external diffusion.

6.20 The decomposition of A is studied in a small fixed bed reactor under atmospheric pressure and 713K. The volume of the packed bed is $5\,\mathrm{cm^3}$ and the bed porosity is 0.4. The diameter of the catalyst is 2.4 mm. The effective diffusion coefficient of gas A is $1.2 \times 10^{-2}\,\mathrm{cm^2/s}$. At the outlet of the reactor, the concentration of A is measured as $1.68 \times 10^{-5}\,\mathrm{mol/cm^3}$, the reaction rate is $1.04 \times 10^{-5}\,\mathrm{mol/(cm^3\,bed \cdot s)}$. The reaction is first order and irreversible. Calculate the catalyst internal effectiveness factor.

Chapter 7

Analysis and Design of Heterogeneous Catalytic Reactors

Chapter Outline

Many important chemicals are commercially manufactured using heterogeneous catalytic reactions. Examples include ammonia, sulfuric acid, nitric acid, methanol, formaldehyde, vinyl chloride, acrylonitrile, and many others. For oil refining and other energy industries, heterogeneous catalysis has also been widely used for product processing, such as fluidized catalytic cracking, catalytic reforming, etc. With ever growing public concern on environmental protection, the requirements for treating air pollution by various industries has become more stringent and many gas treating processes involve heterogeneous catalytic reactions. A typical example is the catalytic converter for automobiles. In summary, catalytic reactions have been widely used for both industrial manufacture and environmental protection. About 80% of chemical reactions used in various industries are heterogeneous catalytic reactions.

Reaction Engineering. DOI: http://dx.doi.org/10.1016/B978-0-12-410416-7.00007-0

The objective of this chapter is to discuss the analysis and design of heterogeneous catalytic reactors. We will limit the discussions to reaction processes using solid catalysts. Depending on whether the solid catalysts are stationary or mobile, the catalytic reactors can be divided into two major categories. For fixed bed and trickle bed reactors the catalyst is at a stationary state. On the other hand, catalyst particles are mobile inside fluidized bed reactors, moving bed reactors and slurry reactors. In Chapter 1, Introduction, the structure of these reactors has been briefly discussed. In this chapter we will focus on fixed bed reactors. Fluidized bed reactors, trickle bed reactors, and slurry reactors will be discussed in the following chapters.

7.1   TRANSPORT PHENOMENA INSIDE FIXED BED REACTORS

In the previous chapter the impact of mass transfer between catalyst particles and the fluid around them on reaction rate has been analyzed. We focused on single catalyst particles and discussed both internal and external mass and heat transfer. The catalyst bed in a fixed bed reactor includes many catalyst pellets and of course the transport phenomena discussed in the previous chapter will occur in the catalyst bed. In addition, there are other transport phenomena that also take place inside the catalyst bed. These include dispersion and heat transfer in both the radial and axial directions. Next we will begin by discussing fluid flow inside a fixed bed, and then review other transport phenomena.

7.1.1   Fluid Flow Inside a Fixed Bed

A key parameter for describing fixed bed structure is the void fraction, which depends on particle size and shape, particle size distribution, the ratio of particle diameter to bed diameter, packing characters, etc.

At a given cross-section of a fixed bed the void fraction will not be uniform. For a fixed bed packed with uniform particles the void fraction is highest at a distance of $1 \sim 2$ particle diameters from the wall while lowest at the bed center, as shown in Fig. 7.1 where r is the radial distance from the wall. The influence on void fraction distribution from the wall is called the wall effect. For nonspherical particles, except for the area impacted by the wall effect the void fraction, distribution is uniform. However, for spherical particles, in addition to the wall effect, the void fraction fluctuates around its mean value, as shown in Fig. 7.1. Due to the wall effect, the larger the ratio of bed diameter to particle size, the more uniform the void fraction distribution will be. Usually the bed void fraction is the average value.

Inside the fixed bed reactor, the fluid flows through the distribution plate and enters the catalyst bed. The fluid will navigate the channels

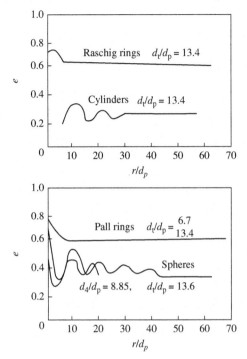

FIGURE 7.1   Radial distribution void fraction in a fixed bed.

between the catalyst particles. Those channels are interconnected to each other and tortuous, and the shape of cross-sections is irregular and the sizes vary a lot. In addition, the number of channels for fluid flow will likely change. Theoretically, the fraction of free area at a cross-section of the catalyst bed should be equal to the void fraction of the bed. As mentioned above, the void fraction is not uniform along the radial direction. Similarly, the free area fraction distribution along the radial direction is also not uniform. As a result, the fluid flow is not uniformly distributed across the bed. From the center of the bed, with the increase of the distance from the center the flow velocity increases. The velocity reaches its maximum at a distance of $1-2$ particle sizes from the wall, it then decreases until it becomes zero at the wall. The smaller the bed diameter to particle diameter ratio, the less uniform the flow velocity distribution will be.

In an empty tube the transition from laminar flow to turbulent flow is very clear. In a packed bed the flow pattern transition from laminar to turbulent flow is gradual. This is because the cross-section areas of different channels are different. At the same volumetric flow rate, fluid flow in some channels will stay in laminar flow status while the fluid flow in other channels becomes turbulent.

The pressure drop across a fixed bed is caused by two factors. One is the drag by the particles, i.e., the friction between the fluid and the solid particles. Another is the resistance caused by changes of cross-areas of the flow channels. Due to the sudden change of flow area and hence the collision between the fluid and the solid particles, additional resistances are introduced. When the flow through the bed is under laminar flow conditions, the former is the main resistance, and the latter becomes dominant when the flow is turbulent.

The fluid flow through a fixed bed is similar to that through an empty tube and the only difference is that for fluid flow through a fixed bed the shape of the channels is irregular. Therefore, the pressure drop equation for an empty tube can be modified to obtain pressure drop equations for a fixed bed. Eq. (7.1) is a commonly used pressure drop correlation for pressure drop of a fixed bed:

$$\Delta P = f \frac{L_r u_0^2 \rho (1 - \varepsilon)}{d_s \varepsilon^3} \tag{7.1}$$

In Eq. (7.1) the particle diameter d_s is defined as the diameter of a sphere that has the same surface area of the solid particle. The friction factor is a function of the Reynold number:

$$f = \frac{150}{Re} + 1.75 \tag{7.2}$$

And

$$Re = \frac{d_s u_0 \rho}{\mu} \cdot \frac{1}{1 - \varepsilon}$$

When $Re < 10$, the fluid flow in the bed is a laminar flow. The second term in Eq. (7.2) is much smaller than the first term and hence can be ignored:

$$f = \frac{150}{Re}$$

When $Re > 1000$, fluid flow in the bed is turbulent flow. The first term of Eq. (7.2) is much smaller than the second term. Under such condition f can be considered as a constant and equal to 1.75.

From Eq. (7.1), the two factors that have most impact on the pressure drop are void fractions of the bed and fluid flow velocity. A small change of either of these two factors will lead to significant changes of pressure drop. Therefore, it is critical to increase bed void fraction. For example, large catalyst particles can be used to reduce the pressure drop. Lower flow velocity can also reduce pressure drop. However, lower flow velocity will have a negative impact on mass and heat transfer. Therefore fluid flow velocity has to be carefully selected to optimize the overall performance.

The pressure drop calculated using the correlation is generally considered as the initial pressure drop of a new catalyst bed. With the increase of online time, the catalyst particle could be damaged or cracked into smaller particles, which will lead to a lower void fraction and a higher pressure drop. Such potential increases of pressure drop have to be considered when selecting the compressor and calculating power consumptions.

Example 7.1
Fe-Cr catalyst has been used in fixed bed reactors for steam reforming reactions at 0.6865 MPa. The catalyst is a cylindrical pellet with diameter of 9 mm and height of 7 mm. The average molecular weight of the feed gas is 18.96 and the mass flow rate (calculated based on an empty tube) is 0.936 kg/(s·m^2). The average temperature of the bed is 689K and the gas viscosity is 2.5×10^{-5} Pa·s. The density of catalyst particle is 2000 kg/m^3 and the bulk density of the bed is 1400 kg/m^3. Please calculate pressure drop per unit bed length.

Solution
The pressure drop can be calculated by using Eq. (7.1). Since it was asked to calculate pressure drop per unit bed length, we can set $L_r = 1$ m.

$$d_s = 6\frac{V_p}{a_p} = 6 \times \frac{0.785 \times 0.009^2 \times 0.007}{2 \times 0.785 \times 0.009^2 + \pi \times 0.009 \times 0.007} = 8.217 \times 10^{-3}\,\text{m}$$

And the bed void fraction is:

$$\varepsilon = 1 - \rho_b/\rho_p = 1 - \frac{1400}{2000} = 0.3$$

$$\rho = \frac{18.96}{22.4 \times \left(\frac{689}{273}\right) \times \left(\frac{0.1031}{0.6865}\right)} = 2.348\,\text{kg/m}^3$$

$$u_0 = \frac{G}{\rho} = \frac{0.9360}{2.348} = 0.3986\,\text{m/s}$$

$$\text{Re} = \frac{d_s G}{\mu(1-\varepsilon)} = \frac{8.217 \times 10^{-3} \times 0.936}{2.50 \times 10^{-5}(1-0.30)} = 439.5$$

Using Eq. (7.2):

$$f = \frac{150}{439.5} + 1.75 = 2.091$$

Using the values in Eq. (7.1) the pressure drop per unit bed length can be calculated:

$$\Delta p = \frac{2.091 \times 2.348 \times 0.3986^2(1-0.3)}{8.217 \times 10^{-3} \times 0.3^3} = 2461\,\text{kg/}(\text{s}^2 \cdot \text{m}^2) = 2461\,\text{Pa/m}$$

7.1.2 Mass and Heat Dispersion Along Axial Direction

The axial dispersion of a single fluid inside a flow reactor has been discussed in detail in Chapter 5, Residence Time Distribution and Flow Models for Reactors. The mass dispersion along the axial direction caused by axial mixing is also called backmixing. In a fixed bed reactor, the existence of solid particles will impact fluid axial mixing. Obviously the particle size will influence the backmixing. For a homogeneous system, the Peclet number has been used to describe the axial dispersion in the reactor. For fixed bed reactors, the Peclet number can also be used, but instead of bed length, L_r, the particle diameter, d_s, is used to define the Peclet number. The Peclet number for mass dispersion in the axial direction is:

$$(Pe_a)_m = \frac{d_p u}{D_a}$$

The Peclet number for heat dispersion in the axial direction is:

$$(Pe_a)_h = \frac{d_p u \rho C_p}{\lambda_{ea}}$$

where λ_{ea} is the effective heat conductivity in the axial direction of the bed. The concept of effective heat conductivity will be discussed later.

Both theoretical and experimental studies have demonstrated that for gas flowing through a fixed bed when the Reynold number $Re = \frac{d_p u \rho}{\mu} > 10$, the Peclet number for mass dispersion in axial direction $(Pe_a)_m = 2$. For liquid, $(Pe_a)_h = 0.3 \sim 1$.

Let's assume the axial mixing status inside a fixed bed with length of L_r is equivalent to that of N CSTR (Continuous Stirred Tank Reactor) reactors with equal volume in series. The length for axial dispersion or diffusion is l:

$$l = \frac{L_r}{N}$$

The diffusion time t_D, axial diffusivity D_a, and axial diffusion length follow the following equations:

$$t_D = \frac{l^2}{2D_a} \tag{7.3}$$

and

$$t_D = \frac{l}{u}$$

therefore

$$l = 2D_a/u = L_r/N$$

or

$$N = \frac{uL_r}{2D_a} = \frac{L_r}{2d_p} \times \frac{d_p u}{D_a} = \frac{L_r}{2d_p}(Pe_a)_m \qquad (7.4)$$

As mentioned above, for gas flowing through a fixed bed, $(Pe_a)_m = 2$, then from Eq. (7.4):

$$N = \frac{L_r}{d_p} \qquad (7.5)$$

We can see that if we want to use a series of equal volume CSTRs to describe the gas flow status inside a fixed bed the number of the CSTRs required is very large. Typically N has to be 50 or larger. The larger the number N is, the closer the flow pattern will be to a plug flow. For most industrial fixed bed reactors, N is much larger than 50 and therefore most of them can be treated as a plug flow reactor. Of course, if the bed length is too short it cannot be treated as a plug flow reactor and the axial dispersion has to be considered.

The above discussions are for isothermal conditions. If the reactor is nonisothermal, we also need to consider heat dispersion. The axial thermal Peclet number $(Pe_a)_h$ for a fixed bed reactor is about 0.6. Due to the difficulty of experimental measurement, so far there is still no good correlation that can be used to calculate axial thermal Peclet number. For a homogeneous system, the axial thermal and mass Peclet numbers can be considered to be the same, since the thermal and mass dispersion are caused by the same mechanism. However, for heterogeneous systems the mass will only transport through the void space between the solid particles while the heat transfer can take place through both the fluid between the solid particles and the solid particle themselves. For nonisothermal fixed bed reactors, the criterion for judging whether they can be treated as a plug flow is to see whether or not $L_r/D_p > 150$. Obviously, most industrial fixed bed reactors will meet this criterion. But for most reactors used in the lab, it is not easy to meet such requirements. Therefore, when designing lab experiments we need to be careful. Otherwise the data analysis may become very challenging.

7.1.3 Mass and Heat Transfer in Radial Direction

In addition to the axial direction, there are also transport processes along the radial direction inside a fixed bed. The concentration distribution over the cross-section that is perpendicular to the fluid flow direction is not uniform. Similarly, the temperature distribution is also not uniform. Fig. 7.2 illustrates the radial temperature distribution for a fixed bed reactor for oxidation of o-xylene. The temperature of feed gas at the reactor entrance

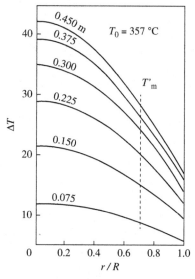

FIGURE 7.2   Radial temperature distributions in a fixed bed reactor.

is 357°C. The ordinate is the difference between bed temperature and entrance temperature and the abscissa is the dimensionless radial position. Since the reaction is exothermic and the reactor is cooled from the wall, the bed temperature is highest at the center and lowest at the wall. If the reaction is endothermic, the situation will be the opposite, i.e., the heat will be transferred from the thermal carrier outside the reactor to the catalyst bed. In Fig. 7.2 the intersection of the dash line and the curves represents the average temperatures at each cross-section.

The resistance for heat transfer in the radial direction can be divided into two parts. One is the bed itself and the other is the boundary layer at the wall. Although there are multiple heat transfer mechanisms involved inside a catalyst bed, the bed can be treated as a pseudo solid and its thermal resistance can be described by effective radial conductivity λ_{er}. When the fluid inside the bed is static, heat can be transferred through the fluid in the void space by conduction and radiation. In addition, heat can be transferred through solid particles and the mechanisms include: (1) conduction at the interface between particles; (2) conduction through the boundary layer around the particles; (3) radiation between particles; and (4) conduction inside the particles. In reality, the fluid flows through the bed, so we also need to add the contribution of convection. The overall heat transfer through both fluid and solid particles can be expressed by using effective radial conductivity. There are correlations for the λ_{er} calculation available in the literature. Thermal conduction at the interface between particles can be ignored unless it is under high vacuum conditions. Also thermal radiation can be ignored unless the temperature is very high.

The resistances for heat transfer between the solid bed and reaction wall are mainly concentrated at the boundary layer. From radial velocity distribution we know that at the wall the velocity is zero, and it is obvious that there is a boundary layer at the wall. The resistance can be described by the heat transfer coefficient h_w. Unfortunately the data available for h_w falls in a very wide range, and different authors reported very different experimental values. Therefore there is no widely accepted correlation available for h_w calculation. In addition, most experimental measurements were conducted without actual chemical reactions, and there were reports in the literature that the heat transfer coefficients obtained with and without chemical reaction were very different.

To simplify the calculation we can also combine the thermal resistances of the bed and the boundary layer and use a bed heat transfer coefficient h_t to describe the overall resistance. For spherical particles:

$$\frac{h_t d_t}{\lambda_f} = 2.03 \, \text{Re}^{0.8} \exp(-6d_p/d_t) \tag{7.6}$$

This correlation can be used when $20 < \text{Re} < 7600$ and $0.05 < d_p/d_t < 0.3$. λ_f is the conductivity of the fluid.

If the particle is cylindrical, and $20 < \text{Re} < 800$; $0.03 < d_p/d_t < 0.2$:

$$\frac{h_t d_t}{\lambda_f} = 1.26 \, \text{Re}^{0.95} \exp(-6d_p/d_t) \tag{7.7}$$

In this equation d_p is the equivalent diameter of the cylindrical particle based on external surface area. These two correlations are obtained by analyzing existing literature data. In most situations, the value of the overall heat transfer coefficient h_t is in the range of $61.2 \sim 320 \, \text{kJ}/(\text{m}^2 \cdot \text{h} \cdot \text{K})$.

After discussing heat transfer along the radial direction, next we will address mass transfer. Due to the distributions of temperature and flow velocity along the radial direction inside the fixed bed reactor, there will be concentration distributions along the radial direction. As a result, the reaction rate will also depend on different radial locations. Even without a chemical reaction, radial dispersion will take place just like the axial diffusion. The radial dispersion can be described by radial Peclet number $\text{Pe}_r = d_p u/D_r$ where D_r is radial diffusion coefficient. In a fixed bed reactor the radial dispersion is caused by redistribution of the fluid as a result of the collision of fluid, which moves up or down through the bed with solid particles and flow direction changes. When colliding with a solid particle, the fluid can move either to the right or left. Therefore, the distance for radial diffusion l_r can be considered to be approximately equal to one half of the particle diameter, i.e., $l_r \approx d_p/2$. The diffusion time is approximately equal to the time required for the fluid flowing through one layer of particles, i.e., $t_D \approx d_p/u$.

Using these two numbers in Eq. (7.3), and also substituting D_a with D_r, we obtain the Peclet number for radial dispersion in a fixed bed:

$$\text{Pe}_r = \frac{d_p u}{D_r} = 8$$

This is an estimation. When $\text{Re} > 20$ the experimentally measured value is $\text{Pe}_r = 10$.

For a fixed bed reactor with diameter of d_t, if the radial dispersion is simulated by using N CSTRs in series:

$$N = \frac{d_t}{l_r} \qquad (7.8)$$

From Eq. (7.3) the diffusion time is:

$$t_D = \frac{l_r^2}{2D_r} = \frac{l_r}{u}$$

Therefore

$$l_r = 2D_r/u$$

Substituting into Eq. (7.8) and rearranging:

$$N = \frac{u d_t}{2D_r} = \frac{d_t}{2d_p}\left(\frac{u d_p}{D_r}\right) = \frac{d_t}{2d_p}(\text{Pe}_r)$$

If $\text{Pe}_r = 10$, the above equation becomes:

$$N = 5\frac{d_t}{d_p}$$

We can see that N cannot be one, therefore the concentration distribution along the radial direction cannot be uniform. There will always be a concentration gradient along the radial direction, but the concentration gradient will decrease with an increase of d_t/d_p. When d_t/d_p is less than 5, the flow inside the bed will be very nonuniform, and the distribution curves for concentration and temperature may not be smooth. On the other hand, with a decrease of d_t/d_p the likelihood of fluid being short-circuited increases. In summary, as long as L_r/d_p is large enough the impact of axial dispersion can be ignored, but it is extremely difficult to achieve uniform distribution of concentration along radial direction by changing d_t/d_p. For adiabatic fixed bed reactors, it is necessary to consider heat and mass transfer in the radial direction.

7.2 MATHEMATICAL MODEL FOR FIXED BED REACTOR

The difference between a fixed bed reactor and a homogeneous tubular reactor is that there are solid catalyst particles in the fixed bed reactor.

Due to the existence of the solid phase, we have to consider transport phenomena between the two phases as well as inside the porous catalyst particles that were discussed in Chapter 6, Chemical Reaction and Transport Phenomena in Heterogeneous System. In addition, transport behaviors along both the axial and radial directions inside a fixed bed reactor are also different from those in a homogeneous reactor. The mathematical model for a fixed bed reactor needs to consider all transport phenomena and will be a group of nonlinear partial differential equations. The dependence of reaction rate on temperature is highly nonlinear, which makes solving these model equations very difficult. Even with modern computers it is still not easy. Therefore, a certain degree of simplification is usually required.

Most fixed bed reactors are cylindrical with two spatial variables along the radial and axial directions. As long as the concentration and temperature distributions are continuous, the variations of concentration and temperature along these two directions can be described by using a group of partial differential equations. As the fluid flows through the bed along the axial direction, the chemical reactions will progress and as a result there are always concentration and temperature gradients along the axial direction. As discussed in the above section, there will also be temperature and concentration distributions along the radial direction. If we consider temperatures and concentrations along the radial direction to be uniform and use average values to represent the temperature and concentration at each cross-section, the model can be simplified as one-dimensional, and the mathematical model becomes a group of ordinary differential equations.

As we discussed above, the fluid flow inside a fixed bed is very close to a plug flow, and therefore the plug flow model can be used. External and internal mass and heat transfers can be treated using the concept of effectiveness factor η_0, as discussed in Chapter 6, Chemical Reaction and Transport Phenomena in Heterogeneous System. Assuming the mass flow rate of the fluid entering the bed is G and the mass fraction of the key component is w_{A0}, a mass balance equation for component A can be written over a small section dZ of the bed:

$$\frac{G w_{A_0}}{M_A} = \frac{dX_A}{dZ} = \eta_0 \rho_b (-\mathscr{R}_A) \tag{7.9}$$

In Eq. (7.9), $\mathscr{R}_A$ is the consumption rate of component A based on per unit weight of catalyst. Since the mass balance is conducted over a small bed volume, the weight based reaction rate has to be converted to volume based by using catalyst bulk density ρ_b. Eq. (7.9) has the same format as the equation used for homogeneous reactor, Eq. (4.27), and the only difference is that Eq. (7.9) includes an effectiveness factor η_0 which describes the impact of internal and external diffusions. For most industrial catalytic reactions, interphase mass transfer usually does not have a big impact.

Thus η_0 can be substituted by η, i.e., only the internal diffusion impact needs to be considered.

It should be noted that although Eq. (7.9) has the same format, it is fundamentally different from the one-dimensional model for pseudohomogeneous reactors discussed in Chapter 4, Tubular Reactor. The pseudohomogeneous model ignores the concentration and temperature differences between the two phases. For example, for heterogeneous catalytic reactions, the pseudohomogeneous model considers the concentrations of reacting components in the fluid bulk phase to have the same values as those inside catalyst particle. The fluid temperature is also the same as that of the catalyst particles. In contrast, Eq. (7.9) considers such differences. Therefore Eq. (7.9) is not a pseudohomogeneous model.

If the thermal dispersion along the axial direction can be ignored, the heat balance over the element volume can be written as:

$$GC_{pt}\frac{dT}{dZ} = \eta_0\rho_b(-\mathscr{R}_A)(-\Delta H_r) - \frac{4U}{d_t}(T - T_c) \qquad (7.10)$$

The derivation procedure of Eq. (7.10) is the same as that of Eq. (4.26). The format of these two equations is also the same. The only difference is the item for reaction heat, i.e., the first term on the right-hand side, as discussed above.

If a significant pressure drop is required for the fluid to flow through the bed, a momentum balance equation is also needed to describe the pressure profile:

$$-\frac{dP}{dZ} = \frac{fG^2(1-\varepsilon)}{\rho d_p\varepsilon^3} \qquad (7.11)$$

The initial conditions for Eqs. (7.9) to (7.11) are:

$$Z = 0, \quad X_A = 0, \quad T = T_0 \quad P = P_0 \qquad (7.12)$$

Eq. (7.10) includes the temperature of the cooling medium T_c. If T_c cannot be treated as a constant, an equation to describe the axial distribution of T_c is also needed:

$$G_cC_{pc}\frac{dT_c}{dZ} = \frac{4U}{d_t}(T - T_c) \qquad (7.13)$$

And the initial condition for Eq. (7.12) is:

$$Z = 0, \quad T_c = T_{c0} \qquad (7.14)$$

G_c is the mass flow rate of the cooling fluid calculated based on the cross-section area of the bed, and C_{pc} is the heat capacity at constant pressure of the cooling fluid.

Solving Eqs. (7.9−7.14) is an initial value problem in ordinary differential equations, which are relatively easy using numerical approaches. By solving these equations, we can answer the most important questions for the design of a fixed bed reactor, i.e., reactor volume needed to achieve a given conversion. Through the simulation we can also understand the status of the reactor, compare various options, and select optimal operation conditions. Of course, in addition to economics, potential social impact must also be considered when selecting the optimal operation conditions, which is beyond the scope of this book.

Among the model equations the most important ones are the mass balance equation Eq. (7.9) and the heat balance equation Eq. (7.10). Unless at high pressure and the pressure drop is large, pressure distribution may not need to be considered, and then Eq. (7.11) is not required. Additional simplifications can be made based on specific situations. For example, under isothermal conditions Eq. (7.10) is not needed.

Eqs. (7.9) and (7.10) were derived for single reactions and need to be modified for multiple reaction systems. Assuming there are a total of M reactions taking place in the bed and there are K key components, then the mass balance for a key component A_i is:

$$\frac{d(u_0 c_i)}{dZ} = \rho_b \sum_{j=1}^{M} \eta_j v_{ij} \bar{r}_j, \quad i = 1, 2, \ldots, k \qquad (7.15)$$

η_j is the effectiveness factor for reaction j, $\bar{r}_j$ the normalized reaction rate for reaction j based on per weight of catalyst.

Correspondingly the heat balance equation is:

$$u_0 C_{pt} \rho_f \frac{dT}{dZ} = \rho_b \sum_{j=1}^{M} \eta_j \left| v_{ij} \right| \bar{r}_j (-\Delta H_r)_j - \frac{4U}{d_t}(T - T_c) \qquad (7.16)$$

And the initial values are:

$$Z = 0, \quad T = T_0, \quad c_i = c_{i_0}, \quad i = 1, 2, \ldots, k, \qquad (7.17)$$

The other equations are the same. In fact, only those terms related to the reaction rate have to be revised. In the above two equations, u_0 instead of G is used, but fundamentally these two are the same.

As pointed out in the previous section, for very shallow beds the assumption of plug flow is not valid. Under this situation we have to consider the axial dispersion. Similar to the axial diffusion model derived in Chapter 5, Residence Time Distribution and Flow Models for Reactors, the model for a fixed bed reactor for a single reaction under constant volume condition can be derived as:

$$\varepsilon D_a \frac{d^2 c_A}{dZ^2} - u_0 \frac{dc_A}{dZ} - \eta_0 \rho_b(-\mathscr{R}_A) = 0 \qquad (7.18)$$

And the heat balance equation is:

$$\lambda_{ea}\frac{d^2T}{dZ^2} - \rho_f u_0 C_{pt}\frac{dT}{dZ} + \eta_0\rho_b(-\mathscr{R}_A)(-\Delta H_r) - \frac{4U}{d_t}(T - T_c) \qquad (7.19)$$

The corresponding boundary conditions are:

$$Z = 0, \quad u_0(c_{A0} - c_A) = -\varepsilon D_a\frac{dc_A}{dZ} \qquad (7.20a)$$

$$u_0\rho_f(T_0 - T) = \lambda_{ea}\frac{dT}{dZ} \qquad (7.20b)$$

$$Z = L, \quad \frac{dc_A}{dZ} = \frac{dT}{dZ} = 0 \qquad (7.20c)$$

We can see that the model equations now are second order ordinary differential equations, i.e., an axial dispersion term has to be added to the equations. Obviously for normal kinetics the backmixing will lead to lower conversions. When axial mass and heat dispersions have to be considered, solving the model equations is a boundary problem in ordinary differential equations, which is much harder. For multiple reaction systems, the model equations need to be modified correspondingly, which will not be discussed here.

In order to use the model equations derived above, three sets of data will be required. (1) Kinetic data, i.e., reaction rate equation. The reaction rate equation depends on the specific reactions involved. In addition, for heterogeneous catalytic reactions even for the same reaction the rate equation may also depend on catalyst used. Usually reaction kinetics have to be measured in the lab since in most cases they cannot be found in the literature. (2) Thermodynamic data, such as reaction heat, specific heat capacity, equilibrium constant, etc. In most situations thermodynamic data can be found in the literature, and there are many such correlations that exist. (3) Transport data, including transport properties of the fluid such as viscosity, diffusivity, thermal conductivity, etc. In addition, structure data of the solid catalyst particles are also needed, such as pore size distribution, particle density, bulk density, specific surface area, etc. Based on these structure data and the operation conditions heat and mass transfer coefficients can be calculated using various correlations. Of course, it is more desirable to directly measure those transfer coefficients since the values will be more reliable.

Eqs. (7.9—7.11) are the most widely used mathematical models for a fixed bed reactor, and Eqs. (7.9) and (7.10) are the most important since in many situations Eq. (7.11) is not necessary. In the following sections, we will use the one-dimensional heterogeneous plug flow model to analyze fixed bed reactors.

7.3   ADIABATIC FIXED BED REACTOR

From reactor design and analysis perspective, it is convenient to classify the fixed bed reactors based on whether there is heat exchange between catalyst bed and the surrounding environment. And based on this criterion, the fixed bed catalytic reactor can be divided into two major categories. One type is that there is no heat exchange between the catalyst bed and the surrounding environment, and is called an adiabatic reactor. Another type is the heat exchanging reactor where heat exchange between the catalyst bed and the environment takes place.

7.3.1   Adiabatic Reactors

Adiabatic fixed bed reactors can be further divided into single- and multistage reactors. For single-stage reactors, the overall reaction takes place under adiabatic conditions. For multistage adiabatic reactors, the reaction takes place adiabatically in each stage. But after one stage of reaction, the reaction mixture temperature is adjusted through heat exchange to meet the reaction conditions needed for the next stage.

Single-stage adiabatic reactors have simple structures and space utilization is high. Therefore, they have the advantage of low cost. However, for reactions with large heat effect the temperature increase in the bed may be too large and the temperature at reactor exit may exceed acceptable levels. For reversible exothermic reactions, the axial temperature profile will be very far away from optimal temperature progression, which will lead to lower productivity. In certain situations, the conversion at reactor exit will not meet the target due to equilibrium limitation. Therefore, single-stage adiabatic reactor can only be applied in limited situations, such as:

(1) Reaction heat is small;
(2) The impact of temperature on the yield of desired product is limited;
(3) Reaction heat is high, but single pass conversion is low or the reaction mixture contains significant amount of inert component so that the temperature increase is limited. One example is ethanol synthesis by hydration of ethylene. The reaction heat is relatively low (44.16 kJ/mol), and single pass conversion is also low (4−5%), therefore single-stage adiabatic reactor is usually used for this reaction.

Multistage adiabatic reactors are commonly used for exothermic reactions, such as ammonia synthesis, methanol synthesis, and oxidation of SO_2, etc. Based on the heat exchange mode, multistage adiabatic reactors can be divided into three types: (1) indirect heat exchange; (2) cold shot cooling with feed; (3) cold shot cooling with nonfeed. The latter two are also known as direct heat exchange. Fig. 7.3 illustrates these three types of reactors.

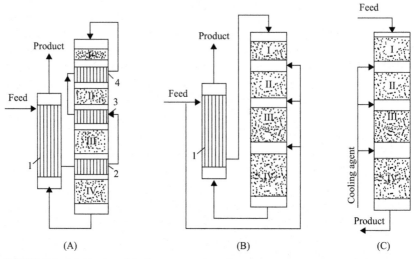

FIGURE 7.3   Multistage adiabatic reactors. (A) Indirect heat exchange; (B) Cold shot cooling with feed; (C) Cold shot cooling with nonfeed.

Fig. 7.3(A) schematically shows a four-stage catalytic reactor with indirect heat exchange. The feed mixture is preheated in heat exchanger 1−4 and then enters the first stage. Since the reaction is exothermic, the temperature of reaction mixture will increase after the first stage. The actual temperature increase depends on feed composition and conversion. The reaction mixture from the first stage will be cooled in the heat exchanger 4 and then enter the second stage. The cooling medium is the feed mixture, so the heat exchangers between stages serve the purpose of both cooling the reaction mixture and preheating the feed so that its temperature will meet the requirement dictated by the catalyst. The reaction mixture from the second stage will be cooled again and enter the third stage, and eventually leave the fourth stage and enter the first preheater to recover the heat. Overall, the reaction and heat exchange take place alternatively, which is a character of multistage adiabatic reactors. This type of reactor is widely used for the oxidation of sulfur dioxide, water gas shift reaction, etc.

For reactors with direct heat exchange or cold shot cooling, the temperature adjustment is achieved by directly adding cold stream. Fig. 7.3(B) is an adiabatic reactor with direct cold shot cooling with feed. It has four-stages, and the cold feed stream is used to adjust the temperature. Only part of the feedstock is preheated in the preheater 1 and the rest is used as cooling medium. After preheating, the feed stream is fed to the first stage and a cold feed stream is added to the reaction mixture to adjust the temperature before entering the second stage. This process is repeated until the final stage, and the product from the final stage is sent to the preheater for heat recovery. If the feed stream is already at high temperature, it is not practical to use the feed stream for cooling. In this case, other cold streams have to be used.

Fig. 7.3(C) shows a four-stage reaction with cold shot with nonfeed. Its operation is very similar to that with feed as the cooling agent. Typically one of the feed components is selected as cooling material. For example, water or steam is used for water gas shift reactions, i.e., water or steam is injected into the reaction mixture between the stages to adjust its temperature. Adding water can also increase water partial pressure in the mixture, which is beneficial for both reaction rate and equilibrium. Similarly, for the oxidation of sulfur dioxide, air is used as the cooling material, and the addition of air is helpful to maintain high oxygen partial pressure.

Compared with using heat exchange between stages, direct cold shot cooling has the advantage of reducing the number of heat exchangers needed. It is relatively easy to control reaction mixture by cold shot cooling, which can be achieved by adjusting the flow rate of the cold stream. However, the catalyst volume is usually higher than that of an indirect heat exchange reactor. In addition, it is also limited by what stream is available or useable as cooling materials. There are fewer limitations when using heat exchangers. Using heat exchangers also has the advantage of achieving heat recovery. More importantly, less catalyst is required. However, when using heat exchangers, the process scheme becomes more complex and the control is more complicated. The use of heat exchangers also means higher capital cost.

It is also possible to combine these two heat exchange modes, e.g., using the feed stream for cooling between the first and second stage and heat exchangers for cooling between other stages. Such an arrangement has been used in the oxidation of sulfur dioxide on vanadium catalyst.

7.3.2 Catalyst Volume for Adiabatic Fixed Bed Reactor

A key task for catalytic reactor design is to decide the amount of catalyst needed to achieve the required production rate. After the amount of catalyst is determined, the height of the catalyst bed can be determined based on the selected diameter of the bed. If the reactor is operated at constant pressure, then Eq. (7.11) is not needed. In order to determine the bed height required to achieve the desired conversion, Eqs. (7.9), (7.10), and (7.12) have to be solved. For adiabatic reactors there is no heat exchange between the bed and the external environment, and therefore the second term on the right side of Eq. (7.10) is zero, and Eq. (7.10) can be simplified as:

$$GC_{pt}\frac{dT}{dZ} = \eta_0\rho_b(-\mathscr{R}_A)(-\Delta H_r)_T \tag{7.21}$$

Dividing Eq. (7.9) by Eq. (7.21):

$$\frac{dT}{dX_A} = \frac{w_{A0}(-\Delta H_r)_T}{M_A C_{pt}} \qquad (7.22)$$

C_{pt} is the heat capacity at constant pressure for the reaction mixture and is a function of temperature and composition. The other parameters are constants. If C_{pt} is substituted with $\overline{C_{pt}}$, which is the heat capacity at average temperature and composition, the right side of Eq. (7.22) becomes a constant, and after integration:

$$T - T_0 = \lambda X_A \qquad (7.23)$$

$$\lambda = \frac{w_{A0}(-\Delta H_r)_{T_0}}{M_A \overline{C_{pt}}}$$

Eq. (7.23) has been derived and used in Chapter 3, Tank Reactor and Chapter 4, Tubular Reactor. Although the format of equation for λ is a little different, the physical meaning is the same. Careful attention needs to be paid to the units of the variables used in the equation. The unit for heat of reaction is J/mol, and the average specific heat capacity for the mixture is kJ/(kg·K).

Integrating Eq. (7.9)

$$L_r = \frac{Gw_{A0}}{M_A \rho_b} \int_{X_{A0}}^{X_{AL}} \frac{dX_A}{\eta_0(X_A, T)[-\mathscr{R}_A(X_A, T)]} \qquad (7.24)$$

Both the effectiveness factor and consumption rate depend on temperature and composition of the reaction mixture, and can be expressed as a function of X_A by using Eq. (7.23). Eq. (7.24) can be integrated. Multiply both sides of Eq. (7.24) with the cross-section area of the bed and the volume of the catalyst needed is:

$$V_r = \frac{F_{A0}}{\rho_b} \int_{X_{A0}}^{X_{AL}} \frac{dX_A}{\eta_0(X_A, T)[-\mathscr{R}_A(X_A, T)]} \qquad (7.25)$$

where F_{A0} is the molar flow rate of key component A at the reactor entrance.

If the specific heat capacity of the reaction mixture changes dramatically with regard to temperature and composition during the reaction process, using $\overline{C_{pt}}$ at average temperature and composition will lead to significant error. Under this situation the relationship between temperature and conversion is not linear, which means Eq. (7.23) is not valid. Then, in order to determine the amount of catalyst needed, Eqs. (7.9) and (7.22) have to be solved numerically.

Example 7.2

Wastewater and gas containing phenol can be treated by catalytically oxidizing the phenol into CO_2 and water so that the stream will meet disposal standards. Now we need to design an adiabatic fixed bed reactor to burn phenol in a gas stream at 0.1013 MPa. The gas steam flow rate is 1200 Standard m^3/h, and the phenol content is 800 mg/kg. The feed is fed to the reactor at 403K, and it is required that the phenol content after treatment has to be 100 mg/kg. A copper oxide catalyst supported on an 8 mm spherical alumina support is used. On this catalyst, the phenol combustion reaction is first order and the rate constant as a function of temperature is:

$$k = 7.03 \times 10^6 \exp\left(-\frac{5000}{T}\right), \text{min}^{-1},$$

Please calculate the amount of catalyst needed (the impact of external diffusion can be ignored).

Catalyst properties:

Specific area = 140 m^2/g;

Particle density = 0.9 g/cm^3;

Pore volume = 0.42 cm^3/g;

Tortuosity factor = 3;

Bed void fraction = 0.38;

Specific heat capacity at constant pressure for the reaction mixture = 30 J/mol;

Reaction heat = −2990 kJ/mol.

Solution

Eq. (7.23) can be used to calculate catalyst volume. Based on Chapter 6, Chemical Reaction and Transport Phenomena in Heterogeneous System, the effectiveness factor for first order irreversible reaction is:

$$\eta = \frac{1}{\phi}\left[\frac{1}{\tan h\,(3\phi)} - \frac{1}{3\phi}\right] \tag{A}$$

The Thiele module is:

$$\phi = \frac{R}{3}\sqrt{\frac{k_p}{D_e}} \tag{B}$$

The rate constant k that is given is defined based on bed volume, and needs to be converted to k_p:

$$k_p = k(1 - \varepsilon)$$

And $\varepsilon = 0.38$, therefore:

$$k_p = \frac{k}{1 - \varepsilon} = 7.03 \times 10^6 \frac{\exp\left(-\frac{5000}{T}\right)}{1 - 0.38} = 1.134 \times 10^7 \exp\left(-\frac{5000}{T}\right), \text{min}^{-1} \tag{C}$$

(Continued)

(Continued)

In order to calculate effective diffusivity D_e, we need to calculate the average pore diameter of the catalyst particle:

$$\langle r_a \rangle = \frac{2V_g}{S_g} = 2 \times \frac{0.42}{140 \times 10^4} = 6 \times 10^{-7} \text{ cm}$$

So the mass transfer inside the catalyst particle is dominated by Knudson diffusion:

$$D_K = 9700 \times 6 \times 10^{-7} \sqrt{T/94} = 6.003 \times 10^{-4} \sqrt{T}$$

In the above equation, 94 is the molecular weight of phenol. Catalyst porosity $\varepsilon_p = V_g \rho_p = 0.42 \times 0.9 = 0.378$, therefore:

$$D_e = \frac{\varepsilon_p}{\tau_m} D_K = \frac{0.378}{3} \times 6.003 \times 10^{-4} \sqrt{T} = 7.654 \times 10^{-5} \sqrt{T}, \text{ cm}^2/\text{s} \quad \text{(D)}$$

Due to temperature change, η is not a constant. The adiabatic temperature increase is:

$$\lambda = \frac{(-\Delta H_r)y_{A0}}{\overline{C_{pt}}} = \frac{2990 \times 0.0008}{30/1000} = 79.73 \text{K}$$

Use this value in Eq. (7.23) and we can obtain the relationship between temperature and conversion:

$$T = 403 + 79.73 \, X_A \quad \text{(E)}$$

Substitute Eq. (E) into (C) and (D), and then substitute Eqs. (C) and (D) into (B) and use $R_p = 0.4$:

$$\phi = 51,626(473 + 79.73X_A)^{-1/4} \exp\left[-2500/(473 + 79.37X_A)\right] \quad \text{(F)}$$

From Eq. (F) when $X_A = 0$, $\phi = 56.07$, and $X_A = 1$ $\phi = 115.6$. Therefore Eq. (A) can be replaced by:

$$\eta = \frac{1}{\phi} \quad \text{(G)}$$

And

$$(-R_A) = 7.03 \times 10^6 \exp\left(-\frac{5000}{T}\right) c_A = 7.03 \times 10^6 \exp\left(-\frac{5000}{T}\right) c_{A0}(1 - X_A)\frac{273}{T}$$

Use Eq. (E) and simplify:

$$(-\mathscr{R}_A) = 1.919 \times 10^9 \frac{c_{A0}(1 - X_A)}{473 + 79.73X_A} \exp\left(-\frac{5000}{473 + 79.73X_A}\right) \quad \text{(H)}$$

And

$$F_{A0} = \frac{1200c_{A0}}{60} = 20c_{A0} \text{ kmol/min} \quad \text{(I)}$$

(Continued)

(Continued)

The required phenol concentration at reactor outlet is 100 mg/kg, therefore

$$X_{AL} = \frac{800 - 100}{800} = 0.875$$

Use Eqs. (G), (H), and (I) in Eq. (7.25) and simplify:

$$V_t = \int_0^{0.875} \frac{5.38 \times 10^{-4}(473 + 79.73X_A)^{3/4} dX_A}{(1 - X_A) \exp\left[-\dfrac{2500}{473 + 79.37X_A}\right]} \tag{J}$$

This equation has to be integrated numerically and the catalyst volume is $V_r = 15.3 \text{ m}^3$. It should be noted that in this problem the reaction rate is defined based on bed volume, therefore it is not necessary to convert the right side of Eq. (7.25), i.e., divided by bulk density ρ_b.

If the required conversion increases, the catalyst volume needed will also increase. For example, if the phenol concentration has to be reduced to 10 mg/kg, the final conversion will be 98.75%, and the catalyst volume calculated by Eq. (J) will be 28.7 m³. We can see that an increase of conversion of 11.25% leads to the catalyst volume almost doubling.

7.3.3 Multistage Adiabatic Reactors

As mentioned above, there are three types of multistage adiabatic reactors. Here we will only discuss the design of multistage adiabatic reactors with an indirect heat exchanger. Fig. 7.4 illustrates the $T-X_A$ relationship for a reversible reaction inside a multistage adiabatic reactor. The dashed line is the optimal temperature profile. If the axial temperature distribution can be controlled to follow this optimal line, the best performance will be achieved, i.e., the overall reaction rate will reach a maximum value. The solid curve is

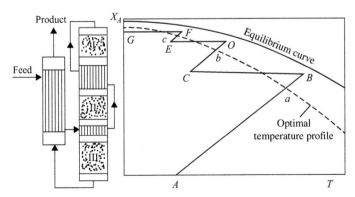

FIGURE 7.4 $T-X_A$ relationship in a three-stage adiabatic reactor with indirect heat exchange.

the equilibrium line, which represents the reaction limit, i.e., at any location inside the reactor the conversion and temperature must be below this line. The straight lines AB, CD, and EF are operation lines, reflecting the relationship between conversion and temperature inside the bed. The operation line can be described by Eq. (7.23). If the adiabatic temperature increase inside each stage is the same, AB//CD//EF. BC, DE, and FG are cooling lines, representing temperature change of the reaction mixture inside heat exchangers. During the heat exchange process the reaction mixture composition does not change, i.e., X_A is a constant, and hence the cooling lines are horizontal.

Fig. 7.4 is the $T - X_A$ plot for a three-stage adiabatic reactor. Similar plots can be developed for adiabatic reactors with more stages. With an increase of stages, the temperature distribution will be closer to the optimal temperature profile, and eventually will overlap the optimal profile when the number of stages approaches infinite. Although the reactor efficiency increases with an increase of stages, in practice at most five to six stages will be used. This is because more stages also mean higher cost, due to an increase in the number of pipes and valves a more complex reactor control will be needed. In addition, the improvement of reactor performance by adding additional stages decreases with the number of reactor stages. Therefore, it does not make economic sense to use too many stages.

For a given number of stages, feed composition, and final conversion, the conversions at the inlet of the first stage and the outlet of the final stage are fixed. However, there will be unlimited possible combinations of conversion and temperature of inlet and outlet of each stage. So determining those values will be an optimization problem, and hence we first need to choose an objective function. For reactor design, the objective function could be minimum catalyst usage, lowest production cost, maximum production, etc. Next we will use minimum catalyst usage as the objective function to decide conversions and temperatures for each stage when the number of stages, feed composition, and final conversion are given.

Assuming X_{Ai} and X'_{Ai} are conversions at the inlet and outlet of stage i, T_i, and T'_i are the temperatures at inlet and outlet of stage i, and the amount of catalyst in this stage is V_{ri}, the total catalyst volume for the reactor is:

$$V_r = \sum_{i=1}^{N} V_{ri} = \min \tag{7.26}$$

Let $\eta_0(X_A, T)\left[-\mathscr{R}_A(X_A, T)\right] = \mathscr{R}_A^*(X_A, T)$, and substitute Eq. (7.25) in Eq. (7.26)

$$V_r = \frac{F_{A0}}{\rho_b}\left[\int_{X_{A1}}^{X'_{A1}} \frac{dX_A}{\mathscr{R}_A(X_A, T)} + \int_{X_{A2}}^{X'_{A2}} \frac{dX_A}{\mathscr{R}_A(X_A, T)} + \cdots + \int_{X_{AN}}^{X'_{AN}} \frac{dX_A}{\mathscr{R}_A(X_A, T)}\right]$$

$$\tag{7.27}$$

For a reactor with N stages, the total number of conversions and temperatures will be $4N$. Since the conversions at the inlet of the first stage (known feed composition) and at the outlet of the last stage are known, the total number of unknown variables is $(4N-2)$. For any given stage the conversion at the outlet equals the conversion at the inlet of the next stage, i.e., $X'_{Ai} = X_{Ai+1}$. Therefore, the number of variables should be reduced by $(N-1)$. For any given stage, the conversions and temperatures at the inlet and outlet should follow Eq. (7.23), one of these four variables is not independent, and hence the total variables should further reduced by N. Therefore, the total independent variables is $(4N-2)-(N-1)-N = 2N-1$. If we choose the inlet temperature of all stages and inlet conversions of all stages except the first stage as independent variables, the total number will also be $(2N-1)$.

Take the partial derivatives for Eq. (7.27) with respect to X_{Ai} and let them equal zero, the following equation can be obtained:

$$\mathscr{R}^*_A(X_{Ai}, T'_{i-1}) = \mathscr{R}_A(X_{Ai}, T_i), \quad i = 2, 3, \ldots, N \qquad (7.28)$$

Similarly, take the partial derivatives of Eq. (7.27) with respect to T_i and let them equal zero:

$$\frac{\partial}{\partial T_i} \int_{X_{Ai}}^{X_{Ai+1}} \frac{dX_A}{\mathscr{R}^*_A(X_A, T)} = 0, \quad i = 1, 2, \ldots, N \qquad (7.29)$$

Eqs. (7.28) and (7.29) include a total of $(2N-1)$ equations. The optimal conversions and temperatures can be determined by solving these equations.

Eq. (7.28) indicates that in order to minimize the amount of catalyst needed, the reaction rate at the outlet of any stage should equal the reaction rate at the inlet of the next stage. The meaning of Eq. (7.29) is that for any stage and given inlet conversion in order to minimize catalyst usage in this stage there will be an optimal inlet temperature. Obviously, the total catalyst usage will be at minimum if the amount of catalyst for each stage has been minimized. The optimal inlet temperature for an adiabatic reaction has been discussed in Section 4.9.

Inside any stage of the reactor, with the progress of the reactions, the temperature at any given point is a linear function of inlet temperature T_i, therefore Eq. (7.29) can be expressed as;

$$\int_{X_{Ai}}^{X_{Ai+1}} \left[\frac{\partial\left(\dfrac{1}{\mathscr{R}^*_A(X_A, T)} \right)}{\partial T} \right]_{X_A} dX_A = 0, \quad i = 1, 2, \ldots, N \qquad (7.30)$$

In many situations using Eq. (7.30) is more convenient.

Neither Eq. (7.28) nor Eq. (7.30) has an analytical solution, and a numerical method has to be used to find the solution. Example 7.3 will illustrate numerical procedure to find conversion and/or temperatures.

For a given catalyst, typically there will be a maximum allowable operation temperature range. Operation temperature in each stage must be above minimum and also not exceed the maximum allowable operation temperature for the catalyst. The optimization procedure discussed above does not consider the temperature constraint. If the allowable temperature range of the catalyst has to be considered, seeking optimal conversion and temperature becomes a constrained optimization.

Example 7.3

A two-stage heat exchange reactor was selected for the water gas shift reaction. The molar fractions for CO, H_2O, CO_2, and H_2 are 0.1267, 0.5833, 0.394, and 0.1575, respectively, and the rest is inert gas. Cylindrical Cr-Fe catalyst has a diameter of 8.9 mm and a height of 7.67 mm. Feed temperature at the inlet of the first stage is 633K, and final conversion of CO is 91.8%. Please determine conversion after the first stage and inlet temperature at the second stage to minimize the catalyst usage.

For the Cr-Fe catalyst used, the overall rate equation (where diffusion impact was already considered) for the water gas shift reaction is:

$$\mathscr{R}_A^* = k^* p_A(1 - \beta)$$

$$\beta = p_C p_D / (p_A p_B K_p)$$

where p_A, p_B, p_C, and p_D are partial pressures of carbon monoxide, steam, carbon dioxide, and hydrogen, respectively. The impact of temperature on the equilibrium constant can be described as:

$$K_p = 0.0165 \exp\left(\frac{4408}{T}\right) \tag{A}$$

And the rate constant as a function of temperature:

$$k^* = 2.172 \times 10^{-4} \exp\left(-\frac{6542}{T}\right) \text{ mol/(g·min·Pa)} \tag{B}$$

Assume adiabatic temperature increase for both stages is 155.2K

Solution

Assuming the conversion for the first stage is 0.85. We already know the temperature at the inlet of the first stage, and from Eq. (7.23) we can obtain the operation equation for the first stage:

$$T = 633 + 155.2 X_A$$

Hence the temperature at the exit of the first stage is:

$$T_i' = 633 + 155.2(0.85) = 765K$$

$$p_A = 1.283 \times 10^4 (1 - X_A)$$

$$p_B = 5.909 \times 10^4 - 1.283 \times 10^4 X_A$$

(Continued)

(Continued)

$$p_C = 3.991 \times 10^4 + 1.283 \times 10^4 X_A$$

$$p_D = 1.595 \times 10^4 + 1.283 \times 10^4 X_A$$

Hence

$$\beta = \frac{\left(3.991 \times 10^3 + 1.283 \times 10^4 X_A\right)\left(1.595 \times 10^4 + 1.283 \times 10^4 X_A\right)}{1.283 \times 10^4 (1 - X_A)\left(5.909 \times 10^4 - 1.283 \times 10^4 X_A\right) K_p} \quad \text{(C)}$$

Based on exit temperature at the first stage, from Eqs. (A) and (B):

$$K_p = 0.0165 \exp\left(\frac{4408}{765}\right) = 5.247$$

$$k^* = 2.172 \times 10^{-4} \exp\left(-\frac{6542}{765}\right) = 4.197 \times 10^{-8} \text{ mol}/(\text{g·min·Pa})$$

When $X_A = 0.85$

$$\beta = \frac{\left(3.991 \times 10^3 + 1.283 \times 10^4 \times 0.85\right)\left(1.595 \times 10^4 + 1.283 \times 10^4 \times 0.85\right)}{\left(1.283 \times 10^4\right)(1 - 0.85)(5.909 \times 10^4 - 1.283 \times 10^4 \times 0.85)5.247} = 0.8225$$

The rate equation can be expressed as a function of conversion:

$$\mathcal{R}_A^* = k^*(p_{A0}(1 - X_A)(1 - \beta)) \quad \text{(D)}$$

Hence the reaction rate at the exit of first stage is:

$$\mathcal{R}_A^* = 4.197 \times 10^{-8} \times 1.283 \times 10^4 (1 - 0.85)(1 - 0.8225) = 1.434 \times 10^{-5} \text{ mol}/(\text{g·min})$$

Since the rate equation already includes the impact of internal diffusion, and for industrial reactors typically the impact of external diffusion can be ignored, there is no need to calculate the effectiveness factor.

Based on Eq. (7.28), the reaction rate at the inlet of the second stage must also be equal to this value, the temperature at the inlet of the second stage must be lower than that at the outlet of the first stage, and the conversion must be equal. Assuming the inlet temperature of the second stage is 633K, from Eqs. (A), (B), and (C):

$$K_p = 12.73$$

$$\beta = 0.339$$

$$k^* = 1.126 \times 10^{-8} \text{ mol}/(\text{g·min·Pa})$$

From Eq. (D) the reaction rate at the exit of the second stage is

$$\mathcal{R}_A^* = 1.126 \times 10^{-8} \times 1.283 \times 10^4 (1 - 0.85)(1 - 0.339) = 1.433 \times 10^{-5} \text{ mol}/(\text{g·min})$$

This value is very close to the value at the exit of the first stage and can be considered equal, which indicates the assumed temperature of 633K at the inlet of the second stage is correct.

(*Continued*)

(Continued)

The conversion at the outlet of the second stage can be calculated using Eq. (7.30). Let $\beta = g(X_A)/K_p$, then rate Eq. (D) can be rewritten as:

$$\mathscr{R}_A^* = p_{A0}(1 - X_A)\left[k^* - k^* g(X_A)/K_p\right]$$

$$\left[\frac{\partial \mathscr{R}_A^*}{\partial T}\right]_{X_A} = p_{A0}(1 - X_A)\left[(1 - \beta)\frac{\partial k^*}{\partial T} + \frac{k^* g(X_A)}{K_p^2}\frac{\partial K_p}{\partial T}\right] \tag{E}$$

Take partial derivative of Eqs. (A) and (B) with respect to temperature T:

$$\frac{\partial k^*}{\partial T} = \frac{6542 k^*}{T^2}$$

and

$$\frac{\partial K_p}{\partial T} = -\frac{4408 K_p}{T^2}$$

Substitute into Eq. (E) and simplify:

$$\left[\frac{\partial \mathscr{R}_A^*}{\partial T}\right]_{X_A} = 6542 k^* p_{A0}(1 - X_A)(1 - 1.674\beta)\frac{1}{T^2} \tag{F}$$

and

$$\left[\frac{\partial(1/\mathscr{R}_A^*)}{\partial T}\right]_{X_A} = -\frac{1}{\mathscr{R}_A^*}\left[\frac{\partial \mathscr{R}_A^*}{\partial T}\right]_{X_A}$$

Use Eqs. (D) and (F):

$$\left[\frac{\partial(1/\mathscr{R}_A^*)}{\partial T}\right]_{X_A} = \frac{6542(1.674\beta - 1)}{k^* p_{A0}(1 - X_A)(1 - \beta)^2 T^2} \tag{G}$$

Based on Eq. (7.23), the operation line for the second stage is:

$$T = 633 + 155.2(X_A - 0.85) = 531 + 155.2 \times X_A \tag{H}$$

Substitute Eq. (B) into Eq. (G), then use Eq. (H) so that it is expressed as a function of conversion. Then from Eq. (7.30):

$$\int_{0.85}^{X_{A2}} \frac{0.09744(1.674\beta - 1)\exp[42.15/(3.421 + X_A)]}{(1 - X_A)(3.421 + X_A)^2(1 - \beta)^2} dX_A = 0 \tag{I}$$

In this equation β is also a function conversion. Using Eqs. (A), (C), and (H):

$$\beta = \frac{478.3(0.0394 + 0.1267 X_A)(0.1575 + 0.1267 X_A)}{(1 - X_A)(0.5833 - 0.1267 X_A)\exp\left[\dfrac{28.4}{3.421 + X_A}\right]} \tag{J}$$

(Continued)

(Continued)

The conversion at the outlet of the second stage, X'_{A2}, can be determined by using Eqs. (J) and (I). Obviously these two equations cannot be solved analytically, a numerical method such as trial and error must be used. Assuming $X'_{A2} = 0.918$, the integration on the left-hand side of Eq. (I) is equal to 0.3654. If $X'_{A2} = 0.9179$, the integration value is −0.1702. Hence when the integration value is zero, X'_{A2} must be between 0.9179 and 0.918. These two values of X'_{A2} are already very close, and it is not necessary to seek a more accurate value. Therefore, the conversion of the second stage can be considered as 0.918, which is equal to the final CO conversion specified in the problem, indicating the assumption of conversion of the first stage of 85% is correct.

In this example, in order to simplify the calculation, the inlet temperature of the first stage was given as 633K. If this value was not given, how can it be determined? Under such circumstances, will the first stage conversion still be 0.85?

7.4 FIXED BED REACTOR WITH INTERNAL HEAT EXCHANGER

The main character of this type of reactor is that during the reaction process, the catalyst bed will exchange heat with external media. Fixed bed reactors with internal heat exchangers are widely used for commercial purposes, such as the epoxidization of ethylene to make ethylene epoxide, acetylene reacting with HCl for make ethylene chloride, ethylbenzene dehydrogenation to make styrene, hydrocarbon steam reforming for syn gas and ammonia synthesis, etc.

7.4.1 Overview

Heat exchanging reactors are also called nonadiabatic reactors, and their structure is similar to heat exchangers. Theoretically, the catalyst can be placed either inside or outside the tubes, but in most cases catalyst is loaded inside the tubes, as illustrated by Fig. 7.5. The feed gas enters the reactor from the top, flows downward through the catalyst bed, and exits the reactor from the bottom. In most industrial applications the reaction fluid will flow downward, although there are some situations where the fluid flows upward. The heat carrier flows outside the tubes, either countercurrently or cocurrently, depending on the need of the specific application. For endothermic reactions, the heat carrier acts as a heat source for the reactions. On the other hand, for exothermic reactions, the heat carrier acts as a cooling agent to remove the heat generated by the reactions. The heat exchange capability must be adequate to maintain the desired reaction temperature.

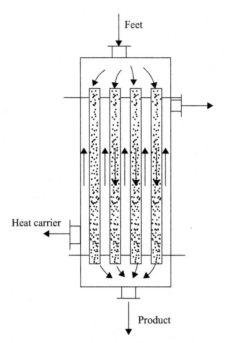

FIGURE 7.5 Fixed bed reactor with internal heat exchanger.

For tubular fixed bed reactors, adequate heat carrier selection is critical to control the reactor temperature and maintain stable operation. It is desirable to maintain a small temperature difference between the catalyst bed and the heat carrier, and at the same time the heat transfer rate must be high enough to remove the heat generated by the reactions. And it is obvious that at a given heat exchange area the heat transfer coefficient must be high. The selection of heat carrier depends on reaction temperature, reaction heat, and how sensitive the reaction is to temperature fluctuation. For reaction temperatures of around 473K, pressurized water is the most commonly used heat carrier; for reaction temperatures in the range of 523–573K, low volatile organic solvents, such as mineral oil, can be used. For reaction temperatures higher than 573K, molten inorganic salts such as potassium nitrate, sodium nitrate, and sodium nitrite mixture can be used. For very high temperatures, such as those higher than 873K, flue gas can be used. Different circulation modes can be used for the heat carrier, such as bubbling bed, forced external circulation, and internal circulation, etc.

The diameter of the tubes inside the reactor is usually pretty small, in the range of 25–30 mm. This is to minimize the radial temperature difference inside the catalyst bed and to increase heat exchange area per unit volume. In the past, due to the constraints of material strength and manufacture capability the maximum numbers of tubes that could be installed inside a reactor was about 3500, which can roughly hold $6-7$ m^3 catalyst. Due to this size limit, for large-scale commercial applications multireactors in parallel have to be used. With the progress of manufacturing technology and improved capability in reactor transportation and installation, over 20,000 tubes can now be installed into one reactor. In addition to reduced cost and increased productivity, using one big reactor instead of multiple reactors is beneficial for both reactor operation and product quality control.

Fixed bed reactors with an internal heat exchanger can be used for either exothermic or endothermic reactions. In reality, it is more commonly used for endothermic reactions. If the catalyst deactivates easily and frequent catalyst change is required, a fixed bed reactor with an internal heat exchanger is not a good choice. If the catalyst does not deactivate very fast and can be regenerated, a fixed bed reactor can still be used, but in this case two reactors will be required, with one being in production while the other is being regenerated.

Compared to adiabatic reactors, the radial temperature distribution inside a heat exchanging fixed bed reactor is more uniform, especially for highly exothermic reactions. Compared to fluidized bed reactors, heat exchanging fixed bed reactors has the advantages of low catalyst attrition, less backmixing, high productivity, and easy scale-up. Reactor behavior in multitube reactors can be studied in a single tube reactor, which makes the scale-up much easier, since higher production can be achieved by adding more reactor tubes. However, multitube fixed bed reactors have the disadvantages of a complex structure, high cost, and labor-intensive loading or unloading of the catalyst.

7.4.2   Analysis for Single Reaction

In this section we will examine the relationship between conversion and temperature inside a heat exchanging fixed bed reactor when only one reaction takes place. For adiabatic reactors we know reaction temperature is a linear function of conversion. The relationship between reaction temperature and conversion inside a heat exchanging fixed bed reactor can be obtained by dividing Eq. (7.9) by Eq. (7.10)

$$\frac{dT}{dX_A} = \frac{w_{A0}(-\Delta H_r)}{M_A C_{pt}} - \frac{4Uw_{A0}(T - T_c)}{d_t M_A C_{pt}\eta_0\rho_b(-\mathcal{R}_A)} \tag{7.31}$$

Using $\overline{C_{pt}}$ to replace C_{pt}, the above equation can be written as:

$$\frac{dT}{dX_A} = \lambda - \frac{4Uw_{A0}(T - T_c)}{d_t M_A \overline{C_{pt}}\eta_0\rho_b(-\mathcal{R}_A)} \tag{7.32}$$

Obviously, for adiabatic reactors the second term is zero and Eq. (7.32) will become Eq. (7.23)

Temperature control is very critical for reactor operations, since reaction rates are very sensitive to reaction temperature. For reversible endothermic reactions and irreversible endothermic or exothermic reactions, it is beneficial to run the reactor at the highest allowed temperature. And of course we need to keep the catalyst temperature as uniform as possible. For reversible exothermic reactions, there is an optimal reaction temperature to achieve maximum reaction rate, as discussed in Chapter 2, Fundamentals of Reaction Kinetics. Therefore, when a heat exchanging reactor is used for a reversible exothermic reaction, the catalyst bed temperature should be maintained as close as possible to the optimal temperature profile. Fig. 7.6 shows the relationship between reaction temperature and conversion. DMN is the equilibrium curve, and PBQ is the optimal temperature profile. The temperature at point H represents the temperature of the cooling medium. Point D, which is the cross point between straight line HD and equilibrium curve DMN, represents the maximum conversion at a given cooling medium temperature. Of course, the actual conversion will be lower than this maximum value.

Points A, E, and G represent the three feed inlet temperatures, and the curves ABD, EFD, and GD represent the relationship between conversion and temperatures for these three inlet temperatures. Let's examine curve ABD first. The reaction temperature first increases with conversion, $dT/dX_A > 0$, and after reaching a maximum value the temperature decreases with conversion, $dT/dX_A < 0$. The curve actually describes the axial temperature distribution inside the reactor.

$$\frac{dT}{dX_A} = \frac{dT}{dZ}\left(\frac{dX_A}{dZ}\right)^{-1} \tag{7.33}$$

Conversion always increases with bed height, i.e., $dX_A/dZ > 0$, hence dT/dZ and dT/dX_A must have the same sign, i.e., follow the same trend. At the early

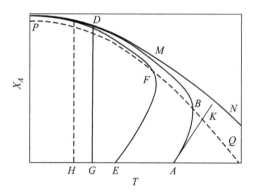

FIGURE 7.6 $T–X_A$ relationship in a heat exchanging fixed bed reactor.

stage the reaction is far from its equilibrium, and the reaction rate is high, which leads to the fact that heat release rate is higher than the heat-removing rate by the cooling medium. As a consequence, the temperature increases with conversion. At later stages the situation will just be the opposite, the heat-removing rate is higher than heat generation rate, and reaction temperature decreases with conversion. The maximum temperature is called the hot spot temperature. At this point $dT/dX_A = 0$. And from Eq. (7.33) dT/dZ is also zero. The location and value of the hot spot is very important for reactor operation and control, and can be used as an indication of reaction status. As long as the hot spot temperature is lower than the given value, the rest of the bed also will not exceed that temperature.

Line AK on Fig. 7.6 represents the adiabatic $T-X_A$ line when inlet temperature is T_A. Obviously, the $T-X_A$ curve ABD under heat exchanging condition is located to the left of adiabatic line AK. At the same conversion, due to the removal of the heat, the temperature under heat exchanging condition is lower than that under adiabatic condition.

When the inlet temperature is at a value that is lower than T_E, the $T-X_A$ curve moves to EFD, which follows the same trend as curve ABD, and at any given conversion the temperature is lower. Of course, this is based on the fact that the heat exchange conditions for these two scenarios are the same. If the heat exchange conditions are different, the temperature of reactors with lower inlet temperatures will not necessarily be lower than that with higher inlet temperature.

If the inlet temperature is further decreased to T_G, the bed temperature will quickly decrease because of the slower heat generation rate caused by low inlet temperature.

Comparing these three curves we can see that their proximities to the optimal temperature profile are different, which is the result of different inlet temperatures. Too high or too low of an inlet temperature will make the operation curve too far away from the optimal temperature profile. Therefore, there is an optimal inlet temperature. For the three temperatures illustrated in Fig. 7.6, point E is the optimal inlet temperature. Of course, this is only a rough and qualitative estimation. During reactor design, many different plans have to be evaluated to determine optimal operation conditions.

It should be noted that for all the curves shown in Fig. 7.6 at the early reaction stage they are all far away from the optimal temperature profile. However, this actually does not have a significant impact on the overall reaction. At early stages the reaction rate is much faster than that at the later stages, and the effect of reaction temperature being far away from optimal value is also much smaller. Therefore when evaluating a $T-X_A$ curve it is more important to examine its proximity to the optimal temperature profile at the mid and late stage. Based on this consideration, curve EFD is better than ABD.

In order to achieve uniform axial temperature distribution, i.e., $dT/dZ = 0$ or $dT/dX_A = 0$, based on Eq. (7.10) the following equation must be satisfied:

$$\eta_0 \rho_b (-\mathscr{R}_A)(-\Delta H_r) = \frac{4U}{d_t}(T - T_c) \qquad (7.34)$$

This means that at any point inside the bed the heat generation rate must equal the heat removal rate. In reality it is extremely difficult if not impossible to satisfy Eq. (7.34) everywhere in the bed, since heat generation intensity, i.e., heat generated per unit bed volume, is not uniform throughout the bed. It is highest at the beginning and gradually decreases along the flowing direction, which is dictated by reaction rate distribution along axial direction.

7.4.3 Analysis of Multiple Reaction Systems

When multiple reactions take place inside a fixed bed reactor with an internal heat exchanger, the reaction behavior becomes much more complex due to the side reactions. We will use the oxidation of o-xylene to make o-phthalic anhydride (PA) as an example to discuss. For o-xylene oxidization on vanadium catalyst the following reactions will take place:

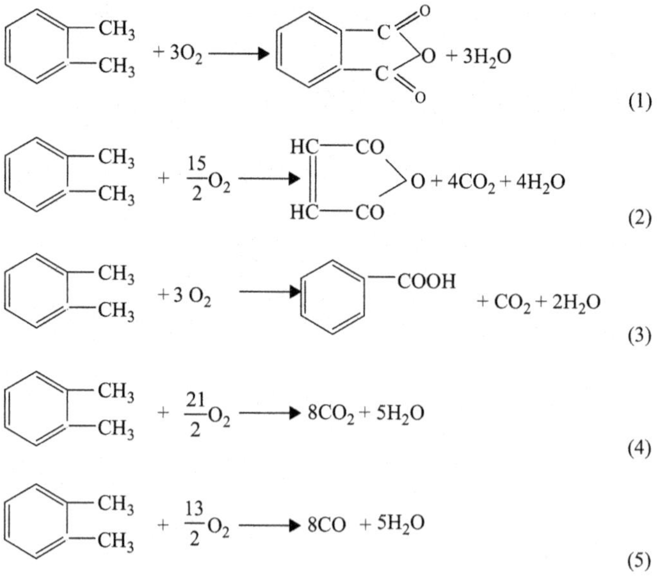

$$\text{(1)}$$

$$\text{(2)}$$

$$\text{(3)}$$

$$\text{(4)}$$

$$\text{(5)}$$

We can see that o-xylene can be oxidized into five different products: PA from the first reaction; maleic anhydride (MA) from the second; benzoic acid from the third; and carbon monoxide and carbon dioxide from reaction (2) to (5). The desired product is PA, so the first reaction is the main reaction, and the other four are side reactions. In addition to the above five reactions, other reactions are also possible, such as further oxidation of PA, MA, and benzoic acid to CO and CO_2.

Considering all the possible reactions will make the reactor analysis extremely complex and is also unnecessary. In practice some side reactions can be ignored. For example, the amounts of MA and benzoic acid formed are very small and hence reactions (2) and (3) can be ignored. If we consider further oxidation of PA to CO and CO_2, the overall reaction system can be simplified as the following reaction network:

$$\text{O-xylene (A)} \xrightarrow{\quad k_1 \quad} \text{PA (B)}$$

with k_3 downward from A and k_2 downward from B to

$$8(CO_2, CO)\ (C)$$

Reaction rate equations for these three reactions are:

$$r_1 = k_1 p_A p_0, \quad k_1 = 0.04017 \exp\left(-\frac{13,500}{T}\right) \tag{7.35}$$

$$r_2 = k_2 p_B p_0, \quad k_2 = 0.1175 \exp\left(-\frac{15,500}{T}\right) \tag{7.36}$$

$$r_3 = k_3 p_A p_0, \quad k_1 = 0.01688 \exp\left(-\frac{14,300}{T}\right) \tag{7.37}$$

The unit of pressure used in the above equations is Pa, and reaction rate is expressed as kmol/(kg·h), that is kmol PA per kg catalyst per hour. The feed is a mixture of air and o-xylene. To ensure safety, the o-xylene concentration must be kept below the explosion limit, which is about 1%. Therefore during the reaction process oxygen concentration can be considered as a constant and the kinetics can be treated as pseudo-first order. Although the number of moles changes for all the reactions during reaction, the reaction process can be treated as constant volume due to a significant excess of inert component (N_2 in the air).

Only two of the three reactions considered are independent. Therefore, the mathematical model for the reactor includes two mass balance equations and one heat balance equation. A pseudo-homogeneous model will be used, and PA yield Y_B, total yield of CO and CO_2 Y_C, and reaction temperature T

are selected as state variables and positioned along the axial direction as control variables to establish the following model equations:

$$\frac{Gy_{A0}}{M_m}\frac{dY_B}{dZ} = \rho_b \mathscr{R}_B \tag{7.38}$$

$$\frac{Gy_{A0}}{M_m}\frac{dY_C}{dZ} = \rho_b \mathscr{R}_C \tag{7.39}$$

$$GC_{pt}\frac{dT}{dZ} = \rho_b \mathscr{R}_B(-\Delta H_r)_B + \rho_b \mathscr{R}_C(-\Delta H_r)_C - \frac{4U}{d_t}(T - T_c) \tag{7.40}$$

$\mathscr{R}_B$ is formation rate of PA, $\mathscr{R}_B = \overline{r}_1 - \overline{r}_2$. Using Eqs. (7.35) and (7.36)

$$\mathscr{R}_B = p_0(k_1 p_A - k_2 p_B) \tag{7.41}$$

Assume initial molar fraction of oxygen is $(y_O)_0$, and if p_0 is considered as a constant, $p_O = p(y_O)_0$. $p_A = py_{A0}(1 - Y_B - Y_C)$, $p_B = py_{A0}(Y_B)$. Substituting these equations into Eq. (7.41) and simplifying:

$$\mathscr{R}_B = y_{A0}(y_O)_0 p^2 \big[k_1(1 - Y_B - Y_C) - k_2 Y_B\big] \tag{7.42}$$

It should be noted that Y_C is the total yield of CO and CO_2, and it is defined as the ratio of mole of o-xylene converted into CO and CO_2 to initial mole of o-xylene. Since one mole of o-xylene will make eight moles of $(CO + CO_2)$, when calculating the amount of CO and CO_2 using Y_C, the result has to be multiplied by eight to obtain the correct values.

CO and CO_2 formation rate is:

$$\begin{aligned}\mathscr{R}_C &= 8\,(\overline{r}_2 + \overline{r}_3) \\ &= y_{A0}(y_O)_0 p^2 \big[k_3(1 - Y_B - Y_C) + k_2 Y_B\big]\end{aligned} \tag{7.43}$$

Substituting Eqs. (7.42) and (7.43) into Eqs. (7.38) to (7.40):

$$\frac{dY_B}{dZ} = \frac{M_m \rho_b}{G}(y_O)_0 p^2 \big[k_1(1 - Y_B - Y_C) - k_2 Y_B\big] \tag{7.44}$$

$$\frac{dY_C}{dZ} = \frac{M_m \rho_b}{G}(y_O)_0 p^2 \big[k_3(1 - Y_B - Y_C) + k_2 Y_B\big] \tag{7.45}$$

$$\begin{aligned}\frac{dT}{dZ} = \frac{\rho_b y_{A0}(y_O)_0 p^2}{GC_{pt}} \Big\{ &\big[k_1(1 - Y_B - Y_C) - k_2 Y_B\big](-\Delta H_r)_B \\ &+ \big[k_3(1 - Y_B - Y_C) - k_2 Y_B\big](-\Delta H_r)_C \Big\} - \frac{4U(T - T_c)}{d_t GC_{pt}}\end{aligned} \tag{7.46}$$

And the initial conditions are:

$$Z = 0, \quad Y_B = 0, \quad Y_C = 0, \quad T = T_0 \tag{7.47}$$

The above reactor model does not consider internal diffusion. The catalyst used for o-xylene oxidation is manufactured by coating vanadium oxide and titanium onto ceramic balls. Since the active components are concentrated at a very thin layer near the catalyst surface, the potential impact of internal diffusion on reaction rates can be ignored. On the other hand, in industrial production, the gas mass flow rate is very high. As a consequence, both concentration and temperature differences between bulk fluid and catalyst surface are negligible. Lide [6] demonstrated the potential errors caused by neglecting external mass and heat transfer resistance would be at most a few percent.

The o-xylene oxidation reactor can be simulated by model equations Eqs. (7.44) to (7.47). O-xylene molar fraction in the feed mixture to the reactor is 0.8432% and the oxygen mole fraction is 20.33%. The average molecular weight of feed mixture is 29.29. Reactor ID is $d_t = 26\,mm$, and operation pressure is $p = 1.274 \times 10^5\,Pa$. Outside the reactor, forced circulated molten salt is used to remove reaction heat, and the molten salt temperature is kept at the same constant temperature as the feed temperature, i.e., $T_c = T_0$. The overall heat transfer coefficient is $U = 508\,kJ/(m^2 \cdot h \cdot K)$. Mass flow rate for gas inside the catalyst bed is $G = 2.948\,kg/(m^2 \cdot s)$. Packing density of the bed is ρ_b $1300\,kg/m^3$. The thermodynamic parameters are:

$$C_{pt} = 1.059 \text{ kJ}/(\text{kg} \cdot K)$$

$$(-\Delta H_r)_B = 1285 \text{ kJ}/\text{mol}$$

$$(-\Delta H_r)_C = 4561 \text{ kJ}/\text{mol}.$$

Using the above data in Eqs. (7.44) to (7.46), then substituting Eqs. (7.35) to (7.37)

$$\frac{dY_B}{dZ} = 4.748 \times 10^8 \left[(1 - Y_B - Y_C)\exp\left(-\frac{135,000}{T}\right) - 2.781 Y_B \exp\left(-\frac{155,000}{T}\right) \right]$$

(7.48)

$$\frac{dY_C}{dZ} = 1.995 \times 10^8 \left[(1 - Y_B - Y_C)\exp\left(-\frac{14,300}{T}\right) + 6.621 Y_B \exp\left(-\frac{155,000}{T}\right) \right]$$

(7.49)

$$\frac{dT}{dZ} = 1.661 \times 10^{11} \left[(1 - Y_B - Y_C)\exp\left(-\frac{135,000}{T}\right) - 2.781 Y_B \exp\left(-\frac{15,500}{T}\right) \right]$$

$$+ 2.477 \times 10^{11} \left[(1 - Y_B - Y_C)\exp\left(-\frac{14,300}{T}\right) + 6.619 Y_B \exp\left(-\frac{15,500}{T}\right) \right]$$

$$- 7.232(T - T_c)$$

(7.50)

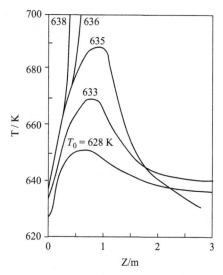

FIGURE 7.7   Axial temperature distributions inside the fixed bed reactor for o-xylene oxidation.

And initial conditions:

$$Z = 0, \quad Y_B = 0, \quad Y_C = 0, \quad \text{and} \quad T = T_0 = T_c$$

Eqs. (7.48) to (7.50) represent an initial value problem for ordinary differential equations, and can be solved numerically.

Figs. 7.7 and 7.8 show simulation results at four different feed temperatures. Fig. 7.7 shows axial temperature distributions. When T_0 is in the range of 628–635K, there is a maximum value on the temperature curve, and this maximum value is called the hot spot temperature, the difference between the hot spot temperature and feed inlet temperature can be several tens of degrees. The higher the feed inlet temperature is, the bigger the temperature increase will be and the further down along the axial direction the hot spot is located. Comparing temperature distributions for $T_0 = 633$ k and $T_0 = 635$K, we can find that 2 degree increases of feed inlet temperature leads to a more than 20 degree increase of hot spot temperatures. Another 1 degree increase of feed inlet temperature ($T_0 = 636$K) will lead to an increase of the hot spot temperature that is so large it will make the reactor unoperable. This phenomenon is called temperature runaway and we will further discuss this in the following section on parameter sensitivity.

The simulation results shown in Fig. 7.7 demonstrated that feed inlet temperature is a very critical design parameter for reactor operation because reactor behavior is very sensitive to its value. Therefore, its value must be carefully selected, since a difference of only a few degrees may lead to a significant increase of bed temperature and in turn cause catalyst damage.

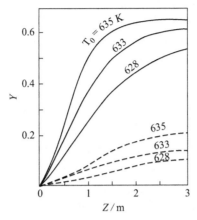

FIGURE 7.8   Axial distribution of o-phthalic anhydride (PA) yield inside the fixed bed reactor for o-xylene oxidation.

These results also reinforce the point discussed above, i.e. managing the hot spot temperature is important for reactor operation.

Fig. 7.8 shows yield distributions. The solid line is PA yield while the dashed line represents CO and CO_2 yield. Both Y_B and Y_C increase along the axial direction. Higher feed inlet temperature also leads to higher PA and (CO + CO_2) yields, as long as the reactions do not run away. PA is the desired product and higher yield means more production. From this perspective, running the reactor at higher feed inlet temperature is beneficial. However, Y_C also increases with feed inlet temperature, which means more feed is converted into undesired by-product (CO and CO_2). Therefore, whether higher feed inlet temperature will be beneficial depends on which yield would increase more. If increasing the feed inlet temperature will lead to a significant increase of Y_B but a lesser increase of Y_C, higher feed inlet temperature will obviously be beneficial. If the temperature increase leads to more of an increase in Y_C than Y_B, it may not be beneficial to increase feed inlet temperature. In fact, we need to consider the impact of temperature on reaction selectivity. Based on our simulations at 628K the selectivity is 83.4%, and will decrease to 80.5% at 633K, i.e., the selectivity decreases with an increase of feed inlet temperature. Lower selectivity means more o-xylene will be consumed per unit PA production. In other words, the feed utilization efficiency is lower. Therefore, both yield and selectivity have to be considered when selecting feed inlet temperature.

7.5 AUTOTHERMAL FIXED BED REACTORS

An autothermal fixed bed reactor is a special type of fixed bed reactor where feed gas is used as the cooling medium to control bed temperature and at the same time is heated to the desired temperature to enter the catalyst bed. Obviously this type of reactor is only suitable for exothermic reactions and a situation where the feed stream has to be heated before entering the reactor.

7.5.1   Feed Flow Direction

For a fixed bed reactor with heat exchange, the relative flow directions of the feed stream and heat exchanging medium will influence the heat exchanging effect. In the previous section, we only discussed a situation when the temperature of heat exchanging medium is constant. In this situation its flow direction has no impact on heat transfer. In this section we will discuss the impact of flow direction by using autothermal reactors as an example.

Fig. 7.9 illustrates temperature distribution for two autothermal fixed bed reactors with different flow patterns. Fig. 7.9(A) shows a countercurrent flow design. The feed stream enters the shell side and flows countercurrently with gas flowing through the catalyst bed inside the tube. The feed stream is heated while flowing through the shell side, and its temperature increases from T_{CLr} at the reactor inlet to T_{c0}. After entering the catalyst bed where both reaction and heat exchange take place, the gas temperature increases from T_0 to the hot spot temperature, and then gradually decreases, as shown by the top curve on Fig. 7.9(A).

Fig. 7.9(B) illustrates a cocurrent flow situation. The temperature distribution seems very similar to that under countercurrent flow conditions. However, a closer examination will reveal that under countercurrent flow conditions, gas temperatures inside the catalyst bed increase quickly to its

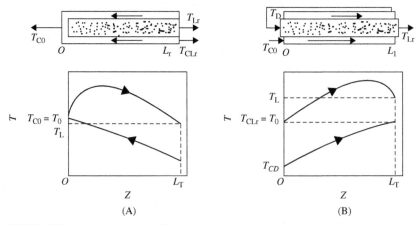

FIGURE 7.9   Axial temperature distribution for two autothermal fixed bed reactors with different flow patterns. (A) Countercurrent flow; (B) Cocurrent flow.

hot spot temperature, and at later stages of the reaction the bed temperature also decreases quickly. Under cocurrent conditions the situation is just the opposite. And the difference between these two operation modes can be explained by considering the temperature difference and heat release rate at different locations inside the catalyst bed. For both countercurrent and cocurrent operations, at the earlier stages, or at a location near the bed entrance, the heat generation rate is high because of high feed concentration and the reaction's large distance from its equilibrium. On the other hand, the gas temperature outside the tube depends on flow pattern. Under countercurrent operation, gas temperature in this region is high, which leads to a small temperature difference acting as the driver for heat transfer. As a consequence, gas temperature inside the tube will increase quickly. Under cocurrent operation, gas outside the tube just enters the reactor and hence is at low temperature, which means a large temperature difference and faster heat transfer rate. Therefore, gas temperature inside the tube will increase slowly. At later stages of the reaction, the amount of heat generated per unit bed volume is smaller, hence it is desirable to have a slow heat removal rate so that the bed temperature won't be too low. Countercurrent flow cannot achieve such a result, because the gas outside the tube in this region is fresh feed just entering the reactor at low temperature. The large temperature difference may lead to an overly high heat removal rate, which in turn leads to unacceptably low bed temperatures. A cocurrent pattern can avoid this phenomenon and maintain relatively high reaction temperatures at later stages. In summary, countercurrent flow operation has the advantage of being able to heat up the feed entering the catalyst bed quickly to reach optimal reaction temperature, but the drawback is the reaction temperature at later stages of the reaction may be too low. The advantage of cocurrent flow operation is that it can maintain bed temperature at later stages of the reaction. The disadvantage is that it may take longer to heat the bed to the desired reaction temperature. It is possible to optimize reactor performance by combining the advantages of these two operation patterns. For example, for cocurrent flow reactors, an adiabatic bed can be used in front of the heat exchanging bed. After being preheated, the feed gas will first enter the adiabatic bed so that the temperature of gas stream will increase quickly and then enter the heat exchanging bed. This design still has the advantage of maintaining bed temperature at later stage of the reaction while overcoming the disadvantage of slow temperature increase at the early stage.

7.5.2 Mathematical Model

Eqs. (7.9) and (7.10) can also be used for autothermal fixed bed reactors. However, since autothermal reactors use feed gas as a cooling agent, the heat balance equation needs to be adjusted accordingly. As discussed, during the preheating process, feed gas temperatures will also be a function of axial location, and it is not an independent variable. It value is dictated by gas temperature and conversion inside the catalyst bed. This relationship will be discussed in this section.

At the entrance of the catalyst bed the temperatures of the reaction mixture and cooling medium are T_0 and T_{c0}, respectively, and at any given cross-section of the bed the reaction gas temperature and conversion are T and X_A, respectively, and the cooling medium temperature is T_c. The heat balance between these two sections can be written as:

$$\frac{Gw_{A0}}{M_A}(-\Delta H_r)X_A = G\overline{C}_{pt}(T - T_0) + G_c\overline{C}_{pt}(T_c - T_{c0}) \qquad (7.51)$$

where G and G_c are the mass flow rate per unit cross-section area for the reaction mixture and cooling medium, respectively.

The left side of Eq. (7.51) represents the amount of heat generated. The first item on the right-hand side is the sensible heat for the gas inside the bed, and the second item is the sensible heat obtained by the cooling medium.

Eq. (7.51) can be rearranged as:

$$T_c = T_{c0} + \frac{G\overline{C}_{pt}}{G_c\overline{C}_{pc}}\left[\frac{w_{A0}(-\Delta H_r)}{M_A\overline{C}_{pt}}X_A - (T - T_0)\right] \qquad (7.52)$$

$$\frac{w_{A0}(-\Delta H_r)}{M_A\overline{C}_{pt}} = \lambda$$

$$\text{Let}\,\frac{G\overline{C}_{pt}}{G_c\overline{C}_{pc}} = \beta$$

Eq. (7.52) can be written as:

$$T_c = T_{c0} + \beta(\lambda X_A - T + T_0) \qquad (7.53)$$

Eq. (7.53) gives the cooling medium temperature at any cross-section as a function of reaction gas temperature and conversion inside the bed. This is a general equation that is applicable for different types of reactors with heat exchange, and the value of β describes different situations. If $\beta = 0$, $T_c = T_{c0}$, which corresponds to the situation of constant cooling medium temperature. If β approaches to infinity, then $G_c = 0$, corresponding adiabatic condition. If $\beta = 1$, then $G\overline{C}_{pt} = G_c\overline{C}_{pc}G$, which means feed gas is used as the cooling medium, i.e., an autothermal reactor. Under this situation:

$$T_c = T_{c0} \pm (\lambda X_A - T + T_0) \qquad (7.54)$$

For cocurrent flows the sign before the parenthesis is positive, and for countercurrent flows it is negative.

Substituting of Eq. (7.54) into Eq. (7.10) (using a positive sign for cocurrent flow operation) we can obtain the differential equation for temperature distribution of a cocurrent flow autothermal reactor:

$$GC_{pt}\frac{dT}{dZ} = (-\Delta H_r)(-\mathscr{R}_A)\rho_b - \frac{4U}{d_t}\left[2T - \lambda X_A - (T_c + T_0)\right] \qquad (7.55)$$

Similarly to the temperature distribution inside a countercurrent flow, autothermal reactors can be described by:

$$GC_{pt}\frac{dT}{dZ} = (-\Delta H_r)(-\mathscr{R}_A)\rho_b - \frac{4U}{d_t}\left[\lambda X_A - (T_{c0} - T_0)\right] \qquad (7.56)$$

For a countercurrent flow autothermal reactor $T_0 = T_{c0}$, and Eq. (7.56) can be simplified as:

$$GC_{pt}\frac{dT}{dZ} = (-\Delta H_r)\rho_b(-\mathscr{R}_A) - \frac{4U\lambda X_A}{d_t} \qquad (7.57)$$

Combining Eq. (7.55) or (7.57) with mass balance equation Eq. (7.10), the autothermal reactor can be simulated. If there is significant pressure drop, Eq. (7.11) will also be included.

It should be noted that for an autothermal reactor the feed inlet temperature T_{c0}, reactor outlet temperature T_{L_r}, and final conversion follows the following equation:

$$T_{L_r} = T_{c0} + \lambda X_{AL_r} \qquad (7.58)$$

This equation can be obtained based on overall heat balance of the reactor, and indicates that for these three variables you can only choose two independently and the third one will be determined by Eq. (7.58).

7.6   PARAMETER SENSITIVITY

When an exothermic reaction takes place in fixed bed reactor, be it adiabatic, with a heat exchanger, or autothermal, there is a maximum temperature inside the reactor bed, i.e., the hot spot. Obviously, for adiabatic reactors, the hot spot is at the outlet of the bed. For a fixed bed reactor with heat exchanger and autothermal reactor, the location of the hot spot depends on many factors. For endothermic reactions, there will be a minimum temperature inside the bed, which can be called a cold spot. However, for reactor control and operation the cold spot is far less important than hot spots, and most studies have focused on the hot spot.

Hot spot temperature T_m should be somewhere between bed inlet temperature T_0 and $(T_0 + \lambda)$, and its value depends on both reaction heat and heat transfer capability of the bed. Higher reaction temperatures may lead to more and/or faster side reactions, lower the yield of desired product, catalyst deactivation, and even explosion. As long as T_m is controlled to be lower than the maximum allowed value, the temperature at other locations inside the bed will not exceed this value. In Section 7.4.3 we simulated the reactor of oxidation of o-xylene, and found that a 1K degree increase of feed inlet temperature can cause significant increase of T_m, and when feed inlet temperature exceeds a certain value the reactor temperature may run away. Therefore, it is necessary to investigate the effects of operation variables, reaction mixture properties, and the heat transfer capability of the catalyst bed on the hot spot temperature. Such an investigation is called a sensitivity analysis.

At a hot spot, $dT/dZ = 0$, hence from Eq. (7.10)

$$\eta_0 \rho_b (-\mathscr{R}_A)(-\Delta H_r) - \frac{4U}{d_t}(T_m - T_c) = 0$$

Or

$$T_m - T_c = \frac{\eta_0 \rho_b (-\mathscr{R}_A)(-\Delta H_r)d_t}{4U} \tag{7.59}$$

In the above equation $(-\mathscr{R}_A)$ and η_0 need to be calculated based on hot spot temperature T_m and reaction mixture composition at that location. Obviously unless we know the reaction mixture composition at the hot spot it is not practical to calculate the hot spot temperature by using Eq. (7.59). However, we can use this equation to analyze the potential impact of various factors on hot spot temperature. From this equation we can see that lower cooling medium temperature, enhanced heat transfer intensity (U/d_t) will help to maintain the hot spot temperature below a given value and hence avoid reactor run away. However, too low cooling medium temperatures will decrease reaction rate, which will in turn lead to an increase of catalyst and reactor volume required to achieve given productivity. This also illustrates why molten salt was selected as the cooling medium for the o-xylene oxidation reactor. One possible way to enhance heat transfer intensity is to reduce d_t. Smaller d_t also has the advantage of more uniform radial temperature distribution, but will cause a larger pressure drop, which is undesirable.

Understanding which operating parameters will have the most significant impact on reactor stability, and more desirably, knowing the most sensitive ranges of those parameters will be very helpful for reactor design and selecting operation conditions. This sensitivity analysis can be performed through reactor simulations. From Eq. (7.32) we can see that w_{A0}, T_c, T_0, and U/d_t are potential sensitive parameters, and a small change of these parameters may lead to unstable conditions. Among them, cooling medium temperature, feed temperature, and concentration are the most critical. Unfortunately, there are no widely applicable and reliable criteria for predicting the stability of fixed bed reactors.

For a first order irreversible exothermic reaction, $(-\mathscr{R}_A) = A$ exp $(-E/RT)c_A$. Using this rate equation in Eq. (7.59) we can obtain component A concentration at the hot spot:

$$c_{Am} = \frac{4U(T_m - T_c)}{A\exp(-E/RT)\rho_b(-\Delta H_r)d_t} \tag{7.60}$$

Here we assumed $\eta_0 = 1$. If cooling medium temperature T_c is a constant and equal to bed inlet temperature T_0, the following equation can be obtained by taking the derivative of Eq. (7.59) with respect to T_m and letting $dc_Am/dT_m = 0$:

$$\frac{E}{RT_{max}^2}(T_{max} - T_c) = 1 \tag{7.61}$$

T_{max} is the temperature when the concentration of component A is at its maximum. The temperature difference in Eq. (7.61) represents the maximum possible temperature difference between the catalyst bed and cooling medium. If the actual temperature difference is higher than this value the reactor can be considered stable. Eq. (7.61) can be rewritten as:

$$\frac{RT_{max}}{E} = \frac{1}{2}\left[1 - \sqrt{1 - \frac{4RT_c}{E}}\right] \tag{7.62}$$

This equation can be used to select the cooling medium temperature. When the cooling medium temperature is lower than T_c calculated by using this equation, the reactor will not run away.

Another empirical criterion that can be used is:

$$\frac{\lambda}{T_{max} - T_c} \le 1 + \sqrt{\frac{N_c}{e}} + \frac{N_c}{e} \tag{7.63}$$

where $N_c = 4U/\rho C_{pt}\, d_t\, k_c$, and k_c is a rate constant calculated using the cooling medium temperature. ρ is the density of the reaction mixture. Eq. (7.63) can be used to select the feed inlet temperature.

Example 7.4

A first order exothermic reaction takes place inside a heat exchange fixed bed reactor. The rate constant is $k = 7.4 \times 10^8 \exp(-13,600/T)$, 1/s; $U = 100$ J/(m$^2 \cdot$ s $\cdot$ K); $\rho C_{pt} = 1300$ J/(m $\cdot$ K), and $(-\Delta H_r) -1300$ kJ/mol. $T_0 = T_c = 635$K.

(1) If reactor tube diameter $d_t = 25$ mm, what is the maximum bed temperature? What is the maximum allowable feed inlet concentration?

(2) If feed inlet concentration is 0.001 kmol/m^3, what is the maximum allowable reactor diameter to ensure stable operation?

(Continued)

(Continued)

Solution

(1) The maximum bed temperature can be calculated by Eq. (7.62)

$$T_{max} = \frac{1}{2} \times 13,600 \left[1 - \sqrt{1 - \frac{4 \times 635}{13,600}} \right] = 667.8K$$

The maximum allowable feed inlet concentration can be calculated by using Eq. (7.63).

$$k_c = 7.4 \times 10^8 \exp(-13,600/635) = 0.37 \; 1/s$$

$$N_c = \frac{4U}{\rho C_{pt} d_t k_c} = \frac{4 \times 100}{1300 \times 0.025 \times 0.37} = 33.3$$

Using Eq. (7.63)

$$\lambda = (667.8 - 635) \left[1 + \sqrt{\frac{33.3}{e}} + \frac{33.3}{e} \right] = 549K$$

We also know

$$\lambda = \frac{(-\Delta H_r)w_{A0}}{M_A C_{pt}} = \frac{(-\Delta H_r)c_{A0}}{\rho C_{pt}}$$

Therefore

$$c_{A0} = \frac{\lambda \rho C_{pt}}{(-\Delta H_r)} = \frac{549 \times 1300}{1300 \times 10^6} = 0.55 \times 10^{-3} mol/m^3$$

Based on the above analysis, as long as the feed inlet concentration is less than 5.5×10^{-4} kmol/m³, the bed temperature will not exceed 667.8K.

(2) We already know $T_{max} = 667.8K$. If feed inlet concentration is 1×10^{-3} kmol/m³, which is higher than the maximum allowable concentration calculated above, reactor diameter has to be changed in order to maintain $T_{max} = 667.8K$.

The adiabatic temperature will also change if feed inlet concentration changes:

$$\lambda = \frac{1300 \times 10^6 \times 1 \times 10^{-3}}{1300} = 1000K$$

From Eq. (7.63)
Solving the above equation:

$$\frac{1000}{667.8 - 635} = 1 + \sqrt{\frac{N_c}{e}} + \frac{N_c}{e}$$

$$N_c = 66.7$$

Based on the definition equation of N_c, the reactor diameter is:

$$d_t = \frac{4U}{\rho C_{pt} k_c N_c} = \frac{4 \times 100}{1300 \times 0.37 \times 66.7} = 12.5 \times 10^{-3} \; m$$

(Continued)

(Continued)

Therefore, if the feed inlet concentration is increased to 1×10^{-3} kmol/m^3, the reactor diameter has to be smaller than 12.5 mm to ensure the maximum bed temperature does not exceed 667.8K. The smaller reactor diameter enhances heat transfer intensity of the catalyst bed, i.e., increases heat transfer area per unit bed volume.

7.7 LABORATORY CATALYTIC REACTOR

The main character of a laboratory reactor is its small volume. Typically, only a few grams or even a few hundred milligram of catalyst are used. For various purposes there are many different types of laboratory reactors. In this section we will only discuss the basic requirements and the major types of laboratory catalytic reactors for reaction kinetic study.

7.7.1 Basic Requirements

The objective of catalytic kinetic study is to obtain reaction rate equations. This objective can be achieved by measuring the reaction rate at different temperatures, pressure, and reaction mixture compositions and then developing reaction rate equations.

For a given catalyst, reaction temperature and reaction mixture composition are the two main factors that will influence reaction rate. For homogeneous reactions, reactant concentrations and temperature can be measured directly. For a gas–solid catalytic reaction system, what can be measured are usually the concentrations and temperature of bulk gas phase. However, the chemical reactions take place on the solid surface, either external surface or internal surface inside the catalyst particles, and in most cases the concentrations and temperature on the catalyst surface are different from those of the bulk gas phase (see Chapter 6 Chemical Reaction and Transport Phenomena in Heterogeneous System). For both concentration and temperature there are interphase differences and distributions inside the catalyst particle. For intrinsic reaction kinetic study we need to measure reaction rate as a function of temperature and reactant concentrations at reaction location, and those differences and internal gradients make the experimental measurement very difficult or impossible. In addition, inside the reactor there will be concentration and temperature distributions along axial and radial direction, which will also have an impact on the accuracy of kinetic measurements. The impact of radial distribution is especially severe.

In summary, for both temperature and concentration, there are four potential distributions. If a reactor can essentially eliminate all eight of these gradients, it is called a gradientless reactor. In a gradientless reactor, the reactions take place under constant temperature and concentration, which make experimental measurement much easier and data processing much simpler. Therefore, gradientless reactors are very useful tools for kinetic study.

The principles of analyzing transport processes, discussed in Chapter 6, Chemical Reaction and Transport Phenomena in Heterogeneous System, as well as earlier in this chapter, can be used to design a reactor that is gradientless or more accurately minimize the gradients inside the reactor. The smaller the particle size the higher the effectiveness factor is. An effectiveness factor equal to one means there is no internal temperature and concentration gradient. Measuring reaction rate using different catalyst particle sizes can decide at what particle size the internal temperature and concentration gradient can be ignored. Using this particle size and measuring reaction rate at different mass flow rates of reactant mixture will allow the determination of minimum mass flow rate when there are no interphase gradients. In summary, the operation conditions to achieve gradientless operation can be determined through experimental study at different catalyst particle sizes and feed mass flow rates.

The axial and radial gradients depend on d_t/d_p, L_r/d_p, fluid dynamics, and physical properties. Generally speaking, the temperature and concentration distributions depend on flow pattern inside the reactor. For a well-mixed reactor, there is obviously no axial or radial gradient, and hence it is a gradientless reactor. Of course this is based on the assumptions that both catalyst particle size and gas mass flow rate meet the requirements. The most effective way to achieve well-mixed flow is to increase recirculation, which will be discussed in more detail later. If well-mixed conditions are not met, there will be both axial and radial gradients. If that is the case, it is desirable to maintain plug plow and isothermal operation to make the data analysis easier. Isothermal plug flow can be achieved by carefully selecting d_t/d_p, L_r/d_p, and proper heat management. Usually reactors with a smaller diameter will help.

In summary, there are two basic requirements for experimental catalytic reactors: (1) isothermal operation and (2) an ideal flow pattern. It is especially important to maintain isothermal operation since the reaction rate is very sensitive to temperature. A small change of temperature may lead to a significant change in reaction rate. To achieve isothermal operation, first we need to make sure the temperature is measured accurately. In addition, temperatures at different locations inside the reactor need to be carefully monitored. If the reaction heat is large, diluting the catalyst and feed gas with inert materials will help reduce the intensity of heat release or absorption by the reaction. In most cases, experimental reactors, whether the reaction taking place inside the reactor is exothermic or endothermic, have to be heated to maintain the reaction temperature. Therefore a constant temperature heat source is needed. When an electrical furnace is used, the reactor needs to be placed in the constant temperature zone of the furnace. In order to meet these two requirements, the reactor needs to be correctly designed and the reaction conditions need to be carefully selected.

7.7.2   Main Types of Experimental Reactor

For lab kinetic study the following types of reactors are most commonly used, and all of them are fixed bed reactors:

(1) Integral reactor
(2) Differential reactor
(3) External recycle reactor
(4) Internal recycle reactor

7.7.2.1   Integral Reactor

Typically it is built using a small diameter (1 cm or less) metal or glass tube that is packed with catalyst particles with proper size. To maintain isothermal operation, the catalyst is usually diluted with inert material with the same particle size. The length of the bed cannot be too long, otherwise it is difficult to maintain the same temperature over the whole length. A thermal couple can be used to measure bed temperature, and in most cases the reaction gas mixture will flow downward through the catalyst bed. The feed gas needs to be preheated to the reaction temperature, and usually a layer of inert material is placed on top of the catalyst bed to ensure uniform flow distribution across the bed cross-section, which also helps to make sure the gas entering the catalyst bed is at the bed temperature. The whole reactor is immerged in a heat source, such as an oil bath, molten salt, or fluidized sand bath.

The main goal of kinetic study is to obtain reactions at different temperatures and concentrations. For a single reaction the reaction rate of key component A can be expressed as:

$$r_A = F_{A0} \frac{dX_A}{dW} \tag{7.64}$$

F_{A0} is the molar flow rate of key component A at the reactor inlet, and can be calculated based on gas flow rate and composition. When integrated over the whole length of the bed, the above equation can be used to describe an integral reactor where the gas composition at reactor outlet is measured. The measured reaction result reflects the overall behavior of the reactor, not the microscopic changes across a thin section of the reactor. For integral reactors, a derivative of the integrated result has to be calculated in order to obtain reaction rate. Therefore, Eq. (7.64) can be rewritten as:

$$r_A = \frac{dX_A}{d(W/F_{A0})} \tag{7.65}$$

The amount of catalyst loaded inside reactor W is a known fixed value; therefore changing W/F_{A0} can be achieved by adjusting F_{A0}. In order to obtain the reaction rate, a series of experiments have to be conducted at different F_{A0}. Based on reaction mixture composition measured at different F_{A0} values, reaction conversions can be calculated. Based on the W/F_{A0} versus X_A correlation measured, the reaction rate can be obtained by using Eq. (7.65). It should be noted that except F_{A0} all other parameters such as reaction temperature and feed composition have to be kept constant.

7.7.2.2   Differential Reactor

The structure of a differential reactor is the same as that of an integral reactor. The only difference is the length of catalyst bed. The main character of a differential reactor is that the composition difference between the reactor inlet and outlet is very small, and so is the conversion of the key component, which is typically less than 5%. In contrast, for an integral reactor there is no limit on conversion and the overall conversion can be quite high. The reason to use a differential reactor is that when the change of W/F_{A0} is very small the X_A change can be treated as linear.

$$r_A = F_{A0} = \frac{dX_A}{dW} = F_{A0}\frac{\Delta X_A}{\Delta W} = \frac{F_{A0}(X_A - X_{A0})}{W} \qquad (7.66)$$

As long as ΔX_A is very small, Eq. (7.66) is valid. Reaction rate calculated using Eq. (7.66) is the reaction rate at average conversion $\overline{X}_A = (X_{A0} + X_A)/2$.

For a differential reactor the composition difference between the reactor inlet and outlet is very small, which means the compositions must be accurately measured, otherwise the reaction rate calculated will not be accurate. In order to measure reaction rate at different compositions, feed streams with different compositions are needed. One way to achieve this is to use an integral reactor in front of a differential reactor. This arrangement will allow the inlet composition to the differential reactor to be adjusted by controlling the conversion of the integral reactor. We can see that the differential reactor can be considered as a thin layer of an integral reactor. If the composition at the inlet and outlet of that thin layer can be measured, the reaction rate can be calculated using Eq. (7.66). The advantage of a differential reactor is the very small net conversion, which leads to small reaction heat and an isothermal operation that is easier to maintain. Another advantage is the ease of data analysis. The main disadvantage is the high requirement of composition analysis.

7.7.2.3   External Recycle Reactor

Both integral and differential reactors have concentration gradient along the axial direction, but the axial concentration for an external recycle reactor is zero or at least very close to zero. The structure of an external recycle reactor is almost identical to a differential reactor with the only difference being the external recycle loop that returns the majority of the reactor outlet flow to the reactor inlet. The constant concentration inside the reactor is achieved by using a large recycle ratio. The recycle reactor has been discussed in Chapter 4, Tubular Reactor, and the following equation can be used:

$$X_{A0} = \frac{\psi X_{Af}}{1 + \psi} \qquad (4.23)$$

where ψ is the recycle ratio, which equals to the ratio of fresh gas to recycle flow. From Eq. (4.23) we can see that when ψ approaches infinity, conversion at the reactor inlet X_{A0} equals the conversion at reactor outlet X_{Af}. In reality, when $\psi = 25-50$ the difference between the reactor inlet and outlet is very small, i.e., the reactor can be considered as being operated at constant concentration. Under such conditions, the reaction rate can be calculated by:

$$r_A = \frac{F_{A0}(X_{Af} - X'_{A0})}{W} \qquad (7.67)$$

It should be noted that X'_{A0} is the conversion of fresh gas or feed gas, not the conversion at the reactor inlet. If there is no product in the feed gas then $X'_{A0} = 0$. Therefore the composition at reactor outlet needs to be measured to calculate the reaction rate. Reaction rate calculation for an external recycle reactor is much easier than that for either an integral reactor or a differential reactor. However, if the contribution of a homogeneous reaction is significant or there is the potential of phase changes caused by temperature fluctuation during the recycling process, using an external recycle reactor may not be practical or even possible.

7.7.2.4 Internal Recycle Reactor

The internal recycle reactor is also a gradientless reactor. The main difference it has from an external recycle reactor is that instead of recycling part of the reactor outlet stream back to the reactor inlet, the internal recycle reactor achieves gradientless operation by employing high-speed mechanical stirring equipment inside the reactor. It is in fact a CSTR reactor. Obviously high-speed stirring, typically at as high as few thousand rpm, is needed.

There are two types of internal recycle reactors. One uses a rotating basket to hold the catalyst, as shown in Fig. 7.10. The catalyst is placed in and rotates with the basket. Another type is illustrated in Fig. 7.11. The catalyst is placed on an annular space between reactor wall and a central tube. A turbine propeller is placed at the bottom and its rotation will suck the gas through the central tube and push the gas toward the edge of the

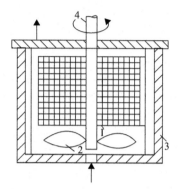

FIGURE 7.10 Rotating basket reactor. (1) Catalyst; (2) Propeller; (3) Baffles; (4) Rotating shaft.

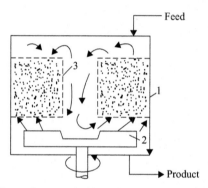

FIGURE 7.11 Integral recycle reactor with stationary catalyst. (1) Catalyst; (2) Propeller; (3) Central tube.

propeller. The gas is forced to flow through the catalyst bed and eventually returns to the central tube. For both of them, the strong mixing is achieved by mechanical stirring.

The internal recycle reactor is a gradientless reactor and hence the reaction rate can be calculated using Eq. (7.67). In this case, X'_{A0} is the conversion of both feed gas and the stream at the reactor inlet. For internal recycle reactors, the feed gas is fed into the reactor directly, while for external recycle reactors the feed gas is first mixed with the recycle stream before entering the reactor.

Among the four types of experimental catalytic reactors, the internal recycle reactor is generally considered to be the best option because both temperature and concentration distributions are more uniform than that in other three types. This is very important to obtain more reliable experimental data. In addition, data analysis for internal recycle reactor is also simpler.

FURTHER READING

Li S. Chemical and catalytic reaction engineering. New York: Chemical Industry Press; 1986.

Froment GF, Bischoff KB. Chemical reactor analysis and design. John Wiley & Son Inc.; 1979. 中译本，反应器分析与设计. 北京: 化学工业出版社, 1985 年.

Carberry JJ. Chemical and catalytic reaction engineering. New York: McGraw-Hill; 1976.

Tarhan MO. Catalytic reactor design. New York: McGraw-Hill; 1983.

Missen RW, Mims CA, Saville BA. Introduction to chemical reaction engineering and kinetics. New York: John Wiley & Sons; 1999.

David R. Lide, ed. CRC handbook of chemistry and physics. 90th Edition. CRC Press/Taylor and Francis, Boca Raton, FL, 2010.

PROBLEMS

7.1 If the superficial velocity (calculated based on empty bed) of a gas flowing through a fixed bed is 0.2 m/s, what is the real velocity of the gas? The bulk density of the solid particle is 1.2 g/cm^3, and particle density is 1.8 g/cm^3.

7.2 In order to determine the Shape Factor of an iron catalyst with irregular shape for ammonia synthesis, the catalyst is packed in a vessel with ID = 98 mm, and the bed height is 1 m. When air flows through the bed at 1 m^3/h, the pressure drop across the bed is measured as 101.3 Pa. The experimental temperature is 298K. Find the Shape Factor for this catalyst. The volume equivalent diameter of the catalyst particle is 4 mm, the bulk density is 1.45 g/cm^3, and the particle density is 2.6 g/cm^3.

7.3 A first order irreversible reaction takes place isothermally in a fixed bed reactor packed with 3 mm diameter spherical catalyst particles. The rate constant based on unit volume of catalyst is 0.8 s^{-1}, the effective diffusivity is 0.013 cm^2/s. The desired conversion can be achieved when bed height is 2 m. In order to reduce pressure drop, 6 mm diameter spherical catalyst is used. If all the other conditions are the same and it is laminar flow inside the bed, find:

1. The bed height required;

2. Percentage of pressure drop reduction.

7.4 A multistage reactor with indirect heat exchange is selected for sulfur dioxide oxidation. The feed flow rate is 35,000 m^3 (standard conditions), and the molar fraction of SO_2, O_2, and N_2 in the feed is 7.5%, 10.5%, and 82%, respectively. A cylindrical vanadium catalyst with diameter of 5 mm and height of 10 mm is used, and the catalyst volume is 80 m^3. Determine the diameter and height of the bed to maintain pressure drop to be less than 4052 Pa.

To simply the calculation, average operating pressure of 0.1216 MPa and average temperature of 733K can be used. The viscosity of the gas mixed is 3.4×10^{-5} Pa · s, and the density is approximately equal to air density.

7.5 The composition of the feed stream to a multistage cold shot reactor for ammonia synthesis is:

Component	NH$_3$	N$_2$	H$_2$	CH$_4$	Ar
Molar fraction (%)	2.09	21.82	66.00	7.63	2.45

1. Find the feed composition on ammonia-less base;
2. If the temperature of gas entering the first stage is 407°C, derive the adiabatic operating equation using both ammonia conversion and ammonia concentration to describe the composition. The average molar specific heat capacity of the feed gas is 33.08 J/(mol · K), The reaction heat $\Delta H_r = -53581$ J/mol NH$_3$;
3. Find the temperature at bed outlet when ammonia content at the outlet is 10% under the following conditions (1) Considering the change of total numbers of moles; and (2) ignoring the change of total numbers of moles. Compare the two results.

7.6 An adiabatic catalytic reactor is used for oxidation of sulfur dioxide. The inlet temperature is 420°C, the SO$_2$ molar fraction at the reactor inlet is 7%, the reactor outlet temperature is 590°C, the SO$_2$ molar fraction at reactor outlet is 2.1%. Three measurements inside the catalyst bed were taken:

1. The temperature at point A was measured as 620°C, do you think this value is correct? Why?
2. The conversion at point B was measured as 80%, do you think this value is correct? Why?
3. The conversion at point C was measured as 50%, and such measurement was repeated three times and it was confirmed to be correct, please estimate temperature at point C.

7.7 The reaction equation for acetylene hydration to make acetone is:

$$2C_2H_2 + 3H_2O \rightarrow CH_3COCH_3 + CO_2 + H_2$$

The reaction rate equation for this reaction on the ZnO-Fe$_2$O$_3$ catalyst is:

$$r_A = 7.06 \times 10^7 \exp\left(-\frac{7413}{T}\right) c_A \, \text{kmol}/(\text{h·m}^3\text{bed})$$

where c_A is acetylene concentration.

The feed contains 3% acetylene and the flow rate is 1000 m^3/h (standard condition), the gas inlet temperature is 380°C, find the amount of catalyst needed to achieve acetylene conversion of 68%. Assume the influence of diffusion can be ignored. The reaction heat is −178 kJ/mol, and the gas average specific heat capacity under constant pressure of 36.4 J/(mol · K).

7.8 The acetylene hydration reaction described in Problem 7.7 takes place in an adiabatic reactor, and the product stream is used to preheat the feed, as shown in Fig. 7A. The heat exchange area of the preheater is 50 m^2; the feed gas contains 3% mol acetylene, enters the preheater at a flow rate of 1000 m^3 (standard condition), and is heated from 100°C to a certain value before entering the catalyst bed. The bed volume is 1 m^3, and the rate equation is the same as given in Problem 7.7. The overall heat exchange coefficient in the preheater is 32.5 J/(m$^2 \cdot$ s $\cdot$ K), and the heat capacity of the reaction mixture is 36.4 J/(mol $\cdot$ K). Please calculate:

1. The adiabatic temperature increases (the changes of total numbers of moles of the reaction mixture can be ignored);
2. Acetylene conversion at the reactor outlet (list all the equations for the calculation and explain solving process).

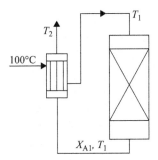

FIGURE 7A

7.9 A two-stage adiabatic reactor with inter-stage heat exchanger is used for the following reaction at atmospheric pressure:

$$CO + H_2O \leftrightharpoons CO_2 + H_2$$

The reaction heat is $\Delta H_r = -41,030$ J/mol. The feed includes semiwater gas and steam and has semiwater gas-to-steam molar ratio of 1:1.4. The composition of semiwater gas is given in the table below:

Component	CO	H$_2$	CO$_2$	N$_2$	CH$_4$	Others	Total
Molar (%)	30.4	37.8	9.46	21.3	0.79	0.25	100

The flow diagram and some of the operation conditions are shown in Fig. 7B. Assume the specific heat of all steams has the same value of 33.5 J/(mol $\cdot$ K), please find the temperatures and CO conversion at the inlet and outlet of the second adiabatic bed. The reactor heat loss to the environment can be ignored.

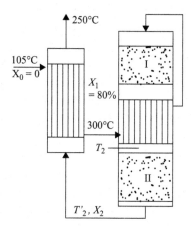

FIGURE 7B

7.10 An alumina catalyst is used for acetonitrile synthesis:

$$C_2H_2 + NH_3 \rightarrow CH_3CN + H_2, \quad \Delta H_r = -92.2 \text{ kJ/mol}$$

The molar ratio in the feed stream is $C_2H_2:NH_3:H_2 = 1:2.2:1$, and a three-stage adiabatic reactor with interstage heat exchangers is used. The temperature of the outlet of each stage is controlled at 550°C, and the inlet temperature of each stage is also kept at the same value. The reaction rate can be described by:

$$r_A = 3.08 \times 10^4 \exp\left(-\frac{7960}{T}\right) \cdot (1 - X_A) \text{ kmol } C_2H_2/(h \cdot kg)$$

where X_A is acetylene conversion. The average heat capacity of the fluid is:

$$\overline{C_p} = 128 \text{ J/(mol·K)}$$

Please find the amount of catalyst needed in order to achieve 92% conversion of acetylene and acetonnitrile productivity of 20 ton per day.

7.11 For the two-stage adiabatic reactor for water gas shift reaction described in Example 7.3, if the CO conversion at the outlet of the first stage is 84%, what should the inlet temperature of the first stage be in order to minimize the amount of catalyst needed? Please use the data given in Example 7.3, and compare the result with the inlet temperature given in the example.

7.12 Figs. 7C and 7D are $T-X$ plots for two chemical reactions. AB is the equilibrium curve, NP the optimal temperature curve, AM the

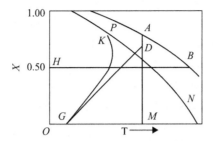

FIGURE 7C

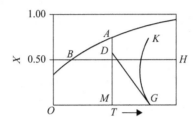

FIGURE 7D

isothermal curve, GD the adiabatic curve, GK the nonadiabatic operating curve, and HB the constant conversion line.

1. Compare the difference of these two plots and analyze potential reasons for such differences;
2. If the reaction illustrated in Fig. 7C is conducted in a fixed bed reactor, and the final conversion is 50%, please compare the amount of catalyst needed when the reaction proceeds following operating line MA, GD, and GK, and explain why;
3. Do the same as (2) for reaction illustrated in Fig. 7D;
4. Do you think the conclusions from (2) to (3) are generally applicable or only valid for a specific situation?

7.13 A multitube reactor is used for o-xylene oxidation to make phthalic anhydride. A cylindrical vanadium pentoxide catalyst with diameter of 5 mm is used and the catalyst particles are placed inside the tube. A molten salt at 370°C is used as coolant. The reaction kinetic equation is:

$$r_s = 0.04017 p_A p_B^0 \exp\left(-\frac{13,636}{T}\right) \text{kmol}/(\text{kg·h})$$

where p_A is the partial pressure of o-xylene, Pa; p_B^0 is the initial partial pressure of O_2, Pa. The reaction heat is $\Delta H_r = -1285 \text{ kJ/mol}$. The ID of reactor tube is 25 mm. The feed gas flow rate is 9200 kg/(m²·h), and contains 0.9 mol% of o-xylene and 99.1 mol% of air. The average molecular weight of the mixture is 29.45, and the average heat capacity at constant pressure is 1.072 kJ/(kg · K). The inlet temperature is 370°C. The bulk density of the bed is 1300 kg/m^3. The operating pressure is 0.1013 MPa(absolute), and the overall heat transfer coefficient is 251 kJ/(m^2 · h · K). Please calculate axial temperature distribution using one dimensional plug flow model and the bed length needed to achieve 73.5% conversion. All the side reactions can be ignored.

7.14 Please evaluate whether the following statements are correct:

1. For a zero order reaction with no reaction heat taking place in an adiabatic reactor, the conversion is a linear function of reactor length;

2. In an adiabatic reactor, the reaction temperature and conversion follows a linear correlation only when the reaction is a first order reaction;

3. To minimize catalyst use in a multistage adiabatic reactor, the catalyst must be evenly distributed in each stage.

7.15 A spherical alumina catalyst with diameter of 6 mm is used for acetonitrile synthesis, and the operation conditions are the same as that in Problem 7.10. The internal diffusion has to be considered while the external diffusion can be ignored. The physical properties of the alumina catalyst are: pore volume $= 0.45$ cm^3/g, particle density $= 1.1$ g/cm^3, specific area $= 180$ m^2, and tortuosity factor $= 3.2$. Please calculate the amount of catalyst in the first stage.

7.16 A nth order irreversible exothermic reaction takes place in a three-stage adiabatic reactor. The amount of catalyst in each stage is the same, and the reactant temperature at the inlet of each stage is also controlled at the same value. If $n > 0$:

1. Which stage will have the highest and lowest conversion?

2. If the cold shot method is used to control reaction temperature, please compare the amount of cooling agent needed for temperature control between the first and the second stage and between the second and the third stage and explain why.

7.17 Naphthalene oxidation reaction takes place in a fixed bed reactor with ID $= 5.1$ cm.

A spherical vanadium catalyst with diameter of 0.318 cm is used. The reaction can be treated as a first order reaction and the reaction rate constant based on bed volume is

$$k = 5.74 \times 10^{13} \exp\left(-\frac{19,000}{T}\right), \text{ s}^{-1}$$

The reaction heat is $\Delta H_r = -1796\ \text{kJ/mol}$. The void fraction of the bed is 0.4. A molten salt at 630K is used for external cooling, and the overall heat transfer coefficient between the catalyst bed and the coolant is $180\ \text{J/(m}^2 \cdot \text{s} \cdot \text{K)}$. The diffusivity of naphthalene inside the vanadium catalyst is $1.2 \times 10^{-3}\ \text{cm}^2/\text{s}$. If the hot spot temperature is 641K, please find the naphthalene concentration at the hot spot location. The impact of external diffusion and all the side reaction can be ignored.

7.18 In Problem 7.17 the feed stream entering the naphthalene oxidation reactor is a mixture of naphthalene and air, and the average heat capacity at constant pressure of the feed mixture is $1.3\ \text{kJ/(kg} \cdot \text{K)}$. If the feed inlet temperature is the same as the temperature of the molten salt, and the maximum allowable operation temperature of the catalyst is 700K,

1. What is the maximum allowable molten salt temperature?
2. If the temperature of the molten salt is selected to be 641K, what is the maximum allowable naphthalene concentration in the feed?

7.19 Toluene hydrogenation reaction takes place inside a $10\ \text{m}^3$ adiabatic fixed bed reactor:

$$C_6H_5CH_3 + H_2 \rightarrow C_6H_6 + CH_4$$

The feed contains 3.85% C_6H_6, 3.18% $C_6H_5CH_3$, 23% CH_4, and 69.97% H_2. The feed temperature is 863K, and the pressure is 6.08 MPa. If the space velocity under standard conditions is $1000\ \text{m}^3(\text{h} \cdot \text{m}^3$ catalyst), please find the gas composition at reactor outlet. The rate equation for the reaction is:

$$r_T = 5.73 \times 10^6 \exp\left(-\frac{17,800}{T}\right) c_T c_H^{0.5}$$

where c_T and c_H are concentrations of toluene and hydrogen in kmol/m^3, respectively. The unit for toluene consumption rate is kmol/(m$^3 \cdot$ s). The heat of reaction is $-49,974$ J/mol. The reactant mixture can be treated as ideal gas, and the heat capacity under constant pressure can be considered as a constant and equals to $42.3\ \text{J/(mol} \cdot \text{K)}$.

7.20 A adiabatic reactor is operated with the inlet temperature of 460°C. When fresh catalyst is used the outlet temperature is 437°C , and the conversion can meet the requirement. After few months of operation the catalyst activity decreased. In order to maintain the conversion the inlet temperature has to be raised to 470°C, and as a results the outlet temperature increases to 448°C. If the reaction activation energy is 83.7 kJ/mol, please estimate the percentage of catalyst activity decrease.

7.21 A second order reaction $2A \rightarrow P + Q$ was studied in the lab using an external recycle reactor. The feed is pure A, and $kc_{A0}\tau = 1$, please calculate conversion of A when:

1. Recycle ratio $= 5$;
2. Recycle ratio $= 30$;
3. The reactor can be treated as a CSTR.
4. Compare and discuss calculation results for the above three situations.

7.22 A multitube reactor is used for benzene oxidation. The ID of the reactor tube is 34 mm. Vanadium catalyst is packed inside the tubes, and the bed length is 1.6 m. A benzene-air mixture with benzene molar fraction of 1.2% is fed into the catalyst bed at velocity of 0.4 m/s. The feed temperature is 673K, and a 673K molten salt is used outside the tube to cool the bed. The overall heat transfer coefficient between the catalyst bed and the molten salt is $100 \, J/(m^2 \cdot s \cdot K)$. The operating pressure is 0.203 MPa. The reaction model and kinetic equation for benzene oxidation are given in Problem 4.16. Please calculate gas composition at reactor outlet and the yield of maleic anhydride.

Chapter 8

Fluidized Bed Reactor

8.1 INTRODUCTION

For some chemical processes, fixed-bed reactors have major disadvantages. If the reactions are fast and highly exothermic, hot spots will form in the fixed beds, which may shorten the catalyst life and increase by-product production. Compared with fixed-bed reactors, fluidized beds have many advantages, such as excellent heat transfer performance, ability of on-line catalyst replacement, feasibility of reaction−regeneration coupled operation, and a high effectiveness factor of fine catalyst particles. However, fluidized bed reactors also face some challenges such as nonuniform gas distribution, intensive backmixing of solid particles, bypassing of the gas, and attrition and entrainment of catalyst particles. Gas−solid fluidized bed reactors have been widely used in many industries, and some typical examples are listed in Table 8.1.

8.2 FLUIDIZATION

8.2.1 Fluidization Phenomenon

The contact regimes of the gas−solid systems can be classified on the basis of solids motion. For a batch-solids system, the gas at a low velocity merely flows through the voids between packed particles while the particles remain motionless. The solids in this case are in the fixed-bed state.

Reaction Engineering. DOI: http://dx.doi.org/10.1016/B978-0-12-410416-7.00008-2
369

TABLE 8.1 Examples of Applications of Gas—Solid Fluidized Bed Reactors

Physical Operations	Chemical Syntheses	Metallurgical and Mineral Processes	Other Applications
Heat exchange	Phthalic anhydride synthesis	Uranium processing	Coal combustion
Solid blending	Acrylonitrile synthesis	Reduction of iron oxide	Coal gasification
Particle coating	Ethylene dichloride synthesis	Pyrolysis of oil shale	Fluidized catalytic cracking
Drying	Methanol to olefin	Roasting of sulfide ores	Incineration of solid waste
Adsorption	Vinyl acetate synthesis	Crystalline silicon production	Cement clinker production
Solidification and granulation	Olefin polymerization	Titanium dioxide production	Microorganism cultivation
Particle growth	Methanol to aromatics	Calcination	Pyrolysis of biomass

With an increase of gas velocity, particles move apart and become fluidized; the bed then enters the gas fluidization state. Under relatively low gas velocities, a dense gas—solid suspension characterizes the bed, and on the bed surface some solid entrainment occurs. Gas bubbles can usually be clearly distinguished from the dense suspension, and bubbling fluidized bed behaves like a fluid. The analogy between the bubble behavior in gas—solid fluidized beds and that in gas—liquid bubble columns is often applied.

8.2.2 Particle Classifications

The fluidization phenomena of gas—solids systems strongly depend on the types of particles. According to Geldart's work, particles can be classified into groups A, B, C, and D based on their fluidization behaviors, as shown in Fig. 8.1. This classification is made in terms of the particle size and the density difference between particles and gas. Fig. 8.1 was obtained empirically and has been widely used in fundamental research and design of fluidized beds.

Group A particles have a typical size range of 30—100 μm and are easy to fluidize. In this case, hydrodynamic forces are more important than the

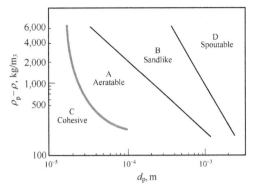

FIGURE 8.1 Geldart's classification of fluidized particles.

interparticle contact forces. Group A particles show a limited tendency to form bubbles and generally exhibit considerable bed expansion between the minimum fluidization gas velocity u_{mf} and the minimum bubbling gas velocity u_{mb}. Thus for Group A particles the minimum fluidization gas velocity u_{mf} is smaller than the minimum bubbling gas velocity u_{mb}. The fluidized beds with Group A particles can be operated in both particulate fluidization regime where bubbles are not present and bubbling fluidization regime where bubbles are present.

Group B particles are also easy to fluidize and tend to form bubbles that grow readily by coalescence. There is no particulate fluidization regime and the bed expansion is small. As a result, u_{mf} is approximately equal to u_{mb}. The fluidized bed with Group B particles collapses more quickly than that with Group A particles when the gas supply is shut off.

Group C particles are small and cohesive, and difficult to fluidize. In fluidization with Group C particles the interparticle contact forces, such as the van der Waals force and electrostatic forces, play an important role. In this case, solid slugging and gas channeling are the most common characteristics. The fluidization quality can be improved by exerting mechanical vibration or acoustic field.

Group D particles have large particle size and/or high solid particle density. When Group D particles are fluidized, the bed expansion is low and the particle mixing is not as good as that for Groups A and B particles. Because the particle size is large, a high fluidizing gas velocity is required, and stable spouted beds can be easily formed.

8.2.3 Fluidization Parameters

8.2.3.1 Minimum Fluidization

When the gas flows upward through a particle bed, there is a drag exerted on the solid particles by the flowing gas. At low gas velocities, the pressure

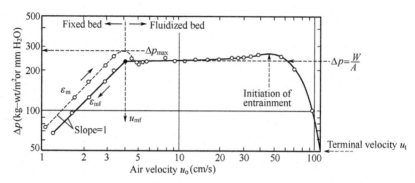

FIGURE 8.2   Relationship between superficial gas velocity and pressure drop through the bed.

drop ΔP resulting from this drag follows the Ergun equation (8.1). As this stage, the pressure drop increases with the gas velocity, as shown in Fig. 8.2.

$$\Delta P = \rho_g u^2 h_s \left[\frac{150(1 - \varepsilon_s)}{Re_p \phi_s} + 1.75 \right] \frac{(1 - \varepsilon_s)}{\phi_s d_p \varepsilon_s^3} \tag{8.1}$$

where ρ_g is the gas density, u is the superficial gas velocity, h_s is height of the settled bed, ε_s is the void fraction of the settled bed, d_p is the particle diameter, Re_p is the particle Reynolds number $\rho_g d_p u / \mu$, and ϕ_s is the particle sphericity.

Fluidization begins at a gas velocity at which the weight of the solid or the gravitational force exerted on the particles equals the drag on the particles from the rising gas. If ρ_p is the density of solid particles, A_c is the cross-sectional area, h_s is the initial settled height of the bed, h is the height of the bed at any time, and ε_s and ε are corresponding viodages of the settled and expanded bed, respectively, then the weight of solid particles in the bed, W_{bed}, is:

$$W_{bed} = \rho_p g A_c h_s (1 - \varepsilon_s) = \rho_p g A_c h (1 - \varepsilon) \tag{8.2}$$

At the point of minimum fluidization, the weight of the bed equals the pressure drop across the bed:

$$W_{bed} = \Delta P A_c \tag{8.3}$$

$$\rho_p g A_c h (1 - \varepsilon) = \rho_g A_c h u^2 \left[\frac{150(1 - \varepsilon)}{Re_p \phi_s} + 1.75 \right] \frac{(1 - \varepsilon)}{\phi_s d_p \varepsilon^3} \tag{8.4}$$

For fine particles, the second term in the right-hand side of Eq. (8.4) can be neglected when $Re_p < 10$, thus the minimum fluidization velocity u_{mf} can be expressed as:

$$u_{mf} = \frac{(\phi_s d_p)^2 (\rho_s - \rho_g) g}{150 \quad \mu} \left(\frac{\varepsilon_{mf}^3}{1 - \varepsilon_{mf}} \right) \tag{8.5}$$

For large particles, the first term in the right-hand side of Eq. (8.4) can be neglected when $Re_p > 1000$, thus the expression for u_{mf} is simplified as:

$$u_{mf} = \left[\frac{\phi_s d_p}{1.75} \frac{(\rho_s - \rho_g)g\, \varepsilon_{mf}^3}{\rho_g} \right]^{1/2} \tag{8.6}$$

When the drag force is greater than the gravitational force, the particles begin to fluidize, the bed expands and the bed void fraction increases, which will reduce the overall drag until it is again balanced by the total gravitational force, as shown in Fig. 8.2.

The range of velocities over which the Ergun equation applies can be fairly large. On the other hand, the difference between the velocity at which the bed starts to expand and the velocity at which the bubbles start to appear can be extremely small and sometimes nonexistent. This means the first evidence of bed expansion may be the appearance of gas bubbles in the bed and the movement of solids. At low gas velocities in the range of fluidization, the rising bubbles contain very few solid particles. The remainder of the bed has a much higher solids fraction and is called as the emulsion phase. The bubbles are shown as the bubble phase. The cloud phase is an intermediate phase between the bubble and emulsion phases.

8.2.3.2   Void Fraction at Minimum Fluidization (ε_{mf})

The void fraction at minimum fluidization, ε_{mf}, appears in many equations describing fluidized bed characteristics. The following correlation can be used to calculate ε_{mf} when the particles in the fluidized bed are fairly small:

$$\varepsilon_{mf} = 0.586\phi_s^{-0.72} \left(\frac{\mu^2}{\rho_g(\rho_s - \rho_g)gd_p^3} \right)^{0.029} \left(\frac{\rho_g}{\rho_p} \right)^{0.021} \tag{8.7}$$

Another commonly used correlation is:

$$\frac{1}{\phi_s \varepsilon_{mf}^3} \cong 14, \frac{1 - \varepsilon_{mf}}{\phi_s^2 \varepsilon_{mf}^3} \cong 11 \tag{8.8}$$

Using Eq. (8.8), u_{mf} can be calculated by:

$$\frac{d_p u_{mf} \rho_g}{\mu} = \left[33.7^2 + 0.0408 \frac{d_p^3 \rho_g (\rho_s - \rho_f)g}{\mu^2} \right]^{1/2} - 33.7 \tag{8.9}$$

When $Re_p < 10$ (fine particles):

$$u_{mf} = \frac{d_p^2 (\rho_p - \rho_g)g}{1650\mu} \tag{8.10}$$

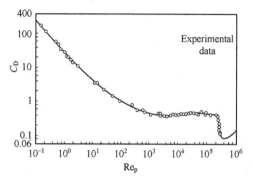

FIGURE 8.3 Drag coefficient for a spherical particle as a function of Re_p.

When $Re_p > 1000$ (large particles):

$$u_{mf}^2 = \frac{d_p(\rho_p - \rho_g)g}{24.5\rho_f} \tag{8.11}$$

8.2.3.3 Particle Terminal Velocity u_t

Individual particles are blown out of the bed when the gas velocity exceeds what is called the particle terminal velocity, u_t. This velocity for spherical particles can be calculated as:

$$\frac{\pi}{6}d_p^3(\rho_p - \rho_g)g = \frac{1}{2}C_D\rho_g\left(\frac{\pi d_p^2}{4}\right)u_t^2 \tag{8.12}$$

where C_D is the drag coefficient. The relationship between C_D and Re_p for a sphere is given by Fig. 8.3. Mathematically, it can be expressed by:

$$C_D = \frac{24}{Re_p} \quad Re_p < 0.4$$

$$C_D = \frac{10}{Re_p^{0.5}} \quad 0.4 \leq Re_p \leq 500 \tag{8.13}$$

$$C_D = 0.44 \quad 500 < Re_p < 2 \times 10^5$$

The three correlations in Eq. (8.13), in order from top to bottom, are known as Stokes's, Allen's, and Newton's equations, respectively. Combining these equations with Eq. (8.12), the terminal velocity of a spherical particle is related to its diameter by:

$$u_t = \frac{d_p^2(\rho_g - \rho_g)g}{18\mu} \quad Re_p < 0.4$$

$$u_t = \left[\frac{4}{225}\frac{(\rho_g - \rho_g)^2 g^2}{\rho\mu}\right]^{1/3} \quad 0.4 \leq Re_p \leq 500 \tag{8.14}$$

$$u_t = \left[\frac{3.03 d_p(\rho_g - \rho_g)g}{\rho}\right]^{1/2} \quad 500 < Re_p < 2 \times 10^5$$

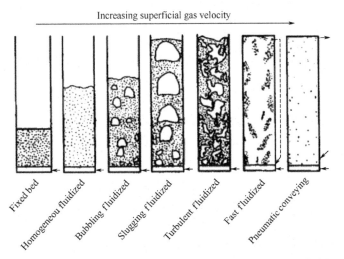

Increasing superficial gas velocity

FIGURE 8.4. Gas−solid contacting regime from low to high superficial gas velocities.

The calculated u_t must be substituted into Re_p to check if it is in the proper range for the equation.

From the above equations for u_{mf} and u_t, the ratio of u_t to u_{mf} can be calculated. For fine particles ($Re_p < 0.4$), $u_t/u_{mf} = 91.6$, and for large particles ($Re_p > 1000$), $u_t/u_{mf} = 8.72$. It can be seen that u_t/u_{mf} is roughly in the range of $10 \sim 90$. The ratio is larger for finer particles, indicating that gas velocity range from minimum fluidization to maximum fluidization is wider. This also explains why small particles are more suitable for fluidization.

8.2.4 Fluidization Regimes

As shown in Fig. 8.4, with increasing superficial velocity the gas−solid contact mode changes. Fluidization can be divided into dense-phase fluidization and lean-phase fluidization. The dense-fluidization regimes include homogeneous or particulate fluidization, bubbling fluidization, turbulent fluidization, and in a broad sense, slugging, spouting, and channeling regimes. The regimes for dense-phase fluidization can be classified based on the bubble behaviors. The lean-phase fluidization, which occurs at higher superficial velocities, covers fast fluidization and pneumatic conveying or dilute transport.

The particulate fluidization regime or homogeneous fluidization regime is bounded by the minimum fluidization velocity u_{mf} and the minimum bubbling velocity u_{mb}. In this regime the gas phase will not form bubbles, and all the gas will pass through the void space between fluidized solid particles. The bed therefore appears homogeneous. This regime exists over a narrow range of gas velocity for Group A particles. At high pressures or with gases of high density, the operating range of this regime expands. For Group B and D particles, bubbles will form as soon as gas velocity reaches u_{mf}, thus the particulate regime does not exist for these particles.

When the gas velocity increases beyond u_{mb}, separated bubble phase will form and the bubbling fluidization regime is reached. Therefore there will be two distinct phases, i.e., the bubble phase and the emulsion phase, in this regime. The bubbles will coalesce and breakup, which will cause vigorous movement of solid particles. Obviously the tendency of bubble coalescence is enhanced at a high gas velocity.

For a fluidized bed of a small bed diameter or large height/diameter ratio, bubbles may grow to a size that is comparable to the bed diameter and hence form slugs. The slugs will break up under sufficiently high gas velocities, and the slugging regime will transform to the turbulent fluidization regime or fast fluidization regime

Further increasing the gas velocity beyond the bubbling fluidization regime will lead to turbulent fluidization. In this regime, bubble breakup is enhanced due to high gas velocity, resulting smaller bubbles, less distinguishable bubble and emulsion phases, and more uniform suspension. In addition the surface of the bed becomes very diffused.

At an even higher gas velocity, the bed enters the fast fluidization regime. One characteristic of transition to fast fluidization is significant increase of solids entrainment. In this regime there will be a dense region at the bottom and a dilute region on the top. The solid movement in the lower region becomes less chaotic and forms a core-annular structure, i.e., a lean core surrounded by a denser annulus. In the upper dilute region the solid holdup gradually decreases and can be approximated by exponential decay.

Finally at the highest gas velocities the bed is operated in a pneumatic conveying regime. This transition velocity depends on the solid flow rate. In the pneumatic conveying regime particles are well distributed in the reactor, with no wall or downflow zone, but solid fraction decreases slightly with height. So we can assume plug flow for both solids and gas in the vessel.

8.3 BUBBLING FLUIDIZED BED

Bubbling fluidized bed reactors have broad applications. As described above, the transition to bubbling fluidization regime depends on property of solid particles. For Group B and Group D particles, the bed transforms from a fixed bed into a bubbling fluidized bed when the gas velocity is increased beyond the minimum fluidization velocity u_{mf}. For Group A particles, the bubbles only appear when the gas velocity is increased beyond u_{mb}. Thus, the transition point from fixed bed to bubbling fluidization regime is u_{mf} for Group B and Group D particles, while for Group A particles, it is u_{mb}. The bubble behaviors play an important role in the hydrodynamics and performance of bubbling fluidized bed reactors. Therefore in this section we will first review bubble behaviors before discussing the mathematical model of a bubbling fluidized bed reactor.

8.3.1   Bubble Behaviors

8.3.1.1   Structure of the Bubble

To describe particle and gas flows around bubbles in a fluidized bed, Davidson and Harrison model is commonly used because of its fundamental importance and relative simplicity. According to this model, the rise velocity of a bubble u_b depends only on the bubble size and the gas behavior in the vicinity of the bubble depends only on the relative velocity of the rising bubble and that of gas rising in the emulsion u_e. The structure of the bubble depends on the bubble's rising velocity, as shown in Fig. 8.5.

The clouded or fast bubble ($u_{br} > u_e$): In this case, emulsion gas enters the lower part of the bubble and leaves at the top. However, because the bubble is rising faster than the emulsion gas, the gas leaving the top is swept around and returns to the bottom of the bubble, forming a circulating region around the bubble which is called the cloud. The rest of the gas in the bed does not mix with that in the cloud.

The cloudless or slow bubble ($u_{br} < u_e$): In this case the emulsion gas rises faster than the bubble, and hence will enter the bubble from bottom and leaves from the top. For the emulsion gas the bubble acts as a convenient shortcut.

For a clouded bubble, the thickness of the cloud decreases with increasing bubble velocity, and its size is given by:

$$\frac{R_c^3}{R_b^3} = \frac{u_{br} + 2u_f}{u_{br} - u_f} \tag{8.15}$$

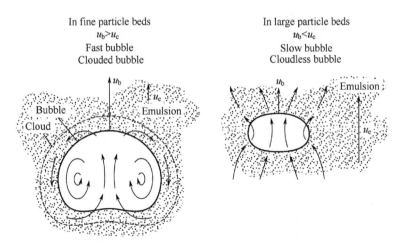

In fine particle beds
$u_b > u_e$
Fast bubble
Clouded bubble

In large particle beds
$u_b < u_e$
Slow bubble
Cloudless bubble

FIGURE 8.5   Gas flow in the vicinity of rising gas bubbles in bubbling fluidized beds.

where $u_f = u_{mf}/\varepsilon_{mf}$ is the upward velocity of gas at minimum fluidizing conditions, and R_c and R_b is the radius of cloud and bubble, respectively. Hence the ratio of cloud to bubble volume is:

$$f_c = \frac{\text{volume of cloud}}{\text{volume of bubble}} = \frac{V_c}{V_b} = \frac{3u_f}{u_{br} - u_f} = \frac{3u_{mf}/\varepsilon_{mf}}{u_{br} - u_{mf}/\varepsilon_{mf}} \tag{8.16}$$

In reality a typical bubble will not be perfectly spherical but instead have a flat or even concave bottom. A wake region will be formed just below the bubble because the pressure near the bottom of the bubble is lower than that in the nearby emulsion phase. The pressure difference will drive the gas into the bubble, which will make the bubble less stable and even partially collapse. For fast clouded bubbles, this gas flow will lead to the leakage of circulating bubble gas into the wake, and also result in solids being sucked toward the region just behind the bubble and forming the wake region. The wake fraction, defined as $f_w = V_w/V_b$, where V_w and V_b are the volumes of the wake and the bubble, respectively, is shown for various particles in Fig. 8.6. The wake angle decreases with particle size, meaning that bubbles are flatter for small particles. The value of f_w has been observed experimentally to vary between 0.25 and 1.0, with typical values close to 0.4.

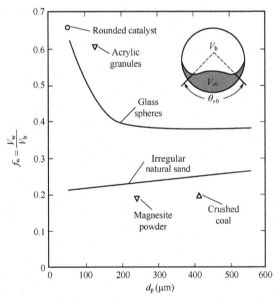

FIGURE 8.6 Wake fraction and wake angle of three-dimensional bubbles at ambient conditions.

8.3.1.2   Bubble Rising Velocity

The rising velocity of a single bubble can be correlated to the bubble size by:

$$u_b = 0.71(gd_b)^{1/2} \tag{8.17}$$

For a group of bubbles, the rising velocity is affected by other factors. The more bubbles present, the less drag on an individual bubble because the bubbles would carry each other up through the bed. The number of bubbles increases with increasing gas velocity u_0. Therefore, bubble rising velocity also increases with increasing u_0. Other factors that should affect bubble rising velocity are the gas viscosity and particle size and density. All these parameters affect the minimum fluidization velocity, which appears in the relationship for the rising velocity of multiple bubbles. The bubble rising velocity in a fluidized bed could be represented by simply adding and subtracting these terms such as the correlation by Davidson and Harrison:

$$u_b = u_0 - u_{mf} + 0.71(gd_b)^{1/2} \tag{8.18}$$

8.3.1.3   Bubble Size

Many properties of the bubble, such as bubble rising velocity, are functions of the bubble diameter. Bubble size depends on many parameters such as bed height, bed diameter, and gas velocity. Bubble size also depends strongly upon reactor internals such as the type and number of baffles and heat exchange tubes within the reactor. The design of the distributor plate used to introduce the gas to the bottom of the bed can also have a pronounced effect on bubble diameter. Most studies on bubble diameter focus on fluidized beds without internals and have used rather small beds. Under these conditions the bubbles grow as they rise through the bed. Based on experimental data covering bed diameters of $7 - 130$ cm, minimum fluidization velocities of $0.5 - 20$ cm/s, and particle diameters of $60 - 450$ μm, a relationship between the bubble diameter and the height in the column is obtained as:

$$\frac{d_{bm} - d_b}{d_{bm} - d_{b0}} = \exp\left(-\frac{0.3h}{D}\right) \tag{8.19}$$

In this equation, d_b is the bubble diameter in a bed of diameter D, observed at a height h above the distributor plate; d_{b0} is the diameter of the bubble formed initially just above the distributor plate, and d_{bm} is the maximum stable bubble diameter.

The maximum stable bubble diameter, d_{bm}, has been observed to follow the relationship:

$$d_{bm} = 0.652[A_c(u_0 - u_{mf})]^{0.4} \tag{8.20}$$

The initial bubble diameter depends on the type of distributor plate. For porous plate distributors, d_{b0} can be calculated by:

$$d_{b0} = 3.76 \times 10^{-3} (u_0 - u_{mf})^2 \tag{8.21}$$

And for perforated plate distributors, d_{b0} can be calculated by:

$$d_{b0} = 0.347 \left[A_c (u_0 - u_{mf}) / n_d \right]^{0.4} \tag{8.22}$$

where n_d is the number of perforations.

Another widely used correlation for calculating bubble size is:

$$d_b = 0.835 [1 + 0.272 (u_0 - u_{mf})]^{\frac{1}{3}} (1 + 0.0684 h)^{1.21} \tag{8.23}$$

Note that in Eqs. (8.19)−(8.23) gas velocities u_0 and u_{mf} have the unit of cm/s, the height above the gas distributor h has the unit of cm, the bubble size d_{bm} and d_{b0} have the unit of cm, and A_c has the unit of cm^2.

8.3.2 Mathematical Model of Bubbling Fluidized Bed

The reaction conversion in a bubbling fluidized bed varies significantly, from as high as that in a plug flow reactor to well below that in a mixed flow reactor, as shown in Fig. 8.7. For many years estimating the performance of a new fluidized bed reactor remains extremely difficult. It was soon recognized that this difficulty was caused by a lack of knowledge on

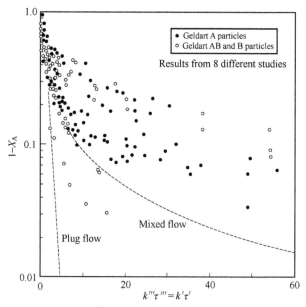

FIGURE 8.7 Conversion in fluidized beds compared with mixed flow and plug flow.

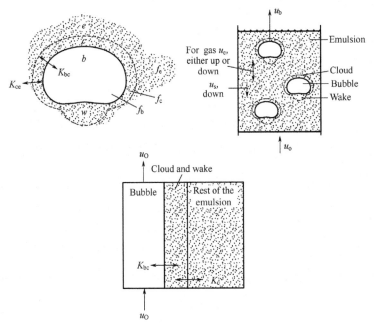

FIGURE 8.8   Model and symbols used to describe the K–L bubbling gas fluidized bed.

contacting and flow pattern in the bed. The flow behaviors inside a bubbling fluidized bed deviates significantly from ideal flow models. A wide variety of approaches have been tried, such as the dispersion and tanks in series model, RTD (residence time distribution) model, contact time distribution model, and two-region model. One model that is widely used was developed by Kunii−Levenspiel and will be described in this session.

In the Kunii−Levenspiel model, the following assumptions are made (Fig. 8.8):

a. The bubbles are all spherical, have the same size d_b, and follow the Davidson's model. Thus the bed contains bubbles surrounded by thin clouds rising through an emulsion. The upflow of gas through the cloud can be ignored for $u_b \gg u_e$ because the cloud volume is much smaller than that of the bubble.

b. The downward flow of particles in the emulsion phase can be considered as plug flow.

c. The emulsion phase is at minimum fluidizing conditions. The particles in emulsion phase move downward, and the minimum fluidizing velocity refers to the gas velocity relative to that of moving particles, $u_e = u_{mf}/\varepsilon_{mf} - u_p$. The velocity of moving particles, u_p, is positive in the downward direction. The velocity of gas in the emulsion, u_e, is positive in the upward direction, but note that it can be negative under some conditions.

d. The concentration of particles in the wakes equal to that in the emulsion phase, and the void fraction is equal to ε_{mf}. The particles and gas in the wake are assumed to move together at a velocity that is equal to the rising velocity of the bubbles.

8.3.2.1 Fraction of Bed in the Different Phases

Using the Kunii–Levenspiel model, the fraction of the bed occupied by the bubbles and wakes can be estimated by material balances on the solid particles and the gas flows. The parameter δ is the fraction of the total bed occupied by the part of the bubbles that does not include the wake, and f_w is the volume of wake per volume of bubble. The bed fraction in the wakes is therefore $f_w\delta$. The bed fraction in the emulsion phase, which includes the clouds, is $(1 - \delta - f_w\delta)$. Letting A_c and ρ_p represent the cross-sectional area of the bed and the density of the solid particles, respectively, a material balance on the solids gives:

(Particles flowing downward in emulsion) = (Particles flowing upward in wakes)
$$A_c\rho_p(1 - \delta - f_w\delta)u_p \qquad = \qquad A_c\rho_p f_w\delta u_b$$
$$(8.24)$$

This gives the velocity of solid particles:

$$u_p = \frac{f_w\delta u_b}{1 - \delta - f_w\delta} \tag{8.25}$$

A mass balance on the gas phase flows gives:

$$\begin{pmatrix} \text{Total gas} \\ \text{flow rate} \end{pmatrix} = \begin{pmatrix} \text{Gas flow} \\ \text{in bubbles} \end{pmatrix} + \begin{pmatrix} \text{Gas flow} \\ \text{in wakes} \end{pmatrix} + \begin{pmatrix} \text{Gas flow} \\ \text{in emulsion} \end{pmatrix}$$
$$A_c u_0 \qquad = \qquad A_c\delta u_b \quad + \quad A_c\varepsilon_{mf}f_w\delta u_b \quad + \quad A_c\varepsilon_{mf}(1 - \delta - f_w\delta)u_e$$
$$(8.26)$$

The rising velocity of gas in the emulsion phase is:

$$u_e = \frac{u_{mf}}{\varepsilon_{mf}} - u_p \tag{8.27}$$

Combining Eqs. (8.25), (8.26), and (8.27) gives:

$$\delta = \frac{u_0 - u_{mf}}{u_b - u_{mf}(1 + f_w)} \tag{8.28}$$

The wake parameter, f_w, is a function of particle size, as shown in Fig. 8.5.

8.3.2.2 Distribution of Particles in the Various Phases

The distribution of solids in the various phases is be defined by:

$$\text{Solids in bubble} = \frac{\text{volume of solids in bubbles}}{\text{bed volume}} \qquad (8.29a)$$

$$\text{Solids in cloud} + \text{wake} = \frac{\text{volume of solids in clouds and wakes}}{\text{bed volume}} \qquad (8.29b)$$

$$\text{Solids in emulsion} = \frac{\text{volume of solids in emulsion phase}}{\text{bed volume}} \qquad (8.29c)$$

These γ values are related by the expression:

$$\delta(\gamma_b + \gamma_{c+w} + \gamma_e) = 1 - \varepsilon_f = (1 - \varepsilon_{mf})(1 - \delta) \qquad (8.30)$$

Rearranging Eq. (8.30) gives:

$$\gamma_e = (1 - \varepsilon_{mf})\left(\frac{1 - \delta}{\delta}\right) - \gamma_{c+w} - \gamma_b \qquad (8.31)$$

With the wake included with the cloud region, we also have

$$\gamma_{c+w} = (1 - \varepsilon_{mf})\left(\frac{3u_{mf}/\varepsilon_{mf}}{u_b - u_{mf}/\varepsilon_{mf}} + f_w\right) \qquad (8.32)$$

Experimental data shows that the value of γ_b ranges between 0.001 and 0.01, with 0.005 being the more typical number.

$$\gamma_b = 0.001 \sim 0.01 \qquad (8.33)$$

8.3.2.3 Mass Transfer Between Phases

Mass transfer between phases in a fluidized bed is illustrated in Fig. 8.9. For gas interchange between bubbles and cloud, the mass transfer coefficient K_{bc}, which has a unit of s^{-1}, is defined in the following manner:

$$W_{Abc} = K_{bc}(C_{Ab} - C_{Ac}) \qquad (8.34)$$

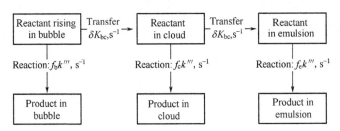

FIGURE 8.9 Mass transfer and reaction in the bubbling fluidized bed.

where C_{Ab} and C_{Ac} in mol/L are the concentration of A in the bubble and cloud, respectively, and W_{Abc} in mol/(L·s) represents the number of moles of A transferred from the bubble to the cloud per unit time and unit volume of bubble.

The mass transfer coefficient K_{bc} based on bubble volume can be calculated as:

$$K_{bc} = 4.5\left(\frac{u_{mf}}{d_b}\right) + 5.85\left(\frac{D_{AB}^{1/2}g^{1/4}}{d_b^{5/4}}\right) \tag{8.35}$$

Similarly, a mass transfer coefficient for gas interchange between the cloud and the emulsion is defined as:

$$W_{Ace} = K_{ce}(C_{Ac} - C_{Ae}) \tag{8.36}$$

Note that even though this mass transfer does not involve the bubbles directly, it is still based on bubble volume. The concept of basing all mass transfer on bubble volume simplifies the calculations markedly.

K_{ce} can be obtained by using analogy of mass transfer from gas to liquid based on penetration theory:

$$K_{ce} = 6.77\left(\frac{\varepsilon_{mf}D_{AB}u_b}{d_b^3}\right)^{1/2} \tag{8.37}$$

where u_b is bubble rising velocity in cm/s. K_{ce} has a typical value of $1\ \text{s}^{-1}$.

8.3.2.4   Reactor Model Equations based on Material Balance

Model equations for a bubbling fluidized bed reactor can be developed by writing material balance equations of each of the three phases inside the reactor. For simplicity, a first order reaction is assumed in the following derivation.

For the bubble phase, the mass balance for species A is:

$$u_b\frac{dC_{Ab}}{dz} = -k_{rb}C_{Ab} - K_{bc}(C_{Ab} - C_{Ac}) \tag{8.38}$$

For the cloud and wake phase, the mass balance for species A is:

$$u_b(f_w + f_c)\delta\frac{dC_{Ac}}{dz} = -k_{rc}C_{Ac} + K_{bc}(C_{Ab} - C_{Ac}) - K_{ce}(C_{Ac} - C_{Ae}) \tag{8.39}$$

For the emulsion phase, the mass balance for species A is:

$$u_e\left(\frac{1 - \delta - f_w\delta - f_c\delta}{\delta}\right)\frac{dC_{Ae}}{dz} = -k_{re}C_{Ae} + K_{ce}(C_{Ac} - C_{Ae}) \tag{8.40}$$

We can see that the reactor model equations are three coupled ordinary differential equations, with one independent variable (z) and three dependent variables (C_{Ab}, C_{Ac}, and C_{Ae}), and have to be solved numerically. The model is further simplified by assuming that for cloud and emulsion phases the derivative terms on the left-hand side of material balance equations are negligible. Using this assumption, and letting $t = z/u_b$ (i.e., the time the bubble has spent in the bed), Eq. (8.38) to Eq. (8.40) can be simplified as:

Bubble phase: $\quad \dfrac{dC_{Ab}}{dt} = -\gamma_b k_{r,cat} C_{Ab} - K_{bc}(C_{Ab} - C_{Ac})$ $\qquad$ (8.41)

Cloud and wake phase: $K_{bc}(C_{Ab} - C_{Ac}) = \gamma_c k_{r,cat} C_{Ac} + K_{ce}(C_{Ac} - C_{Ae})$ (8.42)

Emulsion phase: $\quad K_{ce}(C_{Ac} - C_{Ae}) = \gamma_e k_{r,cat} C_{Ae}$ $\qquad$ (8.43)

To obtain the fluidized-bed design equation for a first order reaction, we express both the concentration of A in the emulsion C_{Ae} and that in the cloud C_{Ac} in terms of the bubble concentration.

Using Eq. (8.42) concentration of A in emulsion phase C_{Ae} can be obtained as a function of the concentration in cloud phase C_{Ac}:

$$C_{Ae} = \frac{K_{ce}}{\gamma_e k_{cat} + K_{ce}} C_{Ac} \qquad (8.44)$$

Substituting Eq. (8.44) into Eq. (8.42) yields:

$$C_{Ac} = \frac{K_{bc}}{k_{cat}\gamma_c + \dfrac{K_{ce}\gamma_e k_{cat}}{\gamma_e k_{cat} + K_{ce}} + K_{bc}} C_{Ab} \qquad (8.45)$$

Then Eq. (8.41) can be expressed as:

$$-\frac{dC_{Ab}}{dt} = \gamma_b k_{cat} C_{Ab} + K_{bc}\left(C_{Ab} - \frac{K_{bc}C_{Ab}}{k_{cat}\gamma_c + \dfrac{K_{ce}\gamma_e k_{cat}}{\gamma_e k_{cat} + K_{ce}} + K_{bc}}\right) \qquad (8.46)$$

$$-\frac{dC_{Ab}}{dt} = k_{cat}C_{Ab}\left[\gamma_b + \frac{K_{bc}(\gamma_e \gamma_c k_{cat} + K_{ce}\gamma_c + K_{ce}\gamma_e)}{k_{cat}(\gamma_e \gamma_c k_{cat} + K_{ce}\gamma_e + K_{ce}\gamma_e) + K_{bc}(\gamma_e k_{cat} + K_{ce})}\right]$$

$$= -k_{cat}K_R C_{Ab} \qquad (8.47)$$

where

$$K_R = \gamma_b + \frac{1}{\dfrac{k_{cat}}{K_{bc}} + \dfrac{1}{\gamma_c + \dfrac{1}{\dfrac{1}{\gamma_e} + \dfrac{k_{cat}}{K_{ce}}}}} \qquad (8.48)$$

C_{Ab} can be expressed as a function of conversion X, and Eq. (8.47) can be rewritten as:

$$\frac{dX}{dt} = k_{cat}K_R(1 - X) \tag{8.49}$$

$$\ln\left(\frac{1}{1-X}\right) = k_{cat}K_R t \tag{8.50}$$

The catalyst weight needed to achieve this conversion is:

$$W = \frac{\rho_s A_c h(1 - \varepsilon_{mf})(1 - \delta)}{k_{cat}K_R}\ln\frac{1}{1-X} \tag{8.51}$$

Example 8.1
Ammonia oxidation takes place in a bubbling fluidized bed reactor at 1.0 atm and $T = 523$ K. The reactor diameter is $D = 12.0$ cm, the gas feed contains 10% NH_3 and 90% O_2, and is fed into the reactor at 800 cm³/s (at reaction conditions). Four kilograms of catalyst with particles size of 100 μm is used in the reactor, and the initial height of settled bed is 40 cm. The catalyst has a sphericity of 0.6 and density $\rho_s = 2.0$ g/cm³. The oxidation reaction is first order with respect to ammonia concentration:

$$-r_A = k_{cat}C_{NH_3} \text{ mol}/(s \text{ cm}^3)$$

$$-k_{cat} = 0.086 \text{ s}^{-1}$$

The gas density is $\rho_g = 0.785 \times 10^{-3}$ g/cm³, viscosity $\mu_g = 2.98 \times 10^{-4}$ g/cm·s, and diffusivity $D_{AB} = 0.618$ cm²/s.
Please estimate ammonia conversion.

Solution
The void fraction of bed at a minimum fluidization, ε_{mf}, is

$$\varepsilon_{mf} = 0.586\phi_s^{0.72}\left(\frac{\mu_g^2}{\rho_g(\rho_s - \rho_g)gd_p^3}\right)^{0.029}\left(\frac{\rho_g}{\rho_s}\right)^{0.021} = 0.661$$

The gas velocity at minimum fluidization is

$$u_{mf} = \frac{(\phi_s d_s)^2}{150}\frac{(\rho_s - \rho_g)g}{\mu_g}\left(\frac{\varepsilon_{mf}^3}{1 - \varepsilon_{mf}}\right) = 1.34 \text{ cm/s}$$

$$u_0 = \frac{v_0}{\frac{\pi}{4}d^2} = \frac{800}{\frac{\pi}{4}12^2} = 7.08 \text{ cm/s}$$

The bubble sizes, d_{b0}, d_{bm}, and d_b are

$$d_{b0} = 3.76 \times 10^{-3}(u_0 - u_{mf}) = 0.12 \text{cm}$$

$$d_{bm} = 0.652[A_c(u_0 - u_{mf})]^{0.4} = 8.67 \text{cm}$$

(Continued)

(Continued)

The initial height of settled bed h_s is 40 cm, and the expanded bed height is usually increased by 40%–50%. Therefore it is estimated that the height of expanded bed is 60 cm. The average bubble size is estimated at the middle of the expended bed, that is $h_{1/2}$ = 30 cm.

$$d_b = d_{bm} - (d_{bm} - d_{b0})\exp\left(\frac{0.3h_{1/2}}{D}\right) = 4.64 \text{ cm}$$

The rising velocity of bubble swarms is

$$u_b = u_0 - u_{mf} + 0.71\left(gd_b\right)^{1/2} = 53.6 \text{ cm/s}$$

The fraction of bed in the bubble phase is

$$\delta = \frac{u_0 - u_{mf}}{u_b - u_{mf}(1 + f_w)} = 0.112$$

where f_w is set a value of 0.4.

The height of expanded bed is

$$h = \frac{W}{A(1 - \delta)(1 - \varepsilon_{mf})\rho_s} = 58.7 \text{ cm}$$

The value is very close to the initial estimated bed height of 60 cm, so we can proceed in the calculations.

The bubble-cloud mass transfer coefficient K_{bc} is

$$K_{bc} = 4.5\left(\frac{u_{mf}}{d_b}\right) + 5.85\left(\frac{D_{AB}^{1/2} g^{1/4}}{d_b^{5/4}}\right) = 5.12 \text{s}^{-1}$$

The cloud-emulsion mass-transfer coefficient K_{ce} is

$$K_{ce} = 46.77\left(\frac{\varepsilon_{mf}D_{AB}u_b}{d_b^3}\right)^{1/2} = 3.17 \text{s}^{-1}$$

The volume of catalysts in the bubble per volume of bubble is assumed as

$$\gamma_b = 0.01$$

The volume of catalyst in the clouds and wakes per cm^3 of bubbles is

$$\gamma_{c+w} = (1 - \varepsilon_{mf})\left(\frac{\dfrac{3u_{mf}}{\varepsilon_{mf}}}{u_b - \dfrac{u_{mf}}{\varepsilon_{mf}}} + f_w\right) = 0.177$$

The volume of catalyst in the emulsion per cm^3 of bubbles is

$$\gamma_e = (1 - \varepsilon_{mf})\left(\frac{1 - \delta}{\delta}\right) - \gamma_{c+w} - \gamma_b = 2.56$$

(Continued)

(Continued)

Now the overall mass transfer coefficient K_R can be calculated as

$$K_R = \gamma_b + \cfrac{1}{\cfrac{k_{cat}}{K_{bc}} + \cfrac{1}{\gamma_c + \cfrac{1}{\cfrac{1}{\gamma_e} + \cfrac{k_{cat}}{K_{ce}}}}} = 2.45$$

And conversion of ammonia X is

$$X = 1 - \exp\left(-\frac{k_{cat}K_R h}{u_b}\right) = 0.21$$

8.4 TURBULENT FLUIDIZED BED

8.4.1 Regime Transition

The turbulent fluidization can be considered as the transition regime from bubbling fluidization to fast fluidization. The intensity of bubble movement increases with increasing gas velocity, and can be measured by pressure fluctuations in the bed. As shown in Fig. 8.10, with the increases of gas velocity, the amplitude of pressure fluctuation will first increase to reach a maximum, and then decrease and gradually level off. This pressure fluctuation variation reflects the changes of flow dynamics in the bed, and can be used to identify the transition from the bubbling to the turbulent regime.

The onset velocity for the transition to the turbulent regime is commonly defined as the gas velocity corresponding to the peak, u_c, and the transition can be directly observed based on bed phenomena, especially the bubble behavior in the bed. During bubbling fluidization ($u < u_c$) the main interaction between bubbles is coalescence. On the other hand, when gas velocity is

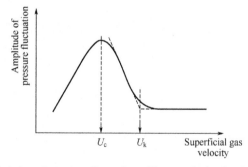

FIGURE 8.10 Variation of pressure fluctuation with gas velocity for dense-phase fluidized beds.

greater than u_c, the dominant bubble behavior is bubble breakup. Therefore there is a significant shift of bubble behavior at gas velocity around u_c. When gas velocity is further increased from u_c to u_k, there is no significant flow structure change. At u_k the bed is usually already highly turbulent and may even be close to the transition to fast fluidization. The plot in Fig. 8.10 for the amplitude of the pressure fluctuation appears to be typical only for Group A particles. For Group B and D particles the leveling-off point for u_k may be nonobvious. Thus for all practical purposes, u_c is defined as the onset velocity for the transition to the turbulent regime.

8.4.2   Hydrodynamic Characteristics

As indicated above the gas phase still forms bubbles in turbulent-fluidized beds. Therefore the hydrodynamic behavior of turbulent fluidization is similar to that of bubbling fluidization, especially at relatively low gas velocity. With increasing gas velocity the differences between bubbling and turbulent fluidization become more obvious. One distinguished behavior of turbulent fluidized bed is that the bubble size tends to decrease with an increase in gas velocity, which is opposite to that exhibited in the bubbling fluidized bed. This is because that in the turbulent regime the dominant bubble behavior is bubble breakup instead of bubble coalescence. However, the impact of operating pressure on bubble size is similar to that in bubbling regime. At a constant gas velocity an increase of operating pressure leads to a decrease in bubble size. The increase of operating pressure promotes gas entry into the emulsion phase, which makes the bubble and emulsion phase less distinguishable, and eventually, lead to fast fluidization. Thus with increase of gas velocity and operating pressure, the bubble size in a turbulent bed will decrease, and the bed will eventually become bubble-free fast fluidized bed.

In general the hydrodynamic characteristics under turbulent fluidization, although different from that under bubbling fluidization, is still largely dominated by bubble behaviors. Under turbulent fluidization the bubbles are small and more irregular in shapes due to frequent bubble breakup. The bubbles move more violently and random, which enhance the interphase exchange (i. e., high heat and mass transfer rate) and also make it difficult to distinguish the emulsion phase from the bubble phase in the bed.

8.5   CIRCULATING FLUIDIZED BED

8.5.1   Introduction

A circulating fluidized bed (CFB) is a fluidized bed system that includes a riser and a down-comer with the solid particles circulating between them. The riser is operated in a fast fluidization regime, with solids carried over from top and returned to bottom of the riser via a standpipe and feeding or control device. The operating variables for a CFB system include both the gas flow rate and solids circulation rate, in contrast to the gas flow rate only in a dense-phase fluidized bed system.

CFB systems have been widely used in the petrochemical industry (e.g., in fluid catalytic cracking (FCC)) and the utility industry (e.g., in coal combustion). Other commercial applications of the CFB systems include maleic anhydride production from n-butane and Fischer−Tropsch synthesis of liquid fuels from carbon monoxide and hydrogen. Typical applications are summarized in Table 8.2. Operating conditions of CFB systems can be significantly different, depending on process applications. For example, for coal combustion, gas velocity and solids flow rate are typically 5−8 m/s and less than 40 kg/m$^2 \cdot$ s, respectively. For FCC, on the other hand, gas velocity at the riser exit and solids flow rate are 15−20 m/s and higher than 300 kg/m$^2 \cdot$ s, respectively.

The advantages of CFBs relative to bubbling beds include:

1. High gas throughputs
2. Limited backmixing of gas
3. Long and controllable residence time of particles
4. Uniform temperature profile without "hot" spots
5. Flexibility in handling particles of widely differing sizes, densities, and shapes

TABLE 8.2 Examples of Applications of CFB Reactors

Applications	Comments
Fluid catalytic cracking	Hundreds of units worldwide; mainstay of petroleum refining
Fischer−Tropsch synthesis	Applied for many years as Synthol process in South Africa
Maleic anhydride	One commercial reactor in Spain
Combustion of coal, biomass, wastes, off-gases	Widespread usage for power generation and boilers in Europe, North America, and Asia
Gasification	Commercial units gaining a foothold, especially in Europe
Calcination (e.g., of aluminum trihydrate and carbonates)	Lurgi units used widely
Reduction of iron ore	Lurgi plant in Trinidad
Smelter off-gas treatment	One plant in Australia supplied by Ahlstrom
Flue gas dry scrubbing of HF, HCl, SO$_2$, dioxins, mercury, etc.	Commercial units since the 1970s, primarily in Europe

6. Effective contacting between gas and particles
7. Lack of bypassing of gas with minimal mass transfer limitations
8. Opportunity for complementary operations, such as catalyst regeneration in the return loop

However, CFBs also have the following disadvantages:

1. Very tall vessel required
2. Substantial backmixing of solid particles
3. Internals not viable because of wear and attribution
4. Heat transfer less favorable than that in dense-phase fluidization
5. Lateral gradients can be significant
6. Losses of particles due to entrainment

8.5.2   Configuration of CFB

A CFB reactor mainly consists of four parts: riser, gas—solid separator, down-comer, and solids flow control device, as shown in Fig. 8.11. In the riser, gas and particles commonly flow cocurrently upward, exit at the top, and flow into the gas—solid separator. The separated particles then flow to the down-comer and return to the riser.

The riser is the main component of a CFB reactor system. In comparing with a fixed bed reactor the design of the entrance and exit of a riser reactor has very significant influences on gas and particle flow behaviors, and therefore is critical for achieving the desired reactor performance. Cyclones are usually used for gas—solid separation, and their efficiency can affect the particle size distribution and solids circulation rate in the system. The down-comer can be either a large reservoir that can help regulating solid recycling

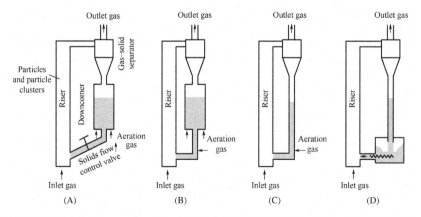

FIGURE 8.11   Various configurations of circulating fluidized bed systems: (A) Mechanical valve with reservoir; (B) Nonmechanical valve with reservoir; (C) Nonmechanical valve without reservoir; (D) Mechanical feeder without reservoir.

rate, as shown in Fig. 8.11A and B, or be a standpipe that simply serve as a pathway for solids to return to the riser, as shown in Fig. 8.11C and D. For reaction applications the down-comer can serve as a heat exchanger or as a catalyst regenerator.

To smoothly operate a CFB system, it is very important to effectively control the solids circulation rate to the riser. The solid flow control device has two major functions: one is preventing gas in riser flows to the down-comer, and the other is controlling solid circulation rate. Both mechanical valves (Fig. 8.12(A) and (D)) and nonmechanical valves (Fig. 8.12(B) and (C)) are used.

8.5.3 Mathematical Models of CFB

As mentioned earlier the riser of a CFB system is typically operated in a fast fluidization regime, and therefore the fluid flow characteristics of fast fluidization beds need to be considered when developing a mathematic model of CFB reactor. Based on experimental observations at macroscale, along the axial direction a fast fluidized bed will have a dense region at the bottom and a dilute region on the top. In the radical direction the solid movement will form a core-annulus structure, i.e., a lean core surrounded by a denser annulus in the radial direction. At mesoscale some solids will form particle clusters. Due to the heterogeneity at both macro- and mesoscales, a complete characterization of the hydrodynamics of a CFB requires detailed description of the void fraction and velocity profiles in both axial and radial directions.

There are a number of mathematical models describing the flow pattern in a CFB. Models commonly contain a series of unproven assumptions and empirical constants. Mechanistic models that have intermediate complexity but can capture major relevant features are often the most useful for engineering design, optimization, and control. In the longer term, more sophisticated models, e.g., those based on multiphase computational fluid dynamic codes, are likely to have increasing impact. Here the one-dimensional model and core−annulus model are introduced.

8.5.3.1 Diffusion-Segregation One-Dimensional Model

Li and Kwauk proposed a one-dimensional model to predict the axial profile of average solids volume fraction based on equilibrium between diffusion and segregation of clusters. In this model, as shown in Fig. 8.12, the fast fluidized bed consists of a dilute suspension phase and a cluster phase. The dense clusters have a volume fraction of f, and the solids concentration in dense clusters is $1 - \varepsilon_a$. The dilute continuum has a volume fraction of $1 - f$, and the solids concentration in it is $1 - \varepsilon^*$. The clusters move upward through a diffusive mechanism from a relatively dense region of the lower section, and then they

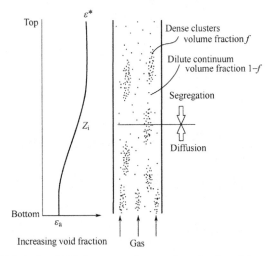

FIGURE 8.12   Schematic of diffusion-segregation one-dimensional model.

tend to fall back to the lower region by virtue of buoyancy when arriving at a higher region, where the average bed density is lower. Dynamic equilibrium calls for equality of the diffusion and buoyancy fluxes.

The diffusion flux is determined by the gradient of strands concentration, $\rho_p f(1 - \varepsilon_a)$, with ξ as the proportionality constant, while the upthrown strands, at their volume concentration of $f(1 - \varepsilon_a)$, are subject to a driving force for segregation, represented by the difference in concentration between the average bed, $(1 - \varepsilon)$, and the strands, $(1 - \varepsilon_a)$, with an analogous proportionality constant ω. Thus,

$$\xi \frac{d}{dz}\left[\rho_s f(1 - \varepsilon_a)\right] = \omega \Delta \rho[(1 - \varepsilon_a) - (1 - \varepsilon)]f(1 - \varepsilon_a) \qquad (8.52)$$

The average solids volume fraction, $(1 - \varepsilon)$, is composed of both the strands and the dilute continuum:

$$1 - \varepsilon = f(1 - \varepsilon_a) + (1 - f)(1 - \varepsilon^*) \qquad (8.53)$$

Combining Eqs. (8.52) and (8.53) gives:

$$-\frac{d\varepsilon}{(\varepsilon^* - \varepsilon)(\varepsilon - \varepsilon_a)} = \frac{\omega \Delta \rho}{\xi \rho_s} dz \qquad (8.54)$$

To find the void fraction ε_i at the point of inflection z_i, set the second derivation of Eq. (8.54) with respect to z to zero, and yielding

$$\varepsilon_i = \frac{(\varepsilon^* + \varepsilon_a)}{2} \qquad (8.55)$$

Integrating Eq. (8.54) from z_i to z gives

$$\frac{\varepsilon - \varepsilon_a}{\varepsilon^* - \varepsilon} = \exp\left(\frac{z - z_i}{Z_0}\right) \tag{8.56}$$

where Z_0 is the characteristic length of fast fluidization and defined as:

$$Z_0 = \frac{\zeta \rho_s}{\omega \Delta \rho} \frac{1}{\varepsilon^* - \varepsilon_a} \tag{8.57}$$

Eq. (8.56) gives an S-shaped axial distribution of the void fraction, trending toward ε^* at the top as $z \to -\infty$ and toward ε_a at the bottom as $z \to +\infty$. The characteristic length Z_0 represents how rapidly the dense region at the bottom merges into the dilute region at the top. Thus, $Z_0 \to 0$ indicates a clear interface between the dense and dilute region, while $Z_0 \to \infty$ implies a uniform axial profile of the void fraction. The value of Z_0, expressed in meters, can be correlated by:

$$Z_0 = 500 \exp[-69(\varepsilon^* - \varepsilon_a)] \tag{8.58}$$

where ε_a is the asymptotic void fraction in the bottom dense region and ε^* is the asymptotic void fraction in the top dilute region. Their values can be calculated from the following empirical correlations:

$$1 - \varepsilon_a = 0.2513\left(\frac{18Re_a + 2.7Re_a^{1.687}}{Ar}\right)^{-0.4037}$$

$$Re_a = \frac{\rho_f d_p}{\mu}\left(u_0 - u_d \frac{\varepsilon_a}{1 - \varepsilon_a}\right) \tag{8.59}$$

$$1 - \varepsilon^* = 0.05547\left(\frac{18Re_a + 2.7Re_a^{1.687}}{Ar}\right)^{-0.6222}$$

$$Re^* = \frac{\rho_f d_p}{\mu}\left(u_0 - u_d \frac{\varepsilon^*}{1 - \varepsilon^*}\right) \tag{8.60}$$

where Ar is the Archimedes number defined as:

$$Ar \equiv \frac{d_p^3 \rho_f g(\rho_p - \rho_f)}{\mu^2} \tag{8.61}$$

The parameter z_i is determined by equating the average solid fraction in the riser calculated from pressure balance through the CFB loop to that calculated by integrating Eq. (8.54) along the riser.

Eqs. (8.59) and (8.60) were obtained on the basis of experimental data for FCC catalyst, fine alumina, coarse alumina, pyrite cinder, and iron ore concentrate, ranging from Group A to weakly Group B particles. These correlations fit the experimental data well in the range of $\varepsilon_a = 0.85 - 0.93$ and $\varepsilon^* = 0.97 - 0.993$.

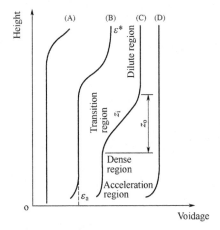

FIGURE 8.13   Typical axial profiles of void fraction for Group A particles.

The Li and Kuawk model can predict the different axial profiles of the void fraction in fast fluidization, which have been experimentally observed at different operating conditions. The axial profile of the cross-sectional averaged void fraction is typically S-shaped, as shown in Fig. 8.13 for Group A particles. This profile reflects an axial solids concentration distribution with a dense region at the bottom and a dilute region at the top of the riser. The boundary between the two regions is marked by the inflection point in the profile. An increase in gas flow rate at a given solids circulation rate reduces the dense region (from (A) to (C) in Fig. 8.13), whereas an increase in solids circulation rate at a given gas flow rate results in an expansion of the dense region (from (C) to (A) in Fig. 8.13). When the solids circulation rate is very low and/or the gas velocity is very high, the dilute region covers the entire riser, as shown in Fig. 8.13(D).

8.5.3.2 Core−Annulus Model

Various models based on the core−annular flow structure have been proposed. The model by Bolton and Davidson is simple to apply and is feasible in its mechanistic account of the large gas−solid slip velocity in the system. Fig. 8.14 shows the conceptual configuration of Bolton and Davidson's model. In the model the flow structure is described as a dilute core region surrounded by a dense annular region near the wall. As shown in Fig. 8.14, the outer diameter of the annular region is D and the inner diameter is D_c. In the annular region, solid particles flow downward with a velocity of u_{pw} and a volume fraction of ε_{sw}. In the core region both gas and solid flow upward with a solid velocity of u_{pc}, a solid volume fraction of ε_{sc}, and a gas velocity u_{fc}. For one-dimensional model, both velocity and volume fraction are considered as uniform in the radial direction. The transport of solid particles from the dilute core region to dense annular

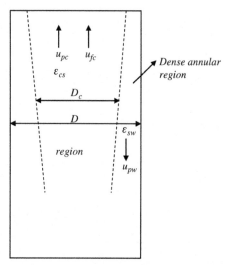

FIGURE 8.14 Conceptual configuration of Bolton and Davidson's model.

region can be described by turbulent diffusion. Therefore a mass balance equation of solid particles in the core region can be derived as:

$$\frac{1}{4} D u_{pc} \frac{d\varepsilon_{sc}}{dz} = -k_d(\varepsilon_{sc} - \varepsilon_{sc\infty}) \tag{8.62}$$

where k_d is called deposition coefficient which describes the turbulent diffusion of particles from the core region to the annular region; $\varepsilon_{sc\infty}$ is an equilibrium solids concentration in the core region. Eq. (8.62) is developed by assuming that D_c can be approximated by D. The changes in particle velocity in the core region are usually small. Therefore, it is reasonable to assume that u_{pc} is independent of the bed height. By further assuming that particles do not interfere with each other significantly, the upward particle velocity in the core region can be calculated by:

$$u_{pc} = u_{fc} - u_{pt} \tag{8.63}$$

Eq. (8.62) can be solved with respect to ε_{sc} to yield:

$$\varepsilon_{sc} = \varepsilon_{sc\infty} + (\varepsilon_{sc0} - \varepsilon_{sc\infty})\exp(-k_d z) \tag{8.64}$$

where $k_d = 4 k_d/Du_{pc}$, ε_{sc0} is the solids concentration in the core region at $z = 0$, and $\varepsilon_{sc\infty}$ is the asymptotic concentration at large z. The upward flow of solids, W_e, is defined as the total flow rate through the cross-section of the core region, can be expressed as:

$$W_e = \frac{\pi}{4} D_c^2 u_{pc} \varepsilon_{sc} \tag{8.65}$$

Combining Eqs. (8.64) and (8.65) yields:

$$W_e = W_{e\infty} + (W_{e0} - W_{e\infty})\exp(-k_d z) \tag{8.66}$$

On the other hand, based on mass balance across a cross-section of the riser, the upward solids flow rate W_e can also be expressed by:

$$\frac{\pi}{4} D^2 J_p = W_e - W_w = W_{e0} - W_{w0} = W_{e\infty} - W_{w\infty} \qquad (8.67)$$

where J_p is the average solids flow rate of the whole cross-section, and W_w is the downward solids flow rate through the cross-section of the annular region. Substituting W_w in Eq. (8.67) into Eq. (8.66) yields:

$$W_w = W_{w\infty} + (W_{e0} - W_{e\infty})\exp(-k_d z) \qquad (8.68)$$

Eq. (8.68) implies that the particle downward flow rate along the wall decays exponentially with bed height. The deposition coefficient, k_d, can be related to the amplitude of the gas velocity fluctuations as:

$$k_d = \frac{0.1\sqrt{\pi}u'}{1 + \frac{St}{12}} \qquad (8.69)$$

where u' is the amplitude of turbulent velocity fluctuations, which can be correlated by $u(1 - 2.8\,\mathrm{Re}^{-1/8})$ with $\mathrm{Re} = \rho_p D u/\mu$, St is the Stokes number, defined as $\rho_p d_p^2 u/(18\,\mu D)$. For small particles, the Stokes number is small, and u' is approximately $0.1\,u$ for a pipe flow. These simplifications give:

$$k_d = 0.01\sqrt{\pi}u \qquad (8.70)$$

Thus, k_d can be given as

$$k_d = \frac{0.04\sqrt{\pi}}{D} \frac{u}{u_{pc}} \qquad (8.71)$$

Once W_e and W_w are known, the cross-sectional averaged solid volume fraction, ε_s, can be estimated by using the following equation:

$$\varepsilon_s = \frac{1}{\rho_p A}\left(\frac{W_e}{u_{pc}} + \frac{W_w}{u_{pw}}\right) \qquad (8.72)$$

The falling velocity of particles in the annular region, u_{pw}, is about 1.8 m/s over a range of gas velocities from 1.5 to 5.0 m/s and is not sensitive to the solids circulation rate.

Example 8.2
A riser is of 0.15 m in diameter and 8 m in height. Particles with a mean diameter of 200 μm and a density of 384 kg/m³ are used in the riser, which operates at $u_0 = 2.21$ m/s and $J_p = 3.45$ kg/m²s. The gas used is air. The particle downward velocity u_{pw} is 0.5 m/s and the particle downward flow rate W_w is 0.2 kg/s in the annular region. Assume that the solid volume fraction in the central core region,

(Continued)

(Continued)

ε_{sc}, is 0.015. Calculate the cross-section averaged solid holdup and the decay constant, k_d, defined in Eq. (8.71) in terms of the core—annular model.

Solution
Assuming that the slip velocity between particles and air in the core region is equal to the particle terminal velocity and D_c can be approximated by D, the particle velocity in the core region can be determined by:

$$u_{pc} = u_{fc} - u_{pt} = \frac{U}{1 - \varepsilon_{pc}} - u_{pt} = \frac{2.21}{1 - 0.015} - 0.331 = 1.91 \text{ m/s}$$

where u_{pt} is calculated from Eq. (8.14).
W_e can be calculated as:

$$W_e = \frac{\pi}{4}D^2 J_p + W_w = \frac{\pi}{4}0.15^2 \times 3.45 + 0.2 = 0.261 \text{ kg/s}$$

The cross-sectional averaged solids holdup ε_s is calculated as:

$$\varepsilon_s = \frac{1}{\rho_p A}\left(\frac{W_e}{u_{pc}} + \frac{W_w}{u_{pw}}\right) = \frac{1}{384 \times 0.0176}\left(\frac{0.261}{u_{pc}} + \frac{0.2}{0.5}\right) = 0.079$$

The decay constant k_d is then calculated as

$$k_d = \frac{0.04\sqrt{\pi}}{D}\frac{U}{u_{pc}} = \frac{0.04\sqrt{\pi}}{0.15}\frac{2.21}{1.91} = 0.55 \text{ m}^{-1}$$

8.6 DOWNER REACTOR

In conventional gas—solid fluidized bed reactors, particles are suspended by upflowing gas stream against the force of gravity. This gives many advantages, such as enhanced mass and heat transfer and improved interphase contact, but also leads to heterogeneous flow structure and significant backmixing, which limits the conversion and selectivity to intermediate product. To achieve high conversion and selectivity for a process with fast and series reactions, a reactor allowing for very short contact time between phases and precisely controlled gas and solid catalyst residence times would be advantageous. The concept of the gas—solid downer reactor was proposed for such a purpose (Fig. 8.15).

The ultrashort contact times between phases in the whole downer impose stringent demands on the solids feeding and rapid gas—solids mixing at the downer inlet. In other words, the solids must be rapidly dispersed in the gas flow to have intimate contact with gas. Otherwise, the nonideal distribution of highly concentrated gas—solids flows at the initial stage would cause fast reaction rates at local positions, leading to worse product distribution.

In principle, the inlet distributor of the downer should provide uniform distribution of phases, quick acceleration of solids, and excellent control of

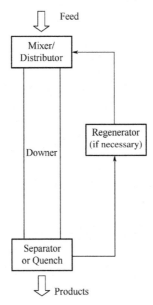

FIGURE 8.15 Conceptual design of a downer reactor.

gas—solids mixing to take the advantages of downer reactors. However, the major challenge to promote the initial gas—solids mixing comes from the gravity acting on the particles. Entering solids tend to drop immediately downward and cannot be refluxed or circulated to enhance the initial heat and mass transfer with gas. Gas usually flows along a less resistant path, i.e., following the region of low solids density. Also, gases can be easily distributed in comparison with solids. Therefore, most efforts have been focused on the design of solids distributors. Fig. 8.16 summarizes several representative inlet geometries reported in the literature.

Because gas and solids move downwards cocurrently in a downer reactor, the backmixing of phases is effectively reduced. This in turn creates a new flow regime in operating the gas—solids fluidized systems. Downer has become a unique reactor that has a plug-flow operation for fast reaction processes among multiphase reactors.

Early works on downers have focused on the significantly different hydrodynamics in the fully developed regions in the riser and downer. Because the flow direction is against the force of gravity in riser but along the force of gravity in downer, the radial profiles of hydrodynamic parameters, such as the solids fraction and velocities, are very different between riser and downer. As shown in Fig. 8.17, the downer has a rather uniform radial flow structure compared with the riser. While the unique feature of dense-ring distribution of solids in the downer is still of academic interest and the physical mechanism is still somewhat an unsolved problem, in the

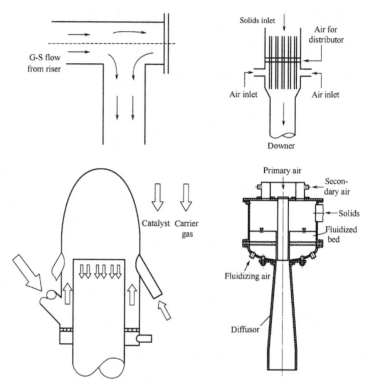

FIGURE 8.16 Inlet configurations of downer reactors in the literature.

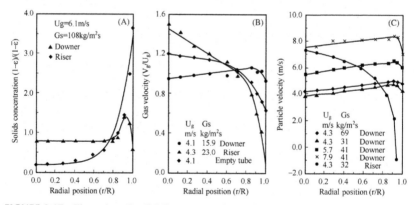

FIGURE 8.17 Illustration of radial flow structure in the fully developed region of the downer:
(A) Solids concentration; (B) Gas velocity; (C) Particle velocity.

downer the radial profile of the solids velocity is very consistent with that
of the solids fraction. In the riser, the clusters forming near the wall result
in slower motion of collected particles, and even cause backflow along the
wall. However, the agglomerated particles in the downer assemble together

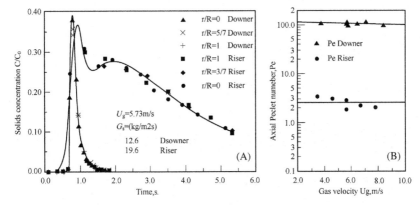

FIGURE 8.18 Typical solids RTDs and Peclet numbers in the riser and downer: (A) Solids RTDs; (B) Axial Peclet number.

very loosely and move downwards faster than the dispersed particles and the gas in the fully developed region.

The downer and riser have very different gas–solid mixing behaviors. The downer has a narrow RTD, and demonstrates a plug flow behavior. In contrast, the riser has a bimodal RTD with a long tail, showing a strong backmixing of solids. Fig. 8.18 shows the typical solids RTDs measured in the riser and downer and the calculated Peclet numbers in these two reactors, which provided solid experimental evidence on the major advantage of downers over risers, i.e., the approaching plug flow behavior due to the effect of flow direction. It can be further concluded that flow direction has the largest influence on the axial solids mixing in the vertical gas–solids suspension systems among many influencing factors such as operating conditions, bed diameters, particle properties, and bed geometry.

FURTHER READING

Kunii D, Levenspiel O. Fluidization engineering. 2nd ed. Boston: Butterworth-Heinemann; 1991.

Kwauk M, Li YC. Idealized and bubbleless fluidization. Beijing: Science Press; 2009.

Levenspiel O. Chemical reaction engineering. New York: Wiley; 1999.

Fan LS, Zhu C. Principles of gas–solid flows. Cambridge, New York: Cambridge University Press; 1998.

Cheng Y, Wu CN, Zhu JX, Wei F, Jin Y. Downer reactor: From fundamental study to industrial application. Powder Technol 2008;183:364–84.

Fogler HS. Fluidized-bed reactors. Professional Reference Shelf of Elements of Chemical Reaction Engineering, unpublished.

Jin Y, Zhu JX, Wang ZW, Yu ZQ. Fluidization engineering principles. Beijing: Tsinghua University Press; 2001.

Yang WC. Handbook of fluidization and fluid-particle systems. Marcel Dekker, Inc.; 2003.

PROBLEMS

8.1 Determine the minimum fluidization velocity for particles with the following size distribution

d_p, μm	400	315	250	160	100	50
Mass fraction, %	5.80	27.05	27.95	30.07	6.49	3.84

Other data: sphericity $\phi_s = 0.75$, $\varepsilon_{mf} = 0.55$, $\rho_p = 1.30 \times 10^3$ kg/m^3

a. At 120°C and 1 atm, $\rho = 1.453$ kg/m^3, $\mu = 1.368$ Pa·s

b. At 10 atm and 420°C, $\mu = 3.2$ Pa·s

8.2 In Example 8.1 the conversion is low and unsatisfactory. One suggestion for improvement is to insert baffles into the bed and thereby cut down the effective bubble size. Find X, if $d_b = 2.5$ cm.

8.3 In Example 8.1 another suggestion for improving performance is to use a shallower bed, keeping W unchanged, thereby decreasing the superficial gas velocity. Find X, if we double the bed cross-sectional area?

8.4 In Example 8.1 still another suggestion is to use a narrower and taller bed, keeping W unchanged. Following this suggestion find X, if we halve the bed cross-sectional area?

8.5 Mathis and Watson in AIChE J., 2, 518 (1956) reported on the catalytic conversion of cumene to phenol and acetone in both fluidized and packed beds of catalyst

$$C_9H_{12} + O_2 \rightarrow C_6H_5OH + (CH_3)_2O + C$$

In very dilute cumene−air mixtures the kinetics are essentially first order reversible with respect to cumene with an equilibrium conversion of 94%. In packed bed experiments ($H_m = 7.62$ cm) using downflow of gas ($u_0 = 6.4$ cm/s) the conversion of cumene was found to be 60%. However, with upflow of gas at the same flow rate in the same bed, the solids fluidize ($u_{mf} = 0.61$ cm/s) and bubbles of gas were observed of roughly 1.35 cm in diameter. What conversion do you expect to find under these conditions?

Estimated values: $D_{Cumene\text{-}Air} = 2 \times 10^{-5}$ m^2/s

8.6 In a laboratory packed bed reactor ($H_s = 10$ cm and $u_0 = 2$ cm/s), conversion is 97% for the first order reaction A + R.

a. Determine the rate constant k''' for this reaction.

b. What would be the conversion in a larger fluidized bed pilot plant ($H_s = 100$ cm and $u_0 = 20$ cm/s) in which the estimated bubble size is 8 cm?

c. What would be the conversion in a downflow packed bed ($H_s = 100$ cm and $u_0 = 20$ cm/s)?

Data: $u_{mf} = 3.2$ cm/s, $\varepsilon_0 \cong \varepsilon_{mf} = 0.5$, $D_{AB} = 3 \times 10^{-5}$ cm^2/s, $f_w = 0.33$.

8.7 Determine the gas flow rate in the bubble phase and emulsion phase for a fluidized bed combustor under the following two operating conditions:

   a. $T = 1173K$, $P = 1.1$ MPa, $D = 1.2$ m, particle loading $M_p = 1600$ kg, $d_p = 0.51$ mm, $\phi_s = 0.84$, $\rho_p = 2422$ kg/m^3, $u_0 = 1.0$ m/s.

   b. $T = 293K$, $P = 0.1$ MPa, and other conditions are the same as (a).

8.8 Determine rising velocity and the thickness of the cloud of a bubble 8 cm in diameter in a bubbling fluidized bed.

   Data: $u_0 = 8.0$ cm/s, $u_{mf} = 2.5$ cm/s, $\varepsilon_{mf} = 0.5$.

Chapter 9

Multiple-Phase Reactors

Chapter Outline

Commercial production of many chemicals often involves reaction systems with multiple phases, such as gas–liquid, gas–solid, liquid–solid, liquid–liquid, and gas–liquid–solid. Since the system involves two or more phases, the reactions will be strongly influenced by transport phenomena, as discussed in Chapter 6, Chemical Reaction and Transport Phenomena in Heterogeneous System, and Chapter 7, Analysis and Design of Heterogeneous Catalytic Reactors. Building on the previous discussions, we will introduce the gas–liquid and gas–liquid–solid reactions, and briefly discuss the issues related to the design of multiphase reactors.

9.1 GAS–LIQUID REACTIONS

Gas–liquid reactions, encountered frequently in chemical and refining processes, can be categorized into two groups. One group produces a desired product using the gas–liquid reaction, and examples include producing nitric acid from water and NO_2 and producing acetaldehyde by oxidizing ethylene

Reaction Engineering. DOI: http://dx.doi.org/10.1016/B978-0-12-410416-7.00009-4
405

in an aqueous solution of palladium chloride. The former is noncatalytic whereas the latter is a catalytic complexation reaction in the liquid phase. Another group is gas purification through gas−liquid reaction, e.g., removing CO by passing the gas mixture through a copper ammonia solution; H_2S removal from natural gas using caustic or other basic solutions. To distinguish from physical absorption, this type of gas absorption is referred to as chemical absorption since it involves chemical reactions. The fundamental principles that govern both types of gas−liquid reactions are essentially the same—the differences lie in the details of reactor design, operation, and control to accommodate different objectives.

Gas−liquid reactions occur at the interface of two phases, and the interface depends on reactor structure as well as the fluid dynamics and is constantly refreshed. This makes the gas−liquid reactions different from the gas−solid reactions discussed in the previous chapter. On the other hand, both mass transfer and chemical reactions need to be dealt with for gas−liquid reactions, similar to that for gas−solid reaction.

There are several models to describe the gas−liquid reactions, including the so-called "two-film theory" and "surface renewal theory." When applied to practical problems the two models give similar results. Herein, we will focus on the conventional two-film theory which is still widely used in spite of its limitations. Consider a gas−liquid reaction:

$$A_{(g)} + \nu_B B_{(l)} \rightarrow \nu_R R_{(l)}$$

Based on two-film theory, the concentration profile of reactant A in the two phases can be depicted as in Fig. 9.1. The concentration profile of B in the liquid phase is also illustrated in Fig. 9.1, and it is assumed that B is nonvolatile. The reaction proceeds through the following steps:

1. Reactant A diffuses from the gas phase through the gaseous film to the interface while its partial pressure is reduced from p_{AG} of the gas phase to p_{Ai} at the interface.
2. Reactant A transfers from the interface into the liquid film and reacts with reactant B which has diffused into the liquid film from the bulk liquid phase. Reaction and diffusion occur simultaneously in this step.
3. The remaining reactant A continues to diffuse into bulk liquid and reacts with B in bulk liquid.
4. Product R diffuses in the direction of the decreasing concentration gradient (not shown in Fig. 9.1).

Similarly to previous gas−solid catalytic reactions, we can establish the diffusion−reaction equation for a gas−liquid reaction, i.e., the differential equation of the concentration distribution in the liquid film. At a

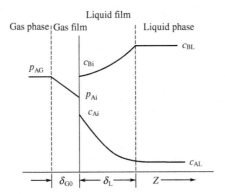

FIGURE 9.1 Illustration of Two-Film Model

distance Z from the interface inside the film, balancing the mass of reactant A within an infinitesimal volume of dZ with a cross-section of unit area, we get,

$$\mathcal{D}_{AL}\frac{d^2 c_A}{dZ^2} = r_A \tag{9.1}$$

at steady-state. Eq. (9.1) takes the same form as the diffusion−reaction equation on a thin-slab catalyst. Similarly, for reactant B,

$$\mathcal{D}_{BL}\frac{d^2 c_B}{dZ^2} = r_B = |v_B|\, r_A \tag{9.2}$$

The boundary conditions are,

$$Z = 0, \quad c_A = c_{Ai}, \quad dc_B/dZ = 0 \tag{9.3}$$

$$Z = \delta_L, \quad c_B = c_{BL}, \quad \mathcal{D}_{AL}a\left(\frac{dc_A}{dZ}\right)_{\delta_L} = r_A(1 - a\delta_L) \tag{9.4}$$

where δ_L is the film thickness, a is the surface area of unit volume liquid, and therefore, $a\delta_L$ corresponds to the volume of the liquid film and $1 - a\delta_L$ is the volume of bulk liquid. Consequently, the last equation in Eq. (9.4) represents that the amount of A reaching the bulk liquid is the same as the amount of A reacted in the bulk liquid.

Solving Eqs. (9.1)−(9.4) gives the concentration profile of A and B within the film, and thereby, the concentration gradient of A at the interface. At steady-state, the rate of A diffusing from the interface to the liquid equals the rate of reaction of A in the liquid phase. Once the gradient of A at the interface is determined, the rate of the reaction can then be calculated by,

$$(-\mathscr{R}_A) = -\mathscr{D}_{AL}\left(\frac{dc_A}{dZ}\right)_{Z=0} \tag{9.5}$$

Note that the consumption rate is calculated here on the basis of interfacial area, i.e., $kmol/(m^2 \cdot s)$, whereas the rate was calculated based on per unit volume or unit mass in the previous chapters.

Solving Eqs. (9.1)–(9.4) is difficult, if at all possible, unless they can be simplified under specific conditions. In what follows, we will solve the equations for a pseudo first order reaction and discuss the implications of the results.

9.1.1 Pseudo First Order Reaction

The reaction between dissolved gaseous component A and liquid component B in the liquid phase can be treated as a second order reaction,

$$r_A = k_2 c_A c_B \tag{9.6}$$

When B is in excess, its concentration inside the film can be considered constant, thus, $k_2 c_B$ is constant. By setting $k = k_2 c_B$, the reaction rate with respect to A can be treated as first order, reducing Eq. (9.6) to,

$$r_A = k c_A \tag{9.7}$$

Substituting Eq. (9.7) to Eq. (9.1) results in,

$$\mathscr{D}_{AL}\frac{d^2 c_A}{dZ^2} = k c_A \tag{9.8}$$

As the concentration of B is constant, we do not have to solve Eq. (9.2). Introducing dimensionless variables $\psi = c_A/c_{Ai}$ and $\varsigma = Z/\delta_L$ and substituting them into Eq. (9.8), we get,

$$\frac{d^2\psi}{d\varsigma^2} = \frac{k\delta_L^2}{\mathscr{D}_{AL}}\psi \tag{9.9}$$

Since $\mathscr{D}_{AL}/\delta_L = k_L$, and letting $k\mathscr{D}_{AL}/k_L^2 = M$, Eq. (9.9) becomes,

$$\frac{d^2\psi}{d\varsigma^2} = M\psi \tag{9.10}$$

The dimensionless forms of Eqs. (9.4) and (9.5) are,

$$\varsigma = 0, \quad \psi = 1 \tag{9.11}$$

$$\varsigma = 1, \quad \left(\frac{d\psi}{d\varsigma}\right)_{\varsigma=1} = M\psi(\alpha - 1) \tag{9.12}$$

where $\alpha = 1/a\delta_L$, the volume ratio of bulk liquid and liquid film.

Applying boundary conditions Eqs. (9.11) and (9.12) to solve Eq. (9.10), we get,

$$\psi = \frac{e^{(1-\varsigma)\sqrt{M}} + e^{-(1-\varsigma)\sqrt{M}} + (\alpha - 1)\sqrt{M}\left[e^{\sqrt{M}(1-\varsigma)} - e^{-\sqrt{M}(1-\varsigma)}\right]}{e^{\sqrt{M}} + e^{-\sqrt{M}} + \sqrt{M}(\alpha - 1)\left(e^{\sqrt{M}} - e^{-\sqrt{M}}\right)}$$

Based on the definition of hyperbolic functions, the above equation can be rewritten as,

$$\psi = \frac{\cosh\left[\sqrt{M}(1 - \varsigma)\right] + \sqrt{M}(\alpha - 1)\sinh\left[\sqrt{M}(1 - \varsigma)\right]}{\cosh\sqrt{M} + \sqrt{M}(\alpha - 1)\sinh\sqrt{M}} \tag{9.13}$$

Differentiating Eq. (9.13) with respect to ς at $\varsigma = 0$ gives,

$$\left(\frac{d\psi}{d\varsigma}\right)_{\varsigma=0} = -\frac{\sqrt{M}\left[\sqrt{M}(\alpha - 1) + \tanh\sqrt{M}\right]}{\sqrt{M}(\alpha - 1)\tanh\sqrt{M} + 1}$$

i.e., the dimensionless concentration gradient at the interface. Recovering its units and substituting them in Eq. (9.5) gives,

$$(-\mathscr{R}_A)\frac{\mathscr{D}_{AL}c_{Ai}\sqrt{M}\left[\sqrt{M}(\alpha - 1) + \tanh\sqrt{M}\right]}{\delta_L\left[\sqrt{M}(\alpha - 1)\tanh\sqrt{M} + 1\right]} \tag{9.14}$$

Setting

$$\beta = \sqrt{M}\frac{\sqrt{M}(\alpha - 1) + \tanh\sqrt{M}}{\sqrt{M}(\alpha - 1)\tanh\sqrt{M} + 1} \tag{9.15}$$

transforms Eq. (9.14) to,

$$(-\mathscr{R}_A) = \beta k_L c_{Ai} \tag{9.16}$$

β represents the enhancement factor of chemical absorption. Since $k_L c_{Ai}$ corresponds to the rate of absorption for physical absorption, the existence of chemical reaction enhances the rate of mass transfer by a factor of β. The magnitude of β depends on α and M. The dimensionless $\sqrt{M}$ is referred to as the Hatta number, an important parameter characterizing chemical absorption and corresponding to the ratio of chemical reaction rate inside the liquid film to physical absorption rate.

When the gas—liquid reaction is used in purifying gas, it is appropriate to use the enhancement factor β to quantify the increase in mass transfer rate due to chemical reactions, as the focus in such a case is the rate of mass transfer. On the other hand, when the reaction is used to produce a desired product, the reaction rate in the liquid phase and conversion of the reactant as well as the impact of interphase mass transfer on the reaction rate will be the major concerns. Similar to treating the gas—solid catalytic reaction, the effectiveness factor was introduced to account for the impact of the mass transfer. The effectiveness factor of the gas−liquid reactions is defined as,

$$\eta = \frac{\text{Rate under mass transfer influence}}{\text{Rate without mass transfer influence}} = \frac{a(-\mathscr{R}_A)}{r_A} \qquad (9.17)$$

Substituting Eq. (9.14) and $r_A = kc_{Ai}$ into Eq. (9.17) gives,

$$\eta = \frac{\sqrt{M}(\alpha - 1) + \tanh\sqrt{M}}{\alpha\sqrt{M}\left[\sqrt{M}(\alpha - 1)\tanh\sqrt{M} + 1\right]} \qquad (9.18)$$

the magnitude of η is a measure of liquid phase utilization. When $\eta = 1$, the entire liquid phase participates in the reaction, and when $\eta < 1$, only a portion of liquid phase participates, meaning that the liquid phase is not fully utilized. Therefore, it is also called liquid phase utilization factor.

On the basis of Eqs. (9.15) and (9.18), β and η are both functions of α and $\sqrt{M}$. In what follows, we will discuss a few specific cases,

1. $M \gg 1$ for a very large value of reaction rate constant k. The property of the hypertangent leads to $\tanh\sqrt{M} \to 1$ for $\sqrt{M} > 3$. Consequently, Eq. (9.15) can be reduced to

$$\beta = \sqrt{M}$$

and substituting into Eq. (9.17), we get

$$(-\mathscr{R}_A) = \sqrt{M}k_L c_{Ai} = \sqrt{k\mathscr{D}_{AL}}c_{Ai} \qquad (9.19)$$

whereas Eq. (9.18) can be reformulated to

$$\eta = \frac{1}{(\alpha\sqrt{M})}$$

Since the volume of the bulk liquid phase is much larger than the volume of the liquid film, $\alpha \gg 1$. If $\sqrt{M} > 1$, the value of η is very small. In such a case, the reaction proceeds predominantly in the liquid film and the concentration of A in bulk liquid approaches 0.

2. $\sqrt{M} \ll 1$ for a very small k value. In such a case, $\tanh\sqrt{M} \to \sqrt{M}$ and Eqs. (9.15) and (9.18) can be reduced to

$$\beta = \frac{\alpha M}{\alpha M - M + 1} \qquad (9.20)$$

$$\eta = \frac{1}{\alpha M - M + 1} \qquad (9.21)$$

Both equations contain the αM term. Based on the definition of α and M,

$$\alpha M = \frac{kD_{AL}}{\alpha\delta_L k_L^2} = \frac{kc_A}{k_L \alpha c_A} = \frac{\text{Reaction rate in liquid phase}}{\text{Reaction rate in liquid film}}$$

therefore, αM represents the relative magnitude of the reaction rate in the liquid phase and the mass transfer rate through the liquid film. Although the result was derived on the basis of $M \ll 1$, αM also depends on the magnitude of α. Consequently, αM can be much larger or smaller than 1, depending on α.

If there is not much liquid in a reactor, as in the case of a bubble reactor, $\alpha \gg 1$, even if the small reaction rate constant makes M small, αM can still be larger than 1. From Eqs. (9.20) and (9.21), we get $\beta \rightarrow 1$, $\eta \rightarrow 1/(\alpha M)$. For example, $M = 0.05$, $\alpha = 1000$, and $\alpha M = 50$, thus, $\beta = 0.98$ and $\eta = 0.0196$. $\beta \rightarrow 1$ indicates that the overall reaction rate depends on the rate of physical absorption. $\eta \rightarrow 1/(\alpha M)$ shows that chemical reaction in the liquid phase still contributes to the overall rate although the utilization of the liquid phase is not high.

If both α and M are small, resulting in $\alpha M \ll 1$. Based on Eqs. (9.20) and (9.21), $\beta \rightarrow \alpha M$, and $\eta \rightarrow 1$. For example, setting $M = 0.01$ and $\alpha = 10$ gives $\alpha M = 0.1$ and $\beta = 0.0917$ whereas $\eta = 1$. In such a case, the reaction is slow and takes place entirely in the liquid phase. The overall rate is determined by the homogeneous reaction rate in the liquid phase.

In summary, gas−liquid reactions have different characteristics, depending on the relative magnitude of the rates of chemical reaction and mass transfer. For reactions with a high reaction rate, the reaction is essentially completed within the liquid film. The overall reaction rate as defined in Eq. (9.19) depends only on k, $\mathscr{D}_{AL}$, and c_{Ai}. In such a case, increasing the reaction temperature (to increase k and $\mathscr{D}_{AL}$) and reducing the resistance in liquid film will benefit the overall rate but increasing the liquid turbulence and reducing the thickness of liquid film will not have much effect on the overall rate. For a process with a slow reaction rate, the reaction takes place primarily in the bulk liquid and, therefore, reactors with large liquid retention are favorable. In the case where the reaction rate in bulk liquid is comparable with the mass transfer (i.e., $\alpha M > 1$), any means that enhance mass transfer will increase the overall reaction rate. On the contrary, where the reaction rate in the bulk liquid is much slower than the rate of mass transfer ($\alpha M < 1$), improving the conditions that are beneficial to the reaction in bulk liquid will increase the overall rate of the process.

Example 9.1
An aqueous NaOH solution was used to absorb CO_2 in a packed bed tower. The rate of the reaction can be written as

$$r_A = k_2 c_A c_B$$

where c_A and c_B are the concentrations of CO_2 and NaOH, respectively. At the interface of a specific location inside the tower, the partial pressure of CO_2 is 2.03×10^{-3} MPa, the concentration of NaOH is 0.5 mol/L and temperature is

(Continued)

(Continued)

20°C. Assuming pseudo first order kinetics, determine the absorption rate at this point. Given $k_L = 1 \times 10^{-4}$ m/s, $k_2 = 10$ m^3/(s·mol), the solubility constant of CO_2 $H_A = 3.85 \times 10^{-7}$ kmol/(m^3·Pa), the diffusion constant of CO_2 in the solution $\mathscr{D}_{AL} = 1.8 \times 10^{-9}$ m^2/s.

Solution

Treating the reaction as pseudo first order, the apparent rate constant

$$k = k_2 c_B = 1 \times 10^4 \times 0.5 = 5 \times 10^3 \text{ s}^{-1}$$

The Hatta number of the absorption is,

$$\sqrt{M} = \frac{\sqrt{k \mathscr{D}_{AL}}}{k_L}$$

$$= \frac{\sqrt{5 \times 10^3 \times 1.8 \times 10^{-9}}}{(1 \times 10^{-4})} = 30 \gg 1$$

i.e., the reaction is fast, and the enhancement factor

$$\beta = \sqrt{M} = 30$$

From Eq. (9.19), we get the rate of absorption,

$$(-\mathscr{R}_A) = \sqrt{M} k_L c_{Ai} = \sqrt{M} k_L H_A p_{At}$$
$$= 30 \times 1 \times 10^{-4} \times 3.85 \times 10^{-7} \times 2.03 \times 10^3 = 2.34 \times 10^{-6} \text{ kmol/(s·m}^2)$$

9.2 GAS–LIQUID REACTORS

9.2.1 Main Types of Reactors

There are two categories of reactors for gas–liquid reactions: column and mechanically stirred tanker reactors. The column reactors can be further divided into four types: packed bed, plate, bubble, and spray columns, schematically shown in Fig. 1.1. A brief description of each type is given below.

1. *Packed bed column.* This type of reactor is suitable for processing a large quantity of gas with a small amount of liquid since such reactors have low fluid flow resistance. In a packed bed reactor, liquid flows on to the surface of the packing materials from top to bottom whereas gas may flow against or in parallel with the liquid. The liquid content in a packed bed reactor is low. Both gas and liquid can be described as plug flow. Packed bed reactors are not suitable for reactions involving solid products.

2. *Spray reactor.* In such reactors, the liquid reactant is nebulized before reacting with the gas. As such, the gas is the continuous phase whereas the liquid is dispersed. Since the spray tower is empty, this type of reactor is suitable for reactions forming solid products. In general, this type of reactor is not widely used.

3. *Plate column.* Contrary to the spray tower, the gas is dispersed whereas the liquid is the continuous phase in this type of reactor. The plate can be in the form of sieves, bubble caps, or any other types. This type of reactor can handle a large quantity of liquid, and is suitable for gas−liquid processes with kinetic control or where the resistance of both mass transfer through liquid film and liquid phase reaction are not negligible. Plate column reactors can also be used in processes where solid precipitates. For example, a plate column has been used in the production of sodium carbonate from ammonium chloride and carbon dioxide in which crystalline sodium carbonate is formed. In the plate column reactor, the gas and liquid flow in opposite directions. The pressure drop across the column is much less than that of the packed bed reactor because the gas only passed through a thin layer of liquid on each plate.

4. *Bubble column reactor.* A bubble column reactor is filled with liquid. Gaseous reactants, introduced to the reactor at the bottom, pass through a gas distributor and rise evenly as bubbles. Bubble column reactors have the advantage of structural simplicity and the ability to handle large quantities of liquid, and are therefore suitable for gas−liquid processes with kinetic control. However, this type of reactor has the smallest interface areas per unit volume and is not appropriate for reactions with mass transfer control. In addition, the pressure drop of gas through the liquid is significant.

5. *Stirred tanker reactor (shortened as Stirred Tanker).* This type of reactor typically has a height to diameter ratio of $1 \sim 3$, and is much smaller than any type of the column reactors. This type of reactor has a mechanical stirrer. Gas is introduced at the bottom through a gas distributor, similar to the bubble column. The presence of the stirrer improves gas distribution by breaking the bubbles. Consequently, both interface area and the liquid content of the stirred tanker reactor are much larger than that of other types of reactors. The stirred tanker is a versatile gas−liquid reactor. On the other hand, the structure of this reactor is complex. In particular, the sealing of the rotating axis under high pressure is challenging. The power consumption is also higher than other types of reactors.

The selection of a gas−liquid reactor requires consideration of the requirements of interface area and liquid content based on the objective and characteristics of the reaction as well as the rate-limiting step of the process. In cases where a reaction takes place predominantly in the liquid film, i.e., mass transfer limits the process, a reactor with a large interface area should be selected as liquid content in the reactor is not critical. In the case where both reaction and mass transfer are important, both interface area and liquid content should be high. For a process limited by the chemical reaction, maximizing liquid content is preferable whereas the interface area becomes less important.

9.2.2 Design of Bubble Column Reactor

In an upright bubble column reactor, gas A reacts with liquid B, and gas and liquid flow counter-currently in the reactor. Given the flow rates and compositions of the gas and liquid, the conversion of the reactants will be determined by the quantity of the gas–liquid mixture in the reactor. If the diameter of the column is known, the conversion will be determined by the height of the mixture inside the column. As such, the design task is reduced to determining the column height. We will discuss this in detail.

Fig. 9.2 shows the symbols used in our calculations. The gas can be treated as a plug flow and liquid as well-mixed, thus, the concentration in the liquid phase inside the column is uniform and equals the concentration at the exit. In addition, the interface boundary area per volume, α, does not vary with position whereas the total pressure linearly changes with the column height, i.e.,

$$p = p_0(1 + \gamma Z) \tag{9.22}$$

where p_0 is the pressure at the gas–liquid interface at the top. If the ratio between the maximum pressure difference and the pressure at the top of the column is less than 0.3, the pressure variation inside the column can be neglected.

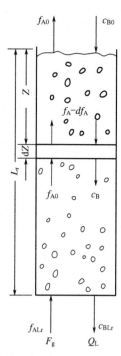

FIGURE 9.2 Schematics of a bubble column

The mass balance of a component in an infinitesimal cell, dZ, at Z from the interface can be written as,

$$F_g df_A = (-\mathscr{R}_A) a A_C dZ \tag{9.23}$$

where F_g represents the molar flow rate of the inert gas, f_A is the molar ratio of component A and the inert gas, or the specific molar number of A. A_C is the area of the column cross-section. Integrating Eq. (9.23), we get,

$$L_r = \int_0^{L_r} dZ = \frac{F_g}{A_C} \int_{f_{A0}}^{f_{AL_r}} \frac{df_A}{a(-\mathscr{R}_A)} \tag{9.24}$$

As $\mathscr{R}_A$ is a function of the concentrations of A and B, the column can only be calculated using Eq. (9.24) when the relationship between f_A and the concentrations is known. The relationship can only be determined for specific situations—there is no universally applicable method.

Assuming the reaction is pseudo first order to A and B, and $\sqrt{M} \gg 1$ the apparent reaction rate can be expressed by Eq. (9.19). Substituting it in Eq. (9.23) results in,

$$F_g df_A = a A_C c_{Ai} \sqrt{k \mathscr{D}_{AL}} dZ \tag{9.25}$$

where $k = k_2 c_B$, and is a function of c_B. Since we assumed the liquid phase is well-mixed and the concentration of B in the column is constant, and under isothermal condition, k_2 is constant, and therefore, k is also constant. Otherwise, c_B as a function of column heights is needed even for the case of pseudo first order reaction.

If the resistance of the gas film can be neglected, the partial pressure at the interface, p_{Ai}, should equal the gas phase partial pressure, p_{AG}. Therefore,

$$c_{Ai} = H_A p_{Ai} = H_A p_{AG} = \frac{H_A P f_A}{1 + f_A} \tag{9.26}$$

Substituting Eqs. (9.22) and (9.26) into Eq. (9.25) and rearranging results in,

$$\frac{1 + f_A}{f_A} df_A = \frac{a A_C p_0 H_A \sqrt{k \mathscr{D}_{AL}}}{F_g} (1 + \gamma Z) dZ \tag{9.27}$$

At $Z = 0, f_A = f_{A0}$, and at $Z = L_r, f_A = f_{AL}$, integrating Eq. (9.27), we get,

$$f_{AL_r} - f_{A0} + \ln \frac{f_{AL_r}}{f_{A0}} = \frac{a A_C p_0 H_A \sqrt{k \mathscr{D}_{AL}}}{F_g} \left(L_r + \frac{\gamma L_r^2}{2} \right) \tag{9.28}$$

Solving this second order equation will provide the height L_r of the gas–liquid mixture.

If the resistance of the gas film is not negligible, the relationship between p_{Ai} and p_{AG} needs to be established. As the rate of mass transfer of A through the gas film equals the rate of its conversion in the liquid phase,

$$k_G(p_{AG} - p_{Ai}) = \sqrt{M} k_L c_{Ai} = \sqrt{M} k_L H_A p_{Ai}$$

and

$$P_{Ai} = \frac{k_G p_{AG}}{k_G + \sqrt{M} k_L H_A} \tag{9.29}$$

or

$$c_{Ai} = H_A p_{Ai} = \frac{k_G H_A p_{AG}}{k_G + \sqrt{M} k_L H_A} \tag{9.30}$$

In a bubble column, the ratio between the height of the gas−liquid mixture and column diameter, L_t/d_t, is generally in the range of 3−12. The diameter depends on the gas velocity, u_G. In fact, the height of the gas−liquid mixture is calculated after u_G has been determined. Since u_G correlates with the rate-determining step, it also determines the gas contents, bubble size, as well as the gas−liquid interfacial area, etc. The apparent reaction rate equation and other parameters can only be determined once u_G is known. As such, the L_t/d_t value calculated under a predetermined u_G should be checked. If the ratio does not meet the requirement of 3−12, it needs to be recalculated after u_G is adjusted.

In addition to the volume occupied by the gas−liquid mixture, gas−liquid separation and heat transfer components also take up space in a bubble column. The overall height of the column should include all three components. The volume used for gas−liquid separation can usually be calculated as one third of the volume of gas−liquid mixture. In the case of high gas velocity, a gas−liquid separation unit would have to be installed, which would in turn lead to a much larger volume for gas−liquid separation.

Example 9.2

A caustic solution of 3 wt.% NaOH is used to absorb CO_2 in air under the iso-thermal condition at 40°C in a bubble column. The molar fraction of CO_2 in air is 0.04% and the air velocity is 23.51 kmol/h. The required purification rate is 94%. The consuming rate of the caustic solution is 21.59 m^3/h and the density of the caustic solution is 1.03 g/cm^3. The pressure at the top of the column 1.013×10^3 Pa determines the diameter and the height of the gas−liquid mixture in the column.

Data: the reaction is first order with respect to both CO_2 (A) and NaOH (B) with a rate constant of 7.75×10^6 cm^3/(s · mol). $\mathscr{D}_{AL} = 1.2 \times 10^{-5}$ cm^2/s, $k_G = 1.036 \times 10^{-8}$ mol/(s · m^2 · Pa), $k_L = 7.75 \times 10^{-3}$ cm/s, $H_A = 2.753 \times 10^{-8}$ mol/(cm^3 · Pa). At $u_G = 18$ cm/s, $a = 18.6$ cm^{-1}; $a \propto u_G^{0.7}$ under other flow rates.

(Continued)

(Continued)

Solution

Assuming gas and liquid flow counter-currently and the concentration of NaOH solution at the entrance is

$$c_{B0} = \frac{\frac{3}{40}}{\frac{100}{1.03}} = 7.725 \times 10^{-4} \text{ mol/cm}^3,$$

and the concentration at the exit is

$$c_{BL} = 7.725 \times 10^{-4} - \frac{23.51 \times 0.0004 \times 0.94 \times 2}{21.59 \times 1000} = 7.717 \times 10^{-4} \text{ mol/cm}^3.$$

Therefore, the difference between the concentrations at the entrance and exit is small and the concentration of NaOH in the column can be treated as constant. In addition, the result also indicates B (NaOH) is in large excess and the reaction can be considered as pseudo first order. The rate constant for the pseudo first order reaction is,

$$k = k_2 c_B = 7.75 \times 10^6 \times 7.717 \times 10^{-4} = 5.981 \times 10^3 \text{ s}^{-1}$$

$$\sqrt{M} = \frac{\sqrt{k \mathscr{D}_{AL}}}{k_L} = \frac{\sqrt{5.981 \times 10^3 \times 1.2 \times 10^{-5}}}{7.75 \times 10^{-3}} = 34.58 \gg 1$$

Therefore, the overall rate of the reaction can be expressed in Eq. (9.19). Based on the given conditions, the diffusion across the gas film cannot be neglected, we need to establish the relationship between c_{Ai} and p_{AG}, i.e.,

$$c_{Ai} = \frac{\left(1.036 \times \frac{10^{-8}}{10^4}\right)\left(2.735 \times 10^{-8}\right)p_{AG}}{\left(1.036 \times \frac{10^{-8}}{10^4}\right) + 34.58\left(7.75 \times 10^{-3} \times 2.753 \times 10^{-8}\right)} \quad \text{(A)}$$

$$= 3.865 \times 10^{-12} \, p_{AG} \text{ mol/cm}^3$$

The interfacial area a, gas content ε_G, and density ρ_m all depend on the velocity of gas u_G, letting $u_G = 20$ cm/s, we get

$$a = 18.6 \left(\frac{20}{18}\right)^{0.7} = 20 \text{ cm}^{-1}$$

Based on the correlations we can determine $\varepsilon_G = 0.20$, hence

$$\rho_m = \rho_L(1 - \varepsilon_G) = 1.03(1 - 0.2) = 0.824 \text{ g/cm}^3$$

Based on the dependence of pressure on height in the column expressed in Eq. (9.22), and

(Continued)

(Continued)

$$\gamma = \frac{0.824}{1033.6} = 7.972 \times 10^{-4} \text{ cm}^{-1}$$

in the equation, (A) can be rewritten as,

$$c_{Ai} = 3.865 \times 10^{-12} \frac{f_A p_0 (1 + \gamma Z)}{1 + f_A} = 3.865 \times 10^{-12} \frac{f_A (1 + 7.972 \times 10^{-4} Z)}{1 + f_A} p_0 \quad \text{(B)}$$

Let us assume the average pressure to be 0.142 MPa, we can verify this after we determine the height of the column. Based on the gas velocity and processing capacity, we can calculate the cross-section area of the column,

$$A_C = \frac{23.51 \times 22.4 \times 313 \times 10^6}{3600 \times 273 \times 1.4 \times 20} = 5990 \text{ cm}^2$$

The flow rate of inert gas is,

$$F_g = \frac{23.51(1 - 0.0004) \times 10^3}{3600} = 6.53 \text{ mol/s}$$

Substituting the value of the parameters and Eq. (B) into Eq. (9.25), we get,

$$6.53 df_A = 20 \times 5990 \sqrt{5.981 \times 10^3 \times 1.2 \times 10^{-5}} \times 3.865 \times 10^{-12}$$

$$\times \frac{f_A (1 + 7.972 \times 10^{-4} Z) p_0}{1 + f_A} dZ$$

or

$$\frac{1 + f_A}{f_A} df_A = 1.925 \times 10^{-3} (1 + 7.972 \times 10^{-4} Z) dZ \quad \text{(C)}$$

Integrating (C) results in,

$$f_{AL_r} - f_{A0} + \ln \frac{f_{AL_r}}{f_{A0}} = 1.925 \times 10^{-3} (L_r + 3.986 \times 10^{-4} L_r^2) \quad \text{(D)}$$

and $f_{AL_r} = \frac{0.04}{99.6} = 4.016 \times 10^{-4}$, $f_{A0} = 0.04 \left(\frac{1 - 0.94}{99.6} \right) = 2.14 \times 10^{-5}$, substituting those values in (D) we get,

$$L_r = 1035 \text{ cm}$$

The diameter of the column $d_t = (4 \times 5990 / \pi)^{1/2} = 87.33$ cm, therefore, $L_r / d_t = 1035 / 87.33 = 11.85$ and satisfies the requirement of $L_r / d_t = 3 \sim 12$. The pressure in the column bottom $p = 1.103 \times 10^5 (1 + 7.972 \times 10^{-4} \times 1035) = 1.849 \times 10^5$ Pa. Therefore, average pressure $= (1.013 + 1.849) \times 10^5 / 2 = 0.143$ MPa. These results confirmed that the assumption is correct and the selected gas velocity $u_G = 20$ cm/s is adequate.

9.2.3   Design of Stirred Tank Reactor

When mechanically stirred tanks are used as gas−liquid reactors, as long as the ratio of the height of the gas−liquid mixture to the diameter of the tank is not too big, both gas and liquid phases can be assumed well-mixed. Applying mass balance to reactive components in gas and liquid phase in turn results in the equations for reactor design.

A schematic diagram of a stirred tank reactor is shown in Fig. 9.3. The symbols used in the calculation are labeled in the figure. A stands for the gas phase component and B stands for the liquid component. Mass balance for A in gas phase leads to,

$$F_g(f_{A0} - f_{AL_r}) = k_L a V_r \beta (c_{Ai} - c_{AL_r}) \tag{9.31}$$

Mass balance for A in liquid phase results in,

$$k_L a V_r \beta (c_{Ai} - c_{AL_r}) = Q_L(c_{AL_r} - c_{A0}) + V_r(1 - \varepsilon_G)r_A \tag{9.32}$$

Similarly, mass balance for B in liquid phase leads to

$$Q_L(c_{B0} - c_{BL_r}) = V_r(1 - \varepsilon_G)|\nu_B|r_A \tag{9.33}$$

Solving Eqs. (9.31)−(9.33) will give the volume to meet the productivity requirement, i.e., the volume of gas−liquid mixture in the reactor. The derivation of Eqs. (9.32) and (9.33) was based on the assumption that the liquid flow rates at the entrance and exit are equal.

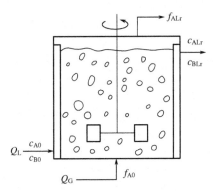

FIGURE 9.3   Schematics of a stirred tank reactor

Example 9.3

Design a stirred tank reactor to produce o-toluic acid from o-xylene oxidation with a production rate of 3650 kg/h. Operating pressure: 1.378 MPa, Temperature: 160°C. The reaction is first order with respect to oxygen and can be treated as pseudo first order. The reaction rate of o-xylene is expressed as,

$$r_B = 2400c_A, \ \text{kmol}/(m^3 \cdot h)$$

where c_A is the concentration of oxygen in liquid phase, kmol/m³. The amount of air is 1.25 times of theoretical quantity. Calculate reaction volume needed to achieve 16% conversion of o-xylene.

Data: the diffusion coefficient of o-xylene $\mathscr{D}_{AL} = 1.44 \times 10^{-5} \ cm^2/s$, the solubility constant of oxygen $H_A = 7.875 \times 10^{-8} \ kmol/(m^3 \cdot Pa)$, gas holdup $\varepsilon_G = 0.2293$, mass transfer coefficient in liquid phase $k_L = 0.07702 \ cm/s$, specific interfacial area $a = 8.574 \ cm^{-1}$, liquid density $\rho_L = 0.75 \ g/cm^3$, neglecting resistance across the gas film.

Solution

O-xylene inlet flow rate $n = 3750/(136 \times 0.16) = 172.3 \ kmol/h$ where 136 is the molecular weight of o-xylene. Converting 1 mol of o-xylene needs 1.5 mol oxygen, thus, the amount of air entering the reactor is

$$\left(\frac{3750}{136}\right) \times 1.5 \times \frac{1.25}{0.21} = 246 \ \text{kmol/h}$$

$$\sqrt{M} = \frac{\sqrt{kD_{AL}}}{k_L} = \frac{\sqrt{\left(\frac{2400}{3600}\right)(1.44 \times 10^{-5})}}{0.07702}$$

$$= 4.023 \times 10^{-2} \ll 1$$

$$\alpha M = \frac{1}{a\delta_L}\frac{k\mathscr{D}_{AL}}{k_L} = \frac{k}{ak_L} = \frac{\frac{2400}{3600}}{8.574 \times 0.07702} = 1.01$$

Substituting into Eq. (9.20) gives the enhancement factor,

$$\beta = \frac{1.01}{1.01 - (4.023 \times 10^{-2})^2 + 1} = 0.5029$$

$$c_{Ai} = \frac{PH_A f_{AL_r}}{1 + f_{AL_r}} = \frac{1.378 \times 10^6 \times 7.875 \times 10^{-8} \times 10^{-3} f_{AL_r}}{1 + f_{AL_r}} = \frac{1.085 \times 10^{-4} f_{AL_r}}{1 + f_{AL_r}}$$

(Continued)

(Continued)

$$F_g = \left(246 \times \frac{10^3}{3600}\right) \times 0.79 = 53.98 \text{ mol/s}$$

$$Q_L = \frac{1.723 \times 10^5 \times 106.6}{3600 \times 0.75} = 6803 \text{ cm}^3/\text{s}$$

$$f_{A0} = \frac{0.21}{0.79} = 0.2658$$

Assuming there is no oxygen in the o-xylene feed, i.e., $c_{A0} = 0$, and applying the data provided and the values determined above in Eqs. (9.31)–(9.33), with simplification, will give (Note: the right-hand side of Eq. (9.33) corresponds to the amount of consumed o-xylene, which can be determined from conversion directly, and there is no need to determine the concentration of o-xylene at the entrance and exit):

$$0.2658 - f_{AL_r} = 6.152 \times 10^{-3} V_r \left[\frac{1.085 \times 10^{-4} f_{AL_r}}{1 + f_{AL_r}} - c_{AL_r}\right]$$

$$(6803 + 0.7707 V_r)c_{AL_r} = 0.3321 V_r \left[\frac{1.085 \times 10^{-4} f_{AL_r}}{1 + f_{AL_r}} - c_{AL_r}\right]$$

$$V_r c_{AL_r} = 14.9$$

Solving the above three equations simultaneously results in:

$$V_r = 9.094 \times 10^6 \text{ cm}^3, \quad f_{AL_r} = 0.0528$$

$$c_{AL_r} = 1.638 \times 10^{-6} \text{ mol/cm}^3$$

Thus, the reaction volume needed is about 9 m³. Based on the oxygen concentration at the exit, we determine absorbance of $O_2 = (0.2658 - 0.0528)/0.2658 \approx 0.8$, i.e., $\sim 80\%$. There is the unreacted oxygen in the liquid discharge at the exit, indicating the reaction is slow and does not reach completion within the liquid film. The utilization of liquid phase is about 60%. The effects of both mass transfer and reaction are pronounced.

9.3 GAS–LIQUID–SOLID REACTIONS

9.3.1 Introduction

Gas–liquid–solid reactions refer to the reaction processes that simultaneously involve all three phases and can be classified into three categories:

(1) three phases involved in the process are either reactants or products. For example, gas and liquid reactants react to produce solid products, such as aqueous ammonia reacting with carbon dioxide to produce crystallized ammonium bicarbonate; (2) a gas−liquid reaction catalyzed by a solid catalyst, as in the case of hydrogenation of benzene to produce cyclohexane catalyzed by Raney nickel; and (3) one of the three phases is inert. Although the inert phase does not participate in the reaction, the process is still considered as a three-phase reaction from an engineering point of view. For example, a gas-stirred liquid−solid reaction, a gas−liquid reaction in a solid-packed bed reactor and a gas−solid reaction with liquid as a heat-carrier all belong to this category.

The presence of a liquid in the three-phase reaction usually makes the reaction temperature moderate. Consequently, the selectivity is relatively high, especially for the production of highly thermal-sensitive chemicals. The three-phase reactor is also suitable for solid catalysts and their supports that only work at low temperatures, such as enzyme and immobilized homogeneous catalysts. Operating at low temperature can avoid catalyst deactivation and provide an opportunity to maximize selectivity by selecting an optimized solvent. In addition, unlike with a catalytic gas−solid reactor, the liquid reactant(s) does not have to be gasified in a three-phase reactor, which would save energy. Furthermore, the large heat capacity of the liquid as well as the latent heat of vaporization make controlling reaction temperature much easier. Consequently, this type of reactor has the advantage of a fast heat transfer rate and high heat transfer efficiency, thereby, good thermal stability and no risk of temperature runaway. However, the three-phase reactor is often plagued by severe corrosion and demands high performance materials. In addition, corrosion is also a concern for the catalysts used.

This chapter will focus on the second and third type of three-phase reactions, i.e., gas−liquid−solid catalytic reactions. Commercial gas−liquid−solid catalytic reactions include hydrogenation, oxidation, hydrorefining, ethynylation, etc. The hydrogenation reactions include the hydrogenation of benzene, fatty acids, butynediol, glucose, aniline, crotonaldehyde, methyl styrene, etc. Oxidation reactions include the oxidation of ethylene, cumene, SO_2, etc. Hydrorefining processes in the petroleum industry, such as hydrodesulfurization and hydrodenitrogenation, are all gas−liquid−solid catalytic reactions.

The two commonly used reactors for gas−liquid−solid catalytic reactions are trickle bed and slurry reactors. A detailed discussion on each type will be provided in turn in the following sections.

9.3.2 Mass Transfer Steps and Rates in Gas−Solid−Liquid Catalytic Reactions

Let gas component A react with liquid B on a solid catalyst through the following reaction,

$$A_{(g)} + \nu_B B_{(1)} \rightarrow \nu_R R_{(1)}$$

and produce nonvolatile liquid product R. The steps involved in the reaction process are depicted in Fig. 9.4. The gas component A diffuses through the gas film to the gas–liquid interface, accompanied by a concentration decrease from C_{AG} to C_{AG}^*. At the interface, the concentration of A on the gas side, C_{AG}^*, is in equilibrium with that on the liquid side, C_{AL}^*. Then A diffuses through the liquid film to the bulk liquid and the concentration decreases from C_{AL}^* to C_{AL}. Since the reaction takes place on the surface of the solid catalyst, A in the bulk liquid needs to be transferred from the bulk liquid to the external surface of the catalyst particles, and then diffuses through the inner pores of the particle to the internal surface and reacts with B to form product R. The readers should think about the transfer steps for B and R, although they were not plotted in Fig. 9.4.

Comparing Figs. 9.4 and 9.1, we can see that the left half of the former is the same as those describing gas–liquid transport steps in the latter. Obviously, gas–liquid–solid catalytic reactions are more complex than gas–liquid or gas–solid reactions.

In what follows, we will use the mass of catalyst as the base to calculate the rates of the mass transfer steps, and let a_L and a_S be the gas–liquid and gas–solid interfacial areas per unit mass of solid catalysts, thus a_S is the specific external surface area of the catalyst particles. The mass transfer rate of A from bulk gas phase to the interface is,

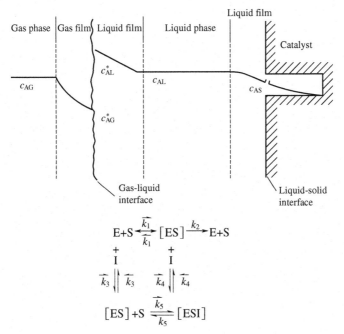

FIGURE 9.4 Steps involved in a catalytic gas-liquid-solid reaction

$$\mathscr{N}_{AG} = k_G a_L (c_{AG} - c_{AG}^*).$$ (9.34)

Its rate from the interface to the bulk liquid is,

$$\mathscr{N}_{AL} = k_L a_L (c_{AL}^* - c_{AL})$$ (9.35)

and the rate from bulk liquid to the external surface of the catalyst particle is,

$$\mathscr{N}_{AS} = k_{LS} a_S (c_{AL} - c_{AS})$$ (9.36)

Assuming the reaction is pseudo first order, the reaction rate including the effect of internal diffusion is,

$$(-\mathscr{R}_A) = \eta k c_{AS}$$ (9.37)

At the gas−liquid interface, component A is in equilibrium. Based on Henry's law,

$$c_{AG}^* = H_A c_{AL}^*$$ (9.38)

When the process reaches the steady-state, the rates of all steps are equal. Eliminating the concentrations of intermediate states results in,

$$(-\mathscr{R}_A) = K_{OG} c_{AG}$$ (9.39)

and

$$\frac{1}{K_{OG}} = H_A \left(\frac{1}{H_A k_G a_L} + \frac{1}{k_L a_L} + \frac{1}{k_{LS} a_S} + \frac{1}{k\eta} \right)$$ (9.40)

where K_{OG} is referred to as the overall mass transfer coefficient or macroscopic reaction rate constant. Eq. (9.39) provides an overall reaction rate and takes into account the resistance of all mass transfer steps and chemical reactions. The inverse of K_{OG} is the resistance of the process. Based on Eq. (9.40), the resistance has five components: the gas and liquid sides of the gas−liquid interfaces, the liquid side of the liquid−solid interface, and diffusion inside the catalyst particle and chemical reaction. The contribution to the overall resistance from each resistance is reflected in the corresponding coefficient. Under certain conditions, some of the resistances may be neglected. If the gas phase contains only one pure component, e.g., pure hydrogen for hydrogenation reaction, the film on the gas phase side does not exist, and therefore, its resistance is zero. In the gas−liquid−solid reaction, the solubility of the gas in liquid is generally small, and the resistance of the gas film is significantly smaller than that of the liquid film and therefore can be neglected. In fact, the resistance of the gas film can be neglected in most cases. Further, if the sizes of the catalyst particles are sufficiently small, where $\eta \simeq 1$, i.e., resistance of internal diffusion is negligible.

Although the above analysis is based on a first order reaction, the methodology and concept are applicable to other reactions, albeit with more complicated and difficult mathematical manipulations.

9.4 TRICKLE BED REACTORS

9.4.1 Introduction

Trickle bed reactors are similar to fixed bed reactors for the catalytic gas–solid reactions discussed in the previous chapter—the major difference is that a single phase fluid flows in the latter whereas the fluid flows in two phases (gas and liquid) in the former. Obviously, the situation in the two-phase flow is more complicated than a single phase. In principle, the two flowing phases can be either cocurrent or countercurrent, although cocurrent flow is more commonly used in commercial applications. Cocurrent flow reactors can be further divided into upward and downward. The selection of flowing direction depends on processing capacity and heat recovery as well as the driving forces of mass transfer and chemical reaction. Countercurrent reactors are subject to the limitation of liquid flooding whereas cocurrent reactors do not have this restriction, and therefore, are suitable for larger flow rates.

The flow patterns of two-phase fluid in the trickle bed reactors, which can be divided into various regimes, are different from that of the single phase fluid in a fixed bed reactor. The flow patterns depend on the sizes of particles and bed, the flow rates and physical properties of gas and liquid. Fig. 9.5 shows different flow regimes as a function of gas and liquid flow rates in a downward cocurrent flow trickle bed reactor. As shown in the figure, there are four flow regimes at different gas and liquid flow rates. At low gas and liquid flow rates, the system is in the trickle zone, characterized by the fact that the gas is a continuous phase whereas the liquid is dispersed and flows along the external surface of the catalyst particles or drips down as droplets. When the gas flow rate is increased to a threshold, the liquid will be completely broken apart and flow in the form of spray, i.e., the so-called "spray regime." Increasing both gas and liquid flow rates will result in the so-called "pulsing regime." The key character of this regime is that the gas velocity is so high that the drag of gas on the liquid will cause turbulence in the liquid. In this regime, two flow zones alternate: gas dominant and liquid dominant, resulting in the gas and liquid flowing downward in sections. When liquid flow rate is high and gas flow rate is low, the reactor is in the so-called "bubbling regime," with liquid being the continuous phase and gas is dispersed as bubbles in the liquid layer.

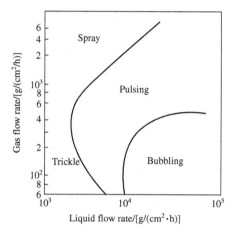

FIGURE 9.5 Flow regimes in a downward cocurrent flow trickle bed reactor

The degree of wetting of the solid catalyst is a key consideration in the design and operation of the trickle bed reactors. Only the gaseous reactants that are dissolved in the liquid phase can participate in the reaction on the catalyst surface. Obviously, the wetting efficiency will critically affect the conversion of the reaction. Wetting involves both the external and internal surfaces of the catalyst particles. Incomplete wetting can be a result of ill-designed liquid distribution, low liquid loading, or high liquid–solid interfacial tension. In order to achieve complete wetting, a minimum liquid loading should be $10 \sim 30$ m^3/(m^2 h). Generally, the capillary effect will lead to the internal pores being liquid filled, i.e., complete wetting, although there are exceptions. For example, liquid in the internal pores may be evaporated under highly exothermic conditions, causing incomplete wetting of the internal surface. In the latter case, a gas–solid reaction may occur on the dry catalyst surface, complicating the process. Although wetting efficiency is extremely important, our understanding and knowledge are very limited.

The amounts of gas and liquid in a trickle bed reactor directly affects the mean residence time of the reactants, which in turn determines reaction conversions. Holdups are usually used to represent the relative amount of each phase. For both gas and liquid, their holdups include contributions from internal pores and voids between particles. Most of the fluid surrounding a particle will flow, while a small fraction will be either partially stationary or completely stationary, equivalent to the stagnant zone (dead area), and contribute little to the overall reaction. The holdups also influence the wetting efficiency of the catalyst and the pressure drop across the reactor.

We now have briefly introduced the concepts of two-phase flow direction, flow patterns and regimes, and holdups in a trickle bed reactor. The reaction and mass transfer have been discussed in the previous section. Next we will discuss the mathematical model of trickle bed reactors.

9.4.2 Mathematical Model

Herein, we focus on the mathematical model of an ideal trickle bed reactor. The assumptions for an ideal trickle bed reactor include: the flows of both phases are in the trickling regime, gas and liquid distribute uniformly, catalyst particles are completely wet, liquid is nonvolatile, and the gas, liquid, and solid phases are at the same temperature. Due to uniform gas and liquid distribution, there is no concentration gradient in the radial direction, and the flow of both the liquid and gas phase can be treated using the plug flow model. In such a multiphase system, mass balance equations have to be derived individually for each phase. For gas A reacting with liquid B, a mass balance equation for A in gas phase can be derived based on the assumptions given above:

$$\frac{d(u_{0G}c_{AG})}{dZ} = -\rho_b k_{LA} a_L (c_{AG}/H - c_{AL}) \tag{9.41}$$

where u_{0G} is the superficial velocity of gas phase. It is also assumed that the resistance of gas film is negligible. Letting u_{0L} be the superficial velocity of the liquid phase, the mass balance equation for A in liquid phase is

$$u_{0L}\frac{dc_{AL}}{dZ} = -\rho_b \left[k_{LS,A} a_S (c_{AL} - c_{AS}) - k_{LA} a_L \left(\frac{c_{AG}}{H} - c_{AL} \right) \right] \tag{9.42}$$

The mass balance for B is

$$u_{0L}\frac{dc_{BL}}{dZ} = -\rho_b k_{LS,B} a_S (c_{BL} - c_{BS}) \tag{9.43}$$

Under steady-state, the mass transferred from bulk liquid to the catalyst equals the amount reacted on the surface of the catalyst, consequently,

$$k_{LS,A} a_S (c_{AL} - c_{AS}) = \eta k c_{AS} c_{BS} \tag{9.44}$$

$$k_{LS,B} a_S (c_{BL} - c_{BS}) = |v_B| \eta k c_{AS} c_{BS} \tag{9.45}$$

Obviously, the reaction is assumed to be first order with respect to both A and B. The effectiveness coefficient can be determined by following the procedures discussed in Chapter 6, Chemical Reaction and Transport Phenomena in Heterogeneous System. However, the rate equation has to be reformulated so that it is a function of the concentration of one specific component.

Therefore we need to find the relationship between concentrations of various reacting components. In Chapter 6, Chemical Reaction and Transport Phenomena in Heterogeneous System, we derived the differential equations describing concentration distribution of reactant A in a slab catalyst:

$$D_{eA} \frac{d^2 c_A}{dZ^2} = r_A$$

The concentration distribution of reactant B is

$$D_{eA} \frac{d^2 c_A}{dZ^2} = |\nu_B| r_A$$

Combining the above equations results in,

$$|\nu_B| D_{eA} \frac{d^2 c_A}{dZ^2} = D_{eB} \frac{d^2 c_B}{dZ^2} \qquad (9.46)$$

Assuming the effective diffusion coefficients of A and B, D_{eA} and D_{eB}, are constants, and integrating Eq. (9.46) leads to,

$$c_B = c_{BS} - \frac{|\nu_B| D_{eA}}{D_{eB}} (c_{AS} - c_A) \qquad (9.47)$$

Hence, the relationship of concentrations of the reacting components inside the catalyst particle is linear and c_B can be expressed as a function of c_A using Eq. (9.47). The rate equation can then be transformed into,

$$r_A = k c_A c_B = k c_A \left[c_{BS} - \frac{|\nu_B| D_{eA}}{D_{eB}} (c_{AS} - c_A) \right]$$

and is a function of c_A, the concentration of reactant A, allowing the determination of η using methods previously discussed.

When $Z = 0$, $c_{AG} = c_{AG}^0$, $c_{AL} = 0$, $c_{BL} = c_{BL}^0$, solving Eqs. (9.41)–(9.45) gives the height of the catalyst bed needed to achieve certain conversion. Obviously, these equations can only be solved numerically.

If superficial velocity is low, the effect of axial diffusion needs to be considered. In such a case, an axial diffusion term needs to be added to Eqs. (9.41)–(9.43), making it an axial diffusion model. When the reaction is irreversible with a reaction order of α, the magnitude of the axial diffusion can be evaluated based on the following equation. When

$$\frac{L_r}{d_P} > \frac{20}{(Pe)_L} a \ln \frac{1}{1 - X_1}, \qquad (9.48)$$

the influence of axial diffusion can be neglected. In Eq. (9.48), $(Pe)_L = u_{0L} d_P / D_{aL}$, and D_{aL} is the axial diffusion coefficient of the liquid. From Eq. (9.48), we know that the assumption of plug flow is not valid when liquid superficial velocity is low.

If

$$\frac{L_r}{d_P} > \frac{4}{(Pe)_L},$$ (9.49)

it can be treated as a continuous stirred tank reactor.

If the reaction rate is only a function of concentration of reactant A, Eqs. (9.43) and (9.45) can be eliminated. c_B in the right-hand side of Eq. (9.46) will disappear, further simplifying the problem. If the reaction is first order or can be treated as pseudo first order, only one equation, i.e., the mass balance of A, is needed.

$$-\frac{d(u_{0G}c_{AG})}{dZ} = \rho_b K_{OG} c_{AG}$$ (9.50)

If u_{0G} can be considered as constant, integrating Eq. (9.50) will provide the relationship between gas phase conversion and catalyst bed heights.

Example 9.4

Hydrogenation of unsaturated hydrocarbons occurs in an isothermal trickle bed reactor at 400 K. Hydrogen with a purity of 50% (mixed with inert gas) enters the reactor at 3.04 MPa with a flow rate of 36 mol/s. The reaction is first order with respect to hydrogen and zeroth order with respect to the hydrocarbons, and the rate constant at 400K is 2.5×10^{-5} m^3/(kg·s). The diameter of the catalyst particles is 0.4 cm and tortuosity is 1.9. Determine the height of the catalyst bed needed to achieve a conversion of 60% in a reactor with a diameter of 2 m.

  Data: Diffusion coefficient of hydrogen in liquid $= 7 \times 10^{-9}$ m^2/s, $k_L a_L = 5.01 \times 10^{-6}$ m^3/(kg·s), $k_{LS} a_S = 3.19 \times 10^{-5}$ m^3/(kg·s), negligible gas film resistance. The solubility index of hydrogen in liquid $= 6.13$. Density of catalyst particles $= 1.6$ g/cm^3, porosity $= 0.45$, and bulk density $= 0.96$ g/cm^3. Pressure drop $= 2.0 \times 10^{-2}$ MPa/m.

Solution

Since the reaction is first order, the bed height can be determined using Eq. (9.50) although u_{0G} cannot be treated as constant due to the fact that the quantity of gas phase reactant decreases significantly and the pressure varies across the bed. The equation is rewritten as,

$$-\frac{dF_A}{dZ} = A_c \rho_b K_{OG} c_{AG}$$ (A)

where A_c is the cross-section area of the bed. The pressure at height Z,

$$p = 3.04 - 2.0 \times 10^{-2} Z$$ (B)

The molar fraction of component A (hydrogen) in gas phase is

$$y_A = \frac{F_A}{[F_0 - (F_{A0} - F_A)]}$$ (C)

(Continued)

(Continued)

where F_0 is the gas molar flow rate at the entrance, F_{A0} is the molar flow rate of hydrogen at the entrance. Since $c_{AG} = p y_A / RT$, substituting (B) and (C) into (A) and simplifying results in

$$c_{AG} = \frac{F_A \left(3.04 - 2.0 \times 10^{-2} Z \right)}{RT \left(F_0 - F_{A0} + F_A \right)} \tag{D}$$

Substituting (D) into (A) and integrating gives

$$-\int_{F_{A0}}^{F_A} \left(1 + \frac{F_0 - F_{A0}}{F_A} \right) dF_A = \frac{A_c \rho_b K_{OG}}{RT} \int_0^{L_r} \left(3.04 - 2.0 \times 10^{-2} Z \right) dZ$$

or

$$\left(F_0 - F_{A0} \right) \ln \frac{F_{A0}}{F_A} + F_{A0} - F_A = \frac{A_c \rho_b K_{OG}}{RT} \left(3.04 L_r - 1.0 \times 10^{-2} L_r^2 \right) \tag{E}$$

Solving (E) gives the height of the catalyst bed. K_{OG} can be calculated by using Eq. (9.40), and η can be determined following the procedure discussed in Chapter 6, Chemical Reaction and Transport Phenomena in Heterogeneous System. Thiele modulus,

$$\phi = \frac{R_p}{3} \sqrt{\frac{k_p}{D_e}} \tag{F}$$

whereas

$$D_e = \frac{\varepsilon_p}{\tau_m} D = \frac{0.45}{1.9} \times 7 \times 10^{-9} = 1.658 \times 10^{-9} \text{ m}^2/\text{s}$$

$$k_p = \rho_p k = 2.5 \times 10^{-5} \times 1600 = 0.04 \text{ s}^{-1}$$

Substituting these values into (F) results in

$$\phi = \frac{0.002}{3} \sqrt{\frac{0.04}{1.658 \times 10^{-9}}} = 3.275.$$

Hence,

$$\eta = \frac{1}{\phi} \left[\frac{1}{\tanh(3\phi)} - \frac{1}{3\phi} \right] = \frac{1}{3.275} \left[\frac{1}{\tanh(3 \times 3.275)} - \frac{1}{3 \times 3.275} \right] = 0.2743$$

Using all the values in Eq. (9.40) leads to

$$\frac{1}{K_{OG}} = 6.13 \left(\frac{1}{5.01 \times 10^{-6}} + \frac{1}{3.19 \times 10^{-5}} + \frac{1}{0.2743 \times 2.5 \times 10^{-5}} \right)$$

and

$$K_{OG} = 4.33 \times 10^{-7} \text{ m}^3/(\text{kg} \cdot \text{s})$$

(Continued)

(Continued)

Given $F_0 = 0.036\,\text{kmol/s}$, thus $F_{A0} = 0.5 \times 0.036 = 0.018\,\text{kmol/s}$. The targeted H_2 conversion is 60%, therefore, $F_A = 0.018(1 - 0.6) = 0.0072\,\text{kmol/s}$. $A_c = (\pi/4)2^2 = \pi\text{m}^2$. Substituting all these values into (E) gives

$$(0.036 - 0.018)\ln\frac{0.018}{0.0072} + 0.018 - 0.0072$$

$$= \frac{\pi \times 960 \times 4.33 \times 10^{-7}}{400 \times 8.314 \times 10^{-3}}(3.04L_r - 1.0 \times 10^{-2}\,L_r^2).$$

After simplifying, we get

$$1.0 \times 10^{-2}L_r^2 - 3.04L_r + 69.505 = 0$$

Solving the above equation gives $L_r = 24.9$ m. As this height is too high, three reactors with a height of 8.3 m should be used in series to meet the requirements. If the pressure drop across the reactor is neglected and the pressure at the entrance is used, the height of the bed is calculated as 22.86 m. Why is this value lower?

9.5 SLURRY REACTOR

The key difference between a slurry reactor and a trickle bed reactor is that the solid catalyst particles move in the former whereas they are still in the latter. In addition, gas phase is dispersed in the former and continuous in the latter. Slurry reactors find wide applications in processes such as hydrogenation, oxidation, halogenation, polymerization, and fermentation.

9.5.1 Types of Reactors

There are, mainly, four types of slurry reactors, i.e., mechanically stirred tank reactor, loop reactor, bubble column reactor, and three-phase fluidized bed reactor, as sketched in Fig. 9.6. The mechanically stirred tank and bubble column reactors used for gas—liquid—solid reactions are fundamentally the same as those used for gas—liquid reactions, and the only added complexity is the presence of catalyst particles suspended in the liquid. The advantage of loop reactor shown in Fig. 9.6(B) is that the internal circulating tube makes the fluid circulate at a high velocity of 20 m/s or higher, which significantly enhances mass transfer. Fig. 9.6(D) illustrates a three-phase fluidized bed reactor where the liquid is introduced to the reactor

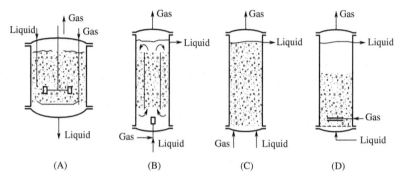

FIGURE 9.6   Slurry Reactors: (1) Stirred tank reactor; (2) Loop reactor; (3) Bubble column reactor; (4) Three-phase fluidized bed reactor

through a distributor at the bottom and fluidizes the catalyst particles. Similarly to a gas−solid fluidized bed, the height of the fluidized region expands as the velocity of the liquid increases. There is a clear liquid region at the top of the bed and the interface between this region and the fluidizing region can be clearly identified. If a gas stream is introduced together with the liquid, the height of the fluidized region will be lower. At low liquid velocity, the solid catalysts cannot be fluidized by simply increasing the gas velocity. In three-phase fluidized reactors, the presence of the gas enhances solid particle movements and blur the boundary at the top of the fluidized bed.

The liquid phase is responsible for the suspension of the catalyst particles in the three slurry reactors shown in Fig. 9.6(B)−(D). Due to differences in reactor structure and operating gas and liquid velocity, the fluid dynamics inside the reactors are different. For a mechanically stirred tank reactor shown in Fig. 9.6(A), it is the mechanical forces that keep the particles suspended.

The slurry reactors have the benefits of high space-time yield, low mass transfer resistance, high heat transfer rate, being suitable for both continuous or semicontinuous operation, and allowing continuous regeneration of the catalysts. Their limitations include severe backmixing, high attrition of catalysts, difficulty and high cost of separating catalysts from the products, and high probability of side reactions in liquid phase due to high liquid/solid ratio.

9.5.2   Mass Transfer and Reaction

The mass transfer and reaction steps discussed previously for the gas−solid−liquid reaction processes are applicable to the slurry reactors. For a first order reaction, Eq. (9.39) can represent the overall reaction rate in a slurry reactor. However, the rate in Eq. (9.39) is calculated on the basis of per unit mass of catalyst. For slurry reactors it is more convenient to use the rate of reaction per unit reaction volume (sum of volumes of gas, liquid, and solid). Thus, letting the mass of the catalyst per unit reaction volume be W_S, Eq. (9.39) can be reformulated to,

$$(-\mathscr{R}_A) = c^*_{AL} \left(\frac{1}{k_L a'_L} + \frac{1}{W_S k_{LS} a_S} + \frac{1}{W_S k \eta} \right)^{-1} \tag{9.51}$$

whereas a'_L is the gas–liquid interfacial area per unit reaction volume. In most cases, the resistance of the gas film can be neglected, and therefore was not taken into account in Eq. (9.51). Eq. (9.51) can be rewritten as,

$$\frac{c^*_{AL}}{(-\mathscr{R}_A)} = \frac{1}{k_L a'_L} + \frac{1}{W_S} \left(\frac{1}{k_{LS} a_S} + \frac{1}{k \eta} \right). \tag{9.52}$$

By maintaining other conditions the same, the overall reaction rate can be measured against the concentration of the catalyst, W_S. The plot of $c^*_{AL}/(-\mathscr{R}_A)$ against $1/W_S$ should be linear. The intercept corresponds to $1/k_L a'_L$, i.e., the resistance of absorption, and the slope is $(1/k_{LS} a_S + 1/k \eta)$, i.e., the overall resistance of the internal and external diffusion and the chemical reaction. Fig. 9.7 shows lines plotted for various sizes of the catalyst particles. As shown in the figure, reducing the diameter, d_p, results in a decrease of the slope, owing to the increased η and specific surface area a_S at a smaller d_p, which reduces the resistances for both internal and external diffusion. When d_p is sufficiently small, $\eta = 1$, $1/k_{LS} a_S \to 0$, the slope of the line equals $1/k$. Further reducing the size of the catalyst particle will not change the slope of the line, and therefore, will not affect the rate of the process.

The resistance of absorption can be determined by following the procedure outlined above. Combining the resistances of liquid–solid mass transfer, internal diffusion and chemical reaction gives the overall resistance represented by the slope. We use the following analysis to illustrate the relative contribution of each step.

If the resistances of liquid–solid mass transfer and internal diffusion can be neglected, plotting $(1/k_{LS} a_S + 1/k \eta)$ against d_p would result in a straight line with a slope of zero.

If the resistance of the liquid–solid mass transfer is negligible but the internal diffusion is significant, we have

$$\eta = \frac{1}{\phi} = \frac{6}{d_p} \sqrt{\frac{D_e}{k \rho_p}}$$

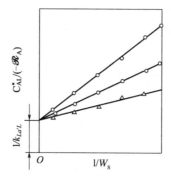

FIGURE 9.7 $C^*_{AL}/\mathscr{R}_A - 1/W_S$ plot for different catalyst particle sizes

Thus,

$$\frac{1}{k\eta} = \frac{1}{6}\sqrt{\frac{\rho_p}{kD_e}}\,d_p \qquad (9.53)$$

Consequently, if the slope of $\ln(1/k_{LS}a_S + 1/k\eta) \sim \ln d_p$ plot is 1, the liquid–solid mass transfer can be neglected and the main resistance is inside the catalyst particle.

If the particles move with the liquid motion so that there is no shear force between the particles and liquid, the liquid–solid mass transfer is equivalent to that of stagnant liquid and solid. Under such conditions, the Sherwood number is 2, i.e.,

$$\frac{k_{LS}d_p}{\mathscr{D}} = 2 \qquad (9.54)$$

or $k_{LS} = 2\mathscr{D}/d_p$, and a_S is the external surface area of per unit mass of catalyst particles, hence,

$$a_S = \frac{\pi d_p^2}{(\pi/6)d_p^3\rho_p} = \frac{6}{d_p\rho_p}$$

and

$$\frac{1}{k_{LS}a_S} = \frac{\rho_p}{12\mathscr{D}}d_p^2 \qquad (9.55)$$

Therefore, if the plot of $\ln(1/k_{LS}a_S + 1/k\eta)$ against $\ln d_p$ results in a straight line with a slope of 2, it indicates that liquid mass transfer, i.e., external diffusion, plays an important role in the process. In such a case, increasing the stirring would have no effect in increasing the overall rate as the particles move with the liquid and there is no shear force between the solid particle and liquid.

If there is shear force between particles and liquid, the mass transfer correlation is

$$\frac{k_{LS}d_p}{\mathscr{D}} = 2 + 0.6 \times \left(\frac{d_p u \rho}{\mu}\right)^{1/2} \left(\frac{\mu}{\rho \mathscr{D}}\right)^{1/3} \tag{9.56}$$

This can be approximated by $k_{LS} \propto d_p^{-0.5}$ as the second term on the right-hand side of Eq. (9.56) is much larger than 2. Hence,

$$\frac{1}{k_{LS}a_S} = \frac{1}{bd_p^{-0.5}(6/d_p\rho_p)} = \frac{\rho_p}{6b}d_p^{1.5} \tag{9.57}$$

where b is a proportionality constant. When $\ln(1/k_{LS}a_S + 1/k\eta)$ is plotted against $\ln d_p$, the liquid−solid mass transfer plays a significant role if the slope of the straight line is 1.5. In this situation, increasing stirring will increase the overall reaction rate.

In summary, the experimental data obtained by systematically varying the catalyst loading and particle size are extremely useful in providing information on the relative contribution of resistance of individual steps to the overall process, and the information can be used to develop effective measures to optimize the entire process.

Example 9.5
The catalytic hydrogenation of methyl linoleate was carried out in a stirred slurry reactor isothermally at 25°C. The data from the experiment is tabulated below, with A being hydrogen.

No.	p_A(MPa)	c_{A^*L}(kmol/m³)	$(-\mathscr{R}_A)$ (kmol/m³ per min)	W_S(kg/m³)	d_p (μm)
1	0.303	0.007	0.014	3.0	12
2	1.82	0.042	0.014	0.5	50
3	0.303	0.007	0.0023	1.5	50
4	0.303	0.007	0.007	2.0	750

The reaction is first order, and the catalyst is spherical and has a density of 2 g/cm³. Calculate: (1) the intrinsic reaction rate constant; (2) the liquid−solid mass transfer coefficient for the catalyst at 750 μm with a Thiele modulus of 3; and (3) discuss and suggest means to intensify the process.

Solution
Based on the given data, we first calculate $c_{AL}^*/(-\mathscr{R}_A)$ and $1/W_S$,

No	1	2	4	5
$c_{AL}^*/(-\mathscr{R}_A)$/min	0.5	3	1	3
$1/W_S$/(m³/kg)	1/3	2	2/3	1/2

(Continued)

(Continued)

We then plot $c_{AL}^*/(-R_A)$ against $1/W_S$, as shown in Fig. 9A. As shown in the plot, the results from 12 to 50 μm catalysts are on the same line, indicating that for these two particle sizes the total resistances of liquid–solid mass transfer and reaction inside the particle are the same. In addition, the comparison of the first group and second group of data further supports the conclusion—the catalyst loading of the former is six times that of the latter and c_{AL}^* is 1/6 whereas the overall reaction rates are the same, which cannot be true unless the total resistances of liquid–solid mass transfer and reaction inside the particle are the same.

1. Since the reaction is not affected by the particle sizes, the resistance of internal diffusion can be neglected, i.e., $\eta = 1$. For the particle with a size of 12 μm, $1/k_{LS}a_S \to 0$ should be valid, as will be demonstrated by results shown later. Therefore, the intrinsic reaction rate constant, k, can be determined from the slope of the line. From the plot, we measured a slope of 1.5 kg·min/m³. Based on $1/k\eta = 1/k = 1.5$ kg·min/m³, we get $k = 2/3$ m³/(kg·min).

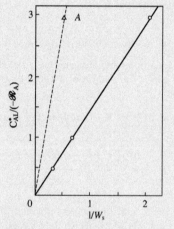

FIGURE 9A $C_{AL}^*/\mathscr{R}_A - 1/W_S$ plot

2. The solid line in Fig. 9A passes through the origin, indicating that the resistance of absorption in the early three groups of experiments is zero but the fourth group is not on the line, as shown by the point labeled A. Is the resistance of absorption also zero in this case? Comparing the data in No. 1, 3, and 4, we find the hydrogen partial pressure is the same, thus the hydrogen concentration in the liquid phase should be the same, which will result in the same absorption resistance. Consequently, the slope of OA, 6 kg·min/m³, corresponds to the resistance of the overall process with a catalyst of 750 μm, i.e.,

(Continued)

(Continued)

$$\frac{1}{k_{LS}a_S} + \frac{1}{k\eta} = 6 \tag{A}$$

Given $\phi = 3$, we get $\eta = 1/\phi = 1/3$. As $a_S = 6/d_p\rho_p = 6/(750 \times 10^{-6} \times 2 \times 10^3) = 4 \, m^2/kg$ Using all the data in (A), we get $1/4k_{LS} + 1/[(2/3)(1/3)] = 3$.

Solving this equation, we get the liquid–solid mass transfer coefficient,

$$k_{LS} = \frac{1}{6} \, m/min.$$

From Eq. (9.56) we have $k_{LS} \propto d_p^{-0.5}$. Based on this we can deduce the liquid–solid mass transfer coefficient for the 12 μm particle as $k_{LS} = (1/6)(750/12)^{0.5} = 1.32 \, m/min$. Since $a_S = 6/(12 \times 10^{-6} \times 2 \times 10^3) = 250 \, m^2/kg$, we get

$$\frac{1}{k_{LS}a_S} + \frac{1}{k\eta} = \frac{1}{(1.32 \times 250)} + \frac{1}{(2/3)} = \frac{3}{2}$$

These results show that neglecting the liquid–solid mass transfer resistance to get the intrinsic rate constant k is a valid assumption.

3. Based on the above analysis, the absorption resistance can be neglected. Consequently, any attempt at enhancing gas–liquid contact such as reducing bubble size and improving gas distribution will not produce any meaningful impact on the reaction process. For particle sizes of $12 \sim 50$ μm, the resistance of internal diffusion is negligible. Under such conditions, the larger particles will have the advantage of easy separation of catalysts from the product. On the other hand, a further increase in particle size will result in increased resistances of both internal diffusion and liquid–solid mass transfer. In the latter case, increasing stirring can intensify the process.

9.5.3 Design of Mechanically Stirred Slurry Tank Reactor

Mechanically stirred slurry tank reactors are widely used for three-phase reactions. The size of the particles is typically in the range of $100 \sim 200$ μm with a loading of $10 \sim 20 \, kg/m^3$. The energy to sustain particle suspension is provided by the mechanical stirrer. Therefore, the stirrer should be maintained at a sufficiently high speed to maintain uniform particle suspension and movement in the liquid phase. The threshold stirring speed depends on the type and size of the stirrer as well as diameter/height ratio of the tank, density and viscosity of the liquid, liquid/solid ratio, particle size, etc. Low gas velocity and high stirring rate can both make the particles suspend. Of course, the higher the stirring speed, the larger the power consumption.

Stirring not only makes the particles suspend but also improves the gas distribution, resulting in better gas—liquid contact.

There are restrictions to the gas loading in the stirred slurry tank reactor. Beyond the limit, the gas distribution will not be affected by stirring. The gas holdup will decrease while the average bubble size will increase, reducing the gas—liquid contact area. Generally, the gas velocity is 0.5 m/s or slightly higher and the gas holdup is 0.2 ∼ 0.4. When the stirring rate is high enough, the gas velocity and the type of gas distributor do not have much impact on the fluid dynamics. On the contrary, when the stirring rate is low, the effects on fluid dynamics will be significant.

The height/diameter ratio of a typical stirred slurry tank reactor is 1. The optimal diameter of the stirrer is 1/3 of reactor diameter. The best performance is achieved when the space between the stirrer and the bottom of the reactor is 1/6 of the reactor diameter. In practice, the height/diameter ratio is larger than 1 and can be as large as 2.5 or bigger. Under such circumstances, multiple stirrers on the same axis should be installed. The stirred slurry reactors are typically operated at a pressure below 10.13 MPa. Higher pressures make reactor design and manufacture difficult.

The presence of multiphase in the stirred slurry tank reactor for gas—liquid—solid catalytic reactions makes the flow patterns complex due to many factors. There is limited research in this area with limited data available. As an approximation for reactor design, the liquid phase is usually considered well-mixed. The gas phase component can also be considered well-mixed when the stirring is intense, which causes rapid coalescence and redistribution of bubbles. At moderate stirring, the gas will flow through the liquid in plug flow. Fortunately, for gases that are either highly soluble or highly insoluble, gas residence time distribution does not have a significant impact on reactor performance, and hence assuming plug flow for the gas phase is appropriate. For gases with intermediate solubility the situation is different and has to be treated more rigorously.

Neglecting the gas film resistance, the mass balance of the gaseous reactant is,

$$-u_{0G}\frac{dc_{AG}}{dZ} = k_L a_L(c_{AG}/H_A - c_{AL}) \tag{9.58}$$

The equation is per unit reaction volume based. Since the liquid phase is well-mixed, the concentration of A in liquid phase is constant and does not vary with the height. Thus, integrating Eq. (9.58) gives the concentration of A at the exit for a reactor with height of the reactive mixtures L_r,

$$(c_{AG})_{L_r} = (c_{AG})\exp(-\beta L_r) + [1 - \exp(-\beta L_r)]c_{AL}H_A \tag{9.59}$$

where $(c_{AG})_0$ is the concentration of A at the entrance and $\beta = k_L a_L / u_{0G} H_A$. The integration of Eq. (9.58) is based on u_{0G} being constant. This assumption is appropriate for gas with a small solubility but not for highly soluble gases. In the latter case, the resistance of gas–liquid mass transfer is significantly smaller than other resistances, and therefore can be neglected.

If the liquid does not have A, the mass balance for A in the reactor is

$$\frac{V_r u_{0G}}{L_r}\left[(c_{AG})_0 - (c_{AG})_{L_r}\right] = Q_L c_{AL} + V_r \varepsilon_S \eta r_A(c_{AS}, c_{BS}) \tag{9.60}$$

where Q_L is liquid flow rate, ε_S is volume fraction of the catalyst in the reactive mixture, and r_A is the reaction rate based on per unit volume catalyst.

If the liquid is nonvolatile, the mass balance of the liquid reactant B is

$$Q_L\left[(c_{BL})_0 - c_{BL}\right] = |\nu_B| V_r \varepsilon_S \eta r_A(c_{AS}, c_{BS}) \tag{9.61}$$

At the liquid–solid interface, the following equations are valid,

$$(k_{LS}a_S)_A[c_{AL} - c_{AS}] = \eta r_A(c_{AS}, c_{BS}) \tag{9.62}$$

$$(k_{LS}a_S)_B[c_{BL} - c_{BS}] = |\nu_B| \eta r_A(c_{AS}, c_{BS}) \tag{9.63}$$

Eqs. (9.59)–(9.63) are the equations for designing the stirred tank slurry reactor. As long as the liquid phase is well-mixed, gas phase is plug flow and liquid–solid shear is weak, these equations are applicable for designing a slurry reactor. Since the sizes of the catalyst particle are small, the effectiveness factor $\eta = 1$.

Example 9.6
Butynediol (B) hydrogenation was carried out in a stirred tank slurry reactor at 1.479 MPa and 35°C on a Pd catalyst of diameter 2×10^{-3} cm:

$$CH_2OH - C \equiv C - CH_2OH + H_2 \rightarrow CH_2OH - CH = CH - CH_2OH$$

The flow rate of hydrogen (A) containing gas with a hydrogen molar fraction of 80% is 5000 m³(STP)/h. The concentration of butynediol at the entrance is 2.5 kmol/m³ and flow rate is 1 m³/h. The reaction is first order in both hydrogen and butynediol. Assume the liquid is well-mixed and gas is in plug flow. The catalyst loading is 8 kg/m³. Calculate the reaction volume required to achieve a conversion of 90% butynediol.

Data: $k = 5 \times 10^{-5}$ m⁶/(mol·s·kg); $k_L a_L = 0.277$ s⁻¹; $k_{LS,A} = k_{LS,B} = 6.9 \times 10^{-4}$ m/s; $\rho_p = 1.45$ g/cm³; $H_A = 56.8$ cm³(gas)/cm³(liquid).

Solution
Solving Eqs. (9.59)–(9.63) simultaneously gives the reaction volume. In Eq. (9.59),

$$\beta L_r = \frac{k_L a_L L_r}{u_{0G} H_A} = \frac{k_L a_L V_r}{Q_G H_A}$$

(Continued)

(Continued)

$$Q_G = \frac{5000}{3600}\left(\frac{1.013 \times 10^5}{1.479 \times 10^6}\right)\left(\frac{308}{273}\right) = 0.1073 \ \text{m}^3/\text{s}$$

Therefore,

$$\beta L_r = \frac{0.277 V_r}{0.1073 \times 56.8} = 0.04545 V_r$$

$$(c_{AG})_0 = \frac{1.479 \times 0.8}{(8.314 \times 10^{-3} \times 308)} = 0.4625 \ \text{kmol}/\text{m}^3$$

Substituting the values of βL_r and $(c_{AG})_0$ into Eq. (9.59) results in,

$$(c_{AG})_{L_r} = 0.4625 \exp(-0.04545 V_r) + 56.8\left[1 - \exp(-0.04545 V_r)\right]c_{AL} \quad \text{(A)}$$

The conversion of butynediol is 0.9, thus $c_{BL} = 2.5(1 - 0.9) = 0.25 \ \text{kmol}/\text{m}^3$. Substituting c_{BL} and other available values into Eq. (9.61) results in,

$$\left(\frac{1}{3600}\right)(2.5 - 0.25) = 8V_r(0.05)c_{AS}c_{BS}$$

After simplification, we get

$$V_r c_{AS} c_{BS} = 1.5625 \times 10^{-3} \quad \text{(B)}$$

From Eq. (9.60), we get

$$0.1073\left[0.4625 - (c_{AG})_{L_r}\right] = \left(\frac{1}{3600}\right)c_{AL} + 8 \times 0.05 V_r c_{AS} c_{BS} \quad \text{(C)}$$

$$a_S = \frac{\pi\left(2 \times 10^{-5}\right)^2}{1450 \times \frac{\pi\left(2 \times 10^{-5}\right)^3}{6}} = 206.9 \ \text{m}^2/\text{kg}$$

Substituting the values of needed parameters into Eqs. (9.62) and (9.63) gives,

$$6.9 \times 10^{-4} \times 206.9(c_{AL} - c_{AS}) = 0.05 c_{AS} c_{BS} \quad \text{(D)}$$

$$6.9 \times 10^{-4} \times 206.9(0.25 - c_{BS}) = 0.05 c_{AS} c_{BS} \quad \text{(E)}$$

Solving Eqs. (A)–(E) simultaneously determines the values of V_r, c_{AL}, $(c_{AG})_{L_r}$, c_{AS}, and c_{BS}. Eliminating several variables will make the calculations easier. Dividing (D) by (E) and rearranging results in

$$c_{BS} = 0.25 - c_{AL} + c_{AS} \quad \text{(F)}$$

Substituting (F) into (D) gives,

$$c_{AS}^2 + (3.105 - c_{AL})c_{AS} - 2.855 c_{AL} = 0$$

(Continued)

(Continued)

Converting c_{AS} as a function of c_{AL},

$$c_{AS} = \frac{1}{2}\left[c_{AL} - 3.105 + \left(c_{AL}^2 + 5.21 c_{AL} + 9.641\right)^{\frac{1}{2}}\right] \tag{G}$$

Substituting (A) and (B) into (C) and rearranging results in

$$c_{AS}^2 = \frac{0.4567 - 0.4625 \exp(-0.04545 V_r)}{2.589 \times 10^{-3} + 56.8\left[1 - \exp(-0.04545 V_r)\right]} \tag{H}$$

Substituting (F) and (G) into (B) and further reduction gives,

$$V_r\left[\left(c_{AL}^2 + 5.21 c_{AL} + 90,641\right)^{1/2} - 2.605 - c_{AL}\right]\left[\left(c_{AL}^2 + 5.21 c_{AL} + 9.641\right)^{1/2} + c_{AL} - 3.105\right]$$

$$= 6.25 \times 10^{-3}$$

$$\tag{I}$$

Substituting (H) into (I) leads to an equation for V_r and solving it gives $V_r = 1.12 \text{ m}^3$.

FURTHER READING

Li S. Chemical and catalytic chemical reaction engineering. Beijing: Chemical Industry Press; 1986.

Danckerts PV. Gas–liquid reactions. New York: McGraw-Hill; 1970.

Carberry JJ. Chemical and catalytic reaction engineering. New York: McGraw-Hill; 1976.

Fogler HS. Elements of chemical reaction engineering. Englewood Cliffs, NJ: Prentice-Hall; 1986.

Shan YT. Gas–liquid–solid reactor. New York: McGraw-Hill; 1979.

PROBLEMS

9.1 Pure CO_2 reacts with aqueous sodium hydroxide solution. Neglecting the vapor pressure above the liquid phase, sketch the concentration distribution of CO_2 in gas and liquid phases based on the two-film theory.

9.2 A 1.2 mol/m^3 hydrous ammonia was used for absorbing CO_2 at the exit of a production unit. If the partial pressure of CO_2 in the mainstream of flowing gas is 1.013×10^{-3} MPa, what would be the corresponding CO_2 absorption rate?

Data: The diffusion coefficients of CO_2 and NH_3 in liquid are both $3.5 \times 10^{-5} \text{ cm}^2/\text{s}$, the reaction constant for the second order reaction is $38.6 \times 10^5 \text{ cm}^3/(\text{mol s})$. The solubility constant of CO_2 is $1.53 \times 10^{-10} \text{ mol/(cm}^3 \cdot \text{Pa)}$, $k_L = 0.04 \text{ cm/s}$, $k_G = 3.22 \times 10^{-10}$ mol/ $(\text{cm}^2 \cdot \text{s} \cdot \text{Pa})$, specific interfacial area $a_L = 2.0 \text{ cm}^2/\text{cm}^3$.

9.3 Gas A and liquid B react irreversibly. The reaction is first order to A and zero order to B. Given the rate constant k, mass transfer coefficient on the liquid side k_L, the diffusion coefficient of A in liquid phase, $\mathscr{D}_{AL}$, and the volume ratio of bulk liquid and liquid film α under three conditions,

$k\ (s^{-1})$	$k_L\ (cm/s)$	$\mathscr{D}_{AL} \times 10^5\ (cm^2/s)$	α
400	0.001	1.6	40
400	0.04	1.6	40
1	0.04	1.6	40

Determine the values of the enhancement and effectiveness factor in all three conditions and discuss the results.

9.4 Isothermal oxidation of cyclohexane was carried out in a stirred tank reactor at 0.891 MPa and 155°C. The concentration of cyclohexane in the liquid mixture at entrance is 7.74 kmol/m^3 and the mixture does not contain oxygen. The liquid and gas were fed at rates of 0.76 and 161 m^3/h, respectively. The concentration of cyclohexane in liquid is required to reach 6.76 kmol/m^3. Given that both gas and liquid are well-mixed and the resistance of gas film is negligible, calculate the necessary reaction volume.

Data: The reaction is first order to both oxygen and cyclohexane, the rate constant under the reaction condition $k = 0.2$ m^3/(s · mol), oxygen solubility = 1.115×10^{-7} kmol/(m^3 · Pa). Liquid side mass transfer coefficient $k_L = 0.416$ cm/s. Specific interfacial area = 6.75 cm^{-1}, gas holdup $\varepsilon_G = 0.139$, diffusion coefficient of oxygen in liquid $\mathscr{D}_{AL} = 2.22 \times 10^{-4}$ cm^2/s.

9.5 The oxidation of cyclohexane is carried out in a bubble column with a 1 m diameter. The air pressure at the bottom entrance is 0.912 MPa, other conditions are the same as those of Problem 9.4. Assuming gas is in plug flow and liquid is well-mixed, determine the height of the column.

Data: $a = 2.6$ cm^{-1}; $k_L = 0.12$ cm/s, $\varepsilon_G = 0.12$. For other data see Problem 9.4.

9.6 Isothermal hydrogenation of a styrene was carried out in a trickle bed reactor packed with Pd-Al2O3 catalysts of 4 mm diameter at 0.1013 MPa and 50°C. The concentration of a styrene in the feeding liquid is 4.3×10^{-4} mol/cm^3. The superficial velocities of pure hydrogen and liquid mixture at the entrance are 12 and 1 cm/s, respectively. Determine the height of the packed bed to reach a styrene conversion of 90%, given that both gas and liquid phases are in plug flow and the pressure drop across the packed bed is negligible.

Data: the reaction is first order in hydrogen and zero order in a styrene. Under the operating condition, $k = 16.8$ s^{-1}; $k_L a_L = 0.203$ s^{-1};

$k_{LS}a_S = 0.203 \text{ s}^{-1}$. The effective diffusivity of hydrogen in the catalyst = $1.02 \times 10^{-6} \text{ cm}^2/\text{s}$. The solubility constant of hydrogen in the liquid = $0.6 \text{ cm}^3\text{gas/cm}^3\text{liquid}$. bed voidage $\varepsilon = 0.42$.

9.7 In a stirred reactor of 5 m³ to hydrogenate methyl linoleate at 25°C on a catalyst of 0.9 mm diameter. The hydrogen partial pressure is maintained at 0.81 MPa. The processing rate of methyl linoleate is 1.5 kmol/min. Determine the necessary amount of catalysts to reach a conversion of 40% based on the data and results available in *Example 9.5*.

9.8 The hydrogenation of aniline is first order in hydrogen and zero order in aniline. Aniline reacts with pure hydrogen on a catalyst of 4 mm diameter at 1.01 MPa and 130°C, $k = 51.5 \text{ cm}^3/(\text{g} \cdot \text{s})$, $k_{LS} = 0.008 \text{ cm/s}$, $k_L a_L = 0.12 \text{ s}^{-1}$, the solubility of hydrogen is $3.56 \times 10^{-6} \text{ mol/cm}^3$, the effective diffusion coefficient of hydrogen in the catalyst is $8.35 \times 10^{-6} \text{ cm}^2/\text{s}$, calculate the overall reaction rate.

9.9 Hydrogenation of butynediol to butenediol with pure hydrogen was carried out in a slurry reactor. The reaction is first order in both hydrogen and butynediol. The concentration of butynediol at the entrance is 2.5 kmol/m³. Under the reaction condition, $k = 4.8 \times 10^{-5} \text{ m}^3/(\text{kg} \cdot \text{s} \cdot \text{mol})$, $k_L a_L' = 0.3 \text{ s}^{-1}$, $k_{LS} = 0.005 \text{ cm/s}$, solubility of hydrogen in liquid = 0.01 kmol/m³. Catalyst loading in liquid is 0.1 kg/m³, $\rho_p = 1.5 \text{ g/cm}^3$, $a_S' = 40 \text{ m}^2/\text{g}$. Both gas and liquid phases are in plug flow.

1. Neglecting internal diffusion resistance, determine the percentage contributions of each step to the overall resistance to achieve a liquid conversion of 95%.

2. Determine the liquid space velocity to achieve a liquid conversion of 95%.

3. Discuss possible means to increase the liquid space velocity while maintaining the overall conversion.

9.10 To purify the air contaminated by SO_2, a trickle bed reactor with activated carbon as catalyst and water as liquid medium at 0.101 MPa and 25°C is used to oxidize SO_2 to SO_3. SO_3 is then dissolved in water to forms diluted sulfuric acid which is discharged at reactor bottom. The rate-limiting step is the adsorption of oxygen on the catalyst surface, the rate of the reaction can be expressed as

$$r_A = \eta \rho_b k c_{AS} \text{ mol}/(\text{s} \cdot \text{cm}^3 \text{ bed})$$

where r_A is O_2-based reaction rate, c_{AS} is the concentration of oxygen on the catalyst surface in the unit of mol/cm³.

Given internal diffusion effectiveness factor $\eta = 0.6$, bulk density $\rho_b = 1.0 \text{ g/cm}^3$, rate constant for the first order reaction $k = 0.06 \text{ cm}^3/(\text{g} \cdot \text{s})$, the voidage of the packed bed $\varepsilon = 0.3$. $k_{LS}a_S = 0.30 \text{ s}^{-1}$, $k_L a_L = 0.03 \text{ s}^{-1}$, gas flow rate = 100 cm³/s, the Henry's law constant for oxygen in water $H = 5.0$. The diameter of the reactor is 10 cm, the gas composition at the top entrance: SO_2 2%, O_2 19%, N_2 79%. Determine the height of the packed bed needed to reach 80% conversion of SO_2.

9.11 Hydrogenation of ethylene was carried out in a slurry reactor with toluene as liquid media and Raney Ni as the catalyst. The reaction conditions are: $p = 1.52\,MPa$, $T = 50°C$, entrance $C_2H_4/H_2 = 1/1.5$ (moles), the entrance hydrogen flow rate is 1.386 mol/s. Given $a_L = 1.5\,cm^2/cm^3$, $k_L = 0.015\,cm/s$, $a_S = 200\,m^2/kg$, $k_{LS} = 0.2\,cm/s$, $W_S = 0.1\,kg/m^3$, the Henry's law constant of H_2 in toluene is 9.4. The liquid phase is well-mixed and gas phase is in plug flow. The rate of the process depends on the rate of hydrogen transfer from the gas−liquid interface to the liquid−solid interface. Determine the necessary reaction volume to reach a 50% ethylene conversion.

Chapter 10

Fluid–Solid Noncatalytic Reaction Kinetics and Reactors

Chapter Outline

A fluid–solid noncatalytic reaction refers to a class of important noncatalytic reactions where both the fluid phase and solid phase participate in the reactions, and either reactants or products, or both, are solid. The difference from fluid–solid catalytic reactions is that fluid–solid noncatalytic reactions

Reaction Engineering. DOI: http://dx.doi.org/10.1016/B978-0-12-410416-7.00010-0

do not involve a catalytic process and either produce or consume a solid during the reaction process. Thus, the property and quantity of the reaction components in the solid phase will change as the reactions progress and have to be treated as a function of time. Fluid−solid noncatalytic reactions have a longer history than catalytic reactions. Due to the revolutionary changes brought to the chemical industry by catalyst technology, fluid−solid noncatalytic reactions appear to be less important. However, with the continuous emergence of new materials, especially the synthesis of inorganic materials, fluid−solid noncatalytic reactions play more prominent roles and hence have been attracting new attention again.

This chapter will discuss fluid−solid noncatalytic reactions and their applications. We will first focus on the analysis of the reaction kinetics of solid particles and then introduce commonly used industrial reactors and design methods. The overall objective of this chapter is to provide readers with basic concepts on fluid−solid noncatalytic reactions and their applications in new material synthesis.

10.1 FLUID−SOLID NONCATALYTIC REACTIONS AND THEIR APPLICATIONS

According to the property of the fluid, fluid−solid noncatalytic reactions can be divided into gas−solid noncatalytic reactions and liquid−solid noncatalytic reactions. According to the function of the solids in the reactions, they can also be divided into solid consuming reactions, solid producing reactions, or a combination of both. Specifically, they can be expressed by the following types of reactions:

(1)	$A_{(s)} \rightarrow B_{(s)} + D_{(f)}$
(2)	$A_{(f)} \rightarrow D_{(s)} + E_{(f)}$
(3)	$A_{(s)} + B_{(f)} \rightarrow D_{(s)}$
(4)	$A_{(s)} + B_{(f)} \rightarrow D_{(f)}$
(5)	$A_{(s)} + B_{(f)} \rightarrow D_{(s)} + E_{(f)}$
(6)	$A_{(f)} + B_{(f)} \rightarrow D_{(s)} + E_{(f)}$

It can be seen from the above classification that fluid−solid noncatalytic reactions are typically heterogeneous. The modeling method and reactor selection criteria described in previous chapters for heterogeneous reactions can be used as a reference in this chapter.

Fluid−solid noncatalytic reactions belong to thermochemical reactions. Thus, they have the basic characteristics of thermochemical reactions, such as the effects of temperature and particle size on reaction rate. However, they also have their own characters, such as change in solid size during the reaction and the existence of a total reaction time. All of these need to be carefully controlled during the reaction process.

Fluid–solid noncatalytic reactions have long been widely used in chemical industries, and new applications are being developed such as the preparation of ultrafine particles. Fluid–solid noncatalytic reactions are most commonly used in the inorganic chemical industry, such as the reduction and regeneration of catalysts and heat-induced polymerization processes. Other typical processes include:

(1) Coal Chemical Industry

Coal can be used for producing many chemicals. One example is to react with water and air to make semiwater gas:

$$C_{(s)} + H_2O_{(g)} + Air_{(g)} \rightarrow CO_{(g)} + CO_{2(g)} + H_{2(g)} + N_{2(g)}$$

This reaction is used for ammonia production.

Another example is to make syngas:

$$C_{(s)} + H_2O_{(g)} \rightarrow CO_{(g)} + H_{2(g)}$$

The resulting syngas can be used for synthesis of different products such as methanol.

Coal combustion for heating involves the following gas-solid noncatalytic reactions:

$$C_{(s)} + O_{2(g)} \rightarrow CO_{2(g)}$$

$$2C_{(s)} + O_{2(g)} \rightarrow 2CO_{(g)}$$

(2) Inorganic Salt Manufacture

$$CaCO_{3(s)} \rightarrow CaO_{(s)} + CO_{2(g)}$$

$$2NaHCO_{3(s)} \rightarrow Na_2CO_{3(s)} + H_2O(g) + CO_{2(g)}$$

The above are the important reaction steps for sodium carbonate production.

(3) Chemical Metallurgy

Dry processes used in chemical metallurgy involve gas-solid noncatalytic reactions such as:

$$2ZnS_{(s)} + 3O_{2(g)} \rightarrow 2ZnO_{(s)} + 2SO_2(g)$$

$$Fe_2O_{3(s)} + 3H_{2(g)} \rightarrow 2Fe_{(s)} + 3H_2O(g)$$

The wet processing of fluorapatite ore for phosphorous acid production is an example of liquid-solid noncatalytic reaction:

$$Ca_5F(PO_4)_{3(s)} + 5H_2SO_{4(l)} + 10H_2O \rightarrow 5CaSO_4 2H_2O_{(s)} + 3H_3PO_{4(l)} + HF(g)$$

(4) Nuclear Industry

$$(NH_4)_2U_2O_{7(s)} \rightarrow 2UO_{3(s)} + 2NH_{3(g)} + H_2O_{(g)}$$
$$UO_{3(s)} + H_{2(g)} \rightarrow UO_{2(s)} + H_2O_{(g)}$$
$$UO_{2(s)} + 4HF_{(g)} \rightarrow UF_{4(s)} + 2H_2O_{(g)}$$
$$UF_{4(s)} + F_{2(g)} \rightarrow UF_{6(g)}$$

The above steps are the important gas–solid noncatalytic reactions that take place during the dry production process of uranium hexafluoride.

(5) Organic Synthesis

Reactive crystallization is used to make p-dichlorobenzene:

Thermal cracking of methane:

$$CH_{4(g)} \rightarrow C_{(s)} + 2H_{2(g)}$$

The coke-burning regeneration of catalyst deactivated by carbon deposition:

$$C_{(s)} + O_{2(g)} \rightarrow CO_{(g)} + CO_{2(g)}$$

This process has been used in fluidized bed catalytic cracking reactors.

(6) Ion Exchange Reaction

$$Ca^{2+}_{(aq)} + 2NaR_{(s)} \rightarrow 2Na^+_{(aq)} + CaR_{2(s)}$$

The R in the reaction equation represents resin-like carriers. Ion-exchange techniques have been widely used in catalyst preparation, water treatment, product synthesis, and many other fields.

When dealing with fluid–solid noncatalytic reactions from an engineering perspective, we need to consider the same topics as when dealing with other chemical reactions, including reaction kinetics, reactor design and operation. However, due to the diversity of the shape of solid reactants, the solid reaction rate has to be specially defined.

10.2 REACTION RATE OF PARTICLES IN DIFFERENT SHAPES

The shape and size of solid particles, which are reaction components, may remain unchanged, or undergo significant changes during the reaction process. For example, when passing ammonia through sulfuric acid, an acid–base neutralization reaction will occur and ammonia sulfate crystals will be formed

$$2NH_{3(g)} + H_2SO_{4(l)} \rightarrow (NH_4)_2SO_{4(s)}$$

During the reaction, solid particles emerge and gradually grow into grains with a certain shape and size.

The chemical vapor deposition process is similar to this. For the following reaction

$$SiH_{4(g)} + O_{2(g)} \rightarrow SiO_{2(s)} + H_2O_{(g)}$$

The produced solid silicon dioxide can be deposited on the support surface. This method can be used in the preparation of ultrafine particles and inorganic membrane materials.

In contrast to this are the reactions where solid reactants are consumed such as burning carbon particles in air to form carbon dioxide.

Another situation is that the shape and size of the solid particles undergo almost no change in the reaction process, so they can be approximately treated as a constant. Such reactions include:

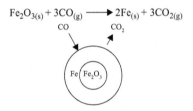

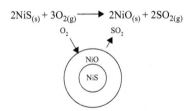

For the homogeneous and fluid–solid catalytic reactions described in previous chapters, the reaction rate is calculated based on unit reaction volume or unit catalyst volume (or weight). However, as the volume of the solid particles may vary, the basis for reaction rate calculation needs to be adjusted accordingly.

For the following reaction:

$$n_A A_{(g)} + n_B B_{(s)} \rightarrow n_D D_{(s)}$$

According to the definition of reaction rate:

$$r_B = -\frac{dn_B}{dt} = \frac{\upsilon_B}{\upsilon_A} r_A = \frac{\upsilon_B}{\upsilon_D} r_D \tag{10.1}$$

Together with the volume change of the solid reactant, the reaction rate can be expressed by a particle size-related function

$$r_B = f\left(k, c, \frac{S_p}{V_p}\right) \tag{10.2}$$

where $f(k,c)$ represents the function of the concentration (c) of the reaction components and the reaction rate constant (k), S_p is the surface area of the particles, and V_p is the volume of the particles.

It can be seen from Eq. (10.2) that the reaction rate of solid particles, r_B, is related to the shape and size of the particles.

When the reaction occurs only on the outer surface of the particles and the reaction rate is not affected by the diffusion of the fluid inside the pores in the particles, the above relationship can be simplified as

$$r_B = \frac{S_p}{V_p} f(k, c) \tag{10.3}$$

In this case, if the reactant concentration and the reaction temperature are constant, the smaller the particle size of reactant B, the faster the reaction rate. The effect of particle size change on the reaction rate is closely related to the shape of the particles. When reactant B is completely consumed, r_B will become zero, i.e.:

$$r_B = \begin{cases} \dfrac{S_p}{V_p} f(k, c) & V_p > 0 \\ 0 & V_p = 0 \end{cases} \tag{10.4}$$

where V_p changes with time, therefore, S_p/V_p contains the time variable implicitly.

For the specific reaction conditions described by Eq. (10.3), reaction rate equations for three different shapes of solid particles can be derived as below.

For spherical particles of radius R_p, the initial reaction rate ($t = 0$) is

$$r_{B0} = \frac{3}{R_p} f(k, c) \tag{10.5}$$

For cylindrical pellets of radius R_p and height H, the initial reaction rate is

$$r_{B0} = \left(\frac{1}{H} + \frac{2}{R_p}\right) f(k, c) \tag{10.6}$$

For very long cylindrical pellets, $H \gg R_p$, Eq. (10.6) can be simplified as

$$r_{B0} = \frac{2}{R_p} f(k, c)$$

For rectangular particles with length L, width W, and height H, the initial reaction rate is

$$r_{B0} = 2\left(\frac{1}{L} + \frac{1}{W} + \frac{1}{H}\right) f(k, c) \tag{10.7}$$

When $W \gg L$ and $H \gg L$, it becomes a thin sheet, and in this case,

$$r_{B0} = \frac{2}{L} f(k, c)$$

At the same volume, the outer area of a sphere is the smallest. Therefore, if the volume of the particles is the same, the reaction rate of spherical particles is the slowest.

10.3 THEORETICAL MODELS OF SOLID REACTIONS

The development of theoretical models for fluid–solid noncatalytic reactions should be based on the analysis of the changes of solid particles during the reactions. The changes in molar amounts in the fluid phase can be obtained based on stoichiometry.

Solid particles can be divided into two different classes: dense and porous. Fluids cannot enter the interior of a dense solid. Therefore, the reaction can only occur on the external surface of the particles. If the reactants are dense solid particles, regardless of whether the product is a solid, the volume of the unreacted part is constantly shrinking as the reaction proceeds. Therefore, the theoretical model for such solid particle reactions is referred to as the shrinking core model, and the unreacted part of the solid is called the unreacted core. For a porous solid, the fluid can diffuse through the pores to the interior of the solid, and the pore size and the reaction rate determine the level of resistance and the impact of the diffusion. When the diffusion resistance of the fluid is very large, the reaction still mainly occurs on the external surface of the solid particle. This situation can also be treated using the shrinking core model. When the diffusion rate of the fluid in the solid is much higher than the reaction rate, the fluid can directly diffuse to the center of the particle and react with the substances at any point inside the particle to varying extents. In this case, there is no longer an unreacted core. This case can be described by the reaction-diffusion equation for gas−solid catalytic reactions, which is also called a continuous model. Between the above two cases is the situation where the pore diffusion resistance affects the reaction. In this case, the large solid particles can be considered as composed of many microparticles aggregated together. The voids between the microparticles are large. The diffusion resistance of the fluid inside the voids is very small, so it can be described by a continuous model. The effect of diffusion on the reaction rate lies in the inside of each microparticle. The reaction inside the microparticles can be described by a shrinking core model. The resulting model combining the shrinking core model and the continuous model is called the microparticle model.

10.4   KINETIC ANALYSIS OF CONTINUOUS MODEL

There are crisscrossing channels within porous solid particles. The fluid flows into and out of the particles, depending on the concentration gradient as the driving force. When the particle is relatively loose, the local density at any point is the same as or very similar to the overall density of the particle. The gas can easily enter the interior of the particle and react with every solid particle. Depending on the relative magnitude of the diffusion rate of the fluid in the porous particle and the fluid−solid noncatalytic reaction rate, there are two different situations. In the first situation, the reaction rate is much smaller than the diffusion rate. The concentration of the reactant is the same throughout the particle (Fig. 10.1A), and the solid reactant is consumed or the solid product is produced at the same reaction rate. This is an ideal state, similar to the gas−solid catalytic reaction where the influence of internal diffusion has been eliminated. In the second situation, the fluid phase has concentration gradients in the particle (Fig. 10.1B). On the surface of the particle, the reaction rate is the highest since the concentration of the fluid reactant is high. The closer to the center, the lower the concentration of the fluid reactant, and the slower the reaction rate is. This situation is similar to the gas−solid catalytic reaction with internal diffusion influence. We can use the solid gasification reaction as an example to illustrate these two scenarios. In the former situation, the solid particles disappear simultaneously from inside to outside. Obviously, this is an ideal situation. In the latter situation, the disappearance rate of the surface layer is faster than that of the inside region, and the particle seems to be becoming smaller. The difference with the shrinking core model is that there are no dense regions in the particle.

The above two cases can both be described by a reaction-diffusion equation, which can be developed by mass balance of the fluid. Based on the calculated changes in fluid composition inside the particle, the

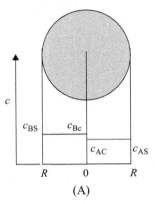
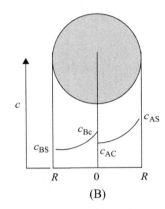

(A)   (B)

FIGURE 10.1 The concentration distribution inside the particle in the continuous model. (A) No internal diffusion influence; (B) Under internal diffusion influence.

reaction rate of the solid can be obtained from the stoichiometry. For solid spherical particles, the reaction-diffusion equation for the fluid phase can be written as:

$$\frac{1}{r^2}\frac{d}{dr}\left(D_e r^2 \frac{dc_i}{dr}\right) = -v_i \bar{r}_i \qquad (10.8)$$

where the subscript i stands for component i in the fluid. For the first order reaction

$$A_{(g)} + B_{(s)} \rightarrow D_{(g)} + E_{(s)}$$

assuming that the shape and size of the particle do not change, and the effective diffusion coefficient D_e is a constant, then

$$\frac{d^2 c_A}{dr^2} + \frac{2}{r}\frac{dc_A}{dr} = \frac{k_p}{D_e}c_A \qquad (10.9)$$

The boundary conditions are

$$r = 0 \quad \frac{dc_A}{dr} = 0$$

$$r = R_p \quad c_A = c_{AS}$$

Solving Eq. (10.9), the following can be obtained

$$c_A = c_{AS}\frac{R_p}{r}\frac{sh\left(\dfrac{3\varphi r}{R_p}\right)}{sh(3\varphi)} \qquad (10.10)$$

where r stands for the radius of the particles containing reactant B.

$$\varphi = \frac{R_p}{3}\sqrt{\frac{k_p}{D_e}}$$

The consumption rate of B in the particle is

$$r_B = r_A = k_p c_A = k_p c_{AS} \frac{R_p}{r} \frac{sh\left(\dfrac{3\varphi r}{R_p}\right)}{sh(3\varphi)} \tag{10.11}$$

The above equation implies that when D_e is a finite number, r_B is a function of particle radius.

At the beginning of the reaction, $r = R_p$, so

$$r_B = r_A = k_p c_{AS} \tag{10.12}$$

When B is completely consumed, $r = 0$, thus

$$r_B = r_A = 0 \tag{10.13}$$

When D_e approaches infinity, the concentration of the fluid components inside the particles is uniform, and $\varphi = 0$. Substituting these results into Eq. (10.10) gives

$$\begin{aligned} c_A &= c_{AS} \\ r &= R_p \end{aligned} \tag{10.14}$$

Substituting Eq. (10.14) into Eq. (10.11) gives

$$r_B = r_A = k_p c_{AS} \quad r > 0 \tag{10.15}$$

Eqs. (10.11) and (10.14) are identical. The reaction continues at constant maximum speed until the particles instantly disappear and the reaction is terminated.

10.5 KINETIC ANALYSIS AT CONSTANT PARTICLE SIZE USING THE SHRINKING CORE MODEL

If the size and shape of the solid particles do not change during the fluid—solid noncatalytic reaction, and there is only a concentration distribution but no temperature gradient inside the particles, the process can be treated as an isothermal process with constant particle diameter. The examples of such cases include reactions where the product is a solid or there is an inert solid residue even though the product is a fluid.

For a first order irreversible reaction conducted on the spherical particle B shown in Fig. 10.2

$$A_{(g)} + bB_{(s)} \rightarrow dD_{(g)} + pP_{(s)}$$

The concentration distribution of the reactant in the particle is shown in Fig. 10.3. There is a layer of stagnant film over the external surface of the

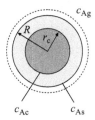

FIGURE 10.2 Shrinking core model.

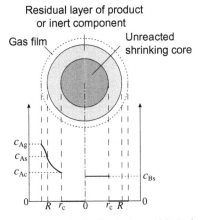

FIGURE 10.3 Reactant concentration distribution in the particle in the shrinking core model.

particle, the external diffusion resistance of the fluid lies mainly in this film. After overcoming this resistance with concentration-difference as the driving force, the fluid will reach the external surface of the spherical particle and the chemical reaction will occur. As the reaction proceeds, the radius of the unreacted core decreases. To react with the solid reactant, the fluid reactant must pass through the porous solid reaction product layer to reach the unreacted core. Compared with the initial reaction state where there is only external diffusion resistance, the fluid now has additional internal diffusion resistance.

In addition, the radius of the unreacted core in the shrinking core model is decreasing slowly. The shrinking rate can be considered negligible compared to the reaction rate and the diffusion rate. Therefore, based on pseudo steady-state assumption, the external diffusion rate of reactant A in the fluid is equal to its internal diffusion rate through the product layer p, and also equal to the chemical reaction rate of the fluid on the solid surface.

10.5.1 Overall Macroreaction Rate

Next we will derive the mass balance equation for gaseous reactant A. First, the external diffusion rate of A through the stagnant film can be calculated with the following equation:

$$\left(-\frac{dn_A}{dt}\right)_1 = 4\pi R^2 k_g (c_{Ag} - c_{AS}) \tag{10.16}$$

The internal diffusion rate of A through the product (or the inert residues) layer is:

$$\left(-\frac{dn_A}{dt}\right)_2 = 4\pi r_c^2 D_e \left(\frac{dc_A}{dr}\right)_{r=r_c} \tag{10.17}$$

where D_e is the effective diffusion coefficient of A.

For a chemical reaction occurring on the surface of a sphere of radius r_c, the consumption rate of A equals:

$$\left(-\frac{dn_A}{dt}\right)_3 = 4\pi r_c^2 k c_{AC} \tag{10.18}$$

where n_A stands for the number of moles of reactant A, k is the reaction rate constant based on the surface area of the unreacted core, and its unit is $m^3/m^2 s^{-1}$.

The reaction rate:

$$r'_A = -\frac{dn_A}{dt} \, \text{mol/s}$$

can be divided by V_p and rewritten as:

$$r_A = \left(-\frac{dc_A}{dt}\right)_3 = \frac{3r_c^2}{R^3} k c_{AC} \, \text{mol} \cdot m^3 \cdot s^{-1} \tag{10.19}$$

The reaction rate expressed in Eq. (10.19) is consistent with the definition of the reaction rate in Section 10.2. However, it can be seen from the following discussions that the final result is the same no matter what reaction rate expression is used. Therefore, reaction rates Eqs. (10.16) to (10.18) will be still used.

According to the pseudo steady-state assumption, the rates of the above three steps should be the same, i.e.,

$$\left(-\frac{dn_A}{dt}\right)_1 = \left(-\frac{dn_A}{dt}\right)_2 = \left(-\frac{dn_A}{dt}\right)_3$$

Thus, there are two equations that constitute an algebraic equation group. The number of the variables $\left(c_{AS}, c_{AC}, \dfrac{dc_A}{dr}\right)$ is 3. In order to solve these equations, one more equation is required. Inside the product layer p, the fluid diffuses through the layer without any reaction, and the process can be described by:

$$\frac{1}{r^2} D_e \frac{d}{dr}\left(r^2 \frac{dc_A}{dr}\right) = 0 \tag{10.20}$$

The boundary conditions are

$$r = R, \quad c_A = c_{AS}$$
$$r = r_c, \quad c_A = c_{AC}$$

Solving Eq. (10.20), the concentration distribution of reactant A in the product layer p can be obtained

$$c_A - c_{AC} = \frac{1 - \dfrac{r_c}{r}}{1 - \dfrac{r_c}{R}}(c_{AS} - c_{AC}) \tag{10.21}$$

Eq. (10.17) requires concentration gradient, which can be obtained by taking the derivative of Eq. (10.21):

$$\left(\frac{dc_A}{dr}\right)_{r=r_c} = \frac{c_{AS} - c_{AC}}{r_c\left(1 - \dfrac{r_c}{R}\right)} \tag{10.22}$$

After all three unknown variables mentioned above are obtained by simultaneously solving Eqs. (10.19) and (10.22), they are substituted into Eq. (10.18) to obtain the surface reaction rate. The macrokinetic equation for the overall reaction can be obtained:

$$-\frac{dn_A}{dt} = \frac{4\pi r_c^2 k c_{Ag}}{1 + \left(\dfrac{r_c}{R}\right)^2\left(\dfrac{k}{k_g}\right) + \left(\dfrac{kr_c}{D_e}\right)\left(1 - \dfrac{r_c}{R}\right)} \tag{10.23}$$

The number of moles of the solid reactant B in the unreacted core with radius r_c can be calculated by the following equation:

$$n_B = \frac{4}{3}\pi r_c^3 \cdot \frac{\rho_B}{M_B} \tag{10.24}$$

where r_c is a function of time:

$$r_c = f(t)$$

Thus, the consumption rate of reactant B is:

$$\frac{dn_B}{dt} = \frac{4\pi r_c^2 \rho_B}{M_B} \cdot \frac{dr_c}{dt} \tag{10.25}$$

From the stoichiometry:

$$\frac{dn_A}{dt} = \frac{1}{b} \cdot \frac{dn_B}{dt} = \frac{4\pi r_c^2 \rho_B}{bM_B} \cdot \frac{dr_c}{dt} \tag{10.26}$$

Combining Eqs. (10.23) and (10.26) gives:

$$-\frac{dr_c}{dt} = \frac{bM_B k \dfrac{c_{Ag}}{\rho_B}}{1 + \left(\dfrac{r_c}{R}\right)^2 \left(\dfrac{k}{k_g}\right) + \left(\dfrac{kr_c}{D_e}\right)\left(1 - \dfrac{r_c}{R}\right)} \tag{10.27}$$

The initial conditions are:

$$t = 0, \quad r_c = R$$

Solving the ordinary differential Eq. (10.27), the relationship between the radius of the unreacted core and the reaction time can be obtained:

$$t = \frac{\rho_B R}{bM_B c_{Ag}} \left\{ \frac{1}{k} + \frac{1}{3k_g}\left[1 + \frac{r_c}{R} + \left(\frac{r_c}{R}\right)^2\right] + \frac{R}{6D_e}\left[1 + \frac{r_c}{R} - 2\left(\frac{r_c}{R}\right)^2\right] \right\}\left(1 - \frac{r_c}{R}\right)$$

Reaction term   External diffusion term   Internal diffusion term

$$\tag{10.28}$$

Eq. (10.28) contains the contributions of three relatively independent terms, i.e., reaction, external diffusion, and internal diffusion.

Because Eq. (10.28) is an implicit function form of $r_c = f(t)$, it is not convenient to use. Nonlinear algebraic equations need to be solved to obtain r_c as a function of time. Otherwise, the solid conversion at a certain reaction time cannot be obtained. In addition, r_c is not measurable, hence it should be converted into a measurable variable such as the conversion of reactant A in the gas phase, x_A.

From the stoichiometry:

$$\frac{n_{B0} - n_B}{b} = n_{A0} - n_A$$

Thus,

$$x_B = \frac{bn_{A0}}{n_{B0}} x_A \tag{10.29}$$

From conversion x_A of gas reactant A, conversion x_B of solid reactant B can be calculated using Eq. (10.29). However, if the measurement using analytical instruments is not wanted and the various parameters are known, the calculation of the conversion of A still requires the kinetic equation. In other words, Eq. (10.28) still needs to be solved. On the contrary, in the case that the various parameters are unknown, the relationship between x_A and reaction time t has to be measured experimentally. Then, by the following calculation process, supplemented by the method for parameter estimation, the various model parameters in Eq. (10.28) can be obtained.

From Eq. (10.29), we know that x_B is indirectly measurable. Converting an unmeasurable quantity r_c into a function of an indirectly measurable quantity x_B will make Eq. (10.28) more usable.

The conversion of reactant B is defined as:

$$x_B = \frac{n_{B0} - n_B}{n_{B0}} = \frac{\frac{4}{3}\pi\rho_B\left(R^3 - r_c^3\right)}{\frac{4}{3}\pi\rho_B R^3} = 1 - \left(\frac{r_c}{R}\right)^3 \qquad (10.30)$$

$$\frac{r_c}{R} = (1 - x_B)^{\frac{1}{3}} \qquad (10.31)$$

Substituting Eq. (10.31) into Eq. (10.28) gives the $x_B \sim t$ relationship

$$t = \frac{\rho_B R}{b M_B c_{Ag}}\left\{\frac{1}{k}\left[1 - (1 - x_B)^{\frac{1}{3}}\right] + \frac{1}{3k_g}x_B + \frac{R}{6D_e}\left[1 - 3(1 - x_B)^{\frac{2}{3}} + 2(1 - x_B)\right]\right\} \qquad (10.32)$$

When $r_c = 0$, $x_B = 1$. At this moment the solid reactant just disappears. The reaction time corresponding to this moment is defined as the total reaction time t^*. Substituting $x_B = 1$ into Eq. (10.32) we obtain:

$$t^* = \frac{\rho_B R}{b M_B c_{Ag}}\left(\frac{1}{k} + \frac{1}{3k_g} + \frac{R}{6D_e}\right) \qquad (10.33)$$

10.5.2 Macroreaction Rate Under Internal Diffusion Control

From Eq. (10.28) we can see that the reaction time required to reach a certain conversion includes three parts corresponding to external diffusion, internal diffusion, and surface reaction, respectively. Based on the assumption of the rate-determining step, when the value of the internal diffusion resistance term is much larger than the other two terms, the overall macroreaction rate can be considered as being controlled by internal diffusion, and hence approximately equals the rate of the internal diffusion. This process can also be interpreted that D_e is very small compared to k and k_g. Therefore, retaining only the internal diffusion resistance term containing D_e, the following equation is obtained:

$$t = \frac{\rho_B R^2}{6 b M_B c_{Ag} D_e}\left[2\left(\frac{r_c}{R}\right)^3 - 3\left(\frac{r_c}{R}\right)^2 + 1\right]$$

$$= \frac{\rho_B R^2}{6 b M_B c_{Ag} D_e}\left[1 - 3(1 - x_B)^{\frac{2}{3}} + 2(1 - x_B)\right] \qquad (10.34)$$

When the solid reactant is consumed completely, $x_B = 1$. Substituting $x_B = 1$ into Eq. (10.34) we can obtain total reaction time t^* when reaction is controlled by internal diffusion

$$t^* = \frac{\rho_B R^2}{6bM_B c_{Ag} D_e} \tag{10.35}$$

So

$$t = t^* \left[2\left(\frac{r_c}{R}\right)^3 - 3\left(\frac{r_c}{R}\right)^2 + 1 \right]$$

$$= t^* \left[1 - 3(1-x_B)^{\frac{2}{3}} + 2(1-x_B) \right] \tag{10.36}$$

In the case of internal diffusion control, reducing t^* is an effective way to shorten the reaction time. The specific approaches to reduce t^* include increasing bulk concentration of the gas-phase reactant c_{Ag}, increasing effective internal diffusivity D_e, and reducing radius of the particle R. One of the methods to increase D_e is to enlarge the pore size of the porous solid, and at the same time to reduce the particle density ρ_B. Both measures are conducive to shortening the reaction time.

10.5.3 Macroreaction Rate Under External Diffusion Control

When controlled by external diffusion, only the external diffusion resistance term in Eq. (10.28) needs to be retained

$$t = \frac{\rho_B R}{3bM_B k_g c_{Ag}} \left[\left(\frac{r_c}{R}\right)^2 + \left(\frac{r_c}{R}\right) + 1 \right] \left(1 - \frac{r_c}{R} \right)$$

$$= \frac{\rho_B R}{3bM_B k_g c_{Ag}} \left[1 - \left(\frac{r_c}{R}\right)^3 \right] \tag{10.37}$$

$$= \frac{\rho_B R}{3bM_B k_g c_{Ag}} \cdot x_B$$

The total reaction time can be calculated from $x_B = 1$

$$t^* = \frac{\rho_B R}{3bM_B k_g c_{Ag}} \tag{10.38}$$

The total reaction time when controlled by external diffusion is directly proportional to particle radius, while the total reaction time when controlled by internal diffusion is proportional to the square of particle radius. Obviously, the effect of the particle size on the overall macroreaction rate is much smaller when controlled by external diffusion than that when controlled by internal diffusion.

Combining Eqs. (10.37) and (10.38), we can obtain:

$$t = t^* \left[1 - \left(\frac{r_c}{R} \right)^3 \right]$$

$$= t^* x_B$$

(10.39)

10.5.4 Intrinsic Reaction Rate Under Surface Reaction Control

When controlled by the chemical reaction, only the chemical reaction term in Eq. (10.28) needs to be retained

$$t = \frac{\rho_B R}{b M_B k c_{Ag}} \left(1 - \frac{r_c}{R} \right)$$

$$= \frac{\rho_B R}{b M_B k c_{Ag}} \left[1 - (1 - x_B)^{\frac{1}{3}} \right]$$

(10.40)

The total reaction time equals:

$$t^* = \frac{\rho_B R}{b M_B k c_{Ag}}$$

(10.41)

Therefore, Eq. (10.40) can be rewritten as

$$t = t^* \left(1 - \frac{r_c}{R} \right)$$

$$= t^* \left[1 - (1 - x_B)^{\frac{1}{3}} \right]$$

(10.42)

In the above analysis, only spherical particles are considered. The overall macroreaction rate equation is derived based on the pseudo steady-state assumption, and the integration of this equation yields the correlational relationship between the reaction time and the radius of the unreacted core or the conversion of the solid under various control conditions. It should be noted that the foregoing derivation is based on the overall expression, which considers internal diffusion, external diffusion, and surface reaction. The overall expression can be decomposed into three separate terms. When one term is the rate-controlling step, this term is retained but the other two are neglected. The results are identical to the results obtained by doing a separate mass balance to the rate-controlling step.

Fig. 10.4 shows reactant concentration distributions inside the particle for the three rate-controlling steps. It can be clearly seen from the figure that different rate-controlling steps generate different concentration profiles.

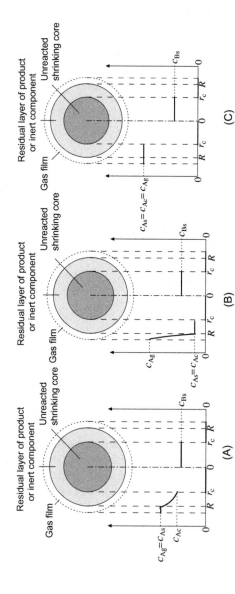

FIGURE 10.4 The reactant concentration distribution inside the particle in the shrinking core model for different rate-controlling steps. (A) Internal diffusion control, (B) External diffusion control, and (C) Chemical reaction control.

10.5.5   Comparison and Differentiation of Rate-Controlling Steps at Constant Particle Size

For spherical particles, Figs. 10.5 and 10.6 can be obtained using Eqs. (10.36), (10.39), and (10.42).

From Fig. 10.5:

1. When controlled by chemical reaction, t/t^* is a linear function of r_c/R.
2. When controlled by external diffusion, the relationship between t/t^* and r_c/R is nonlinear, and the curve is concave downward and intersects the diagonal, which represents the situation of chemical reaction control, at two points. On the other hand, The $t/t^* \sim r_c/R$ curve intersects the diagonal at three points when controlled by internal diffusion.

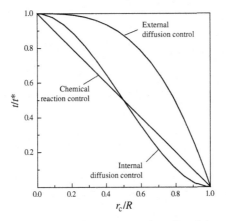

FIGURE 10.5   Relationship between reaction time and the radius of the unreacted core.

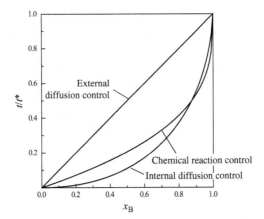

FIGURE 10.6   Relationship between reaction time and the reaction conversion.

Based on the shape of the $t/t^* \sim r_c/R$ curve we can determine the rate-determining step. However, it is difficult to measure r_c directly, and therefore it is not very convenient to use the $t/t^* \sim r_c/R$ curve as the basis for differentiating the rate-determining step.

From Fig. 10.6 we can see that

3. When controlled by external diffusion, t/t^* and x_B have a linear relationship.

4. When the conversion is low, the relative reaction time (t/t^*) required to achieve a given conversion under internal diffusion control is smaller than that under external diffusion control. When the conversion is higher than a certain value, the relative reaction time (t/t^*) required to achieve a given conversion under internal diffusion control larger than that under external diffusion control.

Using the four points discussed above, the rate-determining step for a gas−solid noncatalytic reaction occurring on spherical particles can be determined based on experimental results.

For other shapes of solid particles, corresponding results can also be derived using the same modeling approach. Typical shapes of solid particles include flat sheet, cylinder, and sphere, and Table 10.1 summarizes the results for these three particle shapes.

TABLE 10.1 Reaction Time Calculation Formula for Three Particle Shapes and Different Rate-Determining Steps

Shape			Flat Sheet	Cylinder	Sphere
		Solid conversion	$x_B = 1 - \dfrac{1}{L}$	$x_B = 1 - \left(\dfrac{r_c}{R}\right)^2$	$x_B = 1 - \left(\dfrac{r_c}{R}\right)^3$
Rate-determining step	Internal diffusion		$\dfrac{t}{t^*} = x_B^2$	$\dfrac{t}{t^*} = x_B + (1 - x_B)\ln(1 - x_B)$	$\dfrac{t}{t^*} = 1 - 3(1 - x_B)^{\frac{2}{3}} + 2(1 - x_B)$
			$t^* = \dfrac{\rho_B L^2}{2bM_B D_e c_{Ag}}$	$t^* = \dfrac{\rho_B R^2}{4bM_B D_e c_{Ag}}$	$t^* = \dfrac{\rho_B R^2}{6bM_B D_e c_{Ag}}$
	External diffusion		$\dfrac{t}{t^*} = x_B$	$\dfrac{t}{t^*} = x_B$	$\dfrac{t}{t^*} = x_B$
			$t^* = \dfrac{\rho_B L}{bM_B k_g c_{Ag}}$	$t^* = \dfrac{\rho_B R}{2bM_B k_g c_{Ag}}$	$t^* = \dfrac{\rho_B R}{3bM_B k_g c_{Ag}}$
	Chemical reaction		$\dfrac{t}{t^*} = x_B$	$\dfrac{t}{t^*} = 1 - (1 - x_B)^{\frac{1}{2}}$	$\dfrac{t}{t^*} = 1 - (1 - x_B)^{\frac{1}{3}}$
			$t^* = \dfrac{\rho_B L}{bM_B k c_{Ag}}$	$t^* = \dfrac{\rho_B R}{bM_B k c_{Ag}}$	$t^* = \dfrac{\rho_B R}{bM_B k c_{Ag}}$

TABLE 10.2 Shape Factors of the Particles

Shape of Particle	Flat Sheet	Cylinder	Sphere
n	0	1	2

It can be seen from Table 10.1 that when the reaction is external diffusion-controlled, the relationships between reaction time and conversion for all three particles shapes are the same, i.e., $t/t^* = x_B$. When it is chemical reaction-controlled, the reaction times for all three particle shapes can be expressed as:

$$\frac{t}{t^*} = 1 - (1 - x_B)^{\frac{1}{n+1}} \tag{10.43}$$

where n is the shape factor of the particles, whose values are shown in Table 10.2.

Example 10.1

At atmospheric pressure and 800°C, the total combustion reaction of spherical carbon particles is carried out in oxygen

$$C_{(s)} + O_{2(g)} \rightarrow CO_{2(g)}$$

It is assumed that the carbon particles are dense. As the reaction proceeds, the inert solid components remain, so the reaction process can be treated by a shrinking core model with constant particle diameter. At first, the reaction is conducted in pure and excessive oxygen, and carbon dioxide in the gas phase is almost undetectable. The changes in carbon conversion x_B with reaction time t are measured by weighing

t/min	0	10	20	30	40	50
x_B	0.0	0.2	0.4	0.6	0.8	1.0

Then, only the concentration of oxygen in the gas phase is changed to 9.0 mol/m³. Please calculate the conversion of the solid reactant at 40 min.

Solution

By analyzing experimental data given in the problem, we can see that x_B is proportional to reaction time t. So, this reaction can be considered to be controlled by external diffusion. From Eq. (10.39), t^* can be calculated

$$t^* = \frac{t}{x_B} = \frac{\rho_B R}{3 M_B c_{Ag} k_g}$$

(Continued)

(Continued)

Under atmospheric pressure, the gas phase can be regarded as an ideal gas. So, the oxygen concentration is

$$c_{Ag1} = \frac{P}{RT} = \frac{101,325}{8.314 \times (800 + 273.15)} = 11.36 \ \text{mol/m}^3$$

Under pure oxygen conditions, $\left(\dfrac{t}{x_B}\right)_1 = \dfrac{\rho_B R}{3 M_B c_{Ag1} k_g}$

At oxygen concentration of 9.0 mol/m³, $\left(\dfrac{t}{x_B}\right)_2 = \dfrac{\rho_B R}{3 M_B c_{Ag2} k_g}$

Division of the formulas for the two different conditions gives:

$$\frac{\left(\dfrac{t}{x_B}\right)_2}{\left(\dfrac{t}{x_B}\right)_1} = \frac{c_{Ag1}}{c_{Ag2}} = \frac{11.36}{9.0} = 1.26$$

As $\left(\dfrac{t}{x_B}\right)_1 = 50$ min, so:

$$\left(\frac{t}{x_B}\right)_2 = 50 \times 1.26 = 63 \ \text{min}$$

Therefore, when $t = 40$ min, $x_B = 63.49\%$.

10.6 KINETIC ANALYSIS WITH CHANGING PARTICLE DIAMETER USING THE SHRINKING CORE MODEL

The following reaction occurs on spherical solid particles

$$A_{(g)} + bB_{(s)} \rightarrow dD_{(g)} + pP_{(s)}$$

The difference from the reaction discussed in the previous section is that the particle size is constantly changing as the reaction proceeds. Still using a spherical particle as the example, this change process is illustrated in Fig. 10.7. Assuming that the initial radius of the spherical particle is R_0, the radius becomes R when the reaction proceeds to time t, and the radius of the unreacted core in the shrinking core model is r_c, a mass balance can be

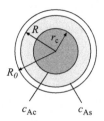

FIGURE 10.7 Shrinking core model with changing particle diameter.

performed for the interval of $0 \sim t$ for the solid reaction component B and product p.

The molar amount of solid B consumed is

$$\Delta n_B = \frac{4}{3} \pi \left(R_0^3 - r_c^3\right) \frac{\rho_B}{M_B} \tag{10.44}$$

At the same time, the molar amount of solid p produced is

$$\Delta n_p = \frac{4}{3} \pi \left(R^3 - r_c^3\right) \frac{\rho_p}{M_p} \tag{10.45}$$

According to the stoichiometric relationship

$$\frac{\Delta n_B}{\Delta n_p} = \frac{b}{p} \tag{10.46}$$

Solving the above three algebraic equations obtains

$$R = R_0 \left[\beta\left(1 - r_e^3\right) + r_e^3\right]^{\frac{1}{3}} \tag{10.47}$$

where

$$r_e = \frac{r_c}{R_0}, \quad \beta = \frac{p \rho_B M_p}{b \rho_p M_B}$$

When $\beta(1 - r_e^3) + r_e^3 > 1$, $R > R_0$. In this reaction, the particle diameter becomes larger.

As $r_e = \frac{r_c}{R_0} \leq 1$, so $1 - r_e^3 \geq 0$. Thus, only at the beginning of the reaction, $r_e = 1$ and $1 - r_e^3 = 0$. Otherwise, $1 - r_e^3 > 0$. When studying the changes in particle diameter at $t > 0$, solving the inequality $\beta(1 - r_e^3) + r_e^3 > 1$ gives

$$\beta > \frac{1 - r_e^3}{1 - r_e^3} = 1$$

Similarly, the following can be deduced

$\beta > 1$	$R > R_0$	Particle diameter increases
$\beta = 1$	$R = R_0$	Particle diameter does not change
$\beta < 1$	$R < R_0$	Particle diameter decreases

The fluid usually reacts simultaneously with many solid particles in the reactor. For a fixed bed reactor, when the particle diameter changes, the bed voidage will also change. When the fluid flows in the bed voids at the same volume flow rate, its flow velocity will change too. This will result in a change in gas film mass transfer coefficient k_g, which should be considered when doing actual calculations.

According to the results of kinetic analysis for constant particle diameter, the total reaction time is the sum of the contributions of chemical reaction, internal diffusion, and external diffusion, i.e.,

$$t = t_e + t_i + t_r \tag{10.48}$$

where the subscripts e, i, and r represent external diffusion, internal diffusion, and surface reaction, respectively.

When the particle diameter changes, the reaction time should also include those three terms. By calculating t_e, t_i, and t_r separately, t can be obtained. This derivation process is significantly simplified, compared to considering all three effects simultaneously.

For the first order irreversible reaction,

$$A_{(g)} + bB_{(s)} \rightarrow dD_{(g)} + pP_{(s)}$$

according to the pseudo steady-state assumption, the consumption rate of gas component A = External diffusion mass transfer rate across the gas film = Internal diffusion mass transfer rate = Surface reaction rate, i.e.,

$$-\frac{dn_A}{dt} = 4\pi R^2 k_g(c_{Ag} - c_{AS}) = 4\pi r_c^2 D_e \frac{dc_A}{dr}\bigg|_{r=r_c} = 4\pi r_c^2 k c_{AC} \tag{10.49}$$

Therefore,

$$n_B = \frac{4}{3}\pi r_c^3 \frac{\rho_B}{M_B}$$

And r_c is a function of time. Taking the derivative of both sides of the above equation with respect to time, the consumption rate of the solid reactant B can be obtained,

$$\frac{dn_B}{dt} = 4\pi r_c^2 \frac{\rho_B}{M_B}\frac{dr_c}{dt} \tag{10.50}$$

From the stoichiometry of the reaction,

$$\frac{dn_A}{dt} = \frac{1}{b}\frac{dn_B}{dt} \tag{10.51}$$

Therefore, the combination of Eqs. (10.50) and (10.51) yields

$$-\frac{dn_A}{dt} = \frac{4\pi r_c^2 \rho_B}{bM_B}\frac{dr_c}{dt} \tag{10.52}$$

In the porous product layer p, it is a pure diffusion process. Therefore, it can be described by the diffusion equation,

$$\frac{d}{dr}\left(D_e r^2 \frac{dc_A}{dr}\right) = 0 \tag{10.53}$$

The boundary conditions can be written as following:

$$r = R, \quad c_A = c_{AS}$$
$$r = r_c, \quad c_A = c_{AC}$$

Solving Eq. (10.53), the concentration distribution of reactant A in the product layer is obtained

$$c_A - c_{AC} = \frac{1 - \dfrac{r_c}{r}}{1 - \dfrac{r_c}{R}}(c_{AS} - c_{AC}) \tag{10.54}$$

In order to use Eq. (10.49), the concentration gradient of A in the product layer is calculated by Eq. (10.54),

$$\frac{dc_A}{dr} = \frac{(c_{AS} - c_{AC})r_c}{r^2\left(1 - \dfrac{r_c}{R}\right)} \tag{10.55}$$

From Eqs. (10.49) and (10.52):

$$-\frac{4\pi r_c^2 \rho_B}{bM_B}\frac{dr_c}{dt} = 4\pi r_c^2 D_e \frac{dc_A}{dr}\bigg|_{r=r_c} \tag{10.56}$$

As

$$\frac{dc_A}{dr}\bigg|_{r=r_c} = \frac{c_{AS} - c_{AC}}{r_c\left(1 - \dfrac{r_c}{R}\right)}$$

It can be substituted into Eq. (10.56), yielding

$$\frac{dr_c}{dt} = -\frac{D_e M_B b}{\rho_B}\frac{c_{AS} - c_{AC}}{r_c\left(1 - \dfrac{r_c}{R}\right)} \tag{10.57}$$

10.6.1 Internal Diffusion Control

When controlled by the internal diffusion in the product layer, the concentration of A in the product layer obeys,

$$c_{AS} \approx c_{Ag} \gg c_{AC}$$

So

$$c_{AS} - c_{AC} \approx c_{Ag}$$

Substituting the above equation into Eq. (10.57), a simplified equation can be obtained

$$\frac{dr_c}{dt} = -\frac{D_e M_B b}{\rho_B}\frac{c_{Ag}}{r_c\left(1 - \dfrac{r_c}{R}\right)} \tag{10.58}$$

Integrating Eq. (10.58) yields

$$\int_{R_0}^{r_c} r_c\left(1 - \frac{r_c}{R}\right)dr_c = -\int_0^{t_i} \frac{D_e M_B b}{\rho_B} c_{Ag} \cdot dt \qquad (10.59)$$

Replacing r_c with $r_c = r_e R_0$, Eq. (10.59) can be rewritten as

$$\int_{R_0}^{r_c} r_c\left(1 - \frac{r_c}{R}\right)dr_c = \int_1^{r_e} R_0^2 r_e\left(1 - r_e\frac{R_0}{R}\right)dr_e$$

$$= \int_1^{r_e} R_0^2 r_e\left[1 - \frac{r_e}{\left(\beta+(1-\beta)r_e^3\right)^{\frac{1}{3}}}\right]dr_e = -D_e M_B b\frac{c_{Ag}}{\rho_B}t_i$$

t_i represents the contribution of internal diffusion to the total reaction time. Integrating the above equation yields

$$t_i = \frac{\rho_B R_0^2}{2b M_B c_{Ag} D_e}\left\{\frac{[\beta+(1-\beta)r_e^3]^{\frac{1}{3}}}{1-\beta} + 1 - r_e^2 - \frac{1}{1-\beta}\right\} \qquad (10.60)$$

10.6.2 External Diffusion Control

When controlled by external diffusion through the stagnant film outside the particle,

$$c_{Ag} \gg c_{AS} \approx c_{AC}$$

Using Eq. (10.49), the material balance for the stagnant film outside the particle can be expressed as:

$$-\frac{dn_A}{dt} = 4\pi R^2 k_g(c_{Ag} - c_{AS}) \approx 4\pi R^2 k_g c_{Ag} \qquad (10.61)$$

Similarly, the material balance of component B combined with reaction stoichiometry can yield

$$-\frac{4\pi r_c^2 \rho_B}{b M_B}\frac{dr_c}{dt} = 4\pi R^2 k_g c_{Ag} \qquad (10.62)$$

As described previously, $k_g = f(R)$, which can be approximately expressed as

$$\frac{k_g}{k_{g0}} = \left(\frac{R_0}{R}\right)^m \quad m > 0$$

At constant particle diameter, factor $m = 0$.

The value of m depends on the reactor bed type and the flow conditions of the fluid outside the particles. For the three special cases, the following values can be used, respectively.

	Fixed-bed	Laminar flow of fluid outside particle	Turbulent flow of fluid outside particle
m	0.5	1	0.25

The integration variable r_c in Eq. (10.62) is replaced with r_e, and R is expressed as a function of β and r_e, then, the following equation is obtained

$$\int_1^{r_e} \frac{R_0 r_e^2 dr_e}{\left[\beta + (1-\beta)r_e^3\right]^{\frac{2-m}{3}}} = -\frac{k_{g0} b M_B c_{Ag}}{\rho_B} \int_0^{t_e} dt$$

After integration,

$$t_e = \frac{\rho_B R_0}{(m+1) b M_B k_{g0} c_{Ag}} \frac{1 - \left[\beta + (1-\beta)r_2^3\right]^{\frac{1}{3}}}{1 - \beta} \tag{10.63}$$

10.6.3 Chemical Reaction Control

When the overall reaction is surface chemical reaction-controlled, from the surface of the unreacted core to the bulk gas phase, the concentration difference of reactant A in the fluid phase is very small, and therefore:

$$c_{Ag} \approx c_{AS} \approx c_{AC}$$

Taking the material balance section for the surface reaction on the unreacted core in Eq. (10.49), i.e.,

$$-\frac{dn_A}{dt} = 4\pi r_c^2 k c_{AC} = 4\pi r_c^2 k c_{Ag} \tag{10.64}$$

and combining Eqs. (10.64) and (10.50) yields:

$$-\frac{4\pi r_c^2 \rho_B}{b M_B} \frac{dr_c}{dt} = 4\pi r_c^2 k c_{Ag}$$

Similarly, substituting the integral variable r_c with r_e gives:

$$-\frac{\rho_B R_0}{b M_B} = \frac{dr_e}{dt} = k c_{Ag} \tag{10.65}$$

Integrating Eq. (10.65)

$$\frac{-\rho_B R_0}{b M_B} \int_1^{r_e} dr_e = k c_{Ag} \int_0^{t_r} dt$$

the following equation can be obtained,

$$t_r = \frac{\rho_B R_0}{b M_B k c_{Ag}}(1 - r_e) \qquad (10.66)$$

Example 10.2

Large silica beads can be prepared by chemical vapor deposition of silane on the silicon seed crystal:

$$SiH_{4(g)} \rightarrow Si_{(s)} + 2H_{2(g)}$$
$$\quad\; A \qquad\; B \qquad\; D$$

Reaction rate is $r_A = kc_A$. When the reaction temperature is 200°C, $k = 0.1$ m/min, the partial pressure of silane $P_A = 0.01$ MPa, the density of silicon $\rho_B = 2.33 \times 10^6$ g/m³, the molecular weight of silicon $M_B = 28$. Assuming that the process is controlled by surface reaction, please calculate the reaction time required to increase the radius of the silicon particle from 1 to 20 μm.

Solution

Since

$$n_B = \frac{4}{3}\pi R^3 \frac{\rho_B}{M_B}$$

Then

$$\frac{dn_B}{dt} = 4\pi R^2 \frac{\rho_B}{M_B}\frac{dR}{dt}$$

$$= 4\pi R^2 k c_{AC}$$

When the process is controlled by the chemical reaction, $c_{AC} \approx c_{Ag}$
So,

$$\frac{dR}{dt} = \frac{M_B}{\rho_B} k c_{Ag}$$

$$R(t) - R_0 = \frac{M_B}{\rho_B} k c_{Ag} t$$

The reason to do the above derivation is that the equations derived in Section 10.6.3 are based on the solid reactant, but the chemical vapor deposition process involves a reaction of a gas phase to form a solid product.

$$c_{Ag} = \frac{P}{RT} = \frac{0.01 \times 10^6}{8.314 \times (200 + 273.15)} = 2.54 \text{ mol/m}^3$$

The time required for the particle growth,

$$t = \frac{\rho_B}{M_B k c_{Ag}}[R(t) - R_0]$$

$$= \frac{2.33 \times 10^6}{28 \times 0.1 \times 2.54}(20 \times 10^{-6} - 10^{-6}) = 6.33 \text{ min}$$

10.6.4 Overall Reaction Time

The relationship between reaction time and the radius of the unreacted core under the conditions of three possible rate-determining steps has been discussed above. The overall reaction time can be calculated according to Eq. (10.48),

$$t = t_i + t_e + t_r = \frac{\rho_B R_0^2}{2b M_B c_{Ag} D_e} \left\{ \frac{\left[\beta + (1-\beta) r_e^3 \right]^{\frac{1}{3}}}{1-\beta} + 1 - r_e^2 - \frac{1}{1-\beta} \right\}$$

$$+ \frac{\rho_B R_0}{(n+1) b M_B k_{g0} c_{Ag}} \frac{1 - \left[\beta + (1-\beta) r_e^3 \right]^{\frac{1}{3}}}{1-\beta}$$

$$+ \frac{\rho_B R_0}{b M_B k c_{Ag}} (1 - r_e)$$

$$(10.67)$$

The r_e in Eq. (10.67) is related to the radius of the unreacted core r_c, and it is very difficult to measure experimentally. Moreover, the conversion has been conventionally used to describe the extent of the reaction. So the solid conversion x_B can be calculated using the following equation,

$$x_B = 1 - r_e^3 \qquad (10.68)$$

By using Eqs. (10.67) and (10.68), the reaction time needed for a given conversion can be calculated and vice versa.

10.7 MICROPARTICLE MODEL

As already mentioned previously, the so-called "microparticle model" is actually a combination of a continuous model and a shrinking core model. Specifically, a solid particle is considered to be composed of many tiny "microparticles." Fig. 10.8 illustrates this basic structure. The transport of the fluid in the gap between the microparticles and the reaction with the solid on the surface of the microparticles obeys the continuous model, while the fluid–solid noncatalytic reaction in each of the microparticles obeys the shrinking core model. This reaction mode is very similar to a fixed bed catalytic reactor. If the fixed bed catalytic reactor is considered to be a large particle, the catalyst is the microparticles. Using the reaction-diffusion equation to describe the diffusion and reaction within the pores of the catalyst corresponds to using the shrinking core model to treat the fluid–solid noncatalytic reaction in the microparticles. Further comparing the two, it is not difficult to find that the pseudo homogeneous model for the fixed bed catalytic

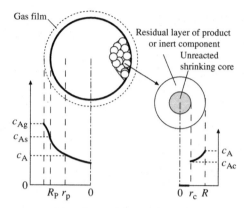

FIGURE 10.8 The structure of the microparticle model and the concentration distribution.

reactor is similar to the continuous model for the fluid–solid noncatalytic reactions. The heterogeneous model for the fixed bed catalytic reactor is similar to the microparticle model for the fluid–solid noncatalytic reaction.

For reaction

$$A_{(g)} + bB_{(s)} \rightarrow dD_{(g)} + eE_{(s)}$$

it is first assumed that the particle is composed of spherical microparticles of the same size. The reaction-diffusion equation of gas component A in the particle is

$$\frac{1}{r_p^2} D_{ep} \frac{d}{dr_p} \left(r_p^2 \frac{dc_A}{dr_p} \right) = r'_A \qquad (10.69)$$

where r_p stands for the radius of the particle, D_{ep} is the effective diffusion coefficient of A in the voids between the microparticles, and r'_A is the reaction rate of component A in the voids of the particle based on the particle volume, rather than the reaction rate on the solid surface r_A. The difference between the two lies in whether the effect of the external and internal diffusion on the surface of the microparticles is considered and whether the reaction rate is based on the surface area of the microparticles or the volume of the particles.

The outer surface area of the unreacted core in each of the microparticle within the particle is:

$$S_i = 4\pi r_c^2$$

If the spherical shell of thickness dr_c is approximately considered as a thin sheet, then the number of moles of component B contained in the spherical shell is:

$$dn_{Bi} = \frac{\rho_B}{M_B} S_i dr_c$$

If a single particle contains m microparticles, then the consumption rate of the solid reactant B in the particle can be written as:

$$\frac{dn_B}{dt} = \sum_{i=1}^{m} \frac{dn_{Bi}}{dt} = \sum_{i=1}^{m} \frac{\rho_B}{M_B} S_i \frac{dr_c}{dt} \tag{10.70}$$

According to the stoichiometry of the reaction,

$$r_A' = \frac{1}{b} r_B = -\frac{1}{b} \frac{dn_B}{dt} \frac{1}{V_p} = -\frac{1}{b} \frac{\sum_{i=1}^{m} S_i}{V_p} \frac{\rho_B}{M_B} \frac{dr_c}{dt} \tag{10.71}$$

The V_p in Eq. (10.71) stands for the volume of the particle, while the combination term can be decomposed into

$$\frac{\sum_{i=1}^{m} S_i}{V_p} = \frac{\text{Total outer surface area of unreacted core}}{\text{Volume of particle}}$$

$$= \frac{\text{Volume of microparticle}}{\text{Volume of particle}} \cdot \frac{\text{Outer surface area of microparticle}}{\text{Volume of microparticle}} \cdot \frac{\text{Total outer surface area of unreacted core}}{\text{Outer surface area of microparticle}} \tag{10.72}$$

The first term in the multiplication on the right-hand side of Eq. (10.72)

$$\frac{\text{Volume of microparticle}}{\text{Volume of particle}} = 1 - \varepsilon \tag{10.73}$$

The second term

$$\frac{\text{Outer surface area of microparticle}}{\text{Volume of microparticle}} = \frac{mS_i}{mV_i} = \frac{3r_c^2}{R^3} \tag{10.74}$$

The last term

$$\frac{\text{Total outer surface area of unreacted core}}{\text{Outer surface area of microparticle}} = \left(\frac{r_c}{R}\right)^n \tag{10.75}$$

In Eq. (10.75), n stands for the shape factor of the particle, whose value is shown in Table 10.2.

Substituting Eqs. (10.72) to (10.75) into Eq. (10.71) yields

$$r_A' = -\frac{1}{b}(1 - \varepsilon)\left(\frac{3r_c^2}{R^3}\right)\left(\frac{r_c}{R}\right)^n \frac{\rho_B}{M_B} \frac{dr_c}{dt} \tag{10.76}$$

Using the shrinking core model for constant particle size to process the first order irreversible reaction, the rate equation of the surface reaction is

$$r_A = kc_A$$

Eq. (10.27) derived previously is substituted into Eq. (10.76), then

$$r'_A = -\frac{kc_A(1-\varepsilon)}{1+\left(\frac{r_c}{R}\right)^2\left(\frac{k}{k_g}\right)+\left(\frac{kr_c}{D_e}\right)\left(1-\frac{r_c}{R}\right)}\left(\frac{3r_c^2}{R^3}\right)\left(\frac{r_c}{R}\right)^n \tag{10.77}$$

where D_e is the effective diffusion coefficient in the microparticle. For spherical particles, $n=2$. Substituting Eq. (10.77) into Eq. (10.69), the microparticle model equation for first order irreversible reaction can be obtained

$$\frac{1}{r_p^2}D_{ep}\frac{d}{dr_p}\left(r_p^2\frac{dc_A}{dr_p}\right)=\frac{3(1-\varepsilon)kc_A\frac{r_c^4}{R^5}}{1+\left(\frac{r_c}{R}\right)^2\left(\frac{k}{k_g}\right)+\left(\frac{kr_c}{D_e}\right)\left(1-\frac{r_c}{R}\right)}$$

$$=\frac{3(1-\varepsilon)c_A\frac{r_c^4}{R^5}}{\frac{1}{k}+\left(\frac{r_c}{R}\right)^2\left(\frac{1}{k_g}\right)+\frac{r_c}{D_e}\left(1-\frac{r_c}{R}\right)} \tag{10.78}$$

The boundary conditions after ignoring the influence of external diffusion are

$$r_p=0 \quad \frac{dc_A}{dr_p}=0$$

$$r_p=R_p \quad c_A=c_{Ag}$$

The radius of the unreacted core r_c in Eq. (10.78) is a function of time, i.e., $r_c=f(t)$ This function can be calculated via Eq. (10.28). Therefore, by simultaneously solving the ordinary differential equation, Eq. (10.78), and the nonlinear algebraic equation, Eq. (10.28), the concentration distribution of component A inside the solid particle at different reaction times can be obtained. Then, the concentration changes of other components can be obtained based on the stoichiometric relationship. Fig. 10.9 shows the calculation results.

It can be seen from Fig. 10.9 that at $t=0$, it is the initial state of the reaction. The concentration distribution within the particles is only caused by the diffusion of component A. As the reaction proceeds, the fluid components diffusing into the particles react continuously, reducing the concentration of component A throughout the particle until the solid reactants disappear at the

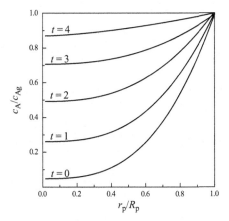

FIGURE 10.9   The concentration distribution of component A in the particle calculated by the microparticle model.

total reaction time. Further increasing the contact time, a pure diffusion equilibrium with uniform concentration within the particle will be reached.

When the reactant concentration distribution in the particle is known, the overall reaction rate in the particle can be calculated using the following equation:

$$\langle r'_A \rangle = \frac{1}{V_p} \iiint_{V_p} r'_A dV \tag{10.79}$$

For spherical particles:

$$\langle r'_A \rangle = \frac{1}{R} \int_0^R r'_A dr_p \tag{10.80}$$

Integrating Eq. (10.80), the reaction rate for the microparticle model can be obtained. Since the microparticle model equation is complicated, a numerical solution is usually obtained.

10.8   CHEMICAL VAPOR DEPOSITION

Chemical vapor deposition, abbreviated as CVD, is a growing area of chemical reaction engineering that has great potential in the field of materials synthesis, and has broad applications in the preparation of optoelectronic equipment, magnetic recording materials, and inorganic membrane materials. Chemical vapor deposition uses a gas phase chemical reaction to produce a solid deposit on the surface of the carrier particles. The mechanism of this process is very similar to gas−solid phase catalysis including the adsorption of the reactants on a solid surface and the formation of a new surface via surface chemical reaction. Some reactions also include the desorption process.

Similarly, Chemical vapor condensation, abbreviated as CVC, is an important method for preparing nanoparticles. The principle is the same as the fluid–solid noncatalytic reaction described previously. It is different from chemical vapor deposition in that the solid particles are formed by condensation, without depositing on the carrier.

Silicon germanium epitaxial film prepared using chemical vapor deposition has attracted more and more attention in the microelectronics industry. Epitaxial germanium film is also an important material for the production of solar cells. Its synthesis mechanism is:

Vapor decomposition	$GeCl_{4(g)} \rightleftharpoons GeCl_{2(g)} + Cl_{2(g)}$
Adsorption	$GeCl_{2(g)} + S \overset{K_G}{\rightleftharpoons} GeCl_2S$
Dissociative adsorption	$H_{2(g)} + 2S \overset{K_H}{\rightleftharpoons} 2HS$
Surface reaction	$GeCl_2S + 2HS \overset{K_S}{\longrightarrow} Ge_{(s)} + 2HCl_{(g)} + 2S$

Using this reaction as an example, we will discuss the method of establishing the kinetics equation of chemical vapor deposition.

The difference from the gas–solid catalytic reaction is that there is no desorption process of the reaction product germanium in the above reaction steps. After a germanium molecule occupies the carrier surface S, it becomes the new carrier, forming layered germanium deposits. Assuming that the surface chemical reaction is the rate-determining step, based on the steady-state assumption, the rate equation of the deposition reaction can be written according to the mass action law.

$$r_G = K_S \theta_{GeCl_2} \theta_H^2 \quad nm/s \qquad (10.81)$$

The vapor decomposition reaction does not occur directly on the surface of the carrier, so it can be treated as an independent reaction, and is not included in the derivation of the kinetics equations.

When the adsorption reaches equilibrium,

$$\theta_{GeCl_2} = \theta_V K_G P_{GeCl_2} \qquad (10.82)$$

The K_G in Eq. (10.82) is the adsorption equilibrium constant.

For the dissociative adsorption of hydrogen, the equilibrium constant is K_H.

$$\theta_H = \theta_V \sqrt{K_H P_{H_2}} \qquad (10.83)$$

In addition,

$$\theta_H + \theta_{GeCl_2} + \theta_V = 1 \qquad (10.84)$$

From Eqs. (10.82–10.84), the following can be obtained

$$\theta_V = \frac{1}{1 + K_G P_{GeCl_2} + \sqrt{K_H P_{H_2}}} \qquad (10.85)$$

Substituting Eqs. (10.85) and (10.83) into Eq. (10.81), the rate equation of the deposition reaction can be obtained

$$r_G = \frac{k' P_{GeCl_2} P_{H_2}}{\left(1 + K_G P_{GeCl_2} + \sqrt{K_H P_{H_2}}\right)^3} \tag{10.86}$$

In Eq. (10.86),

$$k' = k_S K_G K_H$$

Next we will examine the vapor decomposition reaction. As it is not the rate-determining step of the reaction, it is considered to have reached the equilibrium, i.e.,

$$K_d = \frac{P_{GeCl_2} P_{Cl_2}}{P_{GeCl_4}} = \frac{(P_{GeCl_2})^2}{P_{GeCl_4}} \tag{10.87}$$

According to Eq. (10.87), the partial pressure of the decomposition product $GeCl_2$ can be expressed by the partial pressure of the raw material $GeCl_4$. After substituting it into Eq. (10.86), the final reaction rate equation is obtained,

$$r_G = \frac{k \sqrt{P_{GeCl_4}} P_{H_2}}{\left(1 + k_1 \sqrt{P_{GeCl_4}} + \sqrt{K_H P_{H_2}}\right)^3} \tag{10.88}$$

where

$$k = k' K_d^{0.5}, \quad k_1 = K_G K_d^{0.5}$$

10.9 DESIGN OF FLUID–SOLID NONCATALYTIC REACTOR

The three important factors that are critical for designing the fluid–solid noncatalytic reactor are:

1. The reaction kinetics on a single solid particle
2. The particle diameter distribution of the solid material
3. The flowing and mixing state of the solid and fluid in the reactor.

Since the solid is one of the reaction components, its feeding into and removal from the reactor are essential steps. A systematic analysis of the reaction kinetics on a single particle has been discussed in the previous sections, but it is limited to simple reactions. The actual reactions involved in industrial applications are usually much more complicated. For nonisothermal multiple reactions, the kinetics analysis will become very complicated in situations that involve a very wide particle size distribution. Here, only the most commonly used industrial reactors will be introduced, and the design equations for ideal reactors will be developed.

10.9.1 Reactor Types

The reactors commonly used in industrial production are shown in Fig. 10.10. The main operation modes of the reactors are divided into continuous, batch, and semicontinuous operation. A fixed bed reactor is a type of semicontinuous operation reactor and can be used for ion exchange and gas purification, such as ZnO desulfurization. A rotary furnace is the major calcining equipment for sodium carbonate synthesis. Fluidized bed reactors have been used for coal gasification and catalyst regeneration in fluidized bed catalytic cracking. Moving bed reactors can achieve solid movement by mechanical drive or gravity, and have the feature that the solid maintains the same residence time in the bed. It has practical applications in coal gasification, the nuclear fuel processing industry, etc. Blast furnaces are commonly used in the steel-making industry. Stirred bed reactors have a wide range of applications in the field of chemical reaction crystallization, phosphoric acid production, etc., and can be operated in different operation modes. Delivery bed reactors rely on a gas flow to drive the solid into and out of the reactor and are especially suitable for fine particles and fast reactions.

10.9.2 Flowing and Mixing of Reaction Components

When discussing flowing and mixing of the reaction components in a fluid−solid noncatalytic reactor we need to consider two aspects. The first is the flowing behaviors of both fluid and solid reaction components. Since the two phases have an obvious interface, they can flow independently. For example, in a moving bed reactor, the fluid phase can be close to a complete mixing flow while the solid phase complies with a plug flow. The second is the mixing state of the fluid and the solid. The mixing of the fluid and the solid is the necessary condition for a chemical reaction to occur. Because the solid particles are independent clusters composed of a large number of molecules, they exhibit a strong inertness to each other, and no chemical reaction occurs between solid particles. The extent of the fluid−solid reaction depends on the fluid−solid mixing, as well as the residence time of the solid in the reactor, as it is assumed that when the solid particles leave the reactor, the reaction stops immediately. Therefore, a discrete model is used to treat the fluid−solid mixing and to calculate the average conversion of the solid.

For reaction $A_{(g)} + bB_{(s)} \rightarrow dD_{(g)} + eE_{(s)}$
if the diameter of the particles is uniform, then

$$\bar{x}_B = \int_0^\infty x_B(t)E(t)dt = \int_0^{t^*} x_B(t)E(t)dt + \int_{t^*}^\infty x_B(t)E(t)dt \qquad (10.89)$$

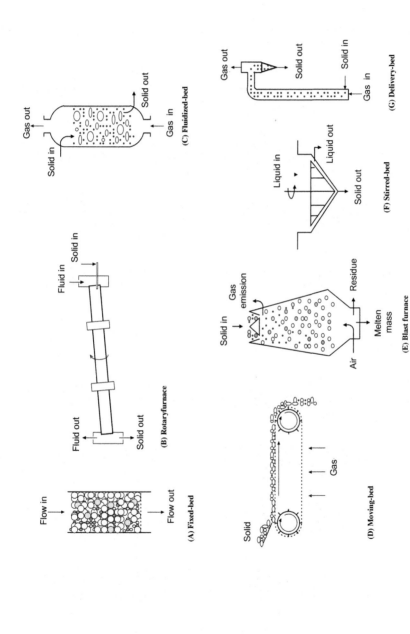

FIGURE 10.10 Main types of fluid—solid noncatalytic reactors. (A) fixed-bed, (B) rotary furnace, (C) fluidized bed, (D) moving-bed, (E) blast furnace, (F) stirred-bed, and (G) delivery-bed.

When the reaction time t is longer than the total reaction time t^*, the conversion of the reaction becomes 1. Therefore,

$$\bar{x}_B = \int_0^\infty x_B(t)E(t)dt = \int_0^{t^*} x_B(t)E(t)dt + \int_{t^*}^\infty E(t)dt \qquad (10.90)$$

Eq. (10.90) needs to be integrated twice, so an appropriate transformation is performed to this equation. Since

$$\int_{t^*}^\infty E(t)dt = 1.0 - \int_0^{t^*} E(t)dt$$

Substituting the above equation into Eq. (10.90), and making appropriate adjustments, an expression containing only one integral can be obtained.

$$1 - \bar{x}_B = \int_0^{t^*} [(1 - x_B(t)]E(t)dt \qquad (10.91)$$

In the above equation, $E(t)$ is the residence time distribution density function of the solid reactant B in the reactor, which is determined by the flowing pattern of the solid. $x_B(t)$ is the conversion of the solid at time t, calculated by the reaction kinetics on a single solid particle.

$\bar{x}_B$ is the average conversion. The upper limit of the integration t^* is the total reaction time.

Typically, the solid particles have a size distribution, expressed by the volume fractions of the different sizes of particles in the solid feed. The fraction of the particles of radius R_i can be calculated by the following equation:

$$f_i = \frac{Q(R_i)}{Q_t} \qquad (10.92)$$

Q_t is the total volume flow rate.

$$Q_t = \sum_{i=1}^m Q(R_i)$$

where m represents the number of particle sizes.

The average conversion of particles with different sizes needs to be calculated by combining Eqs. (10.91) and (10.92):

$$1 - \bar{\bar{x}}_B = \sum_{i=1}^m [1 - \bar{x}_B(R_i)]f_i = \sum_{i=1}^m [1 - \bar{x}_B(R_i)]\frac{Q(R_i)}{Q_t} \qquad (10.93a)$$

or

$$\bar{\bar{x}}_B = \sum_{i=1}^m \bar{x}_B(R_i)f_i = \sum_{i=1}^m \bar{x}_B(R_i)\frac{Q(R_i)}{Q_t} \qquad (10.93b)$$

The above calculation of the solid conversion is based on the assumption that the concentration of the fluid components is uniform in the reactor. When the concentration of the fluid components changes with spatial position, x_B is not only a function of time, but also a function of spatial position. If the change in temperature in the reactor also has to be considered, then the calculation of x_B will become very complex, which will not be discussed here.

10.9.3 Reactor Design When Fluid Is a Complete Mixing Flow and Solid Phase Is a Plug Flow

In the case that the particle size is uniform and does not change, substituting the plug flow residence time distribution density function

$$E(t) = \delta(t - \bar{t})$$

into Eq. (10.91) yields,

$$1 - \bar{x}_B = \int_0^{t^*} [1 - x_B(t)]\delta(t - \bar{t})dt = \begin{cases} 1 - x_B(\bar{t}) & \bar{t} < t^* \\ 0 & \bar{t} \geq t^* \end{cases} \quad (10.94a)$$

or

$$\bar{x}_B = \begin{cases} x_B(\bar{t}) & \bar{t} < t^* \\ 1 & \bar{t} \geq t^* \end{cases} \quad (10.94b)$$

In the presence of a particle size distribution, the total average conversion can be calculated by substituting Eqs. (10.94a, 10.94b) into Eqs. (10.93a, 10.93b).

Example 10.3
A gas–solid noncatalytic reaction is carried out in a moving bed reactor. The particle size distribution of the solid reactant in the feed is

Particle Radius/μm	Volume Fraction
50	0.3
100	0.4
200	0.3

During the reaction process, the particles can be treated by the shrinking core model for constant particle size. The fluid flows as a complete mixing flow, and the solid as a plug flow. The total reaction time for the three sizes of particles is 10, 20, and 30 min, respectively. Please calculate the solid conversion at reaction time of 15 min.

Solution
According to the given data, the total reaction time is proportional to the particle size. It can be determined from Table 10.1 that this reaction may be controlled either by the surface chemical reaction or by the external diffusion.
1. Chemical reaction control
    It is known from Eq. (10.42) that

$$1 - x_B(R_i) = \left(1 - \frac{t}{t^*(R_i)}\right)^3$$

(Continued)

(Continued)

Substituting the above equation into Eqs. (10.94a, 10.94b), and then into Eqs. (10.93a, 10.93b) gives

$$1 - \overline{\overline{x}}_B = \sum_{i=1}^{3} \left(1 - \frac{t}{t^*(R_i)} \right)^3 f_i$$

$$= 0^3 \times 0.3 + \left(1 - \frac{15}{20} \right)^3 \times 0.4 + \left(1 - \frac{15}{30} \right)^3 \times 0.3 = 0.044$$

So

$$\overline{\overline{x}}_B = 0.96$$

2. External diffusion control

It is known from Eq. (10.39) that

$$x_B(R_i) = \frac{t}{t^*(R_i)}$$

Similarly,

$$\overline{\overline{x}}_B = \sum_{i=1}^{3} \frac{t}{t^*(R_i)} f_i$$

$$= 1 \times 0.3 + \frac{15}{20} \times 0.4 + \frac{15}{30} \times 0.3 = 0.75$$

It can be seen from the calculation results that under the circumstances that the relationship between the total reaction time and the particle size is linear but the reaction rate-determining step is unknown the solid conversion cannot be determined since its value depends on what is the rate-determining step. Therefore, when studying reaction kinetics, kinetic experiments and analysis have to be conducted properly to eliminate uncertainty in the rate-determining step.

10.9.4 Reactor Design When Fluid and Solid Phases Can Be Treated as Complete Mixing Flow

The residence time distribution density function for a complete mixing flow is

$$E(t) = \frac{1}{\overline{t}} \exp\left(-\frac{1}{\overline{t}} \right)$$

Therefore, the average conversion of uniform particles can be calculated as follows:

$$1 - \overline{x}_B = \int_0^{t^*} \left[1 - x_B(t) \right] E(t) dt = \int_0^{t^*} \left[1 - x_B(t) \right] \frac{1}{\overline{t}} \exp\left(-\frac{t}{\overline{t}} \right) dt \qquad (10.95)$$

Eq. (10.32), derived from the overall macroreaction rate equation, Eq. (10.23), based on the shrinking core model for constant particle size, is an inverse function of x_B and t. So it cannot be directly substituted into Eq. (10.95) to complete the integration. Instead, numerical methods have to be used to solve and integrate nonlinear algebraic equations. For the $x_B \sim t$ relationships under the three rate-determining steps, Eq. (10.95) can be directly integrated to obtain the analytical solution.

1. Internal diffusion control

   Substituting Eq. (10.36) into Eq. (10.95), and then integrating, the following is obtained

$$\bar{x}_B = 1 - \frac{1}{5}\frac{t^*}{\bar{t}} + \frac{19}{420}\left(\frac{t^*}{\bar{t}}\right)^2 - \frac{41}{4620}\left(\frac{t^*}{\bar{t}}\right)^3 + 0.00149\left(\frac{t^*}{\bar{t}}\right) - \cdots \quad (10.96)$$

2. External diffusion control

   Substituting Eq. (10.39) into Eq. (10.95), and then integrating, the following is obtained

$$\bar{x}_B = \frac{\bar{t}}{t^*}\left(1 - \exp\left(-\frac{t^*}{\bar{t}}\right)\right) \quad (10.97)$$

3. Chemical reaction control

   Substituting Eq. (10.42) into Eq. (10.95), and then integrating, the following is obtained

$$\bar{x}_B = 3\frac{\bar{t}}{t^*} - 6\left(\frac{\bar{t}}{t^*}\right)^2 + 6\left(\frac{\bar{t}}{t^*}\right)^3\left(1 - \exp\left(-\frac{t^*}{\bar{t}}\right)\right) \quad (10.98)$$

Example 10.4

Spherical particles of pyrite are dispersed on the asbestos fibers and roasted. The measured total reaction time t^* is proportional to the particle radius R to the power of 1.5, and the process can be treated according to the shrinking core model for constant particle size. Now uniform particles are placed into a fluidized bed reactor to react at the same conditions as the roasting experiment. Assume that both fluid and solid phases can be treated as complete mixing flows and the effect of the external diffusion has been eliminated. The total reaction time is 30 min. Please calculate the average conversion at the reactor outlet when the average residence time of the particles is 50 min.

Solution

It can be seen from $t^* \propto R^{1.5}$ that this reaction is between an internal diffusion-controlled ($t^* \propto R^2$) and a chemical reaction-controlled ($t^* \propto R$), so both internal diffusion and surface reaction have to be considered simultaneously. However, in order to simplify the calculation, we will first calculate separately the

(Continued)

(Continued)

conversions under internal diffusion control and surface reaction control conditions, and then estimate conversion by averaging the results.

For internal diffusion control

$$\overline{x}_B \approx 1 - \frac{1}{5}\frac{30}{50} + \frac{19}{420}\left(\frac{30}{50}\right)^2 - \frac{41}{4620}\left(\frac{30}{50}\right)^3 + 0.00149\left(\frac{30}{50}\right)^4 = 0.89$$

For reaction control

$$\overline{x}_B = 3\frac{50}{30} - 6\left(\frac{50}{30}\right)^2 + 6\left(\frac{50}{30}\right)^3\left(1 - \exp\left(-\frac{30}{50}\right)\right) = 0.87$$

The real conversion is between 0.87 and 0.89, so the average conversion is 0.88.

When there is a particle size distribution, the average conversion can be calculated according to Eqs. (10.93a, 10.93b), i.e.,

$$1 - \overline{\overline{x}}_B = \sum_{i=1}^{m}\left[1 - \overline{x}_B(R_i)\right]f_i$$

$$= \sum_{i=1}^{m}f_i\int_0^{t^*(R_i)}\left[1 - x_{B,R_i}(t)\right]E(t)dt = \sum_{i=1}^{m}f_i\int_0^{t^*(R_i)}\left[1 - x_{B,R_i}(t)\right]\frac{1}{\bar{t}}\exp\left(-\frac{t}{\bar{t}}\right)dt$$

$$(10.99)$$

Example 10.5

The reaction discussed in Example 10.3 takes place in a fluidized-bed reactor shown in Fig. 10.10(C). Except that the flowing of the solid obeys complete mixing flow, the rest of the conditions remain unchanged. Please calculate the average conversion $\overline{\overline{x}}_B$ at the reactor outlet when the average residence time of the solid is 15 min.

Solution

1. Chemical reaction control

From Eqs. (10.98) and (10.93b), it is obtained that

$$\overline{\overline{x}}_B = \sum_{i=1}^{m}f_i\overline{x}_B(R_i) = 0.3 \times \left[3\frac{15}{10} - 6\left(\frac{15}{10}\right)^2 + 6\left(\frac{15}{10}\right)^3\left(1 - \exp\left(-\frac{10}{15}\right)\right)\right]$$

$$+ 0.4 \times \left[3\frac{15}{20} - 6\left(\frac{15}{20}\right)^2 + 6\left(\frac{15}{20}\right)^3\left(1 - \exp\left(-\frac{20}{15}\right)\right)\right]$$

$$+ 0.3 \times \left[3\frac{15}{30} - 6\left(\frac{15}{30}\right)^2 + 6\left(\frac{15}{30}\right)^3\left(1 - \exp\left(-\frac{30}{15}\right)\right)\right]$$

$$= 0.3 \times 0.85 + 0.4 \times 0.74 + 0.3 \times 0.65 = 0.75$$

(Continued)

(Continued)

2. External diffusion control

Substituting Eq. (10.97) into Eq. (10.93b) gives

$$\overline{\overline{x_B}} = \sum_{i=1}^{m} f_i \overline{x_B}(R_i) = 0.3 \times \frac{15}{10}\left(1 - \exp\left(-\frac{10}{15}\right)\right) + 0.4 \times \frac{15}{20}\left(1 - \exp\left(-\frac{20}{15}\right)\right)$$

$$+ 0.3 \times \frac{15}{30}\left(1 - \exp\left(-\frac{30}{15}\right)\right)$$

$$= 0.3 \times 0.73 + 0.4 \times 0.55 + 0.3 \times 0.43 = 0.57$$

Comparing Example 10.5 with Example 10.3, it can be seen that the attainable conversion is higher when the solid particles flow as a plug flow than that when they flow as a complete mixing flow. This is consistent with the conclusions obtained in the design of homogeneous and heterogeneous catalytic reactors.

FURTHER READING

Missen RW, Mins CA, Saville BA. Introduction to chemical reaction engineering and kinetics. New York: John Wiley & Sons, Inc; 1998.

Schmidt LD. The engineering of chemical reactions. Oxford: Oxford University Press; 1998.

Fogler HS. Elements of chemical reaction engineering. 3rd ed. New Jersey: Prentice Hall PTR; 1999.

Levenspiel O. Chemical reaction engineering. 3rd ed. New York: John Wiley & Sons, Inc; 1998.

Butt JB. Reaction kinetics and reactor design. 2nd ed. New York: Marcel Dekker, Inc; 1999.

Smith J. Chemical engineering kinetics. 3rd ed. New York: McGraw-Hill, Inc; 1990.

Ge QR. Gas-solid reaction kinetics. Beijing: China Atomic Energy Publishing & Media Co., Ltd.; 1991.

Xu HQ. Gas-solid reaction engineering. Beijing: China Atomic Energy Publishing & Media Co., Ltd.; 1993.

PROBLEMS

10.1 For a first order irreversible gas—solid noncatalytic reaction

$$A_{(g)} + bB_{(s)} \rightarrow dD_{(g)} + pP_{(s)}$$

the reactant particles are long cylinders, the particle size does not change during the reaction, and the process is limited by internal diffusion. Please use the shrinking core model to derive the relationship between the conversion x_B of the solid reactant B and the reaction time t.

10.2 The spherical particles of pure B of radius 6 mm are added to a rotary calciner to be heated to react. The gas component reacting with the solid is uniformly distributed in the furnace. The whole process can be considered as an isothermal. It is a first order irreversible reaction for A, and the reaction is conducted according to the following stoichiometric equation:

$$A_{(g)} + B_{(s)} \rightarrow dD_{(g)} + pP_{(s)}$$

B is dense and the product layer is porous. In the furnace, the solid moves at a speed of 2.5 mm/s from the inlet to the outlet in a plug flow. The effect of the external diffusion is negligible.

In a batch stirred tank, at the same gas composition and temperature, the following data are obtained:

Particle Radius, mm	Reaction Time, h	Conversion of B, x_B (%)
3	1.0	87.5
6	2.0	65.7

If the required conversion at the outlet of the rotary calciner, $x_B = 90\%$, please calculate the required furnace length.

10.3 A solid-fluid noncatalytic reaction test is conducted in a small laboratory fluidized bed reactor. The study subject is spherical particles of $\phi = 1.0$ mm. During the reaction, the particle size does not change, the concentration of the gaseous component is uniform, the solid flows in the reactor as a complete mixing flow, and the surface reaction is the rate-determining step. When the feeding rate of the solid is 1.0 kg/min, the amount of solids retained in the reactor is 1.0 kg, and the average conversion of the solid reactant at the reactor outlet is 80%.

It is intended to use the shrinking core model for constant particle size to scale up the processing capacity of the solid reactant to 10 t/h. The particle size, the concentration of the reaction component, the flow pattern, and the reaction temperature are the same as the experimental conditions. Please calculate the required reactor capacity, indicated by the solid retention amount, when the average conversion at the reactor outlet is 95%.

10.4 The particle size distribution of a spherical solid reactant is as follows

Particle Radius, mm	Mass Fraction, %
0.5	10
1.0	30
1.5	50
2.0	10

By single particle experiments at constant temperature and gas phase concentration, the total reaction time for particles of radius 1.5 mm is 1.0 h, and the reaction follows the shrinking core model for constant particle size under internal diffusion control.

Under the same reaction conditions, the conversion measured in a model test reactor is 80%. Now the reaction volume is increased 20 times. The flow pattern in the reactor remains unchanged. Please calculate:

1. In both model and scale-up tests, a moving bed reactor is used. The solid flows following the plug flow model. Please calculate the shortest reaction time needed for 100% conversion of the solid reactant.

2. In both model and scale-up tests, a fluidized bed reactor is used. The solid flows following the complete-mixing model. Please calculate the reaction time needed for an average conversion of 75% for the solid reactant.

3. Please compare the calculation results of the above two questions.

10.5 Natural gas desulfurization can be achieved by the following reaction in a fixed bed reactor loaded with zinc oxide particles,

$$H_2S_{(g)} + ZnO_{(s)} \rightarrow ZnS_{(s)} + H_2O_{(g)}$$

The H_2S content in the feed is 20 mg/kg. After desulfurization it needs to be reduced to 0.2 mg/kg. The reaction is conducted isothermally at 344°C under atmospheric pressure. The gas handling capacity is 3000 kmol/ h. Assuming that the reaction is irreversible and first order to ZnO, please calculate

1. The annual loading of ZnO in cubic meters if the operation time of the reactor is 8000 h per year.

2. The ZnO conversion after one year of reaction.

10.6 The reduction of Fe_3O_4 in H_2 is carried out in a moving-bed reactor at 750°C. The solid reactant flows in a plug-flow, and the gas in a complete mixing flow. Please determine the amount of the solid reactant in the reactor needed to meet the required processing capacity for full reduction of 1000 kg/min Fe_3O_4. Fe_3O_4 is spherical particles of radius 3 mm. The particle density is $\rho_B = 20000$ mol/m³. It is assumed that the reaction is a first order irreversible reaction, the reaction rate constant $k = 1930\exp(-12,100/T)$ m/s, the hydrogen partial pressure is 0.1 MPa, the particle size does not change during the reaction, and the reaction can be treated according to the shrinking core model. The gas effective diffusion coefficient in the product layer is $D_e = 4 \times 10^{-7}$ m²/s, and the effect of the external diffusion has been eliminated.

10.7 Spherical particles with diameter 2.0 mm participate the following reaction:

$$A_{(g)} + 2B_{(s)} \rightarrow D_{(g)} + E_{(s)}$$

During the reaction the particle size decreases. Please calculate the conversion of B at reaction time 20 h according to the shrinking core model with changes in particle size. Assuming:
1. Reaction control
2. Internal diffusion control
3. External diffusion control
4. The three resisting steps act simultaneously

Known: For the two solid reactants $\rho_B/M_B = 1.5$, $\rho_p/M_p = 1.5$, the gas film mass transfer coefficient is $k_g = 9.0 \times 10^{-5}$ m/s, at 450°C the reaction is an irreversible first order reaction $r_A = 8.5 \times 10^{-5} p_A$ m/s, the gas effective diffusion coefficient in the product layer $D_e = 8.5 \times 10^{-7}$ m²/s, the gas partial pressure of component A is 0.2 MPa.

10.8 The following reaction is used for production of boron nitride films via chemical vapor deposition at 1100°C:

$$BF_{3(g)} + NH_{3(g)} \rightarrow BN_{(s)} + 3HF_{(g)}$$

It is assumed that the rate of increase in film thickness obeys

$$r_{BF_3} = 0.01 C_{BF_3} \, m/min$$

The partial pressure of BF_3 in the gas phase is 0.05 MPa. The density of BN is 2.1×10^6 g/m³. Assuming that the reaction is
1. Reaction control
2. Gas film mass transfer control, mass transfer coefficient $k_{g=}2.6 \times 10^{-5}$ m/s

Please calculate the reaction time needed for film thickness to increase by 30 μm.

10.9 The following reaction is used for chemical vapor deposition at a certain temperature:

$$WF_{4(g)} + 2H_{2(g)} \rightarrow W_{(s)} + 4HF_{(g)}$$

Assuming that the reaction includes the following elementary steps:

Adsorption of WF₄	$WF_{4(g)} + S \overset{k_W}{\rightleftharpoons} WF_4S$	
Dissociative adsorption of H₂	$H_{2(g)} + 2S \overset{k_H}{\rightleftharpoons} 2HS$	
Rate-controlling step	$WF_4S + HS \xrightarrow{k_1} WF_3S + HF_{(g)} + S$	
	$WF_3S + HS \xrightarrow{k_2} WF_2S + HF_{(g)} + S$	
	$WF_2S + HS \xrightarrow{k_3} WFS + HF_{(g)} + S$	
	$WFS + HS \xrightarrow{k_4} W_{(s)} + HF_{(g)} + 2S$	

(Chemical reaction)

Please derive the rate expression of Tungsten deposition.

Chapter 11

Fundamentals of Biochemical Reaction Engineering

Chapter Outline

11.1 INTRODUCTION

Biochemical reaction engineering is a branch of biochemistry engineering, and is one of the key technologies for commercializing biotechnology.

Biotechnology is a technique that applies principals of biology, chemistry, and engineering and uses living organisms including microorganisms, animal cells, and plant cells or their components, such as organelles and enzymes, to make useful products or provide certain services for human beings. Biotechnology utilizes the features of biotransformation to manufacture chemical products that are not easily achievable with traditional chemical methods, or to improve existing processes, which help to solve the two long-standing problems: energy crisis and environmental pollution. In recent years, with dramatic developments in modern biotechnology including achievements in gene engineering, cell engineering, protein engineering, enzyme engineering, as well as biochemical engineering, biotechnology applications have attracted more and more attention, and some have already been commercialized. Biotechnology can only benefit human beings and achieve its social and economic impact through commercial application. Biochemical engineering is one of the keys for biotechnology commercialization and biochemical reaction engineering is a branch of biochemical engineering.

Reaction Engineering. DOI: http://dx.doi.org/10.1016/B978-0-12-410416-7.00011-2

From the perspective of chemical engineering, biochemical reaction engineering is a branch of chemical reaction engineering, and is also one of the frontiers in chemical reaction engineering research.

Any kind of biochemical reaction is a process that uses a biocatalyst to make biotechnological products. Biocatalysts can be categorized into two groups: (1) free enzyme or immobilized enzyme and (2) free cell or immobilized cell. A biochemical reaction process is called fermentation when microorganism cells are used as biocatalysts, e.g., different kinds of amino acids, antibiotics, organic acids, vitamins, and methane fermentation. When an enzyme is used as the biocatalyst, the process is called enzyme-catalyzed reaction engineering, and examples include conversion of acrylonitrile and water to acrylamide under acrylonitrile hydratase and dehydrogenation of hydrocotisone to hydroprednisone under hydrocortisone dehydrogenase. In addition, the biochemical reaction process also includes large-scale plant cell and animal cell culture as well as wastewater biotreatment processes.

In general, a biochemical reaction includes three processes: (1) raw material treatment; (2) biocatalyst preparation and biochemical reaction; and (3) product separation and purification. The schematics of these processes are shown in Fig. 11.1. The second part is the core of biochemical reaction process and it is the object of biochemical reaction engineering studies.

The basic components of biochemical reaction engineering can be summarized into two aspects, i.e., biochemical reaction kinetics and bioreactor design and analysis. The difference between biochemical reaction engineering and chemical reaction engineering is that the specific objects of study are different. The former is a process of biochemical reaction and the latter is a process of chemical reaction. So, the two processes have both similarity and individuality. Biochemical reaction engineering is a branch of science that applies the principles and methods of chemical reaction engineering to biochemical reactor design and analysis and to determine the optimal operating conditions.

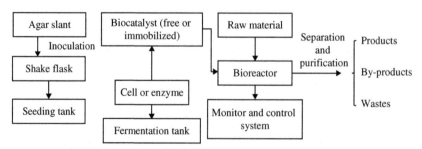

FIGURE 11.1 Schematic diagram of general biochemical reaction processes.

Compared to chemical reaction engineering, biochemical reaction engineering has its specific features that can be categorized as below:

1. Most biocatalysts are complicated except for the single-enzyme system. In general, the multienzyme system of microorganism cells is complicated: cells are various, and morphological forms and physiological characteristics are different too. No matter whether it is a microbial cell, animal cell, or plant cell, they all need to continually absorb nutrition from the external environment, obtain energy and living materials necessary for living through a series of biochemical reaction processes, and discharge metabolites. The mechanism of cell metabolism is very complicated and is often affected by environmental factors such as fluid mechanics and physical chemical properties. Therefore, there are many unmeasured and undetermined factors, which make the study of biochemical reaction kinetics very difficult.

2. A biochemical reaction has mild reaction conditions, strong catalysis specificity and high selectivity. In general, biochemical reactions proceed at room temperature and ambient pressure and usually have low energy consumption, which brings great convenience for biochemical reactor design and operation.

3. The biochemical reaction system is usually complicated. Even for dissociative microbial fermentation, it is always a system of gas, liquid, and solid. The enzymatic catalysis in organic medium that is more widely used in chemical engineering is also a multiphase system. In many of the biochemical reaction systems, liquid is always a viscous non-Newtonian fluid. Its rheological property and other physical properties always change with the progress of biochemical reactions, which will affect the mixing and transport of materials during the reaction. In addition, due to the small size of the cells, and small difference in density between cells and surrounding fluid, the relative movement between the liquid and the cells are small. In addition, cell agglomeration increases the resistance for internal diffusion, which is not good for the mass and momentum transfer between liquid and solid phases. This will bring difficulties to biochemical reactor design and operations.

4. Biocatalysts are subject to the influence of environment and infectious microbes and can even be deactivated. So, biochemical reactions must be operated at sterile conditions. Before the reaction starts, equipment, pipes, and materials must be sterilized, and caution must be executed to prevent the substrate nutrition loss. In addition, biochemical reactions are affected by factors such as pH, temperature, and dissolved oxygen. In order to maintain the activity and life of biocatalyst and yield of the objective product, process conditions have to be more strictly controlled and reactor structure needs to be carefully designed.

5. Although enzymatic catalysis is far more efficient than conventional chemical catalysis, the rate of biochemical reaction is usually limited by substrate (reactant) and product concentrations. It always needs to be operated at lower concentrations and at room temperature and ambient pressure, which slows down the reaction rate and extends the residence time of the materials in the reactor. It thus needs larger reactors. Therefore, scale-up of biochemical reactors is more important.

This chapter will briefly discuss biochemical reaction kinetics and biochemical reactor design and analysis based on principles of the chemical reaction engineering described previously.

11.2 FUNDAMENTALS OF BIOCHEMICAL REACTION KINETICS

Biochemical reaction kinetics studies biochemical reaction and its effect factors. Biochemical reactions include enzyme-catalyzed reaction, microbial reaction, biotreatment of wastewater as well as large-scale culture of animal and plant cells, etc. Kinetics of different processes vary a lot. Due to space limitations, this section only focuses on two basic biochemical reaction kinetics, i.e., enzyme-catalyzed reaction kinetics and microbial reaction kinetics.

11.2.1 Enzyme-Catalyzed Reactions and Its Kinetics

11.2.1.1 Characteristics of Enzymes

An enzyme is a special protein produced by living cells that has catalytic activity and high selectivity. Enzymes can be categorized as simple proteins and binding proteins based on their components. For example, most hydrolytic enzymes belong to enzymes composed of simple protein, while Flavin mononucleotide enzyme is a binding protein composed of zymoprotein and cofactor. The zymoprotein in a binding protein is a protein and the cofactor is nonprotein substance. The binding of zymoprotein and cofactor forms a holoenzyme. Only the holoenzyme has catalytic activity. Although enzymes come from living organisms, enzymes can exist independently from living organisms.

The International Union of Biological Sciences uses the systematic classification system by the enzyme committee to classify enzymes. Enzymes can be classified into six categories based on the forms of the catalytic reactions, i.e., oxidation-reduction enzyme, transferase, hydrolase, lyase, isomerase, and synthetase.

Enzymes not only participate in all kinds of metabolic reactions inside the living body, they also participate in the biochemical reactions outside of the living body. Enzymes not only have properties of conventional catalysis, they also have the characteristics of proteins. Similar to conventional catalysis, enzymes will not change the free energy of the reaction, and thus will not change the equilibrium of the reaction; they can only reduce the activation energy, accelerate the rate to reach equilibrium, and increase the reaction rate. At the end of the reaction, an enzyme itself is not consumed, and is recovered to its original status with no change in quantity and property as long as its protein characteristics did not change.

Let's take the enzyme reaction of single substrate forming product as an example. The reaction mechanism is

$$E + S \leftrightarrow [ES] \rightarrow P + SE$$

The first step is that an enzyme is combined with a substrate by covalent bond to form an intermediate complex [ES]. This process requires that the substrate has enough energy to jump onto a high energy level complex; the second step is that the high energy complex releases energy, forms a product, and releases the enzyme.

Enzymatic catalysis can reduce the activation energy needed between substrate and activated transition state complex, but it will not change the total energy during the reaction. Energy for an enzymatic catalysis and conventional catalysis is compared in Fig. 11.2, which shows that the activation energy of an enzymatic catalysis is much lower than that of conventional catalysis. Therefore, enzymatic catalysis efficiency is much higher than that of conventional catalysis. Catalysis efficiency of enzymes can be described with the activity of enzymes, i.e., the reaction rate of enzymatic catalysis. In general, the number of micromoles of the substrate converted by one mole of enzyme in one minute at the set conditions is used to represent enzyme activity. The amount of enzyme needed to convert one micromole of substrate to product is defined as one enzyme unit.

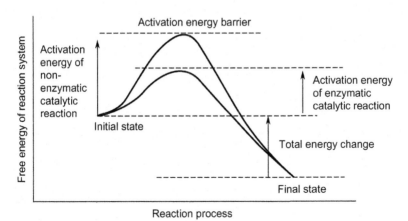

FIGURE 11.2 Diagram of energy changes in enzyme-catalyzed reactions and chemical catalytic reactions.

Compared to catalysis involved in conventional chemical reactions, enzymatic catalysis has the following features:

1. Enzymatic catalysis has high efficiency. In general, the efficiency of enzymatic catalysis is 10^7-10^{13} times higher than the nonenzymatic catalysis; some enzymatic catalysis efficiency, such as urease hydrolyzing urea reaction, can be up to 10^{14} times more efficient.

2. Enzyme-catalyzed reactions have exceptionally high specificity, which includes reaction specificity and substrate specificity. This specificity is determined by the zymoprotein, especially the structure characteristics of its active site. The specificity of enzyme to reaction means that one enzyme can only catalyze one particular reaction of one compound among all thermodynamically feasible reactions. The specificity of enzyme to substrate includes specificity to the structure of the substrate and the stereoscopic specificity. The former specificity means that one enzyme can only catalyze one substrate or one type of substrate and the latter specificity means that one enzyme can only catalyze one stereomer of substrate. When the substrate has an optical isomer, the enzyme can only catalyze one of the optical isomers. In addition, some enzymes have group specificity, i.e., to say they can only catalyze specific groups. For example, alcohol dehydrogenase can only work for alcohols and the alcohol has to be a primary alcohol. Therefore, one enzyme can only catalyze one substrate for a specific reaction.

   Due to high specificity of enzyme, enzyme-catalyzed reactions have very high selectivity and produce fewer by-products, which makes product separation easy. Thus, many reactions unfeasible by conventional catalysis can be accomplished by enzyme catalysis.

3. Enzyme-catalyzed reactions are usually conducted at mild conditions and do not require high temperature and high pressure. Moreover, an enzyme is a protein, and it does not allow a high temperature. Therefore, enzyme-catalyzed reactions are conducted at room temperature and pressure.

4. Enzyme-catalyzed reactions have their feasible temperature, pH, dielectric constant, and ionic strength. As soon as the conditions become inappropriate, the enzyme will change characteristics or even lose activity. This is determined by the fact that an enzyme is a protein. Enzymes are very easily affected by physical and chemical factors (e.g., heat, pressure, ultraviolet rays, acids, bases, and heavy metals) and change their properties. This denaturation will reduce the activity of an enzyme or even result in activity loss, and such deactivation is usually irreversible.

Many factors will affect the reaction rate of enzymatic catalysis, which includes concentration of enzyme, concentration of substrates, concentration of products, temperature, pH values, ionic strength, inhibitors, etc.

11.2.1.2 Single-Substrate Enzyme Kinetics (Michaelis–Menten Equation)

For a typical single-substrate enzyme-catalyzed reaction, e.g.,

$$S \xrightarrow{\ E\ } P$$

The mechanism of reaction can be expressed as

$$E + S \underset{\overleftarrow{k_1}}{\overset{\overrightarrow{k_1}}{\rightleftharpoons}} [ES] \xrightarrow{\ k_2\ } E + P$$

The first step is a reversible reaction, and reaction rate constants of forward and backward reactions are $\overrightarrow{k_1}$ and $\overleftarrow{k_1}$, respectively. The second step is usually irreversible with a reaction rate constant of k_2. Experimental data showed that reaction rate increases with the substrate concentration and when substrate concentration reaches a certain value, reaction rate approaches a constant as shown in Fig. 11.3.

The Michaelis–Menten equation derived by the rapid equilibrium method by Michaelis–Menten or the pseudo-equilibrium assumption by Briggs–Haldane quantitatively describes the relationship between reaction rate and substrate concentration

$$r = \frac{dc_s}{dt} = \frac{dc_p}{dt} = \frac{r_{max} \cdot c_s}{K_m + c_s} \tag{11.1}$$

where c_s is the concentration of substrate s, r_{max} is the maximum reaction rate, c_{E0} is the initial concentration of enzyme, and K_m is usually called the Michaelis constant.

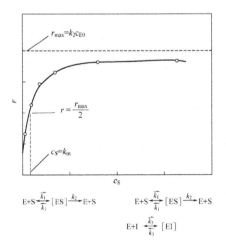

FIGURE 11.3 Relationship between substrate concentration and enzyme-catalyzed reaction rate.

The Michaelis–Menten equation is a hyperbolic function as shown in Fig. 11.3. At given initial enzyme concentrations, the reaction orders are different at different substrate concentrations. When $c_s \ll K_m$, substrate concentration is very low and the reaction is first order. When $c_s \gg K_m$, substrate concentration is very high and the reaction is zero order, i.e., reaction rate is independent of c_s. When c_s is in the middle range, reaction rate is mixed order and reaction rate transits from first order to zero order as substrate concentration increases.

r_{max} and K_m are two important kinetics parameters in the Michaelis–Menten equation. r_{max} represents the maximum reaction rate when almost all enzymes were combined with substrate to form an intermediate complex. The K_m value characterizes the affinity between the substrate and the enzyme. A low K_m value means high affinity between enzyme and substrate and more difficult for [ES] dissociation. A high K_m represents the opposite conditions. The value of K_m depends on the characteristics of the enzyme-catalyzed reaction system and the reaction conditions, so K_m is an intrinsic parameter representing the characteristics of the enzyme-catalyzed reactions. K_m is the substrate concentration required for an enzyme to reach half of its maximum reaction rate r_{max}.

Kinetics parameters r_{max} and K_m can be determined using the differential method or integration method described in Chapter 2, Fundamentals of Reaction Kinetics. Since the Michaelis–Menten equation is a hyperbolic function, it should be linearized first, then the parameters can be estimated by the graphic or linear least square method. There are three differential methods that are commonly used.

1. Lineweaver–Burk method. Plotting $1/r$ versus $1/c_s$ yields a straight line with a slope of K_m/r_{max}, an intercept on the ordinate at $1/r_{max}$ and an intercept on the abscissa at negative $1/K_m$. This method is referred to as the L-B method or double reciprocal graphical method
2. Hanes–Woolf method. Plotting c_s/r versus c_s yields a straight with a slope of $1/r_{max}$ and an intercept on the abscissa at negative K_m
3. Eadie–Hofstee method. Plotting r versus r/c_s yields a straight line with a negative slope of K_m and an intercept on the ordinate at r_{max}

The kinetics experiments of enzyme-catalyzed reactions can be conducted in a batch reactor, continuous stirred tank reactor (CSTR), and plug flow reactor. Experimental data can be fit with the methods described above to get the kinetics parameters.

Example 11.1

At pH 5.1 and constant temperature of 15°C, the initial reaction rate of maltose hydrolysis by glucamylase was measured as a function of maltose concentration as shown in the Table. Determine K_m and r_{max} of this hydrolysis reaction.

Maltose concentration c_s x 10^3/(mol/L)	11.5	14.4	17.2	20.1	23.0	25.8	34.0
Maltose hydrolysis rate r x10^4/[mol/(L · min)]	2.50	2.79	3.01	3.20	3.37	3.50	3.80

Solution

Linearize the Michaelis–Menten equation with Lineweaver–Burk method. After rearrangement, we got

$$\frac{1}{r} = \frac{1}{r_{max}} + \frac{K_m}{r_{max}}\frac{1}{c_s} \tag{A}$$

Eq. (A) shows that $1/r$ and $1/c_s$ has a linear correlation. $1/r$ and $1/c_s$ values can be calculated from the given data. Results are listed in Table 11A. Substituting data from Table 11A into Eq. (A), regressing with linear least square method gives:

$r_{max} = 1.94 \times 10^3$; $K_m/r_{max} = 23.7$
The correlation coefficient $r = 0.9999$. Then

$$\frac{1}{r_{max}} = 1.94 \times 10^3 \quad \frac{K_m}{r_{max}} = 23.7$$

$$r_{max} = 5.16 \times 10^{-4}\,mol/(L \cdot min) \quad r_{max} = K_m = 1.22 \times 10^{-2}\,mol/L$$

TABLE 11A Results from linearizing the Michalelis-Menten equation with the Lineweave-Burk method

(1/c_s)/(L/mol)	87.0	69.4	58.1	49.8	43.6	38.8	29.4
(1/r) × 10^{-3}/[(L · min)/mol]	4.00	3.58	3.32	3.13	2.97	2.86	2.63

11.2.1.3 Kinetics of Enzyme-Catalyzed Reactions With Inhibition

For enzyme-catalyzed reactions, the presence of some materials will decrease the reaction rate. These materials are called inhibitors and this effect is called inhibition. Inhibitors can be foreign materials, the substrate or products of the reactions. Inhibitors can only affect limited functional groups in the enzyme molecules and they will not change the three-dimensional structure of the enzyme. So, the presence of inhibitors will affect the reaction rates.

Enzyme inhibition falls into two classes, i.e., reversible inhibition and irreversible inhibition. When the inhibitor interacts with an enzyme via covalent associations, the inhibition is called an irreversible inhibition. An irreversible inhibitor reduces the concentration of active enzyme and the enzyme is deactivated completely when the inhibitor concentration exceeds the enzyme concentration. When inhibitors interact with an enzyme via non-covalent associates, the inhibition is called reversible inhibition. The inhibitor maintains dissociation equilibrium with enzyme in the reversible inhibition. For reversible inhibition, inhibitors can be removed with physical methods like dialysis and enzyme activity is regenerated. Based on the inhibition mechanisms, reversible inhibitions can be divided into three categories: competitive inhibition; noncompetitive inhibition; and uncompetitive inhibition.

1. Competitive inhibition. When inhibitors have similar structures as the substrate, they will compete for the same binding position—the active site—which prevents bonding of the substrate with the enzyme, and therefore will reduce the enzyme-catalyzed reaction rate. This kind of inhibition is called competitive inhibition. Letting I represent the inhibitor, the mechanism is

$$\mathrm{E} + \mathrm{S} \underset{\overset{\leftarrow}{k_1}}{\overset{\overset{\rightarrow}{k_1}}{\rightleftharpoons}} [\mathrm{ES}] \xrightarrow{k_2} \mathrm{E} + \mathrm{P}$$

$$\mathrm{E} + \mathrm{I} \underset{\overset{\leftarrow}{k_3}}{\overset{\overset{\rightarrow}{k_3}}{\rightleftharpoons}} [\mathrm{EI}]$$

Based on steady-state approximation and $c_{\mathrm{E0}} = c_{[\mathrm{E}]} + c_{[\mathrm{ES}]} = c_{[\mathrm{EI}]}$, kinetics equation of competitive inhibition can be derived as

$$r = \frac{r_{\max} \cdot c_s}{K_m \left(\frac{1 + c_I}{K_I}\right) + c_s} = \frac{r_{\max} \cdot c_s}{K_{mI} + c_s} \tag{11.2}$$

where $K_m = (\overset{\rightarrow}{k_1} + k_2)/\overset{\rightarrow}{k_1}$ is called Michaelis constant, $K_I = \overset{\rightarrow}{k_3}/\overset{\leftarrow}{k_3}$ is the dissociation constant of complex [EI], $r_{\max} = k_2 c_{\mathrm{E0}}$ is the maximum reaction rate, and $K_{mI} = K_m/(1 + c_I/K_I)$ is Michaelis constant in the presence of competitive inhibition.

Apparently, smaller K_I indicates stronger interaction between inhibitor and enzyme and stronger inhibition to the reaction. In this case, reaction rate can be improved by increasing substrate concentrations.

According to the Lineweaver–Burk method, linearizing Eq. (11.2), rearranging experimental data, and plotting $1/r$ versus $1/r_s$ yields a straight line as shown in Fig. 11.4. The slope of the line is $K_{mI}/r_{\max}$, the intercept on the ordinate is $1/r_{\max}$, and the negative intercept on the

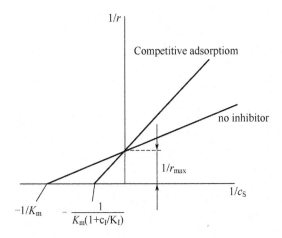

FIGURE 11.4 L-B plot of competitive inhibition.

abscissa is $1/K_{mI}$. Model parameters K_{mI} and r_{max} can be calculated from these values.

When products and substrates have similar structures, the product and enzyme form a complex, which prevents enzyme binding with substrate, and thus reduces the enzyme-catalyzed reaction rate.

2. Noncompetitive inhibition. Some inhibitors usually bind with a nonactive site of the enzyme, and then the enzyme−inhibitor complex binds with the substrate; or the enzyme combines with the substrate to form an enzyme−substrate complex, part of the complex then combines with inhibitors. Although the substrate and inhibitor's binding with enzyme has no competition, the complex they formed with the enzyme cannot form products directly, which reduces the rate of enzyme-catalyzed reactions. Letting I represent the noncompetitive inhibitor, the mechanism is:

$$
\begin{array}{ccc}
\text{E} +\text{S} \underset{\overleftarrow{k_1}}{\overset{\overrightarrow{k_1}}{\rightleftharpoons}} [\text{ES}] \xrightarrow{k_2} \text{E+P} \\
+ \qquad\quad + \\
\text{I} \qquad\quad \text{I} \\
\overleftarrow{k_3} \Big\Vert \overrightarrow{k_3} \qquad \overrightarrow{k_4} \Big\Vert \overleftarrow{k_4} \\
[\text{EI}] + \text{S} \underset{\overleftarrow{k_5}}{\overset{\overrightarrow{k_5}}{\rightleftharpoons}} [\text{SEI}]
\end{array}
$$

Based on steady-state approximation and $c_{E0} = c_{[E]} + c_{[ES]} + c_{[EI]} + c_{[SEI]}$, the kinetics equation of noncompetitive inhibition can be derived as

$$
r = \frac{r_{max}c_s}{\left(\frac{1+c_I}{K_I}\right)(K_m + c_s)} = \frac{r_{I,max}c_s}{K_m + c_s} \tag{11.3}
$$

where $K_m = (\overleftarrow{k_1} + k_2)/\overrightarrow{k_1}$; $r_{max} = k_2 c_{E0}$; and $r_{I,max} = r_{max}/(1 + \frac{c_I}{K_I})$ is the maximum reaction rate with noncompetitive inhibition.

Apparently, due to the presence of the noncompetitive inhibitor, reaction rate was reduced. The maximum reaction rate r_{max} is only $1/(1 + c_I/K_I)$ of that when there is no inhibitor. In this case, even increasing the concentration of substrate will not weaken the effect of the inhibitor on reaction rate. This is the difference between competitive inhibition and noncompetitive inhibition.

Using the Lineweaver–Burk method, Eq. (11.3) can be linearized as:

$$\frac{1}{r} = \frac{\left(1 + \dfrac{c_I}{K_I}\right)}{r_{max}} + \frac{\left(1 + \dfrac{c_I}{K_I}\right)K_m}{r_{max}} \cdot \frac{1}{c_s} \tag{11.4}$$

Plotting $1/r$ versus $1/c_s$ yields a straight line as shown in Fig. 11.5. Model parameters such as K_I, r_{max}, and K_m can be estimated using the slope and intercept on the ordinate and abscissa of the plot.

3. Uncompetitive inhibition. Some inhibitors cannot bind to free enzymes directly, and can only bind with a substrate enzyme complex to form substrate–enzyme–inhibitor complex. The substrate–enzyme–inhibitor complex cannot form products, and as a result the reaction rate decreases. This phenomenon is called uncompetitive inhibition. The mechanism of this inhibition can be described as

$$E + S \underset{k_1}{\overset{\overrightarrow{k_1}}{\longleftrightarrow}} [ES] \overset{k_2}{\longrightarrow} E + P$$

$$[ES] + I \underset{k_3}{\overset{\overrightarrow{k_3}}{\longleftrightarrow}} [SEI]$$

Total concentration of enzyme is $c_{E0} = c_E + c_{[ES]} + c_{[SEI]}$

Kinetics equation of uncompetitive inhibition can be derived based on steady-state approximation

$$r = \frac{r_{max} \cdot c_s}{K_m + \left(\dfrac{1 + c_I}{K_I}\right) \cdot c_s} = \frac{r_{I,max} \cdot c_s}{K'_{mI} + c_s} \tag{11.5a}$$

where $r_{I,max} = r_{max} / \left(1 + \dfrac{c_I}{K_I}\right)$; $K'_{mI} = K_m / \left(1 + \dfrac{c_I}{K_I}\right)$

Based on the L-B method, Eq. (11.5a) is linearized as

$$\frac{1}{r} = \frac{K_m}{r_{max}} \cdot \frac{1}{c_s} + \frac{1}{r_{max}}\left(1 + \frac{c_I}{K_I}\right) \tag{11.5b}$$

Rearranging experimental data, plotting $1/r$ versus $1/c_s$ yields a straight line as shown in Fig. 11.6. Kinetics parameters K_m, r_{max}, and K_I can be estimated from the numbers in the figure and the experiment data.

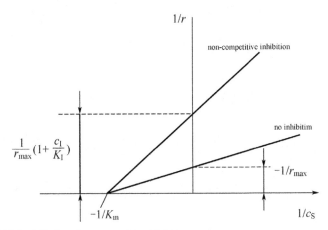

FIGURE 11.5 L-B plot of noncompetitive inhibition.

4. Substrate inhibition. For some enzyme-catalyzed reactions, the relationship between reaction rate and substrate concentration is not a hyperbolic function but a parabolic function as shown in Fig. 11.7.

At the beginning of the reaction, reaction rate increases with substrate concentration; however, after the maximum reaction rate is reached, reaction rate will decrease as substrate concentration increases. This reaction rate reduction caused by high substrate concentration is called substrate inhibition. The reason is that multiple substrate molecules bind to active sites of the enzyme, but the complex they form cannot further decompose to products. Assuming [ES] combines with another substrate molecule, the mechanism is

$$E+S \underset{\overleftarrow{k_1}}{\overset{\overrightarrow{k_1}}{\rightleftharpoons}} [ES] \overset{k_2}{\longrightarrow} P+E$$
$$+$$
$$S$$
$$\overrightarrow{k_3} \Big\Vert \overleftarrow{k_3}$$
$$[SES]$$

The kinetics equation of substrate inhibition can be derived with the method described previously

$$r = \frac{r_{max} \cdot c_s}{K_m + c_s \left(\dfrac{1 + c_s}{K_s} \right)} \tag{11.6}$$

where $K_s = \overleftarrow{k_3}/\overrightarrow{k_3}$ is the dissociation constant of substrate inhibition.
Differentiating Eq. (11.6) with respect to c_s and setting $dr/dc_s = 0$ gives

$$c_{s,opt} = \sqrt{K_m K_s} \tag{11.7}$$

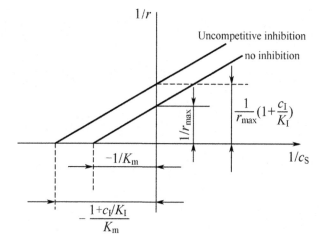

FIGURE 11.6 L-B plot of uncompetitive inhibition.

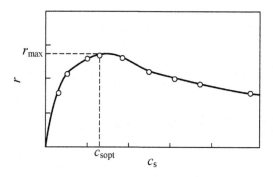

FIGURE 11.7 r_p–c_s curve at substrate inhibition.

This is the optimal substrate concentration, at which the reaction rate reaches its maximum value.

11.2.1.4 Factors Affecting Reaction Rate of Enzyme-Catalyzed Reactions

There are many factors that can affect the reaction rate of enzyme-catalyzed reactions. Besides previously described enzyme concentration, substrate concentration, product concentration, and inhibitor concentrations, temperature, pH value, and ionic strength will affect reaction rate too.

1. Temperature effect. Temperature can have both a positive and negative impact on the reaction rate of enzymatic reactions. On one hand, reaction rate increases with temperature and the dependence of reaction rate on temperature follows the Arrhenius equation. On the other hand, extremely

high temperature will denature zymoprotein and decrease the activity of the enzyme or even deactivate the enzyme. Due to these two effects, enzymatic reactions have optimal temperatures, and the reaction rate versus temperature plot exhibits a bell shape.

2. pH effect. For any enzymatic reactions, zymoprotein, substrate, and intermediate complex have specific dissociation forms and a most favorable pH value, at which reaction rate is maximized. Extremely high or low pH will affect the dissociation of the acidic group, such as carboxyl, or basic group, such as amide, in the active site of the enzyme, reduce the activity of enzyme, or even destroy the structure of enzyme and affect the stability of enzyme.

Therefore, for enzyme-catalyzed reactions, substrate concentrations, product concentrations, temperature, and pH values need to be controlled appropriately.

Example 11.2
An enzyme-catalyzed reaction was carried out at room temperature. At given constant initial enzyme concentration, the initial reaction rate at different substrate concentrations were measured and are listed in Table 11B. Inhibitor was added under the above conditions at a concentration of 1.00×10^{-5} mol/L and initial reaction rate r_{sI} was measured at different substrate concentrations. Results are shown in Table 11B. Please determine the type of inhibition and calculate the kinetic parameters K_m, r_{max}, and K_I based on the experimental data.

TABLE 11B Initial Reaction Rates at Different Substrate Concentrations

$c_s \times 10^3/(\text{mol/L})$	0.333	0.400	0.500	0.667	1.00	2.00
$r_s \times 10^6/(\text{mol} \cdot \text{L/min})$	55.6	62.9	73.0	87.0	107	139
$r_{sI} \times 10^6/(\text{mol} \cdot \text{L/min})$	35.5	41.2	48.8	60.2	78.7	113

Solution
Assuming inhibition is competitive, linearizing Eqs. (11.1) and (11.2) and rearranging gives

$$\frac{1}{r_s} = \frac{1}{r_{max}} + \frac{K_m}{r_{max}} \cdot \frac{1}{c_s} = a + b \cdot \frac{1}{c_s} \tag{A}$$

$$\frac{1}{r_{sI}} = \frac{1}{r_{max}} + \frac{K_{mI}}{r_{max}} \cdot \frac{1}{c_s} = a' + b' \cdot \frac{1}{c_s} \tag{B}$$

From Eqs. (A) and (B), we can tell that both $1/r_s \sim 1/c_s$ and $1/r_{sI} \sim 1/c_s$ are straight lines with the same intercept on ordinate at $1/r_{max}$.

$1/c_s$, $1/r_s$, and $1/r_{sI}$ are calculated with the data in Table 9C, and results are listed in Table 11C

(Continued)

TABLE 11C Results calculated with the data in Table 9C

$1/c_s \times 10^{-3}/$(L/mol)	3.00	2.50	2.00	1.50	1.0	0.5
$1/r_s \times 10^{-3}/$(min·L/mol)	18.0	15.9	13.7	11.5	9.35	7.19
$1/r_{sl} \times 10^{-3}/$(min·L/mol)	28.2	24.3	20.5	16.6	12.7	8.85

Substituting the corresponding data in Table 9C into Eqs. (A) and (B), the following values can be obtained by the least square regression method:

$a = 5.01 \times 10$, $a = 5.01 \times 10^3 3$; $b = 4.34$; correlation coefficient $r = 1.0$

$a' = 4.99 \times 10^3$; $b' = 7.74$; correlation coefficient $r = 1.0$

Results showed that the intercepts on the ordinate of the two lines are very close. Therefore, the assumption of competitive inhibition is valid. From Eqs. (A) and (B):

$$r_{max} = \frac{1}{(a + a')/2} = \frac{1}{(5.01 \times 10^3 + 4.99 \times 10^3)/2} = 2.00 \times 10^4$$

$$K_m = b \cdot r_{max} = 4.34 \times 2.00 \times 10^{-4} = 8.68 \times 10^{-4} \text{mol/L}$$

$$K_{ml} = b' \cdot r_{max} = 7.74 \times 2.00 \times 10^{-4} = 1.55 \times 10^{-3} \text{mol/L}$$

From Eq. (11.2), we get

$$K_I = c_l \left/ \left(\frac{K_{ml}}{K_m} - 1\right)\right. = 1.00 \times \frac{10^5}{\dfrac{1.55 \times 10^{-3}}{8.68 \times 10^{-4}} - 1} = 1.27 \times 10^{-5} \text{mol/L}$$

Fig. 11A shows that for the above two conditions, the vertical intercepts of both straight lines are 5.0×10^3 min·L/mol. Therefore, the vertical intercepts of the L-B line of the competitive inhibition is the same as the intercept when there is no inhibition. The intersection on the x-axis is $(-1/K_{ml})$. The L-B plot of noncompetitive inhibition can be obtained with a similar method. The intersection on the abscissa is same as when there is no inhibition, and the intersection on vertical axis is $1/r_{slmax}$. When the reaction is competitive inhibition, its L-B line is parallel to the line when there is no inhibition.

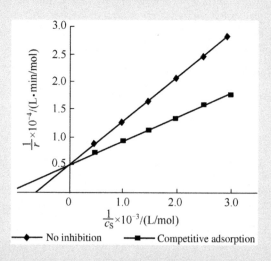

— No inhibition — Competitive adsorption

FIGURE 11A

11.2.2   Kinetics of Microbial Reactions

A microbial reaction is a complicated biochemical process that uses a specific enzyme in the microorganism, and is called the fermentation process. Primary industrial microorganisms are bacteria, yeast, actinomycetes, and mold. Fermentation can be grouped into anaerobic fermentation and aerated fermentation based on the characteristics of the microorganism cells used in the fermentation. For example, alcoholic fermentation, acetone butanol fermentation, and lactic acid fermentation belong to anaerobic fermentation, and antibiotics fermentation and amino acid fermentation belong to aerated fermentation. There are various types of fermentation products, which fall into the following categories: a microorganism cell itself, microorganism metabolites, a microorganism enzyme, biotransformation products, recombinant proteins, microorganism wastewater treatment products, etc.

In the process of microorganism reaction, each microorganism cell behaves like a microbioreactor. The molecules of the substrate, i.e., the nutrients, penetrate into the cell through the cell wall and cell membrane. Under the effect of a complex enzyme system, substrates are converted to the component of the cell and support cell growth and reproduction. On the other hand, part of the cell components are continuously decomposed to metabolites, which are discharged through the cell wall and cell membrane. For genetically engineered bacteria, besides cell growth and reproduction, synthesis of the target protein in gene recombination bacteria also happens. Therefore, microorganism reaction processes include mass transfer, microorganism cell growth, metabolism processes, etc. The process of aerated fermentation also includes gas phase oxygen transferring gradually into the cell and participating in aerated metabolism. Therefore, a microorganism reaction system is a multiphase and multicomponent system. In addition, microorganism cell growth and metabolism is a life activity of a complex group, and degeneration and mutation of biomass occurs during the life cycle. Therefore, it is very complicated to quantitatively describe the reaction rate and influential factors in the process of microorganism reaction.

Due to space limitations, this section will only briefly introduce the bioreaction rate and the influential factors in microorganism reaction-targeting to get metabolite products, including cell growth rate, substrate consumption rate, and product formation rate.

11.2.2.1   Cell Growth Kinetics

Cell growth is affected by environmental conditions such as water, moisture, temperature, nutrient, pH, and oxygen. At given conditions, cell growth follows certain rules. The cell growth process can be described with a cell concentration change in conjunction with balanced growth model.

Cell growth rate, r_x, is defined as the number of cells (biomass) formed per unit time and unit volume of culture solution, i.e.,

$$r_x = \frac{1}{V} \cdot \frac{dm_x}{dt} \tag{11.8}$$

where V is the volume of culture solution and m_x is the mass of cell. For a constant volume process, cell growth rate can be defined as

$$r_x = \frac{dc_x}{dt} \tag{11.9}$$

where c_x is cell concentration, which is usually represented by cell dry weight in unit volume of culture solution.

Balanced growth is similar to a first order autocatalytic reaction. Growth rate based on an increase of cell dry weight is directly proportional to cell concentration and the proportionality coefficient is μ as shown below

$$\mu = \frac{r_x}{c_x} \tag{11.10}$$

μ represents cell growth rate per unit biomass concentration. It is a very important parameter in describing cell growth, which is called specific growth rate.

The specific growth rate in batch culture is

$$\mu = \frac{1}{c_x} \cdot \frac{dc_x}{dt} \tag{11.11}$$

When cell growth is in the exponential growth phase, μ is usually a constant, so

$$\mu = \frac{1}{t} \ln \frac{c_x}{c_{x0}} \tag{11.12}$$

where c_{x0} is initial biomass concentration. The value of specific growth rate μ reflects the potential of cell growth, which is affected by strain and many other physical and chemical factors.

For a specific strain, at constant temperature and the pH, the relationship between cell-specific growth rate and growth-limiting substrate is shown in Fig. 11.8, and can be described with the Monod equation as shown below

$$\mu = \frac{\mu_{max} \cdot c_s}{K_s + c_s} \tag{11.13}$$

where c_s is the growth-limiting substrate concentration (g/L), μ_{max} is the maximum specific growth rate (h^{-1}), and K_s is the saturation coefficient (g/L), which is also called the Monod constant. K_s is the growth-limiting substrate concentration at half of the maximum specific growth rate. Although it does not have a specific physical meaning like the Michaelis constant, it is a constant that characterizes the dependency between cell growth rate and certain growth-limiting substrate.

The Monod equation was obtained empirically, and is a typical balanced growth model. The Monod equation was developed based on the following assumptions: (1) cell growth is balanced growth, which can be described with cell concentration change; (2) there is only one substrate in the culture that will limit cell growth. All the other substrates are excessive and their concentration change will not affect cell growth; and (3) cell growth is a simple reaction and the cell yield with respect to substrate, $Y_{x/s}$, is constant. $Y_{x/s}$ is cell mass synthesized per unit of substrate mass consumed.

When $c_s << K_s$, $\mu \approx \mu_{max}/K_s \times c_s$, kinetics is first order; when $c_s >> K_s$, $\mu \approx \mu_{max}$, kinetics is zero order.

Substituting Eq. (11.13) into Eq. (11.10) gives cell growth rate

$$r_s = \frac{\mu_{max} \cdot c_s}{(K_s + c_s)} \cdot c_x \tag{11.14}$$

The Monod equation was widely used in many microorganism cell growth processes. However, due to the complication of cell growth, e.g., substrate or product inhibition or non-Newton type of fermentation liquid, there are always differences between experimental data and Eq. (11.14). Therefore, there are modifications to the Monod equation reported in the literature (Fig. 11.8).

11.2.2.2 Substrate Utilization Kinetics

Substrate utilization rate refers to substrate mass consumed per unit time and unit volume of culture solution. Substrate includes carbon sources, nitrogen sources, and oxygen in the culture solution. We only briefly discuss the single growth-limiting substrate kinetics herein.

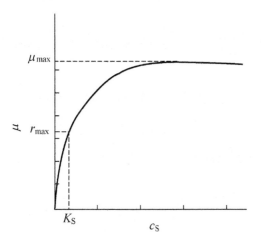

FIGURE 11.8 Cell-specific growth rate as a function of substrate concentration.

In batch culture, the substrate utilization rate is

$$r_s = -\frac{dc_s}{dt} \tag{11.15}$$

Specific substrate utilization rate q_s and specific product formation rate q_p are two very important parameters characterizing substrate utilization rate. q_s is substrate utilization rate per unit cell concentration

$$q_s = \frac{r_s}{c_x} = -\frac{1}{c_x}\frac{dc_p}{dt} \tag{11.16}$$

q_p is product formation rate per unit cell concentration

$$q_p = \frac{r_p}{c_x} = \frac{1}{c_x}\frac{dc_p}{dt} \tag{11.17}$$

where r_p is product formation rate.

In batch fermentation, substrate is consumed for three purposes: cell growth and division, cell activity, and product formation. So, the substrate utilization rate is

$$r_s = \frac{r_x}{Y^*_{x/s}} + mc_x + \frac{r_p}{Y_{p/s}} \tag{11.18}$$

where $Y^*_{x/s}$ is the cell yield coefficient in the absence of metabolism, which is also called the maximum cell yield coefficient; m is the biomass maintenance coefficient, which is the mass of substrate consumed per unit biomass within unit time to maintain routine physiological activity; $Y_{p/s}$ is the yield coefficient of product over substrate, which is the mass of product yielded per unit mass of substrate utilized. Rearranging Eq. (11.18) gives

$$-\frac{dc_s}{dt} = \frac{1}{Y^*_{x/s}}\mu \cdot c_x + m \cdot c_x + \frac{1}{Y_{p/s}}q_p \cdot c_x \tag{11.19}$$

Replacing the expression with substrate utilization rate, Eq. (11.19) becomes

$$q_s = \frac{1}{Y^*_{x/s}}\mu + m + \frac{1}{Y_{p/s}}q_p \tag{11.20}$$

This equation indicates that a specific substrate utilization rate is composed of three parts: rate for cell growth, routine physiological activity maintenance, and product formation.

11.2.2.3 Product Formation Kinetics

Fermentation processes have been classified into three types based on different relationships between product formation and cell growth, i.e., type I, type II, and type III.

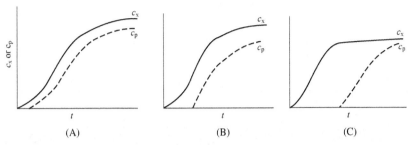

(A)   (B)   (C)

FIGURE 11.9 Relationship between cell growth and product formation in batch reactors. (A) Coupled; (B) Partially coupled; (C) Noncoupled.

Type I: cell growth coupled with production formation. The unique feature of this type is that cell growth is directly associated with product formation. They are synchronous as demonstrated in Fig. 11.9(A). Product formation kinetics of type I is shown below

$$r_p = Y_{p/x} \cdot r_x = Y_{p/x} \cdot \mu c_x \qquad (11.21)$$

$$q_p = Y_{p/x} \cdot \mu \qquad (11.22)$$

This type primarily includes fermentation of the primary by-products of glucose metabolism, e.g., ethanol and lactate.

Type II: cell growth partially coupled with product formation. The feature of this type is that cell growth is partially associated with product formation as shown in Fig. 11.9(B). No product is formed in the early stage of cell growth, but as soon as product is formed, the product formation rate will be related to cell growth and bacterial concentration. Product formation kinetics of type II is

$$r_p = \alpha \cdot r_x + \beta \cdot c_x \qquad (11.23)$$

$$q_p = \alpha \cdot \mu + \beta \qquad (11.24)$$

where α and β are constants. Glutamic acid fermentation and citric acid fermentation fall into this type.

Type III: cell growth is not coupled with product formation. The feature of this type is that product formation has no direct relationship with cell growth, i.e., there is no production accumulation during cell growth, but mass product is synthesized when cell growth stops as illustrated in Fig. 11.9(C). Product formation kinetics of type III is

$$r_p = \beta \cdot c_x \qquad (11.25)$$

$$q_p = \beta \qquad (11.26)$$

Fermentation of secondary metabolites, e.g., antibiotics, falls into this type.

11.2.2.4 Oxygen Consumption Rate

In aerobic microorganism reactions, aeration is needed to provide oxygen as the ultimate electron acceptor for cell aspiration and energy is released through the water formed during reaction.

Oxygen consumption rate or oxygen uptake rate (OUR) is expressed as the amount of oxygen consumed (uptake) by cell per unit time and unit volume of culture solution, i.e.:

$$r_{O_2} = -\frac{dc_{O_2}}{dt} = \frac{r_x}{Y_{x/O_2}} \qquad (11.27)$$

where Y_{x/O_2} is yield coefficient of biomass to oxygen.

An important parameter that describes OUR is the specific oxygen consumption rate q_{O_2}, also known as respiration intensity. It represents oxygen consumption rate per unit biomass concentration.

$$q_{O_2} = \frac{r_{O_2}}{c_x} = \frac{\mu}{Y_{x/O_2}} \qquad (11.28)$$

For a general microorganism reaction, the total oxygen demand is related to the material balance based on combustion reaction, i.e.:

Oxygen consumption = oxygen demand of substrate combustion−oxygen demand of cell combustion−oxygen demand of metabolites.

Substituting with oxygen consumed to form 1 g of cell, then

$$Y_{O_2/x} = \frac{1}{Y_{x/O_2}} = \frac{A}{Y_{x/s}} - B - Y_{p/x} \cdot C \qquad (11.29)$$

where A, B, and C are oxygen demand when 1 g of substrate, cell, and metabolites is completely combusted to form CO_2 and H_2O, respectively. Substituting Eq. (11.29) into Eq. (11.28) gives specific oxygen consumption rate.

$$q_{O_2} = \left(\frac{A}{Y_{x/s}} - B - Y_{p/x} \cdot C\right) \cdot \mu \qquad (11.30)$$

Example 11.3

In a CSTR, *Aspergillus* sp. was cultured under certain conditions and glucose is the limiting substrate. *Aspergillus* specific growth rate was measured at different substrate concentrations and the results are shown in Table 11D.

TABLE 11D Growth rate of Aspergillus measured at different substate concentrations

c_s/(mg/L)	500	250	125	62.5	31.2	15.6	7.8
$\mu \times 10^3/\text{h}^{-1}$	9.54	7.72	5.58	3.59	2.09	1.14	0.60

Solution

Assuming *Aspergillus* sp. growth follows the Monad equation under this condition, determine μ_{max} and K_s.

Linearizing Eq. (11.13) and rearranging gives

$$\frac{1}{\mu} = \frac{1}{\mu_{max}} + \frac{K_s}{\mu_{max}} \cdot \frac{1}{c_x} \qquad (A)$$

It is clear that $1/\mu$ has linear correlation with $1/c_s$. $1/\mu$ and $1/c_s$ are calculated from known c_s and μ values. Results are provided in Table 11E.

TABLE 11E Results of linearizing Eq. 11.13

$(1/c_s)/(\text{L/g})$	2	4	8	16	32.1	64.1	128
$(1/\mu)/\text{h}$	10.5	13.0	17.9	27.9	47.8	87.7	167

Regressing data with Eq. (A) by linear least square method gives:
$\mu_{max} = 0.125\text{h}^{-1}$; $K_s = 0.155\text{g/L}$

11.3 IMMOBILIZED BIOCATALYSTS

11.3.1 Introduction

There are thousands of enzymes in the natural world. There is almost no complicated reaction in the living body that is not catalyzed by enzyme. Enzymatic catalysis has many advantages that are unmatched by other chemical catalysis. However, industrial applications of enzymatic catalysis have been very limited so far. One of the important reasons is that an enzyme is hard to recover due to its hydrophilic feature.

When a free enzyme (or cell) is used as catalyst in a batch reactor the enzyme (or cell) cannot be reused because it is hard to recover an enzyme (or cell) after a reaction. However, in a continuous reactor, the enzyme (or cell) will be carried out with products, which results in gradual concentration reduction in the reactor or even the "wash out" phenomena. Then, the reaction will have to stop and it becomes very difficult to separate enzyme from products and substrates at the exit of the reactor. In addition, the process of enzyme formation and purification is complicated and technically demanding, which makes enzyme production very expensive. Additionally, a free enzyme is prone to the effect of environment temperature, pH, and ionic strength. Therefore, the activity of an enzyme (simplified as enzyme activity thereafter) is not stable. So, the enzyme immobilization technique was developed in the 1960s.

Enzyme immobilization is to attach an enzyme to a carrier or restrict an enzyme to a limited space. Although enzyme immobilization can overcome the limitations of the free enzyme, extraction and purification are needed to obtain the enzyme. In addition, immobilized enzymes can only be applied to simple enzymatic reactions. In fact, most biocatalytic reactions need multienzyme catalysis or participation of coenzymes. Therefore, the cell immobilization technique emerged in the 1970s.

Immobilized enzymes and immobilized cells are generally called immobilized biocatalysts.

Cell immobilization is to attach a cell to a carrier or restrict a cell to a limited space. Based on the psychological status, immobilized cells fall into three categories: immobilized dead cells, immobilized resting cells, and immobilized growing cells. Enzymes in dead cells and resting cells keep their initial activity. Compared to immobilized multienzymes, these two immobilizations are more favorable to improve enzyme stability. Immobilized growing cells still have the function of growth and metabolism. Additionally, if environmental conditions are favorable, cell growth can stay in the equilibrium phase. It will ameliorate the microenvironment of the cell and improve the stability of enzyme activity; this can also increase cell concentration in the reactor and increase reaction rate and reactor productivity. Furthermore, it can simplify the process of product separation and purification, making it feasible for bioreactions to be conducted in fixed bed and fluidized bed reactors, and realize continuous and automatic production. This will be of great importance in enhancing fermentation processes.

11.3.2 Enzyme and Cell Immobilization

There are many methods for enzyme and cell immobilization, and commonly used methods include adsorption, entrapment, covalent binding, and cross-linking.

11.3.2.1 Adsorption

There are two types of adsorption, i.e., surface binding and cell aggregation. In surface binding, an enzyme is fixed onto the adsorbent by nonspecific physical adsorption between enzyme and support or specific adsorption between biological substances. Forces that form nonspecific physical adsorption are Van der Waals force, hydrogen bonding, hydrophobic interaction, electrostatic interaction, etc. The most commonly used adsorbents are activated carbon, bentonite, diatomaceous earth, porous glass, lithium hydride, ion-exchange resin, and polymer materials.

The biggest advantage of surface binding is that the preparation method is simple and it is usually not toxic to cells or enzymes. However, cell or enzyme concentrations on the carrier are low due to the limited adsorption capacity of the porous adsorbents.

Cell aggregation is based on the fact that some cells have the potential to form aggregate or colloids or utilize polyelectrolyte induction to form microorganism cell aggregation to realize cell immobilization.

Some other cells can secrete high-molecular compounds, e.g., mucopolysaccharides, which helps microorganisms adsorb on the surface of adsorbents.

11.3.2.2 Entrapment

Entrapment is the most widely used method for cell or enzyme immobilization. It immobilizes cells or enzymes into the 3-D matrix of high-molecular compounds. There are three types of entrapment, i.e., gel entrapment, microcapsule entrapment, and fiber entrapment.

1. Gel entrapment. There are two types of gels: natural polymer gel and synthetic polymer gel. Natural polymer gel includes calcium algenate, K-carrageenan, agarose gel, gelatin, and chitosan. The advantages of natural polymer gel are: they are nontoxic; the conditions for preparing immobilization biocatalyst are mild; loss of enzyme activity is small; and the microenvironment inside the immobilized cell is suitable for the cell's physiological condition. The biggest limitation of natural polymer gel is that the mechanical strength of the gel particles is weak. The most commonly used gels are calcium algenate and K-carrageenan, which can be easily made into small pellets and are applicable to air lift reactors, fluidized bed reactors, and jet loop reactors. To overcome the disadvantage of the low mechanical strength of the calcium algenate, the gel is always added to glutaraldehyde/polyethyleneimine solution and cross-linked with it. K-carrageenan has better mechanical strength, and it has been used for cell immobilization to produce L-aspartic acid, L-alanine, L-malic acid, and acrylic amide, which have been industrialized.

   The widely used synthetic polymer gels include polyacrylamide, light-cured resin, and polyvinyl alcohol. The biggest advantage of the synthetic polymer gel is its high mechanical strength. But the monomer of polyacrylamide is highly toxic—cells or enzymes are very easily harmed during polymerization. Using polyacrylamide entrapping *Escherichia coli* to produce L-aspartic acid has been applied in industry. Recently, this method has been used to produce L-malic acid, L-lysine, L-alanine, steroid hormones, etc.

The gel entrapment method has high entrapping capacity and adaptability for biomass or enzyme, and it can entrap cells of different types and different physiological status. In addition, it is simple and the stability of the immobilized catalysts is good. However, the internal diffusion resistance is high, which could be problematic, especially when applied to high-molecular substrates. Furthermore, biomass or enzyme may be leaking from the structure, which needs be carefully managed by adjusting preparation conditions.

2. Microcapsule entrapment. Microcapsule entrapment encloses enzymes inside a microcapsule of semipermeable polymer membrane. The diameter of the microcapsule is generally from several micrometers to several hundred micrometers. A typical preparation method includes interfacial polycondensation, liquid-drying, and phase separation. For example, in interfacial polycondensation, hydrophilic monomers with enzymes are emulsified in water, another hydrophobic monomer is dissolved in a water-immiscible organic solvent. Polycondensation of the monomers then occurs at the interface between the aqueous and organic solvent phases in the emulsion to form a thin membrane of polymer, where the enzyme was enclosed. This method can provide very large specific surface areas. In addition, the semipermeable membrane can be made to form a membrane reactor. The small molecular substrates and products can penetrate through the micropores of the membrane, while enzyme and other large molecular compounds cannot pass.

3. Fiber entrapment. In this method, an enzyme solution is emulsified in an organic solvent of cellulose acetate, and sprayed into fibers, which are woven into cloth or form other shapes to meet the structure requirements of different reactors. Since fiber is very thin, the specific area is thus very large, and the capacity of entrapment is high.

11.3.2.3   Covalent Binding

Covalent binding uses the amino group or carboxyl group of an enzyme protein or the aromatic ring of tyrosine and histidine and some organic groups in the carriers to form covalent bonds that immobilize enzyme onto the carrier. In general, the carrier needs to be functionalized using the chemical method. Then it will couple with the corresponding group on the enzyme molecule. The advantage of this method is that the binding between carrier and enzyme is stable and the enzyme will not detach from the carrier easily. However, the preparation method is complicated, preparation conditions are harsh, and it is more likely to cause enzyme deactivation.

11.3.2.4   Cross-Linking Method

In cross-linking, the double functional groups or multifunctional groups in the reagent react with the amino group or carboxyl group of enzyme molecular to make enzyme molecules coupled with each other to form water-insoluble aggregate or make enzyme molecules cross-linked to form a net structure. The commonly used cross-linking reagents are glutaraldehyde, methylene diisocyanate, bis-diazotized benzidine, etc. Most of the time, the cross-linking method was used in combination with entrapment and adsorption methods. Leakage of entrapped cells or enzymes can be avoided by using the former method, and the latter method prevents enzyme detaching from adsorbent.

Other immobilization methods include ionic binding, thermal immobilization method, chemical immobilization, etc.

In summary, there are many methods for enzyme and cell immobilization, novel methods continuously emerge, and industry applications increase. However, so far there are no methods that are ideal and have universal applications. All methods need be tested before use. In general, it is believed that for commercial applications the immobilized biocatalysts should meet the following criteria: (1) the selected carrier should be nontoxic to cell or enzyme; should have appropriate pore diameters, porosity, specific surface areas, and geometric shape; should not only prevent cell leakage or minimize leakage but also have good permeability and low diffusion resistance to substrates and products; and carrier material should be easily available. (2) The immobilization method should be simple; preparation conditions should be mild to minimize enzyme deactivation; the carrier should have good formability to meet the requirements of bioreactors. (3) Should have high cell or enzyme capacity per unit volume to increase reactor's productivity. (4) Should have high mechanical stability and stable activity.

11.3.3   Catalytic Kinetics of Immobilized Biocatalyst

Enzymes or cells become immobilized enzymes or immobilized cells after immobilization. Catalytic activity may be changed due to the effect of the carrier and other factors, and the change is very complicated. Studies on immobilized living cells are still not mature. This section focuses on the kinetics of catalytic reactions of an immobilized enzyme, which can be used as a reference for the study of catalytic kinetics of dead cells (static cells) reactions.

11.3.3.1 Factors Affecting Kinetics of Immobilized Enzyme-Catalyzed Reactions

In general, enzyme activity will drop more or less after immobilization. The following factors will affect enzyme activity decline.

1. Conformational effect. When an enzyme was immobilized by covalent binding or cross-linking method, the active site of enzyme was deformed due to the covalent bond effect between enzyme and carrier. This effect changes the 3-D structure of the active site on the enzyme and weakens the bonding between enzyme and substrate, which reduces enzyme activity. This phenomenon is called the conformational effect.
2. Steric effect. If the pore sizes of the carrier are too small or the carrier was not selected correctly, steric hindrance will exist between the active sites of enzyme and substrate molecules, which will make it very hard for the substrate to be in contact with the active site on the enzyme, and therefore affect enzyme catalysis. This steric effect is also called the shielding effect. It depends on carrier structure, properties, and immobilization methods. It also relates to the substrate molecules' size, characteristics, and shapes.
3. Partition effect. Due to hydrophilic, hydrophobic, and electrostatic interactions of immobilized enzyme, concentrations of substrates or products in the microenvironment of enzyme are different from that in the bulk solution, which affects the reaction rate of enzyme-catalyzed reactions. This effect is called the partition effect and partition coefficient K is usually used to describe partitioning.
4. Diffusion effect. Immobilized enzyme-catalyzed reactions belong to multiphase catalytic reactions. Based on nonhomogenous catalytic reaction principles and analysis methods, immobilized enzyme-catalyzed reactions include five steps, i.e., (1) substrate diffuses from bulk solution to external surface of the immobilized enzyme particles; (2) substrate diffuses from external surface of immobilized enzyme particles to enzyme active site on the inner surface; (3) substrate reacts at active site on inner surface to form products; (4) product diffuses from reaction zone to external surface of immobilized enzyme particles; and (5) product diffuses from external surface to the bulk solution. Apparently, steps (1) and (5) are external diffusion, steps (2) and (4) are internal diffusion, and step (3) is surface reaction. In these procedures, internal diffusion and surface reaction are parallel processes. External diffusion, internal diffusion, and surface reactions happen in series. At steady-state, the reaction rate of each step in the series is the same and equals the reaction rate of the rate-determining step. This reaction rate is called total reaction rate, which is different from the reaction rate of free enzyme. The difference between catalytic reaction rates of immobilized enzyme and free enzyme caused by different diffusions is called diffusion effect.

The above factors are closely interrelated to each other and they combine together to affect the kinetics of immobilized enzyme. The conformation effect and steric effect have a direct impact on enzyme catalysis, but the effect on enzyme kinetics cannot be described quantitatively and has to be determined experimentally. The effect of these factors can be eliminated or improved through selecting the right immobilization conditions, methods, and carrier. The partition effect is usually described with the partition coefficient K, i.e.,

$$K = \frac{\bar{c}'_{si}}{\bar{c}_{SL}} \qquad (11.31)$$

where $\bar{c}'_{si}$ and $\bar{c}_{SL}$ are the average concentration of substrate in the microenvironment inside the interface between liquid and solid, and in the bulk liquid outside of the interface, respectively. Impact of partitioning on kinetics of immobilized enzyme can be described quantitatively with Boltzman distribution law in conjunction with the Michaelis−Menten equation.

The diffusion effect can be quantitatively described with mass transfer theory in a combination of kinetics of enzyme-catalyzed reactions.

Internal and external diffusion in enzyme particles will follow the internal and external diffusion theory and methods of liquid−solid chemical catalysis described in Chapter 6, Chemical Reaction and Transport Phenomena in Heterogeneous System. Compared to chemical catalysis, the only difference is the reaction rate equation.

11.3.3.2   Internal Diffusion Inside Immobilized Enzyme Particles

1. Substrate concentration distribution inside the immobilized enzyme pellet. Assuming enzyme molecules are distributed evenly inside the immobilized enzyme pellet, activity distribution is uniform, and enzymatic reaction kinetics can be described with the Michaelis−Menten equation, the internal diffusion equation of immobilized enzyme pellet with radius of R_p is

$$\frac{d^2 c_s}{dr^2} + \frac{2}{r}\frac{dc_s}{dr} = \frac{1}{D_e}\frac{r'_{max,p} \cdot c_s}{(K'_m + c_s)} \qquad (11.32)$$

where $r'_{max,p}$ and K'_m are the maximal reaction rate of immobilized enzyme particle and m constant, respectively, and $r'_{max,p}$ is based on the unit volume of immobilized enzyme particle.

Introducing the following dimensionless variables

$$\xi = c_s/c_{ss}; \quad \zeta = r/R_p; \quad \beta = c_{ss}/K'_m$$

and Thiele Modulus

$$\phi = \frac{R_p}{3} \sqrt{\frac{r'_{max,p}}{K'_m + D_e}} \tag{11.33}$$

where c_{ss} is substrate concentration at the external surface of the immobilized enzyme particle. Transforming Eq. (11.33) to dimensionless form,

$$\frac{d^2\xi}{d\zeta^2} + \frac{2}{\zeta}\frac{d\xi}{d\zeta} = 9\phi^2 \cdot \frac{\zeta}{(1 + \beta\zeta)} \tag{11.34a}$$

Boundary conditions are

$$\zeta = 1 \quad \xi = 1 \tag{11.34b}$$

$$\zeta = 0 \quad \frac{d\xi}{d\zeta} = 0 \tag{11.34c}$$

Eq. (11.34a) is nonlinear and can only be solved numerically, and the solution describes substrate concentration distribution inside immobilized enzyme particle, which is very similar to Fig. 6.4 in gas–solid chemical catalysis.

When $c_{ss} \ll K'_m$, the Michaelis–Menten equation can be simplified into a first order irreversible reaction and reaction rate equation can be written as

$$r_s = \frac{r'_{max}}{K'_m} \cdot c_s \tag{11.35}$$

The substrate concentration distribution $c_{ss} \ll K'_m$, equation is same as Eq. (6.57), but Thiele Modulus uses Eq. (11.33).

When $c_{ss} \gg K'_m$ it can be treated as a zero order reaction.

For membrane-like immobilized enzyme with thickness L, assuming enzyme molecules with similar activity are distributed uniformly inside the membrane, the diffusion–reaction equation is

$$\frac{d^2c}{dZ^2} = \frac{1}{D_e}\frac{r'_{max,p} \cdot c_s}{(K'_m + c_s)} \tag{11.36}$$

The dimensionless form of Eq. (11.36) can be written as

$$\frac{d^2\xi}{d\zeta^2} = \phi^2 \frac{\xi}{(1 + \beta\xi)} \tag{11.37a}$$

Boundary conditions are:

$$\zeta = 1 \quad \xi = 1 \tag{11.37b}$$

$$\zeta = 0 \quad \frac{d\xi}{d\zeta} = 0 \tag{11.37c}$$

where

$$\zeta = \frac{Z}{L} \quad \phi = L\sqrt{\frac{r'_{\text{max,p}}}{K'_{\text{m}} \cdot D_{\text{e}}}} \tag{11.38}$$

ϕ is Thiele Modulus.

Apparently, Eq. (11.37a) can only be solved numerically too. When $c_{\text{ss}} \ll K'_{\text{m}}$, it also can be treated as a first order irreversible reaction.

Then, the substrate concentration distribution equation is the same as Eq. (6.52) but with Thiele Modulus from Eq. (11.38).

2. Internal effectiveness factor

The definition of internal effectiveness factor for gas−solid chemical catalysis, Eq. (6.53), can be applied to immobilized enzyme particles. For an immobilized enzyme pellet, the internal effectiveness factor in the absence of external diffusion is:

$$\eta = \frac{R^*_{\text{s}}}{r_{\text{s}}} = \frac{4\pi R^2_{\text{p}} D_{\text{e}} \left(\dfrac{dc_{\text{s}}}{dr}\right)_{r=R_{\text{p}}}}{\left(\dfrac{4}{3}\right)\pi R^3_{\text{p}} \cdot [r'_{\text{max,p}} c_{\text{s}}/(K'_{\text{m}} + c_{\text{ss}})]} = \frac{3}{R_{\text{p}}} \cdot \frac{D_{\text{e}} \left(\dfrac{dc_{\text{s}}}{dr}\right)_{r=R_{\text{p}}}}{r'_{\text{max,p}} \cdot c_{\text{ss}}/(K'_{\text{m}} + c_{\text{ss}})}$$

$$\tag{11.39}$$

Eq. (11.39) only has a numerical solution or approximate solution. Several approximate approaches have been developed (Atkinson, 1977).

When $c_{\text{ss}} \ll K'_{\text{m}}$, it can be approximately treated as a first order irreversible reaction. For spherical and membrane type immobilized enzymes, internal effectiveness factor can be referenced to Eqs. (6.60) and (6.54) with Thiele Modulus defined by Eqs. (11.33) and (11.38), respectively.

Mass transfer and reaction mechanisms in immobilized enzymes are much more complicated than those in gas−liquid chemical catalysis reactions especially when substrate and product inhibition exist. Although there are already many publications that are very helpful in understanding and clarifying immobilized enzyme-catalyzed reaction mechanisms and developing kinetics equation, application to reactor design needs to be further investigated.

11.4 BIOREACTORS

Bioreactors are a major piece of equipment in the process of biochemical reactions. Bioreactors are generally similar to chemical reactors based on theory, shape, structure, category, and operation modes. Because bioreactors use enzyme and living cells as catalyst, substrate composition and properties are generally complicated, product types are varied, and they are usually related to cell metabolism processes. Therefore, bioreactors have their specific features. In general, bioreactors should meet the following criteria: (1) provide a good environment for cell growth, enzymatic catalysis reaction and product formation at different scale levels. The reactor is easily sterilized, preventing microorganism contamination, preventing harm to the enzyme, cell, or immobilized biocatalyst inherent characters, and can be easily operated at different conditions so that it can be optimized for a broad range of biochemical reactions. (2) Can provide good mixing and fast mass and heat transfer rates with minimum power consumption. (3) Have good operation flexibility to meet the requirements of biochemical reactions at different stages and/or for producing different types of product.

11.4.1 Types of Bioreactors

Based on the mode of operations, bioreactors fall into three types: batch, semibatch, and continuous reactors. Characteristics of these reactors were described in Chapter 3, Tank Reactor, and Chapter 4, Tubular Reactor. For biochemical reactions, batch reactors also have the advantage of being less prone to contamination, and hence they are the most widely used. Semibatch operation is also referred to as the fed-batch mode. It is usually applied to biochemical processes with substrate and product inhibition or fermentation processes that need to control specific growth rate. Continuous operation is primarily applied to biochemical reaction processes of immobilized biocatalysis. Because continuous reactors are operated continuously and usually for a long time, contamination and biomass mutation are more likely to happen in continuous reactors than other reactors, which limits the application of continuous reactors.

Based on the mode of fluid flowing, material mixing, and backmixing inside the reactors, continuous reactors are further categorized as CSTRs, plug flow reactors and nonideal flow reactors.

The oldest and most classic continuous bioreactor is the fermentation tank for microbial fermentation. With the development of biochemical engineering, there have been many different types of bioreactors, e.g., enzyme reactors for free enzyme or immobilized biocatalysis reactions, reactors for animal cell culture, reactors for plant cell culture, and biochemical equipment for wastewater treatment. In addition, there is a large variety of bioreactors that are used for photosynthesis, new energy (hydrogen or bio-battery), bacteria metallurgy, etc. Here we only introduce a couple of different types of reactors.

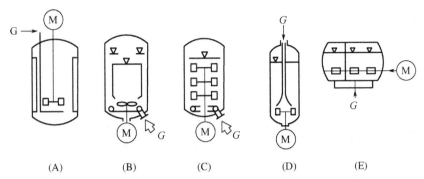

FIGURE 11.10   Mechanically stirred reactors. (A) Classical stirred tank; (B) Waldhof stirred tank; (C) Multilayer paddle stirred tank; (D) Gas self-suction stirred tank; (E) Horizontal stirred tank.

1. Mechanically stirred reactors. Mechanically stirred reactors are the most widely used bioreactor in industrial production, as shown in Fig. 11.10(A). The biggest advantage of this bioreactor is its great operation flexibility and strong adaptability to different systems and processes. But its efficiency is low, energy consumption is high, and scale-up is difficult.

To overcome the limitations described above, a large variety of novel and efficient stirred reactors were developed, e.g., the Waldhof stirred tank, multilayer paddle stirred tank, gas self-suction stirred tank, and horizontal stirred tank, as shown in Fig. 11.9 (B)–(E), respectively. Comparatively speaking, the multilayer paddle stirred tank has high energy consumption and low mass transfer coefficient. The self-suction stirred tank has low energy consumption and high oxygen transfer rate and has been applied in industry. The horizontal stirred tank has the best performance but has a very complicated structure.

2. Gas-lift reactors. Gas-lift bioreactors use the power created by gas injection and the density difference between gas–liquid mixture and liquid to circulate the gas–liquid mixture. This will enhance mass transfer, heat transfer and mixing. There are a great variety of gas-lift reactors, as illustrated in Fig. 11.11. The internal airlift loop reactor has a compacted structure, and the draft tube can be made into multiple stages to strengthen overall and local circulation. In addition, a sieve plate can be installed inside the draft tube to improve gas distribution and inhibit the liquid cycling rate. External loop reactors use the heat exchanger installed in the downcomer to improve heat transfer, which is helpful for the cycling and mixing of top and bottom materials in the tower.

This type of bioreactor has the following features: good mass and heat transfer efficiency, easy to scale up, simple structures, uniformly distributed shear stress, and resistant to contamination.

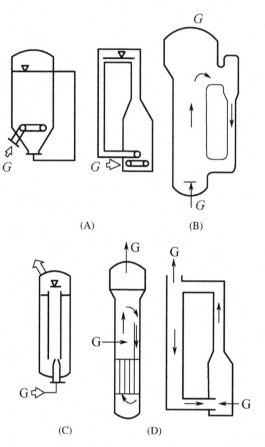

FIGURE 11.11 Gas-lift bioreactors: (A) Bubble column bioreactor; (B) External circulating bioreactors; (C) Internal circulating bioreactors; (D) Differential pressure circulating bioreactors.

3. Liquid jet loop reactor (LJLR). Liquid jet loop reactors have different shapes as shown in Fig. 11.12. They use the jet of pump to circulate the liquid and achieve extensive mixing through the momentum transfer between gas and liquid. LJLR has two types, i.e., up jet loop reactors and reverse jet loop reactors.

    LJLR has the following features: large contact area between gas and liquid, even mixing, good mass and heat transfer property, and easy scale-up.

4. Fixed bed bioreactor. Fixed bed reactors are illustrated in Fig. 11.13. They are mainly used for immobilized enzyme catalysis systems. Based on the flow direction of materials, fixed bed bioreactors can be categorized as up-flow and down-flow reactors.

    Fixed bed reactors have features of continuous operation, small back-mixing, high utilization of substrate, and resistance of wear of immobilized biocatalysts.

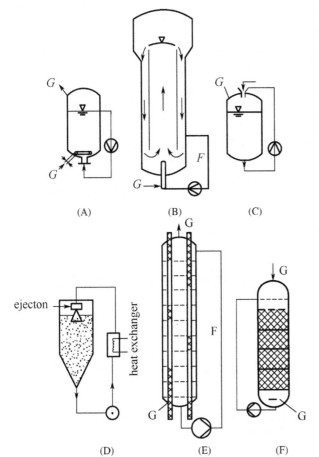

FIGURE 11.12 Liquid jet loop bioreactors: (A) Circular bubble column; (B) Self-aspirated jet flow loop bioreactor; (C) Self-aspirated jet flow bioreactors; (D) External circulating self-aspirated jet flow bioreactors; (E) Multiplate loop reactor; (F) Spray tower reactor.

5. Fluidized bed bioreactor. They are primarily used when substrate is present as solid particles or for reaction systems with immobilized biocatalysis. Due to the efficient mixing, this type of reactor has good heat and mass transfer characteristics but it does not apply to reaction systems with product inhibition. To improve the backmixing in a fluidized bed bioreactor, the magnetically fluidized bed reactor was developed as illustrated in Fig. 11.14, i.e., biocatalyst is mixed with magnetic materials to make the fluidized bed operate under the magnetic field.

6. Membrane reactor. Membrane reactors immobilize enzyme or microorganisms on a porous membrane. When a substrate passes through the membrane, an enzyme-catalyzed reaction will start. Since small molecular product can pass through the membrane and be separated from substrate, it will

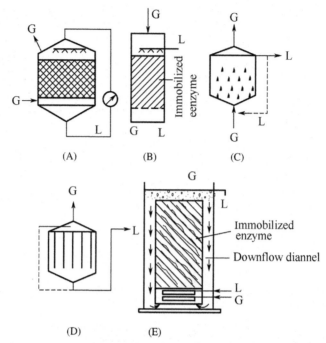

FIGURE 11.13 Fixed bed bioreactors: (A) Pump circulating fixed bed; (B) Trickle bed bioreactor; (C) Up flow; (D) Down flow; (E) Internal circulating fixed bed.

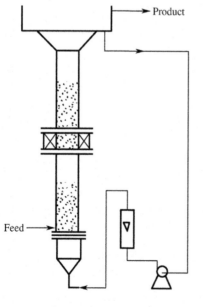

FIGURE 11.14 Two layer magnetically fluidized bed reactor.

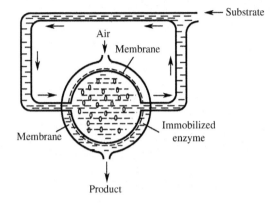

← Substrate   **FIGURE   11.15**   Enzyme-membrane bioreactors.

prevent produce inhibition. This reactor couples the reaction with separation and simplifies the processes. A membrane reactor is illustrated in Fig. 11.15.

In summary, bioreactors have a great variety of types. The proper selection of a bioreactor depends on the characteristics of bioreactions and the requirements of the processes.

11.4.2   Bioreactor Calculations

The basic equations for bioreactor design and calculations, i.e., material balance, heat balance, and momentum balance equations are very similar to those for chemical reactors, but the specific kinetics equations for bioreactions are different from chemical reactions, and are more complicated and more nonlinear than kinetics equations for chemical reactions. Therefore, the analysis and calculation of kinetics equations for bioreactions is more complicated. We only introduce the simplest case here to demonstrate the basic design and calculation methods for bioreactors.

11.4.2.1   Batch Reactor

In batch reactors, if the reaction is controlled by enzyme catalysis and there is no inhibitor and there is only one substrate, the utilization rate of substrate can be described with the Michaelis−Menten Eq. (11.1).

Substituting Eq. (11.1) into the design equation for a batch reactor as described in Chapter 3, Tank Reactor, since enzyme reaction is a constant volume process,

$$t = -\int_{c_{s0}}^{c_s} \frac{dc_s}{r_s} = -\int_{c_{s0}}^{c_s} \frac{dc_s}{\dfrac{k_2 c_{E0} c_s}{(K_m + c_s)}} = \frac{1}{r_{max}} \left[K_m \ln\left(\frac{c_{s0}}{c_s}\right) + (c_{s0} - c_s) \right] \quad (11.40)$$

From Eq. (11.40), the reaction time needed to reach a certain substrate concentration can be obtained. The actual reactor volume can be calculated from Eqs. (3.13) and (3.14).

When an inhibitor is present, the corresponding kinetics equation has to be used in the design equation for batch reactor calculations.

If a multienzyme bioreaction that uses a microorganism as catalysis is conducted in a batch reactor, the process involves biomass growth, metabolism, and product formation. It is thus very complicated. We only discuss the case when a single cell protein is the target product. Assuming biomass growth follows the Monod equation, the substrate is completely consumed for biomass growth, and other consumptions are negligible, biomass growth rate can be written as

$$r_x = \frac{dc_x}{dt} = \frac{\mu_{max}c_s}{K_s + c_s} \cdot c_x \tag{11.41}$$

Substrate utilization rate is related to biomass growth rate as

$$-\frac{dc_s}{dt} = Y_{s/x} \cdot \frac{dc_x}{dt} \tag{11.42}$$

where $Y_{s/x}$ is cell yield rate with respect to substrate. Assuming $Y_{s/x}$ did not change during the process of fermentation, and at $t = 0$, $c_x = c_{x0}$, $c_s = c_{s0}$, then

$$c_s = c_{s0} - Y_{s/x}(c_x - c_{x0}) \tag{11.43}$$

Substituting Eq. (11.43) into Eq. (11.41) yields

$$\frac{dc_x}{dt} = \frac{\mu_{max} \cdot c_x \cdot [c_{s0} - Y_{s/x} \cdot (c_x - c_{x0})]}{K_s + c_{s0} - Y_{s/x} \cdot (c_x - c_{x0})} \tag{11.44}$$

Integrating the above equation by separation of variables, we obtain

$$(c_{s0} + Y_{s/x}c_{x0}) \cdot \mu_{max} \cdot t = (K_s + c_{s0} + Y_{s/x} \cdot c_{x0})\ln\frac{c_x}{c_{x0}} - K_s\ln\frac{c_{s/x} - Y_{s/x}(c_x - c_{x0})}{c_{s0}} \tag{11.45}$$

This equation directly describes the relationship between biomass concentration and fermentation time. The relationship between substrate concentration and reaction time can be derived by combining Eqs. (11.45) and (11.43)

Example 11.4

Maltose hydrolysis reaction was conducted at 15°C isothermally by glucamylase enzyme in a batch reactor. K_m is 1.22×10^{-2} mol/L, and Maltose initial concentration is 2.58×10^{-3} mol/L. After the reaction proceeded for 10 min, the measured Maltose consumption rate was 30%. Please determine the time required to reach 90% conversion.

Solution

At 30% conversion, Maltose concentration is

$$c_{s1} = c_{s0}(1 - x_{s1}) = 2.58 \times 10^{-3}(1 - 0.3) = 1.81 \times 10^{-3} \text{mol/L}$$

Rearranging Eq. (11.40) and substituting the known data gives the maximum cell growth rate for this enzyme-catalyzed reaction:

$$r_{max} = k_2 c_{E0} = \frac{1}{t} \cdot \left[K_m \ln\left(\frac{c_{s0}}{c_{s1}}\right) + (c_{s0} - c_{s1}) \right]$$

$$= \frac{1}{10}\left(1.22 \times 10^{-2} \ln\frac{2.58 \times 10^{-3}}{1.81 \times 10^{-3}} + (2.58 \times 10^{-3} - 1.81 \times 10^{-3}) \right)$$

$$= 5.09 \times 10^{-4} \text{mol/(L·min)}$$

Maltose concentration at 90% conversion is

$$c_{s1} = c_{s0}(1 - x_{s1}) = 2.58 \times 10^{-3}(1 - 0.90) = 0.258 \times 10^{-3} \text{mol/L}$$

Plugging into Eq. (11.40) to obtain the time required to reach 90% conversion

$$t = \frac{1}{5.09 \times 10^{-4}}\left(1.22 \times 10^{-2} \ln\left(\frac{2.58 \times 10^{-3}}{0.258 \times 10^{-3}}\right) + (2.58 \times 10^{-3} - 0.258 \times 10^{-3}) \right) = 59.8 \text{min}$$

11.4.2.2 Continuous Stirred Tank Reactor

1. Enzyme-catalyzed reaction. In a CSTR, if the reaction is controlled by enzyme catalysis and the kinetics equation follows the Michaelis–Menten equation, substituting the kinetics equation directly into Eq. (3.42) gives the residence time

$$\tau = \frac{V_r}{Q_0} = \frac{(c_{s0} - c_s)(K_m + c_s)}{r_{max}c_s} = \frac{(c_{s0} - c_s)(K_m + c_s)}{k_2 c_{E0} c_s} \tag{11.46}$$

Substituting $c_s = c_{s0}(1 - X_s)$ into the above equation and simplifying gives

$$\tau = \frac{1}{k_2 c_{E0}}\left(c_{s0}X_s + \frac{K_m X_s}{1 - X_s} \right) \tag{11.47}$$

When inhibitor is present, we need to substitute the corresponding kinetics equation into Eq. (3.42).

2. Microorganism reaction. In a CSTR, assuming there is no biomass in the feed material, then at steady-state, biomass growth rate in the reactor equals biomass outflow rate from the reactor, i.e.,

$$Q_0 c_x = r_x V_r = \mu c_x V_\tau \tag{11.48}$$

The ratio of feed volumetric flow rate to culture solution volume is defined as dilution rate, i.e., $D = Q_0/V_t$. Substituting it into Eq. (11.48) yields

$$\mu = D \tag{11.49}$$

D represents the extent that feed material was diluted in the reactor, and the dimension is $(\text{time})^{-1}$.

It can be inferred from Eq. (11.49) that if a cell is cultured in a CSTR, at steady-state, cell-specific growth rate equals the dilution rate of the reactor. This is an important feature for cell cultures in CSTRs. We can use this feature to change cell-specific growth rate at steady-state by controlling the feeding rate of the substrate. Therefore, CSTR is also called chemostat when it is used for cell culture. Cell growth characteristics can be conveniently investigated with a chemostat.

In a CSTR, the limiting substrate concentration and biomass concentration are related to the dilution rate. If biomass growth follows the Michaelis–Menten equation, then

$$D = \mu = \frac{\mu_{max} c_s}{K_s + c_s} \tag{11.50}$$

Therefore, in the reactor substrate concentration and dilution rate has the following relationship

$$c_s = \frac{K_s D}{\mu_{max} - D} \tag{11.51}$$

Assuming the limiting substrate is only used for cell growth, then at steady-state

$$Q_0(c_{s0} - c_s) = r_s \cdot V_r \tag{11.52}$$

While

$$r_s = \frac{r_x}{Y_{x/s}} = \frac{\mu \cdot c_s}{Y_{x/s}} \tag{11.53}$$

Substituting Eq. (11.52) into the above equation and combining with Eq. (11.49), we get the cell concentrations in the reactor

$$c_x = Y_{x/s}(c_{s0} - c_s) \tag{11.54}$$

Substituting Eq. (11.51) gives the relationship between cell concentration and dilution rate, i.e.,

$$c_x = Y_{x/s}\left(c_{s0} - \frac{K_s \cdot D}{\mu_{max} - D}\right) \tag{11.55}$$

From Eq. (11.51) we can see that, as D increases, c_s in reactor also increases; when D is large enough to make $c_s = c_{s0}$, the dilution rate becomes the critical dilution rate, i.e.,

$$D_c = \mu_c = \frac{\mu_{max} c_{s0}}{K_s + c_{s0}} \tag{11.56}$$

The dilution rate of the reactor must be less than the critical dilution rate. When $D > D_c$, cell concentration in the reactor will become smaller and smaller, and the cell will be finally "washed" out from reactor, which is definitely not allowed.

Cell yield rate P_x is also cell's growth rate, i.e.,

$$P_x = r_x = \mu c_x = DC_x = DY_{x/s} \left(c_{s0} - \frac{K_s D}{\mu_{max} - D} \right) \tag{11.57}$$

Fig. 11.16 shows cell concentration, limiting substrate concentration and cell formation rate as a function of dilution rate at $c_{s0} = 10$ g/L, $\mu_{max} = 1$ h^{-1}, $Y_{x/s} = 0.5$, and $K_s = 0.2$ g/L. There is a maximum value in the cell formation curve. Setting $dP_x/dD = 0$ gives the optimal dilution rate D_{opt} (Fig. 11.17).

$$D_{opt} = \mu_{max} \left(1 - \sqrt{\frac{K_s}{(K_s + c_{s0})}} \right) \tag{11.58}$$

Then, cell concentration in the reactor is

$$c_x = Y_{x/s} \left(c_{s0} + K_s - \sqrt{K_s(K_s + c_{s0})} \right) \tag{11.59}$$

The maximum cell formation rate, $P_{x,max}$ is

$$P_{x,max} = Y_{x/s} \mu_{max} c_{s0} \left(\sqrt{1 + \frac{K_s}{c_{s0}}} - \sqrt{\frac{K_s}{c_{s0}}} \right)^2 \tag{11.60}$$

FIGURE 11.16

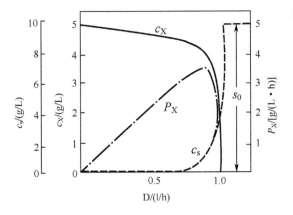

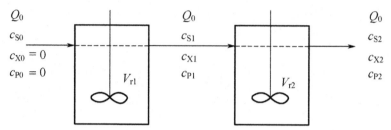

FIGURE 11.17

When $c_{s0} \gg K_s$

$$D_{opt} \approx \mu_{max} \tag{11.61}$$

$$P_{x,max} \approx Y_{x/s}\mu_{max}c_{s0} \tag{11.62}$$

In a CSTR, the relationship between the product formation rate and dilution rate can be derived by performing material balance over the reactor based on the type of product formation and kinetics equation.

If product formation belongs to type I, i.e., growth-associated product formation, performing material balance over product,

$$Q_0c_p - Q_0c_{p0} = r_p \cdot V_L \tag{11.63}$$

Because

$$r_p = q_p \cdot c_x = Y_{p/x}\mu c_x \tag{11.64}$$

Substituting into Eq. (11.63) and assuming there is no product in the feeding material, rearranging Eq. (11.63) gives product concentration

$$c_p = \frac{q_p \cdot c_x}{D} \tag{11.65}$$

Example 11.5
Escherichia coli were cultured in a 10 L CSTR at 30°C. Its kinetics equation follows the Monod equation, where $\mu_{max} = 1.0 \text{ h}^{-1}$ and $K_s = 0.2$ g/L. Glucose's feed concentration is 10 g/L, the feed volumetric flow rate is 4 L/h, and $Y_{x/s} = 0.5$
1. Determine cell concentration and growth rate in the reactor.
2. Calculate the optimal feed rate at maximum cell yield and the maximum cell yield.

Solution
1. Dilution rate of CSTR is
   $D = Q_0/V_r = 4/10 = 0.4 \text{ h}^{-1}$
   So, cell-specific growth rate is
   $\mu = D = 0.4 \text{ h}^{-1}$

(Continued)

(Continued)

From Eq. (11.51), we obtain substrate concentration in the reactor

$$c_s = \frac{K_s D}{\mu_{max} - D} = \frac{0.2 \times 0.4}{1.0 - 0.4} = 0.133 \text{g/L}$$

And cell concentration

$$c_x = Y_{x/s}(c_{s0} - c_s) = 0.5 \cdot (10 - 0.133) = 4.93 \text{g/L}$$

$$P_x = r_x = \mu c_x = D c_x = 0.4 \times 4.93 = 1.97 \text{g/L}$$

2. The optimal feed rate

$$D_{opt} = \mu_{max}\left(1 - \sqrt{\frac{K_s}{K_s + c_{s0}}}\right) = 1.0 \cdot \left(1 - \sqrt{\frac{0.2}{0.2 + 10}}\right) = 0.86 \text{ h}^{-1}$$

$$Q_0 = D_{opt} \cdot V_r = 0.86 \times 10 = 8.6 \text{L/h}$$

Cell concentration in the reactor

$$c_x = Y_{x/s}\left(c_{s0} + K_s - \sqrt{K_s(K_s + c_{s0})}\right) = 0.5\left(10 + 0.2 - \sqrt{0.2(0.2 + 10)}\right) = 4.39 \text{g/L}$$

Maximum cell formation rate

$$P_{x,max} = D_{opt} \cdot c_x = 0.8/6 \times 4.39 = 3.78 \text{g/L} \cdot \text{h}$$

11.4.2.3 CSTR in Series

CSTRs in series are usually operated in three modes when used for cell culture: (1) feed is introduced at the first tank directly; (2) feed is introduced at the first tank, but each tank has continuous supplementary feed; and (3) feed is introduced at the first tank, but part of the effluent from the last tank is recycled back to the first tank.

An example of the first case is two equal volume CSTRs operated in series as shown in Fig. 11.7. The fluid flow follows the model for multi-CSTRs in series. Assuming the reactors have the same operation conditions and yield and there is no product in the feed, then the mass balance of biomass and substrate in the first reactor is

$$c_{x1} = Y_{x/s} \cdot (c_{s0} - c_{s1}) \tag{11.66}$$

$$c_{s1} = \frac{K_s \cdot D}{\mu_{max} - D} \tag{11.67}$$

Then the cell growth rate in the first reactor is

$$r_{x1} = \mu_1 \cdot c_{x1} = D \cdot Y_{x/s} \cdot \left(c_{s0} - \frac{K_s \cdot D}{\mu_{max} - D}\right) \tag{11.68}$$

The mass balance of biomass in the second reactor is

$$Q_0 c_{x1} + \mu_2 c_{x2} V_r = Q_0 c_{x2} \tag{11.69}$$

After rearranging, we get

$$\mu_2 = D \cdot \left(1 - \frac{c_{x1}}{c_{x2}} \right) \tag{11.70}$$

Mass balance of the limiting substrate is

$$Q_0 c_{s1} = Q_0 c_{s2} + r_{s2} V_r \tag{11.71}$$

Substituting Eq. (11.53) and rearranging yields

$$\mu_2 = D \cdot Y_{x/s} \cdot \frac{(c_{s1} - c_{s2})}{c_{x2}} \tag{11.72}$$

Combining with Eq. (11.70) gives

$$c_{x2} = Y_{x/s}(c_{s0} - c_{s2}) \tag{11.73}$$

From Monod equation, we have

$$\mu_2 = \frac{\mu_{max} c_{s2}}{K_s + c_{s2}} \tag{11.74}$$

Combining Eq. (11.72) and substituting Eqs. (11.67) and (11.73), after simplifying we get

$$(\mu_{max} - D) \cdot c_{s2}^2 - \left(\mu_{max} \cdot c_{s0} - \frac{K_s \cdot D^2}{\mu_{max} - D} + K_s \cdot D \right) \cdot c_{s2} + \frac{K_s^2 \cdot D^2}{\mu_{max} - D} = 0 \tag{11.75}$$

Substrate concentration in the effluent of second reactor at different dilution rates, c_{s2}, can be obtained by solving this quadratic equation. Apparently, c_{s2} has to be smaller than c_{s1}.

Cell growth rate in the second reactor can be obtained from c_{x2} and μ derived from Eqs. (11.73) and (11.70), respectively.

$$r_{x2} = \mu_2 \cdot c_{x2} \tag{11.76}$$

Fig. 11.18 shows cell concentration, substrate concentration, and cell formation rate as a function of dilution rate in each of the two CSTR reactors in series.

As shown in the figure, the critical dilution rates in the two-stage equal volume CSTR reactors in series are identical. At $D < D_c$, $c_{x2} > c_{x1}$, even when the dilution rate is close to D_c, c_{s2} is still much smaller than c_{s2}, which indicates that substrate utilization is more complete when two CSTRs are in series. In addition, cell growth rate in the second reactor is much lower than that in the first reactor and so on. When N numbers of equal volume CSTRs are in series, at steady-state, cell concentration, specific growth rate, and substrate concentration in the Pth reactor are

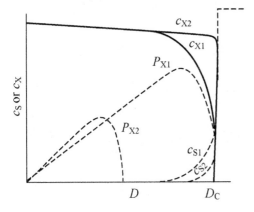

$$c_{x,P} = \frac{D \cdot c_{x,P-1}}{D - \mu_p} \tag{11.77}$$

$$\mu_p = D \cdot \left(1 - \frac{c_{x,P-1}}{c_{x,P}}\right) \tag{11.78}$$

$$c_{s,P} = c_{s,P-1} - \frac{\mu_p \cdot c_{x,P}}{D \cdot Y_{x/s}} \tag{11.79}$$

11.4.2.4   Fixed Bed Reactors

Assuming a fixed bed reactor was uniformly packed with immobilized enzyme and fluid flow through the bed is close to plug flow. Therefore, a plug flow model can be applied to this situation. For enzyme kinetics that follows the Michaelis—Menten equation, under isothermal condition with no external diffusion, the macro kinetics equation is

$$R_A^* = \eta \frac{r'_{max} c_s}{K'_m + c_s} \tag{11.80}$$

where K'_m and r'_{max} are Michaelis constant and maximum reaction rate of the immobilized enzyme intrinsic kinetics, respectively, which can be obtained from the study of immobilized enzyme intrinsic kinetics in the condition of excluding internal and external diffusion.

Since it is an isochoric process, the limiting substrate concentration distribution in the axial direction in the fixed bed reactor can be obtained based on Eq. (4.8)

$$-u_0 \frac{dc_s}{dZ} = \frac{\eta \cdot r'_{max} \cdot c_s}{(K'_m + c_s)} \tag{11.81}$$

Initial condition: at $Z = 0$, $c_s = c_{s0}$

Solving the above equation gives the axial substrate concentration distribution in the fixed bed reactor. Space-time in the fixed bed reactor can be obtained from Eq. (7.25) as shown below.

$$\tau = \frac{V_r}{Q_0} = -\int_{c_{s0}}^{c_s} \frac{dc_s}{\eta \cdot \frac{r'_{max} c_s}{(K'_m + c_s)}} \tag{11.82}$$

The internal effectiveness factor can be calculated from Section 11.3.3. Eq. (11.82) has an analytical solution only when the internal effectiveness factor of immobilized enzyme is a constant, i.e.,

$$\tau = \frac{1}{\eta \cdot r'_{max}} \left[K_m \ln\left(\frac{c_{s0}}{c_s}\right) + (c_{s0} - c_s) \right] \tag{11.83}$$

When $c_s \ll K_m$, it can be treated as a pseudo first order irreversible reaction.

REFERENCE

[1]   Atkinson DE. Cellular energy metabolism and its regulation. New York: Academic Press; 1977.

FURTHER READING

Bailey JE, Ollis DF. Biochemical engineering fundamentals. 2nd ed. New York: McGraw-Hill; 1986.

Nielsen J, Villadsen J. Bioreaction engineering principles. New York: Plenum Press; 1994.

Arkinson B, Mavituna F. Biochemical engineering and biotechnology technology handbook. 2nd ed. New York: Stockton Press; 1991.

Blanch HW, Clark DS. Biochemical engineering. New York: Marcel Dekker; 1996.

Van't Riet K, Tramper J. Basic bioreactor design. New York: Marcel Dekker; 1991.

Fogler HS. Elements of chemical reaction engineering. 3rd ed. Englewood Cliffs, NJ: Prentice-Hall; 1992.

Aiba S, Nagai S. Biochemical engineer—reaction kinetics. Translated by Zhang zhu HU, Changfu Fang, Weijing Wu et al. Beijing: Chemical Industry Press; 1994.

Yamane T. Biochemical reaction engineering. Translated and adapted by Bin Zhou. XiAn: Northwest University Press; 1992.

Yu J, Tang X. Biotechnology. Shanghai: East China Institute of Chemical Technology Press; 1992.

Yizheng QI, Wang S. Bioreaction engineering and bioreactors. Beijing: Chemical Industry Press; 1996.

Guo Y. Enzyme engineering. Beijing: China Light Industry Press; 1994.

PROBLEMS

11.1 For an enzyme-catalyzed reaction the initial reaction rates at different concentrations are given in the table below:

$c_s \times 10^4/(mol/L)$	41.0	9.50	5.20	1.03	0.490	0.106	0.051
$r_s \times 10^6/(mol/(L \cdot min))$	177	173	125	106	80	67	43

Assuming the reaction kinetics follow the Michaelis–Menten equation, please estimate r_{max} and K_m.

11.2 The Monod equation was developed by Monod using four sets of experimental data. One set of the data is given in the table below. In a batch reactor, lactose was used as substrate for cell culture, and concentrations of the substrate (c_s) and cells (c_x) were measured at different times:

No.	Period Δt/h	Substrate concentration c_s/(g/L)	Cell concentration c_x/(g/L)	No.	Period Δt/h	Substrate concentration c_s/(g/L)	Cell concentration c_x/(g/L)
1	0.52	158	15.8~22.8	5	0.36	25	48.5~59.6
2	0.38	124	22.8~29.2	6	0.37	19	59.6~66.5
3	0.32	114	29.2~37.8	7	0.38	2	66.5~67.8
4	0.37	94	37.8~48.5				

Please simulate the experimental data using Monod Equation and estimate model parameters μ_m and K_s.

11.3 An enzyme-catalyzed reaction takes place at a given initial enzyme concentration, and the initial reaction rate was measured at different substrate concentrations (see table below). Under the same conditions, another set of tests with inhibitor concentration of 1.5×10^{-5} mol/L were performed, and the results were also listed in the table.

$c_s \times 10^3$/(mol/L)	0.98	0.20	0.32	0.40	0.50	0.61
$r_s \times 10^6$/(mol/L)	16.2	32.3	38.6	44.2	50.0	55.2
$r_{sl} \times 10^6$/(mol/L)	9.70	17.8	25.8	30.4	35.4	40.2

Please determine the type of inhibition using the data given in the table, and estimate kinetic parameters K_m, r_{max}, and K_I.

11.4 In a 0.5 L CSTR lactose was used as substrate to grow *E. coli*. Inlet concentration of lactose is 160 mg/mL. At different feed flow rates (Q_0) substrate concentration (c_s) and cell concentration (c_x) were measured at reactor outlet:

Q_0/(L/min)	0.1	0.2	0.4	0.5
$c_s \times 10^3$/(mg/mL)	4	10	40	100
$c_x \times 10^3$/(mg/mL)	15.6	15	12	6

Assuming that *E. coli* growth follows the Monod Equation, please develop *E. coli* growth rate equation using the experimental data.

11.5 A reaction catalyzed by a free enzyme take places in a batch reactor with volume of 5 L, and the initial concentration of substrate is 2.77×10^{-4} mol/L. Assume the reaction kinetics can be described by Michaelis–Menten equation with the following parameters:

$K_m = 1.13 \times 10^{-3}$ mol/L

$r_{max} = 2.58 \times 10^{-5}$ mol/L · min

Please calculate:

1. Reaction time needed to achieve 90% substrate conversion

    2. Reaction time needed to achieve 90% substrate conversion if reactor volume if 10 L while all the other conditions are the same as (1)

    3. Reaction time needed to achieve 90% substrate conversion if initial substrate concentration is 2.77×10^{-3} mol/L while all the other conditions are the same as (1).

11.6 A reaction catalyzed by free enzyme takes place in a batch reactor. The initial concentration of substrate is 8.97×10^{-3} mol/L. The reaction kinetics can be described by Michaelis–Menten equation with $K_m = 3.98 \times 10^{-3}$ mol/L. After 53 min 50% of the substrate was converted, please calculate substrate conversions at 80 and 120 min.

11.7 A reaction takes place in a fixed bed reactor with immobilized enzyme catalyst, and the feed flow in the reactor can be considered as plug flow. The apparent reaction kinetic in the reactor can be described by Michaelis–Menten equation with the following kinetic parameters:

$$K_m^{app} = 1.25 \times 10^{-4} \text{mol/L}$$

$$r_{max}^{app} = 2.06 \times 10^{-5} \text{mol/L} \cdot \text{min}$$

    Initial concentration of the substrate is 5.67×10^{-4} mol/L, please calculate space-time needed to achieve 95% conversion at reactor outlet.

11.8 In a CSTR with volume of 1 m^3, *E. coli* was cultured using mannitol as the limiting substrate. The reaction kinetics follows the Michaelis–Menten equation. $\mu_{max} = 1.2$ h^{-1}, $K_s = 2$ g/m^3, $Y_{x/s} = 0.1$ gram cell/gram mannitol. Feed flow rate is 0.88 m^3/h, and concentration of mannitol in the feed is 6 g/m^3. Please calculate:

    1. Cell and mannitol concentrations at reactor outlet;

    2. Optimal feed flow rate to achieve maximum *E. coli* growth rate and the maximum *E. coli* growth rate.

11.9 In a 0.5 L batch reactor glucose was used as the limiting substrate to culture a genetic engineered cell—*Bacillus subtilis* modified with gene of thermally stable α-Amylase—and its host strain TN106. The cell growth rates at different initial glucose concentration were measured and given in the table below:

Initial glucose concentration c_{s0}/(g/L)	0.001	0.002	0.005	0.020	0.500
Growth rate of host strain μ^-/h^{-1}	0.695	0.828	0.889	0.916	0.960
Growth rate of genetic engineered cell μ^+/h^{-1}	0.550	0.711	0.795	0.805	0.876

    Now a 5 L CSTR is used to culture these two cells. If feed flow rate is 4 L/h, please calculate glucose concentration at reactor outlet for these two cells.

11.10 A 50 L CSTR was used for cell culture, and the cell growth rate follows the Monod equation with $\mu_{max} = 1.5$ h^{-1}; $K_s = 1$ g/L;

c_{s0} = 30 g/L; $Y_{x/s}$ = 0.08; and c_{x0} = 0.5 g/L. Under steady-state operation, please determine:

1. Correlation between substrate concentration c_s and dilution rate D
2. Correlation between cell concentration c_x and dilution rate D
3. c_s and c_x at reactor outlet when feed volume flow rate Q_0 is 37.5 L/h.

11.11 An enzyme-catalyzed reaction takes place in a fixed bed reactor packed with 2-mm diameter immobilized enzyme catalyst. The bed voidage is 0.5. The intrinsic kinetic parameters for the immobilized enzyme are:

$$r'_{max} = 0.1 \ \text{mol}/(\text{m}^3 \cdot \text{s})$$

$$K'_m = 100 \ \text{mol}/\text{m}^3$$

Substrate feed flow rate is 1 m³/min, and substrate concentration in the feed is 3 mol/m³. The external diffusion influence can be ignored and the effective diffusion coefficient of substrate in the immobilized enzyme catalyst particles is $D_e = 1 \times 10^{-9}$ m²/s. Please calculate reactor volume needed when substrate concentration is 1 mol/m³.

Chapter 12

Fundamentals of Polymerization Reaction Engineering

Chapter Outline

Polymer products such as synthetic plastics, fibers, rubbers, coatings, adhesives have been applied more and more broadly, and become an group of indispensable material in modern life. Basic human needs such as clothing, food, lodging, and transportation are closely related with polymer products. In this chapter, the following engineering issues of producing these products will be discussed:

1. Polymerization kinetics.
2. Relationship between polymerization rate and reactor and its operation mode.
3. Relationship between degree of polymerization and its distribution and reactor as well as operation mode.

Reaction Engineering. DOI: http://dx.doi.org/10.1016/B978-0-12-410416-7.00012-4
© 2017 Chemical Industry Press. Published by Elsevier Inc. All rights reserved.

4. Heat and mass transfer processes in polymerization.
5. Polymerization reactor design and scale-up as well as other related issues.

12.1 OVERVIEW

Polymerization is a process converting low molecular weight monomer to high molecular weight polymers. Due to their significant performance of plasticity, fiber forming, film forming, high elasticity, etc., which low molar mass monomers do not possess, polymers can be widely used for plastics, fibers, rubber, coatings, adhesives, etc. Polymer compounds consist of one or more of the structural unit (monomer) and are synthesized by repeated reaction of monomers.

It is well known, when comparing to reactions that only involve low molecular weight compounds, polymerization and polymer production have the following characteristics.

(1) Complex Kinetics

Polymerization systems and products are diverse, and so are the reaction mechanisms involved, which leads to very multiple reaction kinetics. In addition, even trace impurities can have significant influence on product quality and may lead to poor reproducibility. As a result, very high purity of raw materials is often required.

(2) Reaction Randomness

Polymers produced in polymerization reactions are often not uniform, which means that many molecules with different sizes and structures are generated. Therefore, in addition to conversion of monomers, it is also necessary to consider the average degree of polymerization and its distribution of the products. In addition, polymer composition, molecular structure, and arrangement, as well as polymer properties are often also needed to be considered. When multiple monomers are copolymerized, the reactions become even more complicated.

(3) High Viscosity of Polymerization Fluids

Most polymers are highly viscous, and the viscosity increases sharply with the increasing of molecular weight of the product during the polymerization. Due to their high viscosity polymer solutions are often nonideal fluids, among which some are non-Newtonian fluids or multiphase systems, and some can be regarded as macroscopic fluids. The fluid nonideality presents new challenges for handling fluid flow and mixing as well as heat and mass transfer. In addition, the structure of the reaction apparatus also has to be specially designed based on the characteristics of the physical system and product performance requirements, which makes transfer processes in the polymerization reactor more complicated and the flow model more complex.

(4)   Rapid Reaction Rate and Strong Exothermicity

Most polymers have a poor thermal conductivity. Even in systems with a large amount of solvent or water, the heat transfer performance declines as the degree of polymerization increases. On the other hand, the degree of polymerization of the polymer is very sensitive to temperature. Therefore, properly managing the balance of heat generation and heat removal during the polymerization process is particularly critical, and often becomes the key factor to be considered when the developing and designing a polymerization reaction process. The heat transfer enhancement and temperature control are very critical for the polymerization process.

From the discussion above, we can see that it is difficult to solve engineering problems associated with polymerization reactions, and the ever growing needs of the great variety of polymer products with special properties and unique performances make polymerization reaction engineering both exciting and challenging. The lack of basic data for some new systems makes the development of suitable mathematical models for design and scale-up more difficult. Until now, design and scale-up of commercial polymerization process have been very challenging, and there is an urgent need of a better understanding of the fundamentals and developing practical methods or theory. However, in the past several decades, to meet the needs of the rapid development of polymeric materials, significant production experience and engineering knowledge of polymerization processes and optimization of polymerization reactor have been accumulated. The single train production capacity can be up to 50 million metric tons per year, and polymerization reactor volume can be up to 200 m³. Various polymerization reactors suitable for the characteristics of a variety of products have been developed.

The design principles, methods, and procedures of polymerization reactors are the same as those for other reactors. The reaction kinetic and transport processes should be considered simultaneously in design of an industrial reactor. The design model has to be established based on experimental data. The inherent characters of the polymerization system should be analyzed carefully and the unique property of the polymerization reaction should also be considered.

12.2 KINETIC ANALYSIS

Polymerization kinetics characterize the dependence of reaction rate on temperature, related component concentrations and other parameters. It reflects the nature of the chemical reaction. But in industrial reactors, the transport processes influence the temperature and concentration distribution and then the polymerization reaction rate. Reaction kinetics is the basis for reactor development and design. It is also the basis of correctly selecting the reactor type and optimizing reactor operation.

The characteristics of the polymerization process determine the main tasks of kinetics study. The primary focus is the reaction rate, i.e., the consumption rate of monomers and the formation rate of product polymer. The degree of polymerization (or molecular weight) and its distribution as well as the structure and composition of the produced polymer are also important and will be discussed in this chapter.

Engineering analysis of polymerization reaction rate relies on experimental data to determine the reaction mechanism and rate equation of each elementary reaction involved. Then overall polymerization rate, degree of polymerization and its distribution, rate constants of elementary reactions, and evolution of concentration of each species with time can be derived based on steady-state approximation and rate-determining step hypotheses. Of course, the kinetics analysis must be based on the specific type of polymerization reaction, and hence next we will first review the main types of polymerization reactions.

12.2.1 Types of Polymerization Reaction

Currently there is no uniformly accepted classification of polymerization reactions. According to the reaction mechanism, the polymerization reaction can be classified as follows.

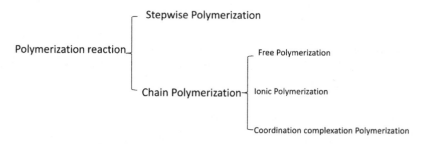

Most polycondensation reactions and polyurethane synthesis reactions belong to stepwise polymerizations characterized by the fact that the reaction is carried out step by step in the process of converting the low molecular weight molecules into the polymer. In the initial stage of reaction, most monomers are polymerized into oligomers such as dimers, trimers, and tetramers. Even at the beginning of the reaction conversion can be very high. Subsequently, the oligomers react with each other, and the molecular weight increases. With that the conversion tends to increase more slowly. Most of the addition reactions of vinyl monomers are chain polymerization reactions, and reactions usually undergo the three stages of chain initiation, propagation, and termination. Chain initiation can be induced by thermal, photo, or radiative approaches, or by using initiators. The most commonly used method is using initiators, and the dosage is generally small—molar concentration is typically between 10^{-4} and 10^{-2} mol/m^3. Firstly, part of the monomer molecules are excited into active molecules (radicals) by initiators, then the active molecules grow by addition with monomer molecules, forming active chains with different sizes before termination. Eventually polymer molecules of the final size (dead polymers) are generated. Active molecules generated during the process also experience propagation and termination from generation to generation, producing dead polymer molecules with different sizes.

Chain polymerization reaction is the most widely used and important class of reaction to produce polymer compounds. According to different polymerization mechanisms, it can be subdivided into radical polymerization (e.g., polyethylene, polyacrylonitrile, and polystyrene), ionic polymerization (initiated by ions generated by a catalyst, carrying out chain propagation in the form of ions, and generally having high degree of selectivity. It can be divided into cationic polymerization and anionic polymerization based on whether the growing ions are positive or negative), coordination complexation polymerization, and so on.

In addition, two or more different monomers can be polymerized by different mechanisms, which is called copolymerization. Monomer molecules with ring structures can polymerize through ring-opening. Whether such polymerization is stepwise polymerization or chain polymerization is determined by reaction conditions.

12.2.2 Degree of Polymerization and Distribution

In comparison with other reactions that only involve low molecular weight molecules, an important feature of polymerization reactions is that the molecular weights of resulting polymers are not uniform, i.e., the number of monomers constituting the product polymer molecules is different. The number of monomers in a polymer chain is defined as degree of polymerization. The random characteristic of the polymerization process results in polymerization degree distribution. For example, for radical polymerization, one end of an active polymer (active chain) may react with another monomer that will increase the degree of polymerization by one, or it may react with another active chain to form an inactive polymer (dead polymer). This is an unpredictable event that is governed by probability.

Obviously, the physical properties of polymer products are closely related with the average value as well as the distribution of degree of polymerization.

In industrial process, the average degree of polymerization and its distribution are usually the main control targets. It has been proved that the average degree of polymerization and its distribution of the product depends on not only the type of reactor used, operating modes, and conditions, but also polymerization methods. Small changes in these conditions will often result in significant differences of product quality.

From different perspectives or using different methods of measuring, the degree of polymerization and its distribution can be defined differently based on different perspectives or measurement methods.

12.2.2.1 Number-Average Value

Generally, molecular weight measured by end-group titration, freezing-point depression, and osmotic pressure measurement is called number-average value

1. Number-average molar mass $\overline{M}_n$

$$\overline{M}_n = \frac{\sum\limits_{j=2}^{\infty} M_j N_j}{\sum\limits_{j=2}^{\infty} N_j} = \frac{W}{N} \tag{12.1}$$

$j = 2, 3 \cdots \infty$, [assuming polymers are dipolymer or higher (the same below)] where N_j is the number of polymer molecules with degree of polymerization of j; M_j is the molar weight of polymer with degree of polymerization of j (polymer j); $N = \sum_{j=2}^{\infty} N_j$ represents total number of polymer molecules; $W = \sum_{j=2}^{\infty} M_j N_j$ represents total mass of all polymer molecules. Therefore, $\overline{M}_n$ is the average molecular weight. This can also be represented by Number-average degree of polymerization $\overline{P}_n$ (i.e., the average number of monomers of polymer molecules),

$$\overline{P}_n = \frac{\sum\limits_{j=2}^{\infty} j N_j}{\sum\limits_{j=2}^{\infty} N_j} \tag{12.2}$$

or

$$\overline{P}_n = \frac{\sum\limits_{j=2}^{\infty} j [P_j]_j}{\sum\limits_{j=2}^{\infty} P_j} = \frac{\sum\limits_{j=2}^{\infty} j [P_j]}{[P]} \tag{12.3}$$

where $[P_j]$ is the concentration of polymer j, $[P_j] = \dfrac{N_j}{V_r}$; $[P]$ is the total concentration of polymers, $[P] = \sum_{j=2}^{\infty} [P_j] = \dfrac{\sum_{j=2}^{\infty} N_j}{V_r}$.

2. Instantaneous number-average degree of polymerization $\bar{P}_n$

The number-average degree of polymerization at a specific moment is called instantaneous number-average degree of polymerization.

$$\bar{P}_n = \frac{\sum\limits_{j=2}^{\infty} j r_{pj}}{\sum\limits_{j=2}^{\infty} r_{pj}} = \frac{r_M}{r_P} \tag{12.4}$$

where r_{pj} is formation rate of polymer j, $r_P = \sum_{j=2}^{\infty} r_{pj}$ is the total formation rate of polymers, and r_M is monomer consumption rate.

In contrast with instantaneous number-average degree of polymerization $\bar{P}_n$, $\overline{P}_n$ is called cumulative number-average degree of polymerization, which is the ratio of numbers of consumed monomer M to total numbers of polymer P accumulated during time period of $0 \sim t$, for a batch reactor with constant volume.

$$\overline{P}_n = \frac{\int_0^t r_M d_t}{\int_0^t r_p d_t} = \frac{\int_0^{[M]} -d[M]}{\int_0^{[P]} d[M]} = \frac{[M]_0 - [M]}{\int_0^x \frac{[M]_0 dX}{\bar{P}_n}} = \frac{X}{\int_0^x \frac{1}{\bar{P}_n} dX} \tag{12.5}$$

Eq. (12.5) gives the relationship between instantaneous number-average degree of polymerization $\bar{P}_n$ and cumulative number-average degree of polymerization $\overline{P}_n$.

The average degree of polymerization tells us the average length of polymer molecular chains. Sometimes polymer properties also depend on the distribution of the degree of polymerization. For this discrete random variable, we can use the fraction of a polymer with a certain degree of polymerization (or molecular weight) in total polymers to characterize the distribution of the degree of polymerization,

3. Number-average distribution function

Just as the degree of polymerization, the number-average distribution function can also be defined as the instantaneous and cumulative number-average distribution functions:

$$f_n(j) = \frac{r_{pj}}{\sum\limits_{j=2}^{\infty} r_{pj}} = \frac{r_{pj}}{r_p} \tag{12.6}$$

$$F_n(j) = \frac{N_j}{\sum\limits_{j=2}^{\infty} N_j} = \frac{[p_j]}{\sum\limits_{j=2}^{\infty} [p_j]} = \frac{[p_j]}{[P]} \tag{12.7}$$

Eq. (12.6) can be adapted as

$$f_n(j) = \frac{d[P_j]}{d[P]} \tag{12.8}$$

When $t = 0$, $[P_j] = 0$, $[P] = 0$, integrating the above equation gives:

$$[P_j] = \int_0^{[P]} f_n(j)d[P] \tag{12.9}$$

Then substituting it into Eq. (12.7) gives

$$F_n(j) = \frac{[P_j]}{[P]} = \frac{\int_0^{[P]} f_n(j)d[P]}{[P]} \tag{12.10}$$

Substituting $[P] = \dfrac{[M]_0 X}{\overline{P}_n}$ or $d[P] = \dfrac{[M]_0 dX}{\overline{P}_n}$ into Eq. (12.10), the relationship between cumulative and instantaneous number fraction distribution can be obtained:

$$F_n(j) = \frac{\overline{P}_n}{X} \int_0^X \frac{f_n(j)}{\overline{P}_n} dX \tag{12.11}$$

12.2.2.2 Weight-Average Method

Molecular weight measured by the light scattering method is defined as the weight-average value (the value in base of weight).

1. Weight-average molecular weight

$$\overline{M}_w = \frac{\sum\limits_{j=2}^{\infty} W_j M_j}{\sum\limits_{j=2}^{\infty} W_j} = \frac{\sum\limits_{j=2}^{\infty} W_j^2 M_j}{\sum\limits_{j=2}^{\infty} W_j N_j} \tag{12.12}$$

Or weight-average degree of polymerization $\overline{P}_w$

$$\overline{P}_w = \frac{\sum\limits_{j=2}^{\infty} j^2[P_j]}{\sum\limits_{j=2}^{\infty} j[P_j]} = \frac{\sum\limits_{j=2}^{\infty} j^2 N_j}{\sum\limits_{j=2}^{\infty} j N_j} \tag{12.13}$$

2. Instantaneous weight-average degree of polymerization $\overline{p}_w$

$$\overline{p}_w = \frac{\sum\limits_{j=2}^{\infty} j^2 r_{pj}}{\sum\limits_{j=2}^{\infty} j r_{pj}} = \frac{\sum\limits_{j=2}^{\infty} j^2 r_{pj}}{r_M} \tag{12.14}$$

The relationship between instantaneous polymerization and cumulative weight-based degree of polymerization can be derived as,

$$\overline{P}_w = \frac{1}{X}\int_0^X \overline{p}_w dX \qquad (12.15)$$

3. Instantaneous weight-average distribution function $f_w(j)$

$$f_w(j) = \frac{jr_{pj}}{\sum\limits_{j=2}^{\infty} jp_{pj}} = \frac{jr_{pj}}{\sum\limits_{j=2}^{\infty} r_{Mj}} = \frac{jr_{pj}}{r_M} \qquad (12.16)$$

4. Cumulative weight-average distribution function $F_w(j)$

$$F_w(j) = \frac{jN_j}{\sum\limits_{j=2}^{\infty} jN_j} = \frac{j[p_j]}{\sum\limits_{j=2}^{\infty} j[p_j]} \qquad (12.17)$$

The relationship between $f_w(j)$ and $F_w(j)$ can be derived as:

$$F_w(j) = \frac{1}{X}\int_0^X f_w(j)dX \qquad (12.18)$$

12.2.2.3 Z-Average Method

The average molecular weight measured by ultracentrifugation is named z-average molecular weight.

1. z-average molecular weight $\overline{M}_z$

$$\overline{M}_z = \frac{\sum\limits_{j=2}^{\infty} M_j^3 N_j}{\sum\limits_{j=2}^{\infty} M_j^2 N_j} \qquad (12.19a)$$

Or z-average degree of polymerization $\overline{P}_z$

$$\overline{P}_z = \frac{\sum\limits_{j=2}^{\infty} j^3 N_j}{\sum\limits_{j=2}^{\infty} j^2 N_j} \qquad (12.19b)$$

2. Instantaneous z-average degree of polymerization

$$\overline{p}_z = \frac{\sum\limits_{j=2}^{\infty} j^3 r_{pj}}{\sum\limits_{j=2}^{\infty} j^2 r_{pj}} \qquad (12.20)$$

12.2.2.4   Viscosity-Average Method

Molecular weight measured by viscosity method is defined as viscosity-average molecular weight $\overline{M}_v$,

$$\overline{M}_v = \left[\frac{\sum_{j=2}^{\infty} M_j^{\alpha+1} N_j}{\sum_{j=2}^{\infty} M_j N_j} \right]^{\frac{1}{\alpha}} \tag{12.21}$$

Or viscosity-average degree of polymerization $\overline{P}_v$

$$\overline{P}_v = \left[\frac{\sum_{j=2}^{\infty} j^{\alpha+1} N_j}{\sum_{j=2}^{\infty} j N_j} \right]^{\frac{1}{\alpha}} \tag{12.22}$$

where α is called the Mark-Houwink Index, a constant related to polymer types. Table 12.1 summarizes the methods for measuring average molecular weight in different ranges. To measure the distribution of degree of polymerization, the samples are usually divided into several grades according to the degree of polymerization so that the average degree

TABLE 12.1 Methods for Measuring Average Molar Mass in Different Ranges

Representation Methods	Measure Methods	Classes	Ranges
Number-average	End-group titration	$\overline{M}_n$	$M < 5 \times 10^4$
	Freezing-point depression		$M < 3 \times 10^4$
	Osmotic pressure measurement		$2 \times 10^4 < M < 10^6$
	Vapor pressure depression		$M < 2 \times 10^4$
	Freezing point ascent		$M < 3 \times 10^4$
Weight-average	Light scattering	$\overline{M}_w$	$10^4 < M < 10^7$
	X-ray small angle scattering		$10^3 < M < 5 \times 10^5$
Z-average	Ultracentrifugation	$\overline{M}_z$	$10^4 < M < 10^7$
Viscosity-average	Viscosimetry	$\overline{M}_v$	Wide range

of polymerization of each grade can be measured using appropriate methods, then the overall distribution of degree of polymerization can be calculated.

Generally speaking, the value of the four average degrees of polymerization as defined above follows the following trend

$$\overline{P}_n \le \overline{P}_v \le \overline{P}_w \le \overline{P}_z \qquad (12.23)$$

If the molar mass of polymer is of monodispersity, i.e., degree of polymerization is the same, the four average degrees of polymerization will have the same value. However, in most cases the distribution of the degree of polymerization follows a certain distribution. The degree of dispersion of the distribution can be characterized by the ratio $\overline{P}_w / \overline{P}_n$, which is called the dispersancy index D.

$$D = \frac{\overline{P}_w}{\overline{P}_n} = \frac{\overline{M}_w}{\overline{M}_n} > 1 \qquad (12.24)$$

For the normal distribution, $\overline{M}_z : \overline{M}_w : \overline{M}_n = 3:2:1$

Example 12.1

The distribution of weight-average degree of polymerization of a polymer is shown in the following table

$j \times 10^{-3}$	0	0.2	0.4	0.6	0.8	1.0	1.5	2.0	2.5	3.0	3.5	4.0
$F_w(j) \times 10^4$	0	2.8	5.1	6.4	6.65	6.2	4.1	2.2	0.8	0.25	0.1	0

What is the number-average, weight-average, and z-average degree of polymerization?

Solution

From Eq. (12.17)

$$F(j) = \frac{jN_j}{\sum\limits_{j=2}^{\infty} jN_j}$$

Firstly, we need to derive between $\overline{P}_n$, $\overline{P}_w$, and $\overline{P}_z$ with $F_w(j)$

$$\overline{P}_n = \frac{\sum\limits_{j=2}^{\infty} jN_j}{\sum\limits_{j=2}^{\infty} N_j} = \frac{1}{\sum\limits_{j=2}^{\infty} \dfrac{jN_j}{\sum\limits_{j=2}^{\infty} jN_j}} = \frac{1}{\sum\limits_{j=2}^{\infty} \dfrac{F_w(j)}{j}} \qquad (A)$$

(Continued)

(Continued)

And the weight-average degree of polymerization and z-average degree of polymerization, respectively, are

$$\overline{P}_w = \frac{\sum\limits_{j=2}^{\infty} j^2 N_j}{\sum\limits_{j=2}^{\infty} N_j} = \frac{\sum\limits_{j=2}^{\infty} j \cdot j N_j}{\sum\limits_{j=2}^{\infty} j N_j} = \sum\limits_{j=2}^{\infty} j F_w(j) \tag{B}$$

$$\overline{P}_z = \frac{\sum\limits_{j=2}^{\infty} j^3 N_j}{\sum\limits_{j=2}^{\infty} N_j} = \frac{\sum\limits_{j=2}^{\infty} j^2 \cdot j N_j}{\sum\limits_{j=2}^{\infty} j \cdot j N_j} = \frac{\sum\limits_{j=2}^{\infty} j^2 F_w(j)}{\sum\limits_{j=2}^{\infty} j F_w(j)} \tag{C}$$

If $F_w(j)$ is a continuous function, we can adopt numerical integration to calculate instead of calculating $\sum_{j=2}^{\infty}$. Taking the calculation of $\overline{P}_n$ as an example, dividing the integral interval into two parts of $j = 0 \sim 1.0 \times 10^3$ and $j = 1.0 \times 10^3 - 4.0 \times 10^3$, and calculating the integration using the trapezoidal method, we can obtain the following table

$j \times 10^{-3}$	0	0.2	0.4	0.6	0.8	1.0	1.5	2.0	2.5	3.0	3.5	4.0
$F_w(j) \times 10^4$	0	2.8	5.1	6.4	6.65	6.2	4.1	2.2	0.8	0.25	0.1	0
$\dfrac{F_w(j)}{j} \times 10^4$	0	14.00	12.75	10.67	8.31	6.2	2.73	1.10	0.32	0.083	0.029	0

Eq. (A) can be transformed as

$$\overline{P}_n = \frac{1}{\int_0^{4.0} \dfrac{F_w(j)}{j} dj} = \left[\int_0^{1.0 \times 10^3} \frac{F_w(j)}{j} dj + \int_{1.0 \times 10^3}^{4.0 \times 10^3} \frac{F_w(j)}{j} dj \right]^{-1}$$

$$= \left\{ 0.2 \times 10^3 \left[\frac{0 + 6.2}{2} + 14.00 + 12.75 + 10.67 + 8.31 \right] \times 10^{-7} + 0.5 \right.$$

$$\left. \times 10^3 \left[\frac{6.20 + 0}{2} + 2.73 + 1.10 + 0.32 + 0.083 + 0.029 \right] \times 10^{-7} \right\}^{-1}$$

$$= 744$$

Following the same procedure, the weight and z-average degree of polymerization can be calculated:

$$\overline{P}_w = 1.13 \times 10^3$$
$$\overline{P}_z = 1.5 \times 10^3$$

12.2.3 Homogeneous Free Radical Polymerization Reaction

Free radical addition polymerization has been used for synthesizing a broad range of products, such as polyethylene, polystyrene, polyacrylonitrile, polyvinyl acetate, and polytetrafluoroethylene. All of them are commercially mass-produced and widely used polymer materials. These free radical polymerization reactions are typical chain reactions, which include chain initiation, propagation, and termination, but may involve different chain initiation methods and chain termination mechanisms. The active chains as reactive intermediates can carry out chain transfer to other substances. Sometimes polymerization inhibitors can be added to control free radical concentrations or stop the polymerization process. For the development of polymerization processes and the design of polymerization devices it is very important to understand the reaction mechanism.

12.2.3.1 Mechanism

For free radical polymerization reactions taking place in a homogeneous-phase and constant-volume reactor, the elementary reaction steps and corresponding rate equations are summarized in Table 12.2.

In Table 12.2, I, M, P, R, S, and Z represent initiator, monomer, polymer, intermediate compound, solvent, and impurity, respectively, while R^*, P^*, S^*, and Z^* represent the free radicals of each corresponding matter; [ ] represents concentrations, e.g., [I] represents the concentration of the initiator; (I) represents intensity of light; f represents the initiation efficiency of the initiator, and its value generally is less than 1; j or i is structure unit number of the monomer, and its value is from 1 to ∞. In addition, except for k_d, rate constants of elementary steps are all based on the products. The rate for coupling termination, r_{tc}, is defined as the reactant consumption rate, which is two times the generation rate, as dictated by reaction stoichiometry.

For an actual polymerization reaction system, reaction rate equations for all elementary steps involved have to be obtained based on the reaction mechanism. Next, we use styrene polymerization with benzoyl peroxide as initiator as an example.

Assuming the reaction mechanism is as follows,

1. Chain initiation
Benzoyl peroxide is decomposed into free radicals

$$C_6H_5\!-\!\underset{\underset{O}{\|}}{C}\!-\!O\!-\!O\!-\!\underset{\underset{O}{\|}}{C}\!-\!C_6H_5 \xrightarrow{\Delta} 2C_6H_5\!-\!\underset{\underset{O}{\|}}{C}\!-\!O* \xrightarrow{part} C_6H_5* + CO_2$$

$$\text{(I)} \qquad\qquad\qquad\qquad \text{(R}*)$$

TABLE 12.2 Elementary Reaction and Rate Equations in a Free Radical Polymerization Mechanism

Stage	Category	Elementary Reaction	Reaction Rate Equation
Monomer initiation (i)	1. Initiator initiation	$I \xrightarrow{k_d} 2R^*$	$r_d = -\dfrac{d[I]}{dt} = k_d[I], \left(\dfrac{d[R^*]}{dt}\right)_d$ $= 2k_d[I]$
		$R^* + M \xrightarrow{k_i} P_1^*$	$r_{i1} = \left(\dfrac{d[p_1^*]}{dt}\right)_{il} = k_i[R^*][M]$ $= 2fk_d[I]$
	2. Photo initiation	$M \to P_1^*$	$r_{i2} = \left(\dfrac{d[p_1^*]}{dt}\right)_{i2} = f(I)$
	3. Thermal initiation bimolecule	$M + M \xrightarrow{k_i} P_1^*$	$r_{i3} = \left(\dfrac{d[p_1^*]}{dt}\right)_{i3} = k_i[M]^2$
	trimolecule	$M + M + M \xrightarrow{k_i} P_1^*$	$r_{i3} = \left(\dfrac{d[p_1^*]}{dt}\right)_{i3} = k_i[M]^3$
Chain propagation (p)		$P_1^* + M \xrightarrow{k_p} P_{j+1}^*$	$r_p = \left(\dfrac{d[M]}{dt}\right)_p = k_p[P^*][M]$
Chain termination (t)	Coupling termination	$P_j^* + P_i^* \xrightarrow{k_{tc}} P_{j+i}$	$r_{tc} = \left(-\dfrac{d[P^*]}{dt}\right)_{tc} = 2k_{tc}[P^*]^2$
	Disproportionation termination	$P_j^* + P_i^* \xrightarrow{k_{td}} P_j + P_i$	$r_{td} = \left(-\dfrac{d[P^*]}{dt}\right)_{td} = 2k_{td}[P^*]^2$
	Mono-radical termination	$P_j^* \xrightarrow{k_{tl}} P_i$	$r_{tl} = \left(-\dfrac{d[P^*]}{dt}\right)_{tl} = 2k_{tl}[P^*]^2$
Chain transfer (f)	Transfer into monomer	$P_j^* + M \xrightarrow{k_{fm}} P_j + P_1^*$	$r_{fm} = \left(\dfrac{d[P]}{dt}\right)_{fm} = k_{fm}[P^*][M]$
	Transfer into solvent	$P_j^* + S \xrightarrow{k_{fs}} P_j + S^*$	$r_{fs} = \left(\dfrac{d[P]}{dt}\right)_{fs} = 2k_{fs}[P^*][S]$
	Transfer into dead-polymer	$P_j^* + P_i^* \xrightarrow{k_{fp}} P_j + P_i^*$	$r_{fp} = \left(-\dfrac{d[P]}{dt}\right)_{fp} = k_{fp}[P^*][P]$
	Transfer into impurity	$P_j^* + Z \xrightarrow{k_{fz}} P_j + Z^*$	$r_{fz} = \left(-\dfrac{d[P]}{dt}\right)_{fz} = k_{fz}[P^*][Z]$

Styrene monomer initiation

$$C_6H_5-CH=CH_2 + C_6H_5-\overset{\overset{\displaystyle O}{\|}}{C}-O^* \longrightarrow C_6H_5-\overset{\overset{\displaystyle O}{\|}}{C}-OCH_2-\underset{\underset{\displaystyle C_6H_5}{|}}{CH}^*$$

$$\text{(M)} \qquad \text{(R*)} \qquad\qquad\qquad \text{(P}_1\text{*)}$$

2. Chain propagation

$$C_6H_5\overset{\overset{\displaystyle O}{\|}}{C}-O-CH_2-\underset{\underset{\displaystyle C_6H_5}{|}}{CH}^* + \underset{\underset{\displaystyle C_6H_5}{|}}{CH}=CH_2 \longrightarrow C_6H_5-\overset{\overset{\displaystyle O}{\|}}{C}-O\Big[CH_2-\underset{\underset{\displaystyle C_6H_5}{|}}{CH}\Big]_j CH_2-\underset{\underset{\displaystyle C_6H_5}{|}}{CH}^*$$

$$\text{(P}_j\text{*)} \qquad\qquad\qquad\qquad\qquad\qquad\qquad \text{(P}_{j+1}\text{*)}$$

3. Chain termination

$$C_6H_5-\overset{\overset{\displaystyle O}{\|}}{C}-O\Big[CH_2-\underset{\underset{\displaystyle C_6H_5}{|}}{CH}\Big]_j^* + \underset{\underset{\displaystyle C_6H_5}{|}}{\overset{\overset{\displaystyle CH-CH_2}{|}}{}} \longrightarrow C_6H_5\overset{\overset{\displaystyle O}{\|}}{C}-O\Big[CH_2-\underset{\underset{\displaystyle C_6H_5}{|}}{CH}\Big]_{j-1} CH_2=\underset{\underset{\displaystyle C_6H_5}{|}}{CH} + CH3-\underset{\underset{\displaystyle C_6H_5}{|}}{CH}^*$$

$$\text{(P}_j\text{*)} \qquad\qquad \text{(M)} \qquad\qquad \text{(P}_j\text{)}$$

4. Chain transfer

$$C_6H_5-\overset{\overset{\displaystyle O}{\|}}{C}-O\Big[CH_2-\underset{\underset{\displaystyle C_6H_5}{|}}{CH}\Big]_j^* + \overset{\overset{\displaystyle CH=CH_2}{}}{C_6H_6} \longrightarrow C_6H_5\overset{\overset{\displaystyle O}{\|}}{C}-O\Big[CH_2-\underset{\underset{\displaystyle C_6H_5}{|}}{CH}\Big]_{j-1} CH_2=\underset{\underset{\displaystyle C_6H_5}{|}}{CH} + CH3-\underset{\underset{\displaystyle C_6H_5}{|}}{CH}^*$$

$$\text{(P}_j\text{*)} \qquad\qquad \text{(M)} \qquad\qquad \text{(P}_j\text{)}$$

Transfer into the solvent (benzene as solvent)

$$C_6H_5-\overset{\overset{\displaystyle O}{\|}}{C}-O\Big[CH_2-\underset{\underset{\displaystyle C_6H_5}{|}}{CH}\Big]_j^* + C_6H_6 \longrightarrow C_6H_5-\overset{\overset{\displaystyle O}{\|}}{C}-O\Big[CH_2-\underset{\underset{\displaystyle C_6H_5}{|}}{CH}\Big]_{j-1} CH_2=\underset{\underset{\displaystyle C_6H_5}{|}}{CH} + C_6H_5^*$$

$$\text{(P}_j\text{*)} \qquad\qquad \text{(S)} \qquad\qquad\qquad \text{(P}_j\text{)} \qquad\qquad \text{(S*)}$$

For the mechanism listed above, the rate equations of each elementary reaction can be obtained according to mass action law,

Reaction	Reaction Rate Equation
1. $I \xrightarrow{k_d} 2R^*$	$r_d = -\dfrac{d[I]}{dt} = k_d[I]$
$R^* + M \xrightarrow{k_i} P_1^*$	$r_{i1} = \left(\dfrac{d[p_1^*]}{dt}\right)_i = k_i[R^*][M] = 2fk_d[I]$
2. $P_1^* + M \xrightarrow{k_p} P_{j+1}^*$	$r_p = \left(-\dfrac{d[M]}{dt}\right)_p = k_p[P^*][M]$

(Continued)

(Continued)

Reaction	Reaction Rate Equation
3. $P_j^* + P_i^* \xrightarrow{k_{tc}} P_{j+i}$	$r_{tc} = \left(-\dfrac{d[P^*]}{dt}\right)_{tc} = 2k_{tc}[P^*]^2$
4. $P_j^* + M \xrightarrow{k_{fm}} P_j + P_1^*$	$r_{fm} = \left(\dfrac{d[P]}{dt}\right)_{fm} = k_{fm}[P^*][M]$
$P_j^* + S \xrightarrow{k_{fs}} P_j + S_1^*$	$r_{fs} = \left(\dfrac{d[P]}{dt}\right)_{fs} = 2k_{fs}[P^*][S]$

It should be mentioned here that $[P^*] = \sum_j[P_j^*]$ represents the total concentration of active chains. The concentration of dead polymers was expressed as $[P] = \sum_j[P_j]$. In the chain transfer step, the chain transfer into monomer and solvent are parallel reactions. The total chain transfer rate can be

$$r_f = r_{fm} + r_{fs} = \left(\frac{d[P]}{df}\right)_f = (k_{fm} + k_{fs})[P^*][M] = k_f[P^*][M] \qquad (12.25)$$

where $k_f = k_{fm} + k_{fs}$. If benzene is used as a solvent, S^* and initiating free radical R^* are the same material by coincidence. They may not be the same if using other initiators or using other solvents.

The mass balance for any one compound or intermediate can be obtained according to the above elementary reaction analysis. If the effect of solvent is ignored, the rate equation can be obtained as

$$\frac{d[R^*]}{dt} = \left(\frac{d[R^*]}{dt}\right)_d - \left(\frac{d[R^*]}{dt}\right)_i = 2fk_d[I] - k_i[R^*][M] \qquad (12.26)$$

Or

$$-\frac{d[M]}{dt} = r_i + r_p + r_{fm} + r_{fs} \qquad (12.27)$$

Because the amounts of monomers consumed in chain initiation and chain transfer are far less than that consumed in chain propagation, r_i, r_{fm} and r_{fs} can be ignored when compared with r_p. Hence the total rate of styrene monomer consumption is approximately equal to the chain propagation rate:

$$-\frac{d[M]}{dt} \cong r_p = k_p[P^*][M] \qquad (12.28)$$

12.2.3.2 Total Reaction Rate

We will continue to use polyethylene polymerization discussed above as an example. Assuming the reaction takes place isothermally in a batch reactor, according to the steady-state approximation and rate-determining steps methods described in Chapter 2, Fundamentals of Reaction Kinetics, the rate equation can be derived from Eq. (12.26)

$$\frac{d[R^*]}{dt} = 2fk_d[I] - k_i[R^*][M] = 0 \tag{12.29}$$

So

$$2fk_d[I] = k_i[R^*][M]$$

For the same system the total free radical concentration in systems is a constant (generally, $r_{fs} \ll r_{tc}$, and can be ignored)

$$\frac{d[P^*]}{dt} = r_i - r_{tc} = 2fk_d[I] - k_{tc}[R^*]^2 = 0$$

So

$$[P^*] = \left(\frac{2fk_d[I]}{2k_{tc}}\right)^{1/2} \tag{12.30}$$

Combining the above equation into Eq. (12.28) we can obtain the total consumption rate of monomers

$$r_M = -\frac{d[M]}{dt} \cong \left(\frac{2fk_d}{2k_{tc}}\right)^{1/2} [I]^{1/2}[M] \tag{12.31}$$

At the initial conditions, $t = 0$, $[M] = [M]_0$, and $[I] = $ constant, then Eq. (12.31) is a simple first order irreversible reaction equation and not difficult to solve. The above result is derived based on the specific mechanism. For a different reaction, or even the same reaction but a different mechanism, the result will be different.

The approaches and procedure of determining the polymerization reaction mechanism and estimating rate constants for each step are the same as those for other reactions. Generally speaking, they have to be based on experimental data, and the experimental methods and data analysis procedures discussed in previous chapters are largely applicable.

Example 12.2

A free radical polymerization takes place isothermally in a batch tank reactor, and its mechanism involves the following steps:

Chain initiation	$I \rightarrow 2R^*$
	$R^* + M \xrightarrow{k_d} P_1^*$
	$r_i = 2fk_d[I]$
Chain propagation	$P_1^* + M \xrightarrow{k_i} P_{j+1}^*$
	$r_p = k_p[P^*][M]$

(Continued)

(Continued)

Coupling termination	$P_j^* + P_i^* \xrightarrow{k_{tc}} P_{j+i}$
	$r_{tc} = 2k_{tc}[P^*]^2$
Transfer into monomer	$P_j^* + M \xrightarrow{k_{fm}} P_j + P_1^*$
	$r_{fm} = k_{fm}[P^*][M]$
Transfer into solvent	$P_j^* + S \xrightarrow{k_{fs}} P_j + S_1^*$
	$r_{fs} = k_{fs}[P^*][S^*]$

It is known that

$$[M]_0 = 7.17 \text{ mol/L}, [S] = 1.32 \text{ mol/L}, f = 0.52, [I] = 10^{-3} \text{ mol/L}.$$

$$k_d = 8.22 \times 10^5 \text{s}^{-1}; \quad k_p = 5.09 \times 10^2 \text{L/(mol} \cdot \text{s)}; \quad k_{tc} = 5.95 \times 10^7 \text{L/(mol} \cdot \text{s)}$$

$$k_{fm} = 0.079 \text{L/(mol} \cdot \text{s)}; \quad k_{fs} = 1.34 \times 10^{-4} \text{L/(mol} \cdot \text{s)}$$

1. Determine monomer concentration $[M]_0$ and conversion X as a function of reaction time?
2. Calculate the reaction time needed to reach conversion = 80%.

Solution
The consumption rate of monomer is:

$$r_m = -\frac{d[M]}{dt} = r_i + r_p + r_{fm} + r_{fs} \cong k_p[P^*][M] \tag{A}$$

According to steady-state approximation assumptions

$$\frac{d[P^*]}{dt} = 0, r_i = r_t$$

We can get $2fk_d[I] = 2k_{tc}[P^*]^2$
So

$$[P^*] = \left(\frac{2fk_d[I]}{2k_{tc}}\right)^{1/2} \tag{B}$$

Combining Eq. (B) into Eq. (A) gives

$$r_M = -\frac{d[M]}{dt} = k_p\left(\frac{2fk_d[I]}{2k_{tc}}\right)^{1/2}[M] = k[M] \tag{C}$$

When $t = 0$, $[M] = [M]_0$, $X_A = 0$, integrating the above equation:

$$[M] = [M]_0 e^{-kt}$$

or

$$X = 1 - e^{-kt}$$

The calculation results are shown in Fig. 12A.

(Continued)

(Continued)

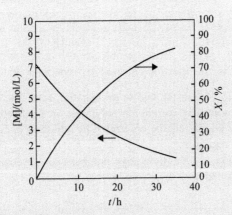

FIGURE 12A The relationship of [M] or X with t.

When X is 80%, the reaction time needed is

$$t = \frac{1}{k}\ln\frac{1}{1-2} = \frac{1}{1.36 \times 10^{-5}}\ln\frac{1}{1-0.8} = 32.9\text{h}$$

12.2.3.3 Average Degree of Polymerization

The polymerization of styrene according to the above mechanism can be represented as follows,

$$\bar{P}_n\text{M} \rightarrow \text{P}$$

where the number-average degree of polymerization P_n is an average stoichiometric coefficient. In a constant-volume batch process, the conversion (or polymerization rate) of monomers is

$$X = \frac{[\text{M}]_0 - [M]}{[M]_0} \tag{12.32}$$

Instantaneous number-average degree of polymerization defined as Eq. (12.4) is

$$\bar{P}_n = \frac{r_\text{M}}{r_\text{p}} \cong \frac{\left(-\dfrac{d[M]}{dt}\right)_\text{p}}{\left(\dfrac{d[P]}{dt}\right)_\text{tc} + \left(\dfrac{d[P]}{dt}\right)_\text{fm}\left(+\dfrac{d[P]}{dt}\right)_\text{fs}} \tag{12.33}$$

$$= \frac{k_\text{p}[\text{M}][\text{P}^*]}{k_\text{tc}[\text{p}^*]^2 + k_\text{fm}[\text{M}][\text{P}^*] + k_\text{fs}[\text{M}][\text{P}^*]}$$

Combining Eq. (12.30) into the above equation

$$\frac{1}{\bar{P}_n} = \frac{[2fk_dk_{tc}]^{1/2}[I]^{1/2}}{k_p[M]} + \frac{k_{fm} + k_{fs}}{k_p} = \frac{[2fk_dk_{tc}]^{1/2}[I]^{1/2}}{k_p[M](1-X)} + k_f \quad (12.34)$$

$k_f = \dfrac{k_{fm} + k_{fs}}{k_p}$ is the transfer constant to monomer and solvent, which represents the ratio of transfer rate to propagation rate. The larger the ratio, the smaller the molecular weight of polymers. In commercial production, product molecular weight can be controlled by adjusting k_f, such as adding a molecular weight regulator.

According to Eq. (12.5), cumulative number-average degree of polymerization can be obtained as

$$\bar{P}_n = \frac{X}{\displaystyle\int_0^X \frac{1}{\bar{P}_n} dX} = \frac{X}{\displaystyle\int_0^X \left[\frac{[2fk_dk_{tc}]^{1/2}[I]^{1/2}}{k_p[M](1-X)} + k_f \right] dX} \quad (12.35)$$

It can be seen that the average degree of polymerization is directly proportional to [M], and inversely proportional to $[I]^{1/2}$. At the same time, we can see from Eq. (12.31) that the polymerization rate is directly proportional to $[I]^{1/2}$. That is to say that the polymerization rate increases with initiator concentration, but the average degree of polymerization is just the opposite, i.e., decreases with initiator concentration. In an industrial process, the dosage of initiator is typically very low. This is because a product with a high molecular weight is usually desired. In addition, due to the limitations imposed by heat transfer and temperature control, it is usually not desirable to have a very fast reaction.

Example 12.3
Using the data and conditions given in Example 12.2, please determine
1. Relationship between number-average degree of polymerization and conversion.
2. Cumulative number-average degree of polymerization when $X = 80\%$.

Solution
1. According to the mechanism given in Example 12.2, the instantaneous number-average degree of polymerization can be written as

$$\bar{P}_n = \frac{r_M}{\sum r_{pj}} = \frac{k_p[P^*][M]}{k_{tc}[P^*]^2 + k_{fm}[P^*][M] + k_{fs}[P^*][S]}$$

$$= \frac{k_p[M]}{k_{tc}\left(\dfrac{2fk_d[I]}{2k_{tc}}\right)^{1/2} + k_{fm}[M] + k_{fs}[S]} \quad (A)$$

(Continued)

(Continued)

Substituting the above equation into number-average degree of polymerization Eq. (12.5) gives

$$\overline{P}_n = \cfrac{X}{\displaystyle\int_0^X \left[\cfrac{k_{tc}\left(\cfrac{2fk_d[I]}{k_{tc}}\right)^{1/2} + k_{fs}[S]}{k_p[M]_0(1-X)} + \cfrac{k_{fm}}{k_p}\right] dX} \tag{B}$$

Integrating Eq. (B) yields

$$\frac{1}{\overline{P}_n} = \frac{k_{fm}}{k_p} X + \left[\cfrac{k_{tc}\left(\cfrac{2fk_d[I]}{k_{tc}}\right)^{1/2} + k_{fs}[S]}{k_p[M]_0(1-X)}\right] \ln\frac{1}{1-X} \tag{C}$$

$$= 1.55 \times 10^{-4} X + 6.18 \times 10^{-4} \ln\frac{1}{1-X}$$

From Eq. (C), the relationship between cumulative number-average degree of polymerization $\overline{P}_n$ to conversion X can be calculated, which was shown in Table 12A and Fig. 12B. The $\overline{P}_n = 900$ when conversion $X = 80\%$.

TABLE 12A Cumulative Number-Average Degree of Polymerization as a Function of Conversion

X	$\overline{P}_n \times 10^{-4}$
0.1	1.24
0.2	0.59
0.3	0.37
0.4	0.26
0.5	0.19
0.6	0.15
0.7	0.12
0.8	0.09

(Continued)

(Continued)

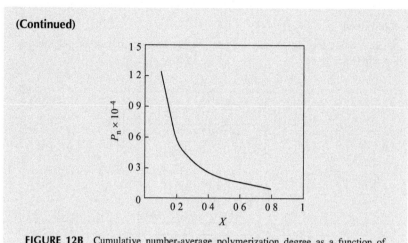

FIGURE 12B Cumulative number-average polymerization degree as a function of conversion X.

12.2.3.4 Polymerization Degree Distribution for Batch Polymerization Process

Assuming the reaction mechanism of a free radical polymerization system as the following:

Chain initiation	$I \xrightarrow{k_d} 2R^*$	
	$R^* + M \xrightarrow{k_i} R_1{}^*$	$r_i = 2fk_d[I]$
Chain propagation	$P_j^* + M \xrightarrow{k_p} P_{j+1}^*$	$r_p = k_p[P^*][M]$
Chain termination	$P_i^* + P_j^* \xrightarrow{k_{ec}} P_{i+j}$	$r_t = 2k_{tc}[P^*]^2$
Chain transfer	$P_i^* + M \xrightarrow{k_{fm}} P_j + P_1^*$	$r_{fm} = k_{fm}[P^*][M]$
	$P_i^* + S \xrightarrow{k_{fs}} P_j + S^*$	$r_{fs} = k_{fs}[P*][S]$

From the definition of instantaneous number-average distribution of degree of polymerization

$$f_n(j) = \frac{r_{pj}}{\sum\limits_{j=2}^{\infty} r_{pj}} = \frac{r_{pj}}{r_p} = \frac{[P_j^*]}{[p^*]}$$

For $f_n(j)$, we have to find the generation rates for different free radicals. Under steady state, the generation rates of free radicals with different sizes should not change with time, thus

$$\frac{d[P_j^*]}{dt} = 0, \ j = 1, 2, \ldots \infty \qquad (12.36)$$

When $j = 1$, the mass balance for $[P_j^*]$ gives

$$\frac{d[P_1^*]}{dt} = 2fk_d[I] + k_{fm}[P^*][M] - k_{fs}[P_1^*][S] - k_p[P_1^*][M] - k_{fm}[P_1^*][M] - 2k_{tc}[P^*][P_1^*]$$

$$= 0$$

So

$$[P_1^*] = \frac{2fk_d[I] + k_{fm}[P^*][M]}{k_p[M] + k_{fm}[M] + k_{fs}[S] + 2k_{tc}[P^*]} \qquad (12.37)$$

Similarly, when $j = 2$

$$\frac{d[P_2^*]}{dt} = k_p[P_1^*][M] - k_p[P_2^*][M] - k_{fm}[P_2^*][M] - k_{fs}[P_2^*][S] - 2k_{tc}[P^*][P_2^*] = 0 \qquad (12.38)$$

Therefore

$$[P_2^*] = \frac{k_p[P_1^*][M]}{k_p[M] + k_{fm}[M] + k_{fs}[S] + 2k_{tc}[P^*]} \qquad (12.39)$$

When $j = j$

$$\frac{d[P_j^*]}{dt} = k_p[P_{j-1}^*][M] - k_p[P_j^*][M] - k_{fm}[P_j^*][M] - k_{fs}[P_j^*][S] - 2k_{tc}[P^*][P_j^*] = 0$$

So

$$[P_j^*] = \frac{k_p[P_{j-1}^*][M]}{k_p[M] + k_{fm}[M] + k_{fs}[S] + 2k_{tc}[P^*]} \qquad (12.40)$$

Since the total concentration of free radicals is constant

$$\frac{d[P^*]}{dt} = r_i - r_t - r_{fs} = 2fk_p[I] - 2k_{tc}[P^*]^2 - k_{fs}[P^*][S] = 0$$

Therefore

$$2fk_d[I] = 2k_{tc}[P^*]^2 + k_{fs}[P^*][S] \qquad (12.41)$$

Or

$$[P^*] = \frac{k_{fs}[S] \pm \left[(k_{fs}[S])^2 - 16k_{tc}fk_d[I]\right]^{1/2}}{2k_{tc}}$$

Substituting the above equation into Eq. (12.37)

$$[P_1^*] = \frac{2k_{tc}[P^*]^2 + k_{fs}[P^*][S] + k_{fm}[P^*][M]}{k_p[M] + k_{fm}[M] + k_{fs}[S] + 2k_{tc}[P^*]} = \left(\frac{1}{1 + v_{tf}}\right)[P^*] \qquad (12.42)$$

Assuming

$$v_{tf} = \frac{k_p[M]}{k_{fm}[M] + k_{fs}[S] + 2k_{tc}[P^*]}$$

$$= \frac{\text{the consumption rate of the monomer in chain growth}}{\text{chain termination rate} + \text{chain transfer rate}} \qquad (12.43)$$

Substituting Eq. (12.55) into (12.52) gives

$$[P_2^*] = \left(\frac{v_{tf}}{1 + v_{tf}}\right)\left(\frac{1}{1 + v_{tf}}\right)[P^*] \qquad (12.44)$$

Similarly, iteration gives

$$[P_j^*] = \left(\frac{v_{tf}}{1 + v_{tf}}\right)^{j-1}\left(\frac{1}{1 + v_{tf}}\right)[P^*] \qquad (12.45)$$

Meanwhile

$$f_n(j) = \frac{[P_j^*]}{[P^*]} = \left(\frac{v_{tf}}{1 + v_{tf}}\right)^{j-1}\left(\frac{1}{1 + v_{tf}}\right) \qquad (12.46)$$

When the value of j is large enough

$$\left(\frac{v_{tf}}{1 + v_{tf}}\right)^{j-1} \cong \left(\frac{v_{tf}}{1 + v_{tf}}\right)^{j} \cong \exp\left(-\frac{j}{v_{tf}}\right)$$

In most cases, $v_{tf} \gtrsim 1$, therefore:

$$\frac{1}{1 + v_{tf}} \cong \frac{1}{v_{tf}}$$

Substituting the above equation into Eq. (12.46) gives

$$f_n(j) = \frac{1}{v_{tf}}\exp\left(-\frac{j}{v_{tf}}\right) \qquad (12.47)$$

It should be noted here that for those reactions which have coupling termination mechanism, by comparing Eq. (12.33) with Eq. (12.43) we can see:

$$\overline{P_n} \neq v_{tf}$$

This is caused by the difference between the coupling termination rate and the polymer formation rate. According to the definition of instantaneous number-average degree of polymerization and v_{tf}, we can know that when the chain transfer rate is small enough compared with the chain termination rate, Eq. (12.33) can be simplified as

$$\overline{P_n} = \frac{r_M}{r_p} = \frac{k_p[M][P^*]}{k_{tc}[P^*]^2} \tag{12.48}$$

And

$$v_{tf} = \frac{k_p[M][P^*]}{2k_{tc}[P^*]^2} \tag{12.49}$$

So, for polymerization involving coupling termination mechanism

$$\overline{P_n} = 2v_{tf} \tag{12.50}$$

For polymerization without coupling mechanism

$$\overline{P_n} = v_{tf} \tag{12.51}$$

We can obtain the cumulative number-average degree of polymerization in batch constant-volume system by substituting Eq. (12.50) into (12.5)

$$\overline{P_n} = \frac{X}{\int_0^X \frac{1}{2v_{tf}} dX} \tag{12.52}$$

The cumulative number-average distribution function of polymerization, $F_n(j)$, can be derived by substituting Eqs. (12.47) and (12.50) into Eq. (12.11).

In addition, from the definition of instantaneous number-average distribution function and weight-average distribution function, we can find out the relationship between the two functions.

$$f_n(j) = \frac{r_{pj}}{\sum\limits_{j=2}^{\infty} r_{pj}} = \frac{r_{pj}}{r_p} = \frac{r_{pj} \cdot \overline{P_n}}{r_M}$$

$$f_w(j) = \frac{jr_{pj}}{\sum\limits_{j=2}^{\infty} jr_{pj}} = \frac{jr_{pj}}{\sum\limits_{j=2}^{\infty} jr_{Mj}} = \frac{jr_{pj}}{r_M} = \frac{jf_n(j)}{\overline{P_n}} \tag{12.53}$$

Therefore, the instantaneous weight-average distribution function can be obtained as long as the $f_n(j)$ is known. From Eq. (12.18):

$$F_w(j) = \frac{1}{X} \int_0^X f_w(j) dX = \frac{1}{X} \int_0^X \frac{jf_n(j)}{\overline{P_n}} dX \tag{12.54}$$

Substituting Eqs. (12.47) and (12.50) into Eq. (12.53) gives

$$f_w(j) = \frac{j}{2v_{tf}^2} \exp\left(-\frac{j}{v_{tf}}\right) \tag{12.55}$$

Substituting the above equation into Eq. (12.54) gives

$$F(j) = \frac{1}{X} \int_0^X \frac{j}{2v_{tf}^2} \exp\left(-\frac{j}{v_{tf}}\right) dX \qquad (12.56)$$

From the discussions above, we can conclude that the polymerization degree distribution is a function of monomer conversion. Different extent of reaction will lead to different polymerization degree distribution of the final product.

Polymerization degree distribution can also be obtained from formation rate of polymer j. Formation rate of polymer j is determined by the rate of chain termination and chain transfer, and can be expressed as:

$$\frac{d[P_j]}{dt} = k_{tc}[P^*]\left[P_j^*\right] + k_{fm}[M]\left[P_j^*\right] + k_{fs}[S]\left[P_j^*\right]$$

Substituting Eqs. (12.41) and (12.45) into above equation gives

$$\frac{d[P_j]}{dt} = k_p[M][P^*]\left(\frac{v_{tf}}{1+v_{tf}}\right)^j$$

And the monomer consumption rate is

$$-\frac{d[M]}{dt} \cong k_p[P^*][M]$$

Dividing the above two equations gives

$$\frac{d[P_j]}{-d[M]} = \left(\frac{v_{tf}}{1+v_{tf}}\right)^j = \phi([M]) = \phi'(X)$$

In the above equation, $\phi([M])$ or $\phi'(X)$ denotes a function of monomer concentration or conversion. $[P_j]$ can be obtained by integrating the above equation. When X is given, number average distribution curve can be derived from the $[P_j] - j$ plot, and weight-average distribution can be derived from the $j[P_j] - j$ plot.

Example 12.4
Using the data and conditions given in Example 12.2, calculate the instantaneous number-average and weight-average distribution of polymerization when conversion $X = 80\%$.

Solution
Eq. (A) in Example 12.2 can be rewritten as

$$\frac{1}{\overline{P}_n} = \frac{k_{fm}}{k_p} + \frac{k_{tc}\left(\frac{2fk_d[I]}{k_{tc}}\right)^{1/2} + k_{tc}}{k_p[M]_0(1-X)} = 1.55 \times 10^{-4} + \frac{6.18 \times 10^{-4}}{1-X}$$

(Continued)

(Continued)

From Eq. (12.50)

$$\frac{1}{v_{tf}} = \frac{2}{\overline{P}_n} = 2\left(1.55 \times 10^{-4} + \frac{6.18 \times 10^{-4}}{1-X}\right)$$

Instantaneous number-average distribution function can be obtained from Eq. (12.47).

$$f_n(j) = \frac{1}{v_{tf}}\exp\left(-\frac{j}{v_{tf}}\right)$$

$$= \left(3.1 \times 10^{-4} + \frac{12.36 \times 10^{-4}}{1-X}\right)\exp\left[(-j)\left(3.1 \times 10^{-4} + \frac{12.36 \times 10^{-4}}{1-X}\right)\right]$$

$$\text{(A)}$$

And instantaneous weight-average distribution function can be obtained from Eq. (12.55)

$$f_w(j) = \frac{jf_n(j)}{\overline{P}_n} = \frac{jf_n(j)}{2v_{tf}} = \frac{j}{2v_{tf}^2}\exp\left(-\frac{j}{v_{tf}}\right)$$

$$= \frac{1}{2}\left(3.1 \times 10^{-4} + \frac{12.36 \times 10^{-4}}{1-X}\right)^2 \exp\left[(-j)\left(3.1 \times 10^{-4} + \frac{12.36 \times 10^{-4}}{1-X}\right)\right]$$

$$\text{(B)}$$

From the discussions above, we can know that polymerization degree distribution is a function of conversion. When the conversion is given, the distribution can be determined. If $X = 80\%$, $f_n(j)$ and $f_w(j)$ can be calculated from Eqs. (A) and (B), as shown in Fig. 12C:

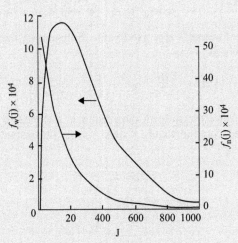

FIGURE 12C Relationship of $f_n(j)$ or $f_w(j)$ with j.

(Continued)

(Continued)

The above results are obtained under batch operation conditions, the concentration of each component is changing with reaction time. Thus the products' average degree of polymerization degree as well as their distribution all change with the reaction time. For a constant-volume system, the above conclusions derived from a batch reactor are also suitable for continuous operation in a plug flow reactor (PFR). Next we will discuss polymerization in a continuous flow stirred tank reactor (CSTR).

12.2.3.5 Continuous Polymerization Reaction in CSTR

The radical polymerization reaction discussed above will be used as a model reaction in this section. The mass balance for this reaction in CSTR is shown in Fig. 12.1.

Mass balance for monomer M in this reactor yields

$$Q_0[\mathrm{M}]_0 = Q[\mathrm{M}] + r_{\mathrm{m}} V_{\mathrm{r}}$$

The average residence time (space-time) is

$$\tau = \frac{V_{\mathrm{r}}}{Q_0} = \frac{[\mathrm{M}]_0 - [\mathrm{M}]}{r_{\mathrm{m}}} \tag{12.57}$$

Monomer consumption rate here can be rewritten based on Eq. (12.31)

$$r_{\mathrm{M}} \cong k_{\mathrm{p}} \left(\frac{2fk_{\mathrm{d}}}{2k_{\mathrm{tc}}} \right)^{1/2} [\mathrm{I}]^{1/2}[\mathrm{M}] = k[\mathrm{M}]$$

where $k = k_{\mathrm{p}} \left(\dfrac{2fk_{\mathrm{d}}}{2k_{\mathrm{tc}}} \right)^{1/2} [\mathrm{I}]^{1/2}$, when [I] can be treated as a constant, monomer consumption rate constant k is a fixed value. Substituting the above equation into Eq. (12.57) yields

$$[\mathrm{M}] = \frac{[\mathrm{M}]_0}{1 + k\tau} \text{ or } \tau = \frac{1}{k}\left[\frac{1}{1 - X} - 1 \right] \tag{12.58}$$

In a CSTR reactor the fluid in the reactor is assumed to be well mixed. Under well mixed condition and at steady state, the instantaneous average

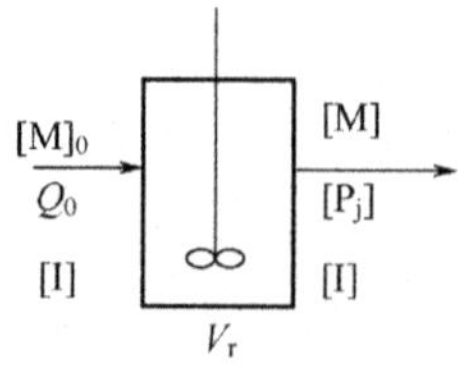

FIGURE 12.1 Schematic of a CSTR.

degree of polymerization and its distribution are equal to the cumulative average polymerization degree and its distribution, respectively

$$\overline{P}_n = \overline{p}_n$$
$$F_n(j) = f_n(j)$$
(12.59)

$$\overline{P}_w = \overline{p}_w$$
$$F_w(j) = f_w(j)$$
(12.60)

For both conversion of monomer and polymerization degree distribution of product, the conclusions derived here are different from those derived for a batch reactor. The conversion difference in CSTR and batch reactors has been discussed in detail in Chapter 3, Tank Reactor, and Chapter 5, Residence Time Distribution and Flow Models for Reactors. The differences of degree of polymerization and its distribution in CSTR and batch reactors are mainly caused by the differences of residence time distribution and concentration distribution. In order to control the degree of polymerization, selecting a suitable operation mode is very important.

12.2.3.6   Tanks in Series Reactor

For commercial polymerization processes, tank reactors in series are widely used. Here the aforesaid radical polymerization is still used as a model reaction. As shown in Fig. 12.2, the polymerization reaction occurred in a series of tank reactors. The feedstock for the first stage is only monomer and the initiator. In the subsequent tanks, the feedstock contains some radicals and polymers with different chain lengths.

Assuming,

1. The volume of each tank is the same and reaction mixture density remains a constant in each tank, the average residence time of each tank is the same

$$\tau_1 = \tau_2 = \cdots = \tau_n = \tau$$

2. Each tank in the reactor system can be described by a perfectly mixed flow model, all tanks are operated isothermally and under steady state, and the initiator concentration [I] remains a constant.

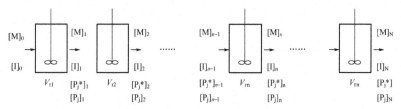

FIGURE 12.2   Schematic diagram for a tanks in series reactor system with multiple stirred tank reactors.

Mass balance of free radical with various chain length in the nth reactor is:

Amount entering the reactor + formation in the reactor

= amount leaving the reactor + converted in the reactor

Under this condition, when $j = 1$, mass balance for $[P_1^*]_n$ gives

$$[P_1^*]_{n-1} + 2fk_{dn}[I]_n \cdot \tau + k_{fm}[P_1^*][M]_n\tau = [P_1^*]_n + k_{pn}[P_1^*]_n[M]_n$$

$$+ 2k_{tcn}[P_1^*]_n[P^*]_n \cdot \tau + k_{fm}[P_1^*]_n[M]_n \cdot \tau + k_{fs}[P_1^*]_n[s]_n \cdot \tau$$

Rearranging yields

$$[P_1^*]_n = \frac{[P_1^*]_{n+1} 2fk_{dn}[I]_n\tau + k_{fm}[P_1^*]_n[M]_n\tau}{k_{pn}[M]_m\tau + 2k_{tcn}[P_1^*]\tau + k_{fm}[M]_n\tau + k_{fs}[S]_n\tau + 1} \tag{12.61}$$

Under steady state

$$\frac{d[P^*]_n}{dt} = r_{in} - r_{tn} - r_{fsn} = 2fk_{dn}[b]_n - 2k_{tcn}[P^*]_n^2 - k_{fs}[P^*]_n[s]_n = 0$$

So

$$2fk_{dn}[I]_n = 2k_{tcn}[P^*]_n^2 + k_{fs}[P^*]_n[s]_n \tag{12.62}$$

And because space-time is much larger than active chain's lifetime τ_s (τ_s is defined as total concentration of free radical divided by chain termination total rate under steady state)

$$\tau \gg \frac{[P^*]_n}{2k_{tcn}[P^*]_n^2 + k_{fm}[M]_n[P^*]_n + k_{fs}[s]_n} = \frac{1}{2k_{tcn}[P^*]_n + k_{fm}[M]_n + k_{fn}[s]_n}$$

So

$$\tau_s = \frac{1}{2k_{tcn}[P^*]_n + k_{fm}[M]_n + k_{fs}[s]_n} \cong 0 \tag{12.63}$$

Substituting Eqs. (12.62) and (12.63) into Eq. (12.61) and rearranging gives

$$[P_1^*] = \left(\frac{1}{1 + v_{tf(n)}}\right)[P^*]_n \tag{12.64}$$

Assuming

$$v_{tf(n)} = \frac{\text{consumption rate of the monomer in chain growth}}{\text{chain termination rate + chain transfer rate}}$$

$$= \frac{k_{pn}[M]_n}{2k_{tcn}[P^*]_n + k_{fm}[M]_n k_{fs}[S]_n} \tag{12.65}$$

When $j = 2$, mass balance for $[P_2^*]_n$ gives

$$[P_2^*]_{n-1} + k_{pn}[P_1^*]_n[M]_n\tau = k_{pn}[P_2^*]_n[M]_n\tau + [P_2^*]_n + 2k_{tcn}[P_2^*]_n[P^*]_n\tau$$

$$+ k_{fm}[P_2^*]_n[M]_n\tau + k_{fs}[P_2^*]_n[S]_n\tau$$

Rearranging gives

$$[P_2^*]_n = \left(\frac{v_{tf(n)}}{1 + v_{tf(n)}}\right)[P_1^*]_n = \left(\frac{1}{1 + v_{tf(n)}}\right)\left(\frac{v_{tf(n)}}{1 + v_{tf(n)}}\right)[P^*]_n \qquad (12.66)$$

For $\left[P_j^*\right]_n$, we can get

$$[P_j^*]_n = \left(\frac{v_{tf(n)}}{1 + v_{tf(n)}}\right)[P_{j-1}^*]_n = \left(\frac{1}{1 + v_{tf}}\right)\left(\frac{v_{tf(n)}}{1 + v_{tf(n)}}\right)^{j-1}[P^*]_n \qquad (12.67)$$

Then according to the definition of instantaneous number-average distribution function of polymerization degree

$$f_{n(n)}(j) = \frac{r_{pjn}}{r_{Pn}} = \frac{d[P_j]_n/dt}{d[P]_n/dt} = \frac{[P_j]_n}{[P^*]_n} = \left(\frac{1}{1 + v_{tf(n)}}\right)\left(\frac{v_{tf(n)}}{1 + v_{tf(n)}}\right)^{j-1} \qquad (12.68)$$

When j is large enough, $v_{tf} \gg 1$, the above equation can be approximately rewritten as

$$f_{n(n)}(j) \cong \frac{1}{v_{tf}}\exp\left(-\frac{j}{v_{tf}}\right) \qquad (12.69)$$

In this case, the instantaneous number-average degree of polymerization is

$$\overline{P}_{n(n)} = \frac{r_{M(n)}}{r_{pn}} = \frac{k_{pn}[P^*]_n[M]_n}{k_{tcn}[P^*]_n^2 + k_{fm}[P^*]_n[M]_n + k_{fs}[P^*]_n[S]_n} \qquad (12.70)$$

$$\neq v_{ef(n)}$$

In the same way with Eq. (12.50)

$$\overline{P}_{n(n)} = 2v_{tf(n)}$$

Instantaneous weight-average distribution function of polymerization degree is

$$f_{w(n)}(j) = \frac{jf_{n(n)}(j)}{\overline{P}_{n(n)}} = \frac{j}{2v_{tf(n)}^2}\exp\left(-\frac{j}{v_{tf(n)}}\right) \qquad (12.71)$$

For the same reaction mechanism, Eqs. (12.69–12.71) for tanks in series system have the same format as Eqs. (12.47), (12.50), and (12.53), respectively. Under steady state, the operation conditions in nth tank for a tanks in series system corresponds to the conditions in a batch system at a specific moment. It should be noted here that distribution function for a tanks in series system is not continuous. Only when the number of tanks in the system approaches to infinite, is the tanks in series system completely equivalent to batch operation. Therefore, although the equations calculating degree of polymerization and their distributions of the two reactor systems have the same format, the actual values of degree of polymerization and their distributions are different.

In a continuous tanks in series system, the concentration and reaction rate in each tank are fixed values. So in each tank in the system

$$\bar{P}_{n(n)} = \bar{p}_{n(n)}$$
$$F_{n(n)}(j) = f_{n(n)}(j)$$

(12.72)

$$\bar{P}_w = \bar{p}_w$$
$$F_{w(n)}(j) = f_{w(n)}(j)$$

(12.73)

From the discussions above, we can conclude that for the same polymerization reaction, the degree of polymerization and its distribution are different for different reactor systems and different operation modes. The overall degree of polymerization and its distribution of the final product in a tanks in series system can be derived by adding the corresponding values of each tank. Here we use the weighted average value

$$\bar{P}_{n(N)} = \sum_{n=1}^{N} \left(\frac{[P]_n - [P]_{n-1}}{[P]_N} \right) \bar{P}_{n(n)}$$

(12.74)

$$F_{n(N)}(j) = \sum_{n=1}^{N} \left(\frac{[P]_n - [P]_{n-1}}{[P]_N} \right) F_{n(N)}$$

(12.75)

$$\bar{P}_{w(n)} = \sum_{n=1}^{N} \left(\frac{X_n - X_{n-1}}{X_N} \right) \bar{P}_{w(n)}$$

(12.76)

$$F_{w(N)}(j) = \sum_{n=1}^{N} \left(\frac{X_n - X_{n-1}}{X_N} \right) F_{w(n)}(j)$$

(12.77)

When $N = 1$, it corresponds to a single CSTR. Eq. (12.59) can be rearranged as

$$\bar{P}_n = \bar{p}_n$$
$$F_n(j) = f_n(j) = \frac{1}{v_{tf}} \exp\left(-\frac{j}{v_{tf}} \right)$$

(12.78)

And weight-average degree of polymerization and its distribution Eq. (12.70) can be rewritten as

$$\bar{P}_w = \bar{p}_w$$
$$F_w(j) = f_w(j) = \frac{j}{2v_{tf}^2} \exp\left(-\frac{j}{v_{ef}} \right)$$

(12.79)

It should be noted that the above conclusions are valid for radical polymerization and the above specific mechanism. For other polymerization reactions with different mechanisms, we can use the same method and procedure to obtain the results needed. In general, the polymerization reaction type is determined by analyzing the experimental data, and then the reaction rate equation of each elementary step can be written according to the reaction mechanism. The reaction rate of a specific component can be derived based on steady-state approximation and the rate-determining step method. After selecting the reactor type and operation mode, a reactor model can be established by mass balance and kinetics analysis, which can be used for reactor design and analysis.

Example 12.5

Assuming that the reaction mechanism and the kinetic rate constant are the same as shown in Example 12.2, please calculate the weight-average distribution of polymerization degree in both a CSTR and an isothermal batch reactor when the consumption rate is 80%, and then compare the results in a figure.

1. CSTR

According to Eqs. (12.59) and (12.60), for a CSTR, the weight-average cumulative distribution of polymerization degree is same as the instantaneous distribution.

$$F_w(j) = f_w(j) = \frac{j}{2}\left(3.1\times10^{-4} + \frac{12.36\times10^{-4}}{1-X}\right)^2$$

$$\exp\left[(-j)\left(3.1\times10^{-4} + \frac{12.36\times10^{-4}}{1-X}\right)\right] \tag{A}$$

2. Batch tank reactor

The cumulative weight-average distribution function of polymerization

$$F_w(j) = \frac{1}{X}\int_0^X f_w(j)dX$$

$$= \frac{1}{X}\int_0^X \frac{j}{2}\left(3.1\times10^{-4} + \frac{12.36\times10^{-4}}{1-X}\right)^2$$

$$\exp\left[(-j)\left(3.1\times10^{-4} + \frac{12.36\times10^{-4}}{1-X}\right)\right]dX \tag{B}$$

Eq. (B) can be integrated numerically by trapezoidal method, and the results are compared with that from Eq. (A) in Table 12B and shown in Fig. 12D.

(Continued)

(Continued)

TABLE 12B Weight-average distribution function of polymerization

j	CSTR, $F_w(j) \times 10^4$	Batch Reactor, $F_w(j) \times 10^4$
50	7.61	2.13
100	11.01	3.44
200	11.50	4.58
250	10.39	4.73
300	9.02	4.74
400	6.28	4.24
500	4.10	4.09
800	0.94	2.84
1000	0.32	2.16

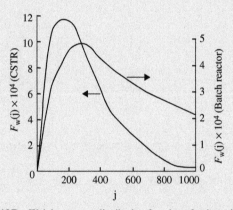

FIGURE 12D Weight-average distribution function of polymerization.

12.2.3.7 Effect of Active Chain Life on Distribution of Polymerization

The degree of polymerization and its distribution are important indicators of product quality, and studying and understanding the factors that influence the degree of polymerization and its distribution are of great theoretical and practical importance.

According to the discussions above, the degree of polymerization and its distribution depend on polymerization reaction types and reaction mechanisms. Even for the same reaction with the same reaction mechanism, they also depend on polymerization method, reactor structure, and operation mode. The operation conditions have also a great influence on the degree of polymerization, and even minor changes can have a significant impact on product quality. In any case, the degree of polymerization and its distribution depend primarily on the microkinetics of the polymerization reaction, which is the inherent character of the reaction. Microkinetics describes the dependence of reaction rate on temperature and reactant concentrations. All the factors that can influence temperature and concentration distribution in the reactor, such as mixing, flow, mass transfer, and heat transfer, may have a great impact on product quality. Therefore, the polymerization process is always studied by analyzing reaction kinetics and transfer processes.

Reactor type and operation modes may have a great effect on the degree of polymerization and the distribution of the products. This is because the transport processes, which influence both residence time distribution and concentration distributions, are very different in different reactors. Whether the residence time or concentration distribution has a stronger effect depends on the polymerization mechanism, especially the life of active chains. The life of the active chain, τ_s, is the time from radical (active chain) generation to radical termination, and under the steady state can be calculated by the free radical total concentration $[P^*]$ divided by overall chain termination rate r_{tf}.

$$\tau_s = \frac{[P^*]}{r_{tf}} \tag{12.80}$$

When the life of active chain is relatively short, such as $\tau_s \ll t$, the effect of residence time distribution is relatively small, since even the reactants with short residence time will have enough for chain termination. In such a case, the concentration distribution will be the determining factor. A CSTR will lead to a narrow distribution of the degree of polymerization due to homogeneous concentration distribution. On the other hand, inside a batch reactor, concentrations of different components change with time, and so does the distribution of instantaneous degree of polymerization. As a result, the distribution of cumulative degree of polymerization will be wide. For reactions involving an active chain with a long life, i.e., $\tau_s \gg t$, residence time distribution will have a greater effect than concentration distribution. In such a case, a CSTR will lead to a wider distribution of the degree of polymerization due to the broader distribution of residence time. But for a batch reactor, there is no distribution of residence time, leading to a relatively narrower molecular weight distribution.

The residence time distribution and concentration distribution in a tanks in series system will be between a CSTR and a batch reactor. Therefore, the distribution of the degree of polymerization of the product in this system will also be between that in those two kinds of reactors. When the number of tanks in the system, n, approaches infinity, the performance is similar to that in a batch reactor, and when $n = 1$ it becomes a CSTR.

It is well known that for a free radical chain reaction the life of active chain is short, generally between 0.1 and 10 s, which is much shorter than average residence time or reaction time. Therefore, for free radical polymerization, a CSTR is usually used to produce product with a narrow distribution of degree of polymerization.

12.2.4 Polycondensation Reaction

Different from chain reactions, polycondensation reactions do not require special active sites to initiate the reaction. All monomers participate in the polymer formation reaction together, and the resulting j-mer still have the functional groups and hence further react with each other It means that the molecular weight of the polymer increases gradually with reaction time. The differences between a polycondensation reaction and a polyaddition reaction are shown in Table 12.3.

Typical examples of polycondensation reactions include manufacturing polyesters, polyamides, polyurethanes, and others. For polyesters, molecular weight molecules such as water will be released during polymerization. For polyamides and polyurethanes synthesis, no low molecular weight molecules will be released. In order to generate polymers, two or more functional groups must be present in each monomer, e.g., bifunctional diamine and diacid monomers can generate a linear polymer. Of course, if there are more than two functional groups in the monomer, it may generate dendrimer polycondensates by branching or cross-linking. Here we use linear polycondensation as examples to illustrate its characteristics and attributes.

12.2.4.1 Kinetics Analysis of Polycondensation

A polycondensation reaction is reversible, and both reaction rate and the degree of polymerization correlate closely with reaction equilibrium. In kinetic studies, to simplify the analysis, the concept of equal reactivity of functional groups, which has been experimentally proved to be a reasonable assumption, is often used. This concept assumes that the activity of functional groups at both ends of monomers, dimers, or oligomers are the same. Under this assumption, all equilibrium constants of each step are the same and the reaction rate constants remain unchanged. When reactants A and B are the only reactants in polycondensation, the reaction rate follows the general equation

$$r = k[A]^{\alpha}[B]^{\beta} \tag{12.81}$$

TABLE 12.3 Comparison Between the Polycondensation and Polyaddition Reaction

Items	Polycondensation Reaction	Polyaddition Reaction
Initiator	Must have	Not needed
Chain propagation mechanism and rate	There are three elementary reactions: Initiation, propagation, and termination, the activation energy of the propagation reaction is small, such as $E_p = 21 \times 10^3$ J/mol, and reaction is very fast, often accomplished in seconds	No so-called initiation, propagation, termination reactions, the activation energy is higher, such as $E_p = 63 \times 10^3$ J/mol for polyester, the reaction rate is slow, often accomplished in hours
Thermal effects and reaction equilibrium	Large thermal effects, $\Delta H_r = 84 \times 10^3$ J/mol, high critical temperature of polymerization, irreversible at normal temperature	Small thermal effects, $\Delta H_r = 21 \times 10^3$ J/mol, low critical temperature of polymerization, reversible at normal temperature
Relationship between monomer conversion and reaction time	Monomers gradually disappear over time	Monomer disappear quickly
Relationship between polymer molecular weight and reaction time	No change after the rapid formation of macromolecules	Macromolecules gradually formed, and the molecular weight gradually increases over time

Obviously, reaction order and rate constant in the above equation can be estimated using the same methods as that for low molecular reactions.

Next we will use preparation of polyester as an example. Assuming that there are two different bifunctional monomers $_A-R_{-A}$ and $_B-R'_{-B}$, the polycondensation reaction occurs through the formation of new bond m and releases low molecular weight substance n,

$$_A-R_{-A} + _B-R'_{-B} \; \underset{\overleftarrow{k}}{\overset{\overrightarrow{k}}{\rightleftharpoons}} \; [R-m-R'-m] + n$$

A—functional groups, such as carboxyl $-COOH$
B—another functional groups, such as hydroxyl $-OH$
m—the new chemical bonds, such as $-O-CO-$
n—low molecular weight substances, such as H_2O

The above reversible reaction usually uses an acid catalyst, and the reaction is first order with respect to all the components involved. The reaction rate for a batch operation is

$$r_M = -\frac{d[-A]}{dt} = -\frac{d[-B]}{dt} = \vec{k}[-A][-B] - \overleftarrow{k}[-M-][n] \qquad (12.82)$$

If the initial concentration of two functional groups is the same, $[-A] = [-B]$, the above equation can be rewritten as

$$\frac{[-A]_0 dX}{dt} = \vec{k}[-A]_0^2(1-X)^2 - \overleftarrow{k}[-A]_0^2 X^2 \qquad (12.83)$$

Integrating Eq. (12.83) gives

$$\frac{[-A]_0}{[-A]} = 1 + \sqrt{K}\frac{1 - \exp\left(-2\vec{k}[-A]_0 t/\sqrt{K}\right)}{1 + \exp\left(-2\vec{k}[-A]_0 t/\sqrt{K}\right)} \qquad (12.84)$$

where K is the equilibrium constant. If there are only two functional materials existing in the system, and no other side reactions, at reaction time t the number of unreacted functional groups is equal to the number of moles of the condensation product N, and N_0 is the total number of monomers structural units incorporated into the polymer chain of the polycondensation product. Therefore, the relationship between the number-average degree of polymerization based on structure and conversion is

$$\overline{P}_n = \frac{N_0}{N} = \frac{[-A]_0}{[-A]} = \frac{1}{1-X} \qquad (12.85)$$

Here, N_0 is the number of initial functional groups, i.e., $N_0 = \dfrac{N_{A0} + N_{B0}}{2}$. When the reaction reaches equilibrium, $r_m = 0$, and Eq. (12.83) becomes

$$K = \frac{X^2}{(1-X)^2}$$

Or

$$X = \frac{\sqrt{K}}{1 + \sqrt{K}} \qquad (12.86)$$

Now substituting Eq. (12.86) into Eq. (12.85), we can obtain number-average degree of polymerization at equilibrium

$$\overline{P}_n = 1 + \sqrt{K} \qquad (12.87)$$

When t approaches to infinity, the reaction approaches to its equilibrium, and using Eq. (12.84) we can obtain the same equation. We can conclude here that the maximum degree of polymerization of the polycondensation reaction depends on its equilibrium constant. For example, in polyester synthesis process, usually $K \approx 1$, the maximum degree of polymerization is

2, which tells us we can't get high molecular weight polymers. Thus, for such polycondensation reactions, in order to increase the degree of polymerization, a high degree of vacuum has to be used to allow continuous and fast release of low molecular substances which in turn increase reaction equilibrium constant and hence the degree of polymerization.

Actually, for polycondensation reactions low molecular substances formed in the reactions are usually removed from the system continuously and the reverse reaction can be ignored. Eq. (12.83) can be simplified to

$$\frac{dX}{dt} = \vec{k}[-A]_0(1-X)^2 \tag{12.88}$$

Integrating yields

$$[-A]_0\vec{k}t = \frac{1}{1-X} - 1 \tag{12.89}$$

In this case the number-average degree of polymerization is

$$\overline{P}_n = [-A]\vec{k}t + 1 \tag{12.90}$$

From Eqs. (12.90) and (12.85) we can see that the degree of polymerization increases with the reaction time and monomer conversion. When $X = 0.90$, the degree of polymerization of the polymer $\overline{P}_n = 10$; if $X = 99\%$, $\overline{P}_n = 100$. That is to say, in order to increase the degree of polymerization and improve the properties of the polymer, it is necessary to ensure the purity of the monomer and control reaction conditions to guarantee high monomer conversion.

It should also be noted that the above conclusion is derived when the initial mole ratios of the two components in the feed are equal, i.e., $\alpha = \frac{[-A]_0}{[-B]_0} = 1$. If $\alpha \neq 1$, and the conversion of the less component is X, the number-average degree of polymerization is

$$\overline{P}_n = \frac{[-A]_0 + [-B]_0}{[-A] + [-B]} = \frac{1+\alpha}{1+\alpha-2\alpha X} \tag{12.91}$$

When $\alpha = 1$, the above formula can be rewritten as

$$\overline{P}_n = \frac{1}{1-X} \tag{12.92}$$

Taking the limit of Eq. (12.91) yields

$$\lim_{X \to 1} \overline{P}_n = \frac{1+\alpha}{1-\alpha} \tag{12.93}$$

It can be seen that under the condition of $\alpha = 1$, when $X \to \infty$, $\overline{P}_n \to \infty$. It means that for a linear polycondensation reaction, to obtain a high degree of polymerization, the molar ratio of the two components must be close to 1, which is another important measure to increase the degree of polymerization.

12.2.4.2 Distribution of the Degree of Polymerization for Linear Polycondensation

Because of the randomness of the polycondensation process, the conversion of functional groups mentioned above indicates the probability that the functional groups are consumed. The probability for functional groups that have not been consumed is $1 - X$. The probability for the generation of j-mer (which have j structure units) is approximately $X^{j-1}(1 - X)$.

In a polymerization reaction system, the number of functional groups is

$$N = N_0(1 - X) \tag{12.94}$$

The number of j-mer is

$$N_j = NX^{j-1}(1 - X) = N_0 X^{j-1}(1 - X)^2 \tag{12.95}$$

The distribution function of number-average degree of polymerization is

$$F_n(j) = \frac{N_j}{N} = X^{j-1}(1 - X) \tag{12.96}$$

The distribution function of weight-average degree of polymerization is

$$F_w(j) = \frac{jN_j}{N_0} = jX^{j-1}(1 - X)^2 \tag{12.97}$$

Eqs. (12.96) and (12.97) show that the distribution of the degree of polymerization depends only on conversion and there is no direct relationship with reaction temperature and other conditions. However, the distribution function of the degree of polymerization depends on operation mode. Fig. 12.3 clearly shows the significant impact of operation mode on the distribution function of weight-average degree of polymerization.

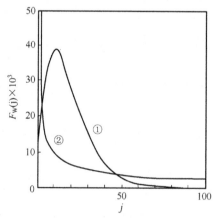

FIGURE 12.3 Effect of operation modes on distribution function of weight-average degree of polymerization for polycondensation ($X = 0.9$). (1) Batch or plug flow reactors (2) CSTR (homogeneous phase)

According to the relationship between the degree of polymerization and its distribution:

$$\overline{P}_n = \sum_{j=2}^{\infty} jN_j/N = \sum_{j=2}^{\infty} jF_n(j) = \sum_{j=2}^{\infty} jX^{j-1}(1-X) = \frac{1-X}{(1-X)^2} = \frac{1}{1-X}$$

(12.98)

where $\sum_{j=2}^{\infty} jX^{j-1} = \frac{1}{(1-X)^2}$. From Eqs. (12.98) and (12.85), we can know that probability and the conversion is actually same. Similarly, the weight-average degree of polymerization is

$$\overline{P}_n = \sum_{j=2}^{\infty} jN_j/N_0 = \sum_{j=2}^{\infty} jF_w(j) = \sum_{j=2}^{\infty} jX^{j-1}(1-X)^2$$

$$= (1-X)^2 \sum_{j=2}^{\infty} j^2 X^{j-1} = \frac{(1-X)^2(1+X)}{(1-X)^3} = \frac{1+X}{1-X}$$

(12.99)

where $\sum_{j=2}^{\infty} j^2 X^{j-1} = \frac{1+X}{(1-X)^3}$. Comparing Eq. (12.98) with Eq. (12.99), we can get the dispersion index D

$$D = \frac{\overline{P}_w}{\overline{P}_n} = 1 + X$$

(12.100)

From the above equation, we can know that the dispersion of the polymerization degree distribution increases with the conversion. The greater the conversion of the polycondensation is, the wider of the molecular weight distribution of the products will be.

Fig. 12.4 shows the calculated results using Eq. (12.97). It clearly shows the change of $F_w(j)$ with X. It quantitatively demonstrates that both the

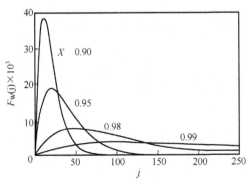

FIGURE 12.4 Distribution of the weight-average degree of polymerization for polycondensation in batch operations.

degree of polymerization and its distribution dispersion increase with conversion. Meanwhile, it also illustrates that monomers disappear quickly with the increase of reaction time and form small polymers, and larger polymers are formed from the polymerization of small polymer molecules.

12.2.5 Factors That Influence Polymerization Rate

The factors that influence reaction rate have been discussed in Chapter 3, Tank Reactor, Chapter 4, Tubular Reactor, and Chapter 5, Residence Time Distribution and Flow Models for Reactors, which include:

1. Kinetics factor, mainly reflected by the impacts of temperature and concentration on rate for a given reaction.
2. Fluid mixing. For normal kinetics, the stronger the fluid backmixing, the slower the reaction rate and the lower the conversion.
3. Micromixing (or the degree of segregation), which depends on reaction orders.
4. Timing of fluid mixing.

The above factors surely have significant impacts on polymerization, and will influence monomer conversion and the average degree of polymerization and its distribution. The impact of micro-backmixing is especially important because the polymerization mixture is highly viscous and its property is close to macrofluid. For instance, for solution polymerization and bulk polymerization, polymers generated can dissolve in monomer or solvent. With the increase of the degree of polymerization, the viscosity of the reaction mixture would become very high, and this fluid of high viscosity would segregate into fluid and solid groups. Generally, this kind of fluid flows like a segregation flow or is called macrofluid. In addition, unlike other reaction systems, the high viscosity of the polymerization mixture can cause a gelling effect, which will in turn lead to an acceleration in the increase of viscosity. This phenomenon is especially common in free radical polymerization and precipitation polymerization. Free radical polymerization of styrene is a typical example, which involves reaction initiation by an initiator, bi-radical termination, and transferring to monomers. The monomer consumption rate can be expressed by Eq. (12.101)

$$r_M = -\frac{d[M]}{dt} \cong k_p \left(\frac{2fk_d[I]}{k_t}\right)^{1/2} [M] = k[M]_0 (1 - X) \tag{12.101}$$

where $k = k_p \left(\frac{2fk_d[I]}{k_t}\right)^{1/2}$. If k is a constant as shown in Fig. 12.5, r_M is a linear function of X. As shown in Fig. 12.5, at low conversion, the experiment result is consistent with the prediction of Eq. (12.31). But at high conversion, the difference between experimental data and the model prediction becomes

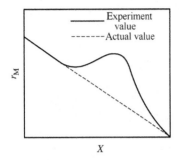

FIGURE 12.5 r_M to X relationship in styrene solution polymerization

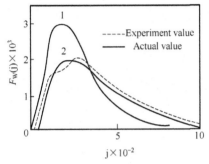

FIGURE 12.6   Polymerization degree distribution in styrene polymerization. (1) No viscosity correction; (2) With viscosity correction.

very large. This phenomenon is generally attributed to the gelling effect. With an increase of conversion, the viscosity of the mixture increases and causes curling of free radical chains which will bury some of the active site. As a result, bi-radical termination will be blocked, leading to the decline of k_t. On the other hand, small monomer diffusion was not influenced at that stage, therefore k_p/k_t increases which will in turn increase reaction rate r_M. When conversion is high and system viscosity is extremely large (such as when $X > 90\%$, the viscosity can reach to hundreds of thousands of Pa·s), monomer diffusion will also be greatly hindered. Therefore k_p decreases, and total rate decreases.

From the above example we know that the reaction rate equation, the degree of polymerization degree and its distribution are all derived based on the microkinetics of free radical polymerization reactions and by assuming that reaction rate constants and initiation efficiency do not change as the reactions progress. However, such assumptions are only valid at low conversion. Fig. 12.6 shows the differences between the experiment result and theoretical value of the degree of polymerization for styrene polymerization. To properly describe kinetic behavior for commercial polymerization processes, the impact of viscosity must be considered. Curve 2 in Fig. 12.6

is the model prediction with viscosity being corrected based on the research work of Shawki and Hamielec [1]. As given by Eq. (12.102), k_t and f are correlated with the viscosity based on experimental data of bulk and solution polymerization of styrene (with benzene as solvent and azodiisobutyronitrile as initiator) when conversion is in the range of 10−60%:

$$\log\left(\frac{k_t}{k_{t0}}\right) = -0.133 \log(1 + \mu \times 10^3) - 0.077 \left[\log(1 + \mu \times 10^3)\right]^2$$

$$\log\left(\frac{f}{f_0}\right) = 0.133 \log(1 + \mu \times 10^3)$$

$$\log\mu = \{17.66 - 0.311 \log(1 + [S] - 7.72 \log T - 10.23 \log(1 - \varpi)$$
$$- 11.82[\log(1 - \varpi)]^2 - 11.22[\log(1 - \varpi)]^3\} - 3 \tag{12.102}$$

where [S] is solvent concentration in mol/L; μ is viscosity of the solution in Pa·s; ϖ is weight fraction of polymers; and f_0 and k_{t0} are initiation efficiency and rate constant of termination reaction, respectively. Eq. (12.102) is applicable for isothermal batch reactors.

Fig. 12.6 shows us that the polymerization rate and the degree of polymerization calculated using Eq. (12.102) are in good agreement with the experimental results. However, Eq. (12.102) is only suitable for conversion lower than 60%. From this example we can see that it is necessary to modify the parameters like k_t, k_p, and f in order to properly describe a polymerization process. The equations like Eq. (12.102), derived based on a large quantity of experimental data, have practical significance for the reactor design, scale-up, and optimization of the operation conditions. Unfortunately, such correlations are still not readily available.

Moreover, just like reactions between low molecular weight compounds, polymerization process can be classified into homogeneous and heterogeneous reactions, and the methods and procedure for kinetic study are similar. Note that even for same polymerization reaction, the kinetics derived can be totally different if a different polymerization process is used, or if the polymerization takes place in different phases. A homogeneous polymerization system includes bulk polymerization and solution polymerization, and a heterogeneous polymerization system includes heterogeneous bulk polymerization, heterogeneous solution polymerization, suspension polymerization, emulsion polymerization, etc.

For bulk polymerization, the starting reactant mixture is generally composed of monomers. If it is a radical polymerization reaction, the starting reactant mixture also contains initiator and other materials. If the polymer generated is mutually miscible with the monomer, the whole reaction system still be homogeneous during the polymerization process. Therefore, it is a homogeneous polymerization system. Styrene or methyl methacrylate polymerization reactions are typical examples of homogeneous bulk radical polymerization. Examples of homogeneous bulk condensation polymerization include synthesis of Polybutylene Terephthalate and Nylon -66.

For bulk polymerization, if the monomer is not mutually miscible with the polymer, the reactant mixture will be heterogeneous, the process belongs to heterogeneous bulk polymerization. A typical example is the high pressure radical polymerization process to make low-density polyethylene.

It should be noted that the same polymer materials can be synthesized using different polymerization processes. For instance, bulk, suspension, and emulsion polymerization processes can be used to produce Polyvinyl chloride (PVC). The bulk polymerization process belongs to a heterogeneous system, since under typical polymerization conditions PVC is not soluble in vinyl chloride and thus will form a separate phase. However, vinyl chloride can totally dissolve in PVC. As a result, the two phases in the PVC polymerization process are a pure monomer phase and a monomer swelling polymer phase. The polymerization can take place in either phase, but the reaction rate is much faster in the polymer rich phase. Therefore, the kinetics in a heterogeneous system is much more complicated than that in a homogeneous system.

12.3 ANALYSIS OF HEAT AND MASS TRANSFERS IN THE POLYMERIZATION PROCESS

12.3.1 Thermal Effect in Polymerization Process

Polymerization processes typically have the characters of fast reaction rate, high viscosity of reaction mixture, and being highly exothermic. Therefore, properly managing heat transfer is very critical for controlling the polymerization reaction. It is also the most important factor for reactor selection and design and selecting the polymerization process and conditions.

Fig. 12.7 shows the relationship between monomer conversion and the viscosity at different temperatures. Obviously, viscosity of the polymerization system increases sharply with monomer conversion. In other words, the viscosity increases sharply with the amount of polymer in the system. For instance, at $100°C$. the viscosity reaches to 10^{10} Pa $\cdot$ s when conversion approaches 100%. This high viscosity will lead to sharp decrease of total heat transfer coefficient. Fig. 12.8 shows polymerization heat as a function of reaction temperature. When reaction temperature is $160°C$, reaction heat reaches 74 kJ/mol. Fig. 12.9 shows the relationship between conversion and reaction time under different temperatures. We can see that at $100°C$ the conversion increases gently with reaction time and the conversion is less than 60% after reacting for 30 h. At $178°C$, conversion increases rapidly and exceeds 90% after 3 h. However, after that the reaction rate declines sharply, and after reaching 98% at 10 h the conversion does not increase any more,

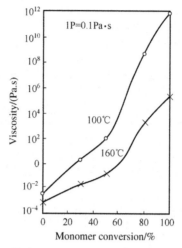

FIGURE 12.7 The relationship between conversion and viscosity.

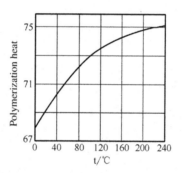

FIGURE 12.8 Styrene polymerization reaction heat.

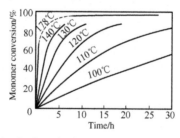

FIGURE 12.9 Relationship of polymerization rate and reaction temperature.

indicating that the reaction rate reduces to 0. This phenomenon is called the autoacceleration effect or gel effect. From Figs. 12.7−12.9, we can know that when the polymerization reaction reaches a certain conversion, an auto-acceleration effect would appear, leading to intensified heat release and viscosity increase. High heat release and high viscosity will cause further temperature and reaction rate increases. These factors influence each other, and form a vicious circle. Therefore, it is very important and challenging to properly address heat and mass transfer issues in a polymerization process.

It was well known that we must control polymerization temperature properly to ensure polymer product quality. Temperature control depends on the relative rates of heat release and heat removal. Balancing the two rates plays a key role in controlling the reaction temperature, polymerization rate, and ultimately product quality. Due to the high viscosity of the reaction mixture, rapid reaction rate, and large reaction heat, removing reaction heat and transporting highly viscous fluid are very challenging issues for the polymerization industry. We will discuss these issues in the following sections.

12.3.2 Heat Transfer and Fluid Flow in Polymerization Process

The heat transfer and fluid flow issues can be addressed from two aspects. One is to properly selecting a polymerization process, and the other is to adequately manage the heat transfer rate.

12.3.2.1 Polymerization Processes

Polymerization processes can be divided into bulk polymerization, solution polymerization, suspension polymerization, and emulsion polymerization. Obviously, when choosing the polymerization process, first we need to ensure the what is most important to guarantee the quality of the products. Here, we mainly compare the advantages and disadvantages of different polymerization processes from the perspectives of heat transfer and fluid flow.

Bulk polymerization is the most commonly used polymerization method process. The reaction can be initiated by a small trace amount of initiator, and generally its monomer conversion is typically very high, and as a result very little or no monomer recovery is needed. In addition, product purity is high, and reactor utilization is also high. However, the most obvious shortcoming of bulk polymerization is the difficulty in managing heat transfer and fluid flow.

Solution polymerization means that the polymerization takes place in a solution. The presence of a solvent prevents the viscosity of the reaction mixture from becoming too high, which is beneficial for fluid flow and heat transfer. Solvent evaporation and reflux can help heat removal from a reaction solution, but at same time will also lead to a larger reactor volume. In addition, it may lead to transferring an active chain to the solvent, which will reduce average molecular weight and narrow the molecular weight distribution of the products.

Suspension polymerization is a special type of bulk polymerization in which the monomer disperses in water as oil-droplets. Its reaction mechanism is the same as that for bulk polymerization. Each dispersed droplet suspended in water can be seen as a reaction system, and therefore the distance for heat transfer shrinks to 0.2−0.4 mm, which is very beneficial for heat transfer. A catalyst can be used to increase the reaction rate and reduce the possibilities of violent polymerization. In suspension polymerization, the reaction temperature can be easily controlled and there is no fluid flow problem. The main disadvantage of suspension polymerization is that it is difficult to achieve continuous operation. In order to maintain a suitable size of monomer droplets, additives such as dispersant have to be added. As a result, product purification, separation, and drying are required as posttreatment process.

In an emulsion polymerization system, in addition to monomers, various additives such as emulsifiers need to be added to the reaction system. It is very difficult to totally remove all the additives from the product. Therefore, emulsion polymerization is only suitable for making products that do not require high purity, such as rubbers, coatings, and artificial leathers.

In summary, we should choose a suitable polymerization process based on the characters of the specific polymerization reaction.

In addition, we can adjust the operation procedure to control the reaction rate and heat release. For instance, compound initiator can be used; monomer and initiator can be added in a batch (batch operation) or alternatively in each tank (continuous operation) or a inhibitor can be used.

12.3.2.2 Heat Transfer Rate

Heat transfer calculation in a polymerization reactor is the same as that for other processes. Heat transfer rate can be expressed as:

$$q = UA_m \Delta t_m \qquad (12.103)$$

where,

U—total heat transfer coefficient in k J/(m$^2 \cdot$ h $\cdot$ °C);
A_m—average heat transfer area across the wall in m^2;
Δt_m—average temperature difference between cold and heat fluid in °C

In order to enhance heat transfer rate in a polymerization reactor that involves highly exothermic reaction and viscous fluid, it is more critical to maintain feed purity. In addition, the internal surface should be smooth and there should be no dead area inside the equipment to avoid fouling and coating. The structure's internal heat exchanger should be as simple as possible to make the cleaning easy and allow installation of a stirring apparatus.

From Eq. (12.103), heat transfer rate can be enhanced by increasing total heat transfer coefficient, heat exchange area, and Δt_m.

12.3.3 Heat and Mass Transfer Coefficients

A key task of heat and mass transfer calculations is to determine the heat transfer coefficient and mass transfer coefficient, and readers can refer to relevant books or other literatures to find suitable correlations. It should be noted that most polymerization reactors include an agitator, and the impact of agitation on the heat transfer coefficient and mass transfer coefficient should be properly considered.

The impact on the heat transfer rate is generally reflected by the total heat transfer coefficient. Usually the heat transfer on the internal side of the reactor is the rate-determining step, and the corresponding heat transfer coefficient α_i is greatly affected by agitation. Obviously, the design of the agitator should first meet the demands for mixing, agitating, dispersing, and suspending during the polymerization process. But agitating can also enhance the heat transfer rate. On the other hand, agitation may generate heat (every kWh electricity is equivalent to 3600 kJ). For polymerization reactions, especially for a highly viscous reactant system, heat generation from agitation usually cannot be ignored since its contribution to total heat removal requirements can be as high as 30−40%.

Similarly, agitation has a significant effect on mass transfer, and its effect is mainly reflected by the change of mass transfer area. Agitation will cause convection and turbulence and hence has an impact on the mass transfer rate. It has been proved that the mass transfer coefficient would not change much after agitation reaches a certain intensity.

The accuracy of heat and mass transfer calculations depend on the quality of physical property data and the correlations used. When selecting a correlation, we must pay special attention to its applicable ranges.

12.4 DESIGN AND ANALYSIS OF POLYMERIZATION REACTOR

12.4.1 Polymerization Reactor and Agitator

According to its structure, a polymerization reactor can be classified into four types: tank, column, tubular, and other special type reactors. However, due to the high viscosity and high exothermicity of a typical polymerization system, special attention has to be paid to heat transfer and fluid flow when designing a polymerization reactor. For instance, most of the polymerization reactors are equipped with agitators. Commonly used tank reactors obviously have stirring capability, and even a tower reactor is usually equipped with multiple blades when it is used for polymerization. The main functions of agitators in a polymerization reactor include:

1. Mixing effect
   When feed contains more than two phases, adequate agitation will ensure uniform concentration, temperature, density, and viscosity in reactor.

2. Agitating effect

    The blades of the agitator exert pressure on the fluid and force the fluid to flow more vigorously. Fluid flow pattern can also be changed by changing blade shapes to enhance heat and mass transfer.

3. Suspension effect

    Agitation can keep solid particles or oil-drops suspended uniformly in a reaction medium. This is especially the case for suspension polymerization processes where water is usually used as the solvent. Maintaining uniform suspension of monomer drops or polymer particles in the water by agitation is very critical for the polymerization process.

4. Dispersion effect

    Gas, liquid, and solid can be dispersed in the liquid medium by agitation, which increases interface area and hence enhances heat and mass transfer rate.

When designing a polymerization reactor, the structure of the agitator blades and rotation speed have to be chosen based on the characters of the specific polymerization reaction to enhance a specific function of agitation. For instance, a high viscosity bulk polymerization process requires intensive mixing and agitating effects. Generally speaking, bulk polymerization and solution polymerization usually require better mixing, and suspension polymerization requires, in addition to good mixing, a uniform and stable suspension of monomer drops. Therefore, agitator design is a very important part of polymerization reactor design.

12.4.2 Mathematical Model

The flow models of the polymerization reaction process can be divided into ideal flow models and nonideal flow models. Polymerization in a batch tank reactor can be described by an ideal batch reactor model, and polymerization in a continuous tank reaction can be treated by a CSTR model. However, polymerization is a complex process and can deviate from the ideal flow model significantly. For instance, when bulk polymerization takes place in a continuous tank reactor, it is possible to achieve complete mixing at the beginning because of low viscosity of the mixture and good agitating effect. But at a later stage of polymerization, the viscosity of reactant mixture increases by several thousand times, or even tens of thousands times. The heat and mass transfer rates decrease sharply. At this stage it is impossible to maintain uniform temperature and concentration distribution inside the reactor. Therefore, using an ideal complete mixing flow model will lead to significant error, and a more realistic model has to be developed.

    Besides the flow model, polymerization kinetics is another issue that has to be properly addressed when designing a polymerization reactor. At later stage of the polymerization, gel effect will not only cause non-ideal flow patterns but also influence reaction kinetics. Correlation like Eq. (12.102) can be used for reactor design and kinetics modeling. Unfortunately, the availability and accuracy of such correlations are far from ideal.

From a reaction engineering perspective, the effect of heat and mass transfer on the reaction kinetics should be considered first. Then a nonideal flow model can be established by modifying ideal flow models. After fine-tuning of model parameters, the model can be used for the design and calculation of polymerization reactors.

12.4.3   Calculation and Analysis of Polymerization Reactors

Polymerization reactor design is an important part of the overall polymerization process development, and a key task for reactor design is to calculate reaction volume. Here we use styrene polymerization as an example to discuss polymerization reactor design.

For reactor design, reactor type and operation conditions are generally determined through lab study, pilot test, and commercial production experience with the overall goal of maximizing economic return.

1. Design bases

Production scale	1000 t/a
Operating time per year	7800 h (i.e. 325 days)
Feed composition	styrene mass fraction is 88%
	toluene (as solvent) mass fraction is 12%
Purity of styrene feed	mass fraction is 99.5%
Yield of reaction	98% by mass
Monomer conversion	80% by mass

Based on the above data, styrene production rate is:

$$w_{\mathrm{p}} = \frac{10,000 \times 10^3}{7800} = 1282 \ \mathrm{kg/h}$$

Total feed mass flow:

$$w = \frac{1282}{0.88 \times 0.995 \times 0.80 \times 0.98} = 1867.52 \ \mathrm{kg/h}$$

which includes styrene:

$$w_{\mathrm{m}} = 1867.52 \times 88\% = 1643.42 \ \mathrm{kg/h}$$

and toluene:

$$w_{\mathrm{s}} = 1867.52 \times 12\% = 224.10 \ \mathrm{kg/h}$$

2. Basic data for reactor design

Thermally initiated styrene polymerization can be treated as a first order reaction. The relationship of reaction rate constant k with

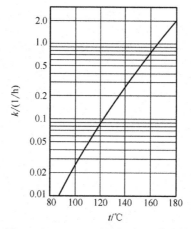

FIGURE 12.10 Relationship reaction rate constant k for thermally initiated styrene polymerization and temperature t.

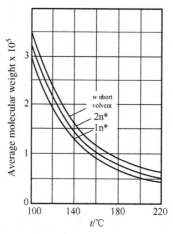

FIGURE 12.11 Average molecular weight of polystyrene and reaction temperature.

temperature is shown in Fig. 12.10. According to the reported reference [2], for thermally initiated styrene polymerization, average molar mass of polystyrene product only depends on reaction temperature and amount of solvent, as shown in Fig. 12.11. Figs. 12.12−12.14 show the dependence of reaction mixture density, specific heat capacity, and viscosity on monomer conversion. The density, specific heat capacity, and viscosity of the mixture are calculated by weighted average based on the corresponding properties of each component in the mixture.

3. The selection of reactor types

In general, the overall goal of reactor selection is to maximize the yield of desired product and minimize reactor volume required. Of course, the

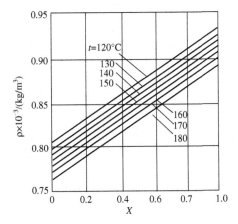

FIGURE 12.12 Reaction mixture density.

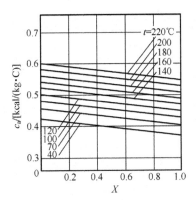

FIGURE 12.13 Reaction mixture specific heat capacity μ.

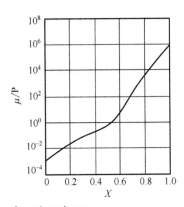

FIGURE 12.14 Viscosity of reaction mixture.

characters of the specific polymerization reaction must be considered. As for styrene bulk polymerization, the two major issues are heat transfer and fluid flow. Especially when conversion is high, the viscosity of the reaction mixture increases dramatically, leading to a sharp drop in the heat transfer rate. The large amount of reaction heat must be removed from the reaction system to control reaction temperature. Therefore, when designing the reactor we must consider temperature control, agitation, and maintaining heat transfer coefficient and heat transfer area.

After considering various factors, we choose four tanks in series for this reaction. Each tank is equipped with an agitator which is properly designed to ensure uniform temperature in the tank, and hence can be treated as a CSTR. Because viscosity of reaction solution in the first tank is less than 0.1 Pa · s, a turbine blade that is suitable for low viscosity is selected. After the second tank, viscosity of reaction solution is between 2 and 200 Pa · s, a stirring blade suitable for high viscosity must be selected. In addition, each tank should have sufficient heat exchange area and adequate heat transfer capacity.

4. Determine reaction temperature

    According to Fig. 12.11, a product's average molecular weight depends on temperature, and a polymer product with a narrow molecular weight distribution can be obtained by controlling the temperature in each tank at the same value. Meanwhile, the impact of reaction temperature on viscosity, reaction heat, and reactor thermal stability should also be considered. This is especially the case for the first tank, because it has a high monomer concentration, high reaction rate, and large heat release. Operating temperature in the first tank cannot be too high, but it cannot be too low either. Low temperature will lead to not only slow reaction rate in the first tank but also lower reaction temperature in the following tanks, which is undesirable since the viscosity will be too high. Based on the above considerations, reaction temperatures in each tank are all selected as 140°C in this design.

5. Kinetics

    Based on literature data, styrene bulk polymerization can be treated as a first order irreversible reaction when the monomer conversion is in the range of $0 \sim 70\%$, i.e., $r_M = k[M]$. Reaction rate constant is shown in Fig. 12.10. In this design, the required monomer conversion is 80%, but we will still approximately treat it as a first order reaction.

6. Mathematical model

    Based on the ideal flow model of a CSTR, mass balance for each tank in steady state can be expressed as:

$$Q_{i-1}[M]_{i-1} = Q_i[M]_i + r_{Mi}V_{ri} \qquad (12.104)$$

And heat balance

$$(-\Delta H_r)_{Tr} \cdot r_{Mi} V_{ri} + q_{ai} = U_i A_i (T_i - T_{ei}) + W\overline{C}_{pt\ i}(T_i - T_{i-1}) \quad (12.105)$$

where

Q_i—volume flow rate of ith tank, m^3/h

V_{ri}—effective reaction volume of ith tank, m^3

$(-\Delta H_r)_{Tr}$—reaction heat under reference temperature, kJ/h

q_{ai}—Heat generation due to stirring in ith tank

U_i—total heat transfer coefficient of ith tank

A_i—total heat transfer area of ith tank

T_{ei}—heat carrier temperature of ith tank

$\overline{C}_{pt\ i}$—average specific heat capacity of reaction mixture in ith tank

W—total mass flow rate, kg/h.

As shown in Fig. 12.12, reaction mixture density changes with progress of the reaction, so

$$[M]_i = \frac{\rho_i}{\rho_0}[M]_0(1 - X_i) \quad (12.106)$$

where $[M]_0$ and ρ_0 are monomer concentration and density of feedstock in the first tank, respectively. Substituting Eq. (12.106) to Eq. (12.104) gives

$$V_{ri} = \frac{W(X_i - X_{i-1})}{k\rho_i(1 - X_i)} \quad (12.107)$$

7. Key tasks and steps for reactor design
   i. Setting reaction temperature for each tank T_i.
   ii. Setting permissible temperature difference Δt_m between reaction temperature T_i and heat carrier temperature T_{ci}.
   iii. Setting conversion for each tank X_1, X_2, X_3.
   iv. Calculating effective reaction volume V_{ri} of each tank based on selected T_i and X_i.
   v. Calculating stirring power P_i and total heat transfer coefficient U_i under specific conditions of each tank.
   vi. Calculating heat transfer area A_i required by each tank based on heat balance equation, and determining heat exchanging mode for each tank.
   vii. Calculating heat transfer rate q_c for each tank based on heat balance.
   viii. Calculating Δt_m by q_c and compare with the value selected in (ii). If it is greater than the permissible temperature difference, it need to reset the X_i in the three tank and repeat the calculations above until the calculated Δt_m is less than the permissible temperature difference.
   ix. If the mechanism of polymerization is known, calculating the average molecular weight and its distribution.

8. Reaction volume calculation

For a first order irreversible reaction, in order to minimize reaction volume of each tank the volume of each tank should be the same, as discussed in Chapter 3, Tank Reactor. Therefore:

$$V_{r1} = V_{r2} = V_{r3} = V_{r4}$$

Mass balance for each tank gives

$$\frac{X_1}{\rho(1 - X_1)} = \frac{X_2 - X_1}{\rho_2(1 - X_2)}$$

$$\frac{X_2 - X_1}{\rho_2(1 - X_2)} = \frac{X_3 - X_2}{\rho_3(1 - X_3)} \tag{12.108}$$

$$\frac{X_3 - X_2}{\rho_3(1 - X_3)} = \frac{X_4 - X_3}{\rho_4(1 - X_4)}$$

From Fig. 12.10, we can find the reaction rate constant at 140°C, i.e., $k = 0.26$. We can also find reactant mixture densities at different monomer conversion X_i from Fig. 12.12. Solving the above three equations we can get

$$X_1 = 32.5\%$$
$$X_2 = 55.1\%$$
$$X_3 = 69.8\%$$

And then from Eq. (12.107), we can obtain the effective volume of each tank

$$V_{ri} = 4.15 \text{ m}^3$$

REFERENCES

[1] Shawki SM, Hamielec AE. The effect of shear rate on the molecular weight determination of acrylamide polymers from intrinsic viscosity measurements. J Appl Polym Sci 1979;23:3323–39.

[2] Duerksen JH, Hamielec AE. Polymer reactors and molecular weight distribution, Part IV: free radical polymerization in a continuous stirred tank reactor train. J Polym Sci Part C 1968;25:155–66.

FURTHER READING

Polymerization Reaction Engineering. Translated by Tianjin University, Beijing: Chemical Industry Press, 1982.

Fundamentals of Polymer Science. Translated by S Haoting Wang, Fu Lng and Jinzhang He, Chemical Industrial Press.

Chen G, et al. Fundamentals of polymerization reaction engineering. Beijing: China Petrochemical Press; 1991.

Polymer Chemistry, Polymer Group, Fudan University, Shanghai: Fudan Press, 1995.

Shi Y. Fundamentals of polymerization reaction engineering. Beijing: Chemical Industry Press; 1991.

Schmidt LD. The engineering of chemical reactions. New York: Oxford University Press; 1998.

PROBLEMS

12.1 According to the relationship between j and $F_w(j)$ given in Example 12.1, derive the weight-average degree of polymerization $\overline{P_w}$ and Z-average degree of polymerization $\overline{P_z}$.

12.2 The reaction described in Example 12.5 is carried out isothermally in a PFR. Please calculate the distribution of weight-average degree of polymerization and compare it with that in a batch reactor.

12.3 Free radical polymerization is initiated by a initiator and terminated by bi-radical termination (coupling and disproportionation also occur). The reaction can be treated as a constant-volume process, the rate constants of each elementary reaction

$$k_p = 5.8 \times 10^3 \text{ m}^3/\text{kmol} \cdot \text{s}$$
$$k_{tc} = 3.0 \text{ m}^3/\text{kmol} \cdot \text{s}$$
$$k_{td} = 1.0 \times 10^4 \text{ m}^3/\text{kmol} \cdot \text{s}$$
$$k_d = 1.2 \times 10^{-5} \text{ s}^{-1}$$

And:

$$[M]_0 = 1.0 \text{ kmol}/\text{m}^3, \ [I] = 1.2 \times 10^{-4} \text{ kmol}/\text{m}^3, \ f = 0.5.$$

a. If this reaction is carried out homogeneously in a PFR, please calculate the distribution of weight-average degree of polymerization of the products when conversion is 50%.

b. If this reaction is carried out homogeneously in CSTR, please calculate the distribution of weight-average degree of polymerization of the products when conversion is 50%.

c. Show and compare the results of above calculations in the same plot.

12.4 Based on conditions and data given in Problem 12.3. Please determine whether it would be possible to obtain $\overline{P_w} = 5.8 \times 10^{-3}$. If it is possible, please calculate the conversions for both PFR and CSTR reactors.

12.5 A solution polymerization process is used for a free radical polymerization reaction, and the reaction mechanism includes initiation by an initiator, termination by disproportionation, and chain transfer can be ignored. Monomer consumption rate

$$r_M = 1.8 \times 10^3 \exp\left(\frac{-11,328}{T}\right)[M], \text{ kmol}/(\text{m}^3 \cdot \text{s})$$

Initial monomer concentration is 2.0mol/L, and the reaction is carried out in a batch tank.

a. Calculate reaction time required to reach conversion of 98% under the conditions of isothermal operation at 65°C.

b. If the reaction is operated adiabatically, initial temperature is 60°C, what is the reaction time required to reach conversion of 98%?

12.6 For the polymerization reaction described in Problem 12.5, assuming when monomer conversion is 0 at 65°C, the instantaneous number-average degree of polymerization is 10×10^4. Initiator concentration stays unchanged during the reaction process. Please calculate the cumulative number-average degree of polymerization for isothermal batch operation at 65°C.

12.7 Using steady-state approximation, derive the reaction rate, instantaneous number-average degree of polymerization and its distribution for a free radical polymerization reaction with initiator initiation, mono-radical termination, and no chain transfer.

12.8 Using steady-state approximation, derive the reaction rate, instantaneous number average degree of polymerization and its distribution for an anionic polymerization reaction with catalytic initiation, no termination, and no chain transfer.

12.9 Methyl methacrylate polymerization is a free radical polymerization process with ABIN as initiator, disproportionation termination, and no chain transfer. Under reaction temperature,

$$k_d = 7.0 \times 10^{-4} \text{min}^{-1}, \ k_t = 5.6 \times 10^8 L/(\text{mol} \cdot \text{min})$$
$$k_p = 2.2 \times 10^4 \ L/(\text{mol} \cdot \text{min}), \ [\text{I}] = 3.0 \times 10^{-3} \ \text{mol/L}, \ [\text{M}]_0 = 0.6 \ \text{mol/L}$$
$$f = 0.52$$

1. The reaction is carried out in a batch tank reactor, please calculate the reaction time required to reach conversion of 40%, and both weight-average and number-average degree of polymerization.

2. The reaction is carried out in a CSTR with space-time equal to the reaction time of the batch reactor to reach conversion of 40%, please calculate both weight-average and number-average degree of polymerization and compare with that in (1).

3. The reaction is carried out in a CSTR and the required conversion is 40%. Calculating the reactor volume required when methyl methacrylate feed rate is 100 kg/h.

4. Under the condition of batch operation, calculate the reaction time required to obtain product $\overline{P_n} = 2.0 \times 10^{-3}$ and the corresponding conversion.

Chapter 13

Introduction to Electrochemical Reaction Engineering

Chapter Outline

13.1 INTRODUCTION

Electrochemical reaction engineering is a branch of chemical engineering, and shares many general concepts and theories. However, some specific issues have to be addressed in electrochemical reaction engineering due to the following characters of electrochemical reactions.

13.1.1 Characters of Electrochemical Reactions

Electrochemical reactions refer to reaction that involves electron exchange taking place at the interface of the first conductor (electronic conductor) and second conductor (ionic conductor). It is heterogeneous and also a redox reaction. At an early stage, most electrochemical reactions occur at the solid (electrode)−aqueous (dilute electrolyte solution) interface. This concept now

Reaction Engineering. DOI: http://dx.doi.org/10.1016/B978-0-12-410416-7.00013-6

has been extended, and electrochemical reactions can occur at the interface between two solids with different charge carriers, such as applications involving solid state electrolytes and conductive polymer electrolytes.

Since the reactants of electrochemical reactions are charged, reactant transport in the system must be accompanied by electron and ion transfers. Therefore, in addition to mass, heat, and momentum transports, charge transfer must also be considered when designing and analyzing an electrochemical reactor. For mass transfer, electrical driven migration (charged particles migration caused by electrical fields) needs to be considered together with diffusion and convection. At the interface, the amount of reactants consumed or products formed must strictly follow a correlation with the quantity of charge transfer across the interface, i.e., Faraday's law.

Electrochemical reaction kinetics, known as electrode kinetics in electrochemistry, demonstrates the following special properties and behaviors:

1. Electrochemical reactions occur at charged phase interfaces (electrical double layer). The electrical structure of this interface (electrical potential difference and its distribution) is especially important, and determines the reaction activity energy and kinetic rate. The reaction rate can be continuously adjusted over a large wide range by changing the electrical potential and overpotential.
2. An electrochemical reaction includes many elementary steps. They generally include reactant particles moving from the bulk phase to the interface, pre-reaction surface transforming process (e.g., absorbing), electron exchanging process, post-reaction surface transforming process, and product particles moving back to the bulk phase. Sometimes, it may include new phase generation (e.g., crystallization, gas generation). According to the rate determining steps, electrochemical reactions can be classified as electrode polarization (electron exchange controlling), concentration polarization (diffusion controlling), or mixed polarization, and each will show different kinetic features and behaviors.
3. The material and surface properties of electrodes are especially important for electrochemical reactions, because they affect the electrocatalytic properties of the electrodes, i.e., the relationship between reaction rate and overpotential.

It is necessary to adopt a new set of parameters to describe electrochemical reactions and analyze electrochemical reactors. For example, based on Faraday's law, the electrical quantity, current, and current density can be used to describe the extent and rate of an electrochemical reaction. For example:

$$i = nFr \tag{13.1}$$

where:

i: current density passing through the electrode, A/m^2
n: number of electrons exchanged during reaction
F: Faraday constant, 96,500 C/mol
r: reaction rate, mol/(s m^2).

13.1.2 Performance Parameters for Electrochemical Reaction Engineering

Besides those commonly used in chemical reaction engineering, the following parameters are also used in describing the performance of electrochemical reaction processes.

13.1.2.1 Current Efficiency

In electrochemical reactions, electrons acts as reactants. When an electrical current passes through the interface between electrode and electrolyte solution, the ratio between the actual amount of product formed and the ideal amount calculated by Faraday's law is defined as the current efficiency. This definition is based on a given electrical quantity. On the other hand, this factor can be defined as the ratio between actual electrical consumption and ideal consumption.

$$G = QK\eta_1 \tag{13.2}$$

$$\eta_1 = \frac{G}{QK} \tag{13.3}$$

where:

G: the mass of reaction product, g or kg
Q: the actual electricity consumed, A h
K: electrochemical equivalent, g/(A h) or kg/(kA h)
η: current efficiency.

$$\text{Or} \quad \eta_1 = \frac{G}{QK} = \frac{Gk}{Q} \tag{13.4}$$

where k is ideal electricity consumption, in the unit of A h/g or kA h/kg.
Obviously:

$$k = \frac{1}{K} \tag{13.5}$$

When the current density is not a constant, the total electricity during a period of time t is

$$Q = \int_0^t I dt \tag{13.6}$$

By inserting this Q in Eq. (13.4), the overall current efficiency can be obtained. The efficiency within a certain period of time, t_1 to t_2, can be calculated by using total electricity consumed during that period of time:

$$Q = \int_{t_1}^{t_2} I dt \tag{13.7}$$

The current efficiency is usually less than 100%, indicating the loss of productivity due to side reactions, or secondary reactions (or reverse reactions).

Example 13.1

NaCl aqueous solution is electrolyzed to produce chlorine and sodium hydroxide. If anode gas contains 98.2% Cl_2 and 1.8% O_2, calculate the current efficiency.

Solution

At the interface of the anode, the main and side reactions are

$$2Cl^- \to Cl_2 + 2e$$
$$4OH^- \to 2H_2 + O_2 + 4e$$

So producing 1 mol of Cl_2 will consume 2 F electricity and producing 1 mol O_2 will consume 4 F electricity. Given that the anode gas consists of 98.2% Cl_2 and 1.8% O_2, their electricity consumptions are

$$98.2 \times 2 = 196.4\, F \quad \text{and} \quad 1.8 \times 4 = 7.2\, F,$$

respectively. Then the current efficiency is

$$\eta = \frac{196.4}{196.4 + 7.2} = 96.46\%$$

13.1.2.2 Operation Voltage of Electrochemical Reactor

The electrochemical reactors are almost always operated under irreversible conditions, and their operation voltages are different from the theoretical decomposition voltage. This phenomenon is mainly caused by electrode polarization. In addition, other electrical resistances in the reactor also cause electrical potential drop. The total operation voltage can be described by the following equation (shown in Fig. 13.1 as well)

$$V = \varphi_{e,A} - \varphi_{e,K} + |\Delta\varphi_A| + |\Delta\varphi_K| + IR_{AL} + IR_{KL} + IR_D + IR_A + IR_K$$

(13.8)

where:

φ_A: anode equilibrium potential
φ_K: cathode equilibrium potential
$|\Delta\varphi_A|$: overpotential of at anode (absolute value)
$|\Delta\varphi_K|$: overpotential of cathode (absolute value)
IR_{AL}: potential drop in electrolyte near anode
IR_{KL}: potential drop in electrolyte near cathode
IR_D: potential drop in diaphragm
IR_A: potential drop across anode and its circuit
IR_K: potential drop across cathode and its circuit

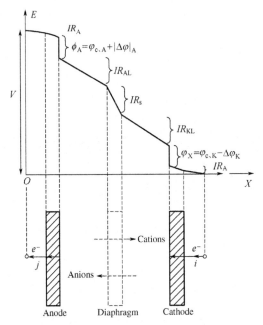

FIGURE 13.1 Components of operation voltage across electrochemical reactors.

Next we will discuss the properties and calculation methods of these parameters.

1. φ_A and φ_K are thermodynamic properties. For electrolyzer type of electrochemical reactors, the difference between these two values is the theoretical decomposition potential E_0, i.e.

$$E_0 = \varphi_{e,A} - \varphi_{e,K} \qquad (13.9)$$

For chemical power (battery), the difference is the electromotive force. When a battery is discharging, the reduction and oxidation is taking place at cathode (electrode with higher electrical potential) and anode (electrode with lower electrical potential), respectively, therefore:

$$E_0 = \varphi_{e,+} - \varphi_{e,-} = \varphi_{e,K} - \varphi_{e,A} \qquad (13.10)$$

Both $\varphi_{e,A}$ and $\varphi_{e,E}$ can be calculated with the Nernst equation.

2. $\Delta\varphi_A$ and $\Delta\varphi_K$ are determined by electrode kinetics, and their calculation methods depend on the electrode polarization mechanism. If the polarization is caused by electrochemical reaction and the overpotential is high, they can be calculated with the Tafel equation (simplified from the Buter−Volmer Equation)

$$\Delta\varphi = a + b\log i \qquad (13.11)$$

where a, b are constants related to reaction mechanism, electrode properties, and reactor conditions.

3. IR_A and IR_K are ohmic potential drops when current passes through an electrolyte solution, and can be calculated by

$$IR_L = \frac{I\rho\Delta}{S} = \frac{i\Delta}{\kappa} \qquad (13.12)$$

where ρ is the electrical resistivity of electrolyte, κ is the electrical conductivity of electrolyte, and Δ is the distance between electrodes or between electrode and diaphragm.

When gas is generated in the reactor during the electrolytic process, the electrical conductivity of the electrolyte solution decreases due to the presence of gas bubbles, and such an effect has to be corrected (see Eq. 13.28).

4. IR_D is the ohmic potential drop over the diaphragm, and depends on electrical resistance of the diaphragm and current density. Diaphragm resistance depends on diaphragm properties (composition, structure, thickness) and operation conditions (electrolyte's composition, concentration, temperature, conductivity, etc.). There are a few formulas that can be used to estimate IR_D, such as:

$$V_D \simeq IR_D = i\left(\rho_0 d + \frac{K_D \cdot d^2}{2}\right) \approx i\rho_0 d \qquad (13.13)$$

where ρ is the electrical resistivity of the diaphragm, d is diaphragm thickness, and K_D is a factor related to diaphragm structure.

5. IR_A and IR_K are ohmic potential drops of electrodes, and can be calculated based on cross-section area, length, and conductivity. Usually, this term is very small due to the high conductivity of metal, and can be neglected when estimating the operation potential of electrochemical reactors.

6. For electrolyzer and electroplating cells the operation voltage is cell voltage, and for a chemical power source (battery) it is recharge or discharge voltage.

Example 13.2
Calculate the operation voltage of diaphragm chlor-alkali electrolytic cells:
1. Electrochemical reaction in the cell:
    Anode reaction: $2Cl^- \rightarrow Cl_2 + 2e$ $\varphi^0_{90°C} = 1.266V$
    Cathode reaction: $2H_2O + 2e \rightarrow 2OH^- + H_2$ $\varphi^0 = -0.828V$
    Overall reaction: $2H_2O + 2NaCl \rightarrow 2NaOH + H_2$
2. The reaction takes place at 90°C and 1500 A/m²
3. Reaction overpotential can be calculated using the Tafel equation, i.e.,
    Anode overpotential $\Delta\varphi_A = 0.025 + 0.027\log i$ (unit of i is A/m²)
    Cathode overpotential $\Delta\varphi_K = 0.113 + 0.056\log i$
4. Diaphragm thickness = 8 mm, resistivity = 0.035 m Ω
5. Distance between electrodes = 9 mm, Conductivity of electrolyte solution = 61.03 m^{-1}/Ω, gas holdup = 0.2
6. Anode electrolyte composition: 320 g/L NaCl (5.7 mol/L)
Cathode electrolyte composition: 100 g/L (2.5 mol/L)

(Continued)

(Continued)

7. Potential drop of electrodes can be neglected.

Solution

Operation voltage of the electrochemical reactor can be calculated using Eq. (13.8):

$$V = \varphi_{e,A} - \varphi_{e,K} + |\Delta\varphi_A| + |\Delta\varphi_K| + IR_L + IR_D + IR_A + IR_K$$

1. Calculate equilibrium potential at anode using the Nernst equation. Assume $p_{Cl_2} = 1$

$$\varphi_{e,A} = \varphi_A^0 + \frac{RT}{2F}\log\frac{p_{Cl_2}}{a_{Cl^-}^2} = 1.226 + \frac{2.3 \times 8.314 \times 363}{96,500}\log\frac{1}{(5.475)^2} = 1.212V$$

2. Calculate equilibrium potential of cathode using the Nernst equation. Assume $a_{H_2O} = 1$, $p_{H_2} = 1$

$$\varphi_{e,K} = \varphi_K^0 + \frac{RT}{2F}\log\frac{a_{H_2O}}{a_{OH^-}^2 \cdot p_{H_2}} = -0.828 - \frac{2.3 \times 8.314 \times 363}{96,500}\log2.5 = -0.856V$$

3. Calculate overpotential

$$\Delta\varphi_A = 0.025 + 0.027\log1500 = 0.1107 \text{ V}$$

$$\Delta\varphi_K = 0.113 + 0.056\log1500 = 0.2901 \text{ V}$$

4. Calculate ohmic potential drop of electrolyte solution

$$IR_i = \frac{i \cdot \Delta}{\kappa_0(1-\varepsilon)^{3/2}} = \frac{1500 \times 0.009}{61.03 \times (1-0.2)^{3/2}} = \frac{1.5 \times 9}{61.03 \times 0.7155} = 0.309V$$

5. Calculate ohmic potential drop across diaphragm: using Eq. (13.13)

$$IR_D \approx i\rho_0 d = 1500 \times 0.035 \times 0.008 = 0.42V.$$

Therefore: $V = 1.212 - (-0.856) + 0.1107 + 0.2901 + 0.309 + 0.42 = 3.1978V$

13.1.2.3 Voltage Efficiency

Voltage efficiency is the ratio between theoretical decomposition potential and reactor operation voltage. For electrolyzers, it is

$$\eta_V = \frac{E}{V} \tag{13.14}$$

For batteries, it is the ratio between working (discharge) voltage and electromotive force

$$\eta_V = \frac{V}{E} \tag{13.15}$$

Obviously, voltage efficiency indicates the reversibility of electrode processes, i.e., the magnitude of overpotential under close circuit conditions.

It can comprehensively reflect the performance of electrochemical reactors, i.e., the overall Ohmic potential drop over all components of the reactor.

13.1.2.4 DC Power Consumption

The DC power consumption of industrial electrolysis process is usually expressed by DC electrical energy consumed per unit production (kg or t)

$$W = \frac{kV}{\eta_I} \qquad (13.16)$$

where:

W: DC power consumption, kW h/t
K: theoretical energy consumption
V: operation voltage, V
η_I: current efficiency, %.

Among these three factors, K is almost constant for a given electrochemical system. The main influencing factors are operation voltage and current efficiency, so reducing operation voltage and improving current efficiency are key approaches to enhance reactor performance.

13.1.2.5 Energy Efficiency

Energy efficiency refers to the ratio of theoretical and actual power consumption per unit production, i.e.,

$$\eta_W = \frac{W_t}{W} \qquad (13.17)$$

$W_t = kE$, and $W = kV/\eta_I$, therefore:

$$\eta_W = \frac{W_t}{W} = \frac{kE}{\frac{kV}{\eta_I}} = \eta_V \cdot \eta_I$$

i.e.,

$$\eta_W = \eta_V \cdot \eta_I \qquad (13.18)$$

We can see that energy efficiency depends on both current and voltage efficiency.

13.1.2.6 Specific Electrode Area

Active electrode area per unit volume of an electrochemical reactor is defined as specific electrode area, A_s. Electrochemical reactions are heterogeneous, so they are greatly influenced by interface areas and their properties. For a constant current density, the greater the interface area is, the higher the current that can pass through. For constant current, the greater the A_s, the lower the actual current density, which can reduce polarization and operation voltage.

$$A_S = \frac{S}{V_R} \qquad (13.19)$$

where:

S: electrode area, m^2
V_R: reactor volume, m^3
A_s: specific electrode area, m^2/m^3 or m^{-1}.

It is difficult to accurately measure A_s, due to the discrepancy between apparent and actual areas of electrodes. A_s depends not only on the electrode structure (macroscopic and microscopic structure, e.g., two-dimensional and three-dimensional electrode), but also on operation conditions. For three-dimensional electrodes, such as porous electrodes prepared by powder metallurgy, packed-bed electrode, fluid bed electrode, the actual effective area of electrodes are determined not only by particle and pore sizes and their distributions, but also by current density, flow pattern of electrolyte solution, and mass transfer behaviors, because these factors affect the depth where reaction occurs (the depth of penetration).

The definition of reactor volume is ambiguous as well. It could be the internal volume of the reactor, or the volume of electrolyte solution under certain circumstances. For reactors whose most internal space are occupied by electrodes, V_R can be represented by volume of electrodes V_E, then

$$A_S = A_E = \frac{S}{V_E} \tag{13.20}$$

No matter of their definition, A_E and A_s should be as high as possible to achieve compact design and high performance in electrochemical reactors.

13.1.2.7 Space-Time Yield

The amount of product generated in a reactor per unit time and volume is defined as space-time yield:

$$Y_{ST} = \frac{G}{tV_R} \tag{13.21}$$

Since, $G = It\eta_I K = iSt\eta_I K$, therefore:

$$Y_{ST} = \frac{iSt\eta_I K}{tV_R} = iA_S\eta_I K \tag{13.22}$$

where K is electrochemical equivalent of the electrochemical reaction, $kg/A \cdot s$.

For a chemical power source, space-time yield is equivalent to power output per unit volume. Since its product is electrical energy instead of materials, the space-time yield is the electrical energy generated per unit time and volume, and its unit is $J/L \cdot s$ or W/L.

Space-time yield of an electrochemical reactor is strongly affected by its design. Compared to other chemical reactors, the space-time yield of electrochemical reactors is generally lower. For industrial reactors space-time yield is typically in the range of $0.2-1$ kg/dm^3 h, but a typical industrial electrolyzer for copper is in the range of $0.05-0.1$ kg/dm^3 h.

13.2 SPECIAL ISSUES IN ELECTROCHEMICAL REACTION ENGINEERING

13.2.1 Electrical Potential and Current Distribution on Electrode Surface

Electrical potential and current distributions on an electrode surface are special and very phenomena in electrochemical engineering. Besides electrolysis, almost all industrial electrochemical processes require uniform distribution of electrical current. An unevenly distributed current could cause the following undesirable consequences:

1. Incomplete utilization of active electrode surface and/or materials, reducing the space-time yield and energy efficiency
2. Reaction at some portion of electrode surface is under "uncontrolled" situation, i.e., the reaction proceeds under unexpected and unreasonable conditions, inducing side reactions and reducing current efficiency and product quality (e.g., purity)
3. Uneven wear, tear, corrosion, and deactivation of electrodes, that shortens their lives
4. Dendritic growth of crystals in electrochemical deposition processes, causing short circuit or diaphragm damage, or local pH change, or undesired formation of oxide deposition, or hydroxide coating.

13.2.1.1 Factors That Affect Current Distributions and Three Major Types of Distribution

The distributions of electrical potential and current density are closely related. Current density distribution depends on potential distribution and localized concentration of reactant and conductivity in electrolyte solution. Potential distribution depends on the shapes of electrodes, their relative positions and distance, and polarization behavior of electrodes. Usually, the electrodes are made of highly conductible materials, so it is reasonable to assume uniform potential distribution on electrode surface.

The equilibrium correlation of cell voltage can be derived by analyzing electrical potential at a specific point on the electrode surface:

$$V_i = \varphi_e + - \varphi_{e^-} + |\Delta\varphi_+| + |\Delta\varphi_-| + \Sigma IR \qquad (13.23)$$

If the point is under positive or anode polarization, its electrode potential is

$$\varphi_+ = \varphi_{e^-} + |\Delta\varphi_+| = V_i + \varphi_{e^-} - |\Delta\varphi_-| - \Sigma IR \qquad (13.24)$$

Terms in Eq. (13.24) have the same meaning as those in previous sections. ΣIR includes ohmic potential of electrolyte solution, diaphragm, etc.

A change in any term in Eq. (13.23) will cause changes of φ_+. Generally, factors that influence potential and current density distributions include:

1. Structures of electrochemical reactors and its electrodes, such as shape, size, positions, and distances
2. Conductivity and its distribution of electrodes and electrolyte solution
3. Polarization processes that cause overpotential, including electrochemical polarization and concentration polarization
4. Other factors, such as conversion steps on electrode surface and surface coating effects.

For an actual electrochemical system, all factors mentioned above may coexist, though their effects are different. When analyzing an electrochemical system some simplification is usually needed. The three main types of current distribution are:

1. Primary current distribution: All overpotentials are neglected and conductivity is uniform
2. Secondary current distribution: Consider overpotential induced by electrochemical polarization but ignore that caused by concentration polarization
3. Tertiary current distribution: Both electrochemical and concentration polarization are considered.

Characters of these three types of distributions are summarized in Table 13.1.

13.2.1.2 Primary Current Distribution

Primary current distribution on electrode surface is determined by the distribution of the electrical potential field in the electrolyte solution. Since at any point the current density is proportional to the potential gradient at that point and media conductivity:

$$i_s = -\kappa \Delta \varphi \tag{13.25}$$

where i_s is the current density in the solution, which is a vector, perpendicular to the potential isosurface, and at a tangent to the electric field line. Current density on an electrode surface is a scalar, and it is the component of i_s along the direction perpendicular to the electrode surface. It can be calculated according to Eq. (13.25) and the partial differential of electrical potential $\left(\frac{\partial \varphi}{\partial n}\right)$ along the direction perpendicular to the electrode surface.

TABLE 13.1 Characters of Different Current Distributions

Current Distribution	Causes for Overpotentials			Major Influencing Factors				
	Electrochemical Polarization	Concentration Polarization	Others	Structure of Reactor and Electrodes	Conductivities of Electrodes and Electrolyte	Electrochemical Polarization	Concentration Polarization	Others
Primary distribution	×	×	×	✓	×	×	×	×
Secondary distribution	✓	×	×	✓	✓	✓	×	×
Tertiary distribution	✓	✓	×	✓	✓	✓	✓	×
Actual distribution	✓	✓	✓	✓	✓	✓	✓	✓

Actually, both potential φ and current density i on the electrode surface can be calculated.

We can see that the first step to quantitatively study current distribution is to calculate potential distribution, $\varphi(X,Y,Z)$, by solving the Laplace equation.

In electrochemical engineering, this equation is derived based on basic mass and charge transfer equations without considering concentration polarization. i.e., the change of concentration of B (c_B) as a given point in the solution $\partial c_B/\partial t$, is determined by three mass transfer processes (diffusion, convection, and electromigration) and the homogeneous reaction that forms B.

In fact, Laplace equations are also used to study heat transfer and mass diffusion under steady state. Similar to heat transfer under temperature gradient and mass transfer under concentration gradient, charge transfer under electrical gradient can be studied with proper boundary conditions.

Both analytical and numerical solving methods can be applied on Laplace equations, such as conformal mapping, Green function, finite difference, and finite element methods.

Next we will use an example illustrated by Fig. 13.2, a system with two parallel plate electrodes, to introduce the conformal mapping method for

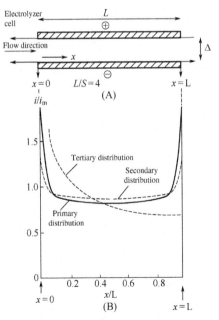

FIGURE 13.2 Current distribution between parallel plates. (A) Reactor; (B) Current distribution.

TABLE 13.2 Current Distribution on Surface of Parallel Electrodes

x/L	i_x/i_m
0	∞
0.001	7.35
0.01	2.39
0.05	1.203
0.1	0.971
0.2	0.857
0.3	0.831
0.4	0.824
0.5	0.823
0.6	0.824
0.7	0.831
0.8	0.857
0.9	0.971
0.95	1.203
0.99	2.39
0.999	7.35
1	∞

solving Laplace equation, and then examine primary current distribution. The calculation results are given in Table 13.2.

When $L/\Delta = 10$, $\varepsilon = 15.7$, $i_{x=L/2}/i_m = 0.958$, for most parallel electrode systems, their length L greatly exceed the distance between electrodes, and therefore the primary current density are relatively uniform. At the electrode edges, current densities may change greatly, as shown in Fig. 13.3.

In summary, the primary current density remains uniform only in systems with parallel electrodes and coaxial cylindrical electrodes with uniform electrode distance and excellent isolation condition. For electrodes with other structures, current densities are not uniform. As shown in Fig. 13.4, current density reaches its highest value at the shortest distance, and approaches zero on the unexposed backside of anodes.

FIGURE 13.3   Primary current distribution at the edge of electrodes. (The thick lines represent electrode surface and the fine lines represent wall of the electrochemical reactor)

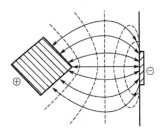

FIGURE 13.4   Primary current distribution on electrode surface.

Current densities are strongly affected by the structure and the shape of electrode as well as the configuration of their circuiting system. For example, current distributes more evenly on surfaces of multiple connected electrodes. Current distribution has to be considered when designing an electrode, such as width−length ratio, location of electric connector, shape of conductive frame. Gas evolution electrodes are more sensitive to flow conditions, electrolyte concentration, and gas production rate.

13.2.1.3   Secondary Current Distribution

When an electric current flows through the electrode surface, electrochemical polarization occurs, which will have an impact on the current density distribution. The current density distribution under such circumstances is referred to as the secondary current distribution.

The fundamental cálculation method is similar to that for primary current distribution, but more complicated. One of the boundary conditions of the Laplace equation is modified, since the potential difference at the electrode−electrolyte solution interface depends on current density. In the electric double layer, even the electric potential is uniform on metal electrodes, electrical potential on the solution side will change with current density:

$$\varphi_s = \varphi_m - \Delta\varphi = f(i)$$

where φ_s is the electric potential on the solution side of double electric layer; φ_m is the potential on the metal side, i.e., potential inside metal electrode; and

$\Delta\varphi$ is the overpotential.

To specify boundary conditions, the relationship between $\Delta\varphi$ and i has to be determined. As discussed previously. The relationship between $\Delta\varphi$ and i can be simplified under two typical conditions:

1. When current density is low, the relationship between $\Delta\varphi$ and i is approximately linear, i.e., $\Delta\varphi = \Delta\varphi^0 + K \cdot i$, K is a constant. Then the boundary conditions for the Laplace equation can be linearized and the equation can be integrated. K reflects the degree of polarization of the electrode reaction.

2. If current density is high, the relationship between $\Delta\varphi$ and i can be approximated using the Tafel equation, $\Delta\varphi = a + b\log i$.

For an intermediate range between the two extreme situations discussed above, fundamental kinetic equations have to be used to determine the $\Delta\varphi - i$ relation and then to specify the boundary conditions.

Ibl reviewed calculation methods and results of secondary current density [1]. The effects of electrochemical polarization can be qualitatively discussed. In an electrochemical system shown in Fig. 13.5, point 1 and 2 are located on the dendritic cathode surface at the distance of Δ_1 and Δ_2 away from the flat anode surface. The ohmic resistance of electrolyte solution is R_1 and R_2, respectively. If electrochemical polarization effects are neglected, $\Delta\varphi_1 = \Delta\varphi_2 = 0$, $i_1 R_1 = i_2 R_2$, $i_1/i_2 = R_2/R_1$; since $\Delta_2 > \Delta_1$, $R_2 > R_1$, so $i_1/i_2 > 1$.

When electrochemical polarization effects are considered, both $\Delta\phi_1$ and $\Delta\phi_2$ are not zero. However, the potential difference between point 1, 2 and anode surface is still equal to each other, i.e.,

$$i_1 R_1 + |\Delta\varphi_1| = i_2 R_2 + |\Delta\varphi_2|.$$

$i_1 > i_2$, therefore $\Delta\phi_1 > \Delta\phi_2$, so, $i_1 R_1 > i_2 R_2$, i.e., $i_1/i_2 < R_1/R_2$.

We can see that although i_1 is still greater than i_2, their difference decreases, implying that the current distribution on the cathode becomes more uniform.

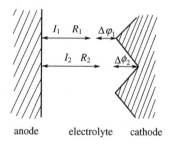

anode   electrolyte   cathode

FIGURE 13.5 Secondary current distribution on electrode surface

From the above discussions we know the three factors that affect secondary current distribution are:

1. Electrolyte conductivity (which determines solution electrical resistance)
2. Degree of electrochemical polarization, $d\varphi/di$ (which determines overpotential $\Delta\varphi$)
3. Distance between electrodes (Δ).

A dimensionless parameter, Wagner number (W_a), is used to characterize secondary current distribution,

$$W_a = \frac{\frac{d(\Delta\varphi)}{di}}{\rho L} = \frac{d(\Delta\varphi)}{di} \cdot \frac{\kappa}{L} \qquad (13.26)$$

Where $d\Delta\varphi/di$: Degree of polarization of the electrode reaction (or polarization resistance);

ρ: electrolyte solution resistivity
κ: electrolyte solution conductivity
L: characteristic length, its definition depends on system design.

W_a represents the ratio of a polarization resistance per unit area to electrolyte resistance per unit area, and can also be interpreted as the ratio between the resistance of the electric double layer to that of the electrolyte solution. The greater the W_a is, the more uniform the current distribution is, which means that the overpotential caused by tiny changes of current density can compensate for the change of potential drop through the electrolyte solution caused by current density variation. If $d\varphi/di \to 0$, $W_a = 0$, there is no electrochemical polarization, and it becomes the primary current distribution.

Generally, the secondary current distribution is more uniform than the primary current distribution, as shown in Fig. 13.2.

Based on parameters included in W_a we can examine the impact of various factors on the value of W_a and the secondary current distribution.

1. Composition of electrolyte solution. Extra electrolytes can increase solution conductivity. On the other hand, some organic additives can increase the value of $d\varphi/di$ of the electrode reaction. Both can increase W_a, and improve current distribution. These techniques are usually used in metal deposition processes (electroplating and metallurgic processes). In contrast, W_a can be decreased by reducing the concentration of electrolytes, which in turn will lead to a more uneven current distribution. This method is usually used in electrolytic machining processes to enhance accuracy. For example using low concentration NaCl and NaNO$_3$ solutions can achieve significant improvement.

2. Current distribution becomes less uniform if current density increases. This is because if the relationship between overpotential and current density follows the Tafel equation

$$\frac{d\Delta\varphi}{di} = \frac{\beta}{i} \qquad (13.27)$$

an increase in current density will lead to decrease of $d\varphi/di$. As a result, W_a becomes smaller and current distribution becomes less uniform.
3. The characteristic length L increases when the reactor or electrode becomes bigger. As a result, W_a decreases, and the current distribution becomes less uniform, as shown in Fig. 13.6. The impact of reactor or electrode size on current distribution has to be considered when scaling up an electrochemical reactor.

Use electrode illustrated in Fig. 13.2 as example, when $L \gg \Delta$, the distribution of secondary current and the impact of W_a value near the electrolyte entrance ($x/L < 0.2$) is shown by Fig. 13.7, in which i_∞ is the current

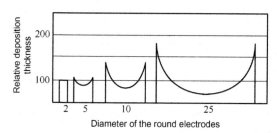

FIGURE 13.6 Effect of electrode size on secondary current distribution. (The numbers on the figure represent the diameter of the round electrodes)

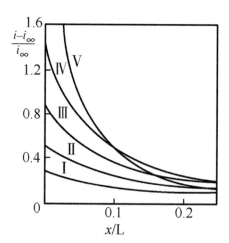

FIGURE 13.7 Secondary current distribution when $L \gg \Delta$ at various W_a. I: 0.63; II: 0.32; III: 0.16; IV: 0.08; V: 0.04. ($W_a = \frac{d\varphi}{di} \cdot \frac{\kappa}{\Delta}$, $L \gg \Delta$, See Fig. 13.2)

density that is far away from the reactor entrance. It is obvious that the smaller W_a is, the less uniform the current distribution at the entrance will be. Similar to Sh and Re used in mass transfer, W_a can be used to quantitatively, or together with i_x/I and x/L, qualitatively describe the current distribution, which is very useful when analyzing experimental results.

13.2.1.4 Tertiary Current Distribution

When concentration polarization, in addition to electrochemical polarization, also has to be considered, the current density distribution is called tertiary current distribution. Concentration polarization will occur when current density is relatively high, reactant concentration decreases dramatically near the electrode surface, and mass transfer resistance cannot be neglected.

The Laplace equation cannot be used to calculate the tertiary current distribution. Instead, the mass transfer equation, which is more complicated to solve, has to be used. As shown in Fig. 13.2, when current density reaches the diffusion-limited current density i_d, current distribution becomes very uneven. Since reactant concentration keeps decreasing along flow path (x positive direction), reaction current decreases correspondingly. When $i < i_d$, the situation is different. The further away i is from i_d, which means the reactant concentration is high enough, the less likely the concentration polarization will occur. Therefore there won't be tertiary current distribution, the system will be governed by secondary current distribution.

The surface profile of electrodes should be considered when studying tertiary current distribution. If the electrodes roughness is significantly larger than the thickness of the diffusion layer, as shown in Fig. 13.8A, the diffusion layer will develop along the surface with almost uniform thickness. As a result, i_d is almost uniform as well. In contrast, if the roughness is lower than δ (Fig. 13.8B), uneven tertiary current distribution develops due to diffusion limit. I_d is greater in convex regions than the value in concave regions. Such phenomena are more common in electroplating, electropolishing and electrochemical production of metal powder.

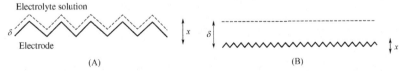

FIGURE 13.8 Effect of surface profile on current distribution. (A) Surface roughness is significantly greater that the thickness of diffusion layer and uniform tertiary current distribution; (B) Surface roughness is significantly smaller that the thickness of diffusion layer and nonuniform tertiary current distribution.

13.2.2 Effects of Gassing

Many industrial electrochemical processes involve gas evolution at the electrodes. Gas formation has significant impact on both the electrode reactions and the performance of the electrochemical reactor, and is usually referred to as gassing. This is most common for electrolysis and examples include hydrogen formation at cathode, oxygen formation at anode, chlorine formation at anode, etc. The main effects of gassing include:

1. Gas curtain is formed at the electrode/electrolyte interface due to the formation and accumulation of gas bubbles. This will reduce active electrode surface area and lead to microscale nonuniform distribution of both potential and current density on electrode surface.
2. In the bulk phase, the existence of gas bubbles will make the electrolyte solution a gas−liquid mixture with lower conductivity. Therefore, ohmic resistance of the electrolyte solution will increase, and so does the working voltage of the electrochemical reactor. In addition, the nonuniform distribution of gas bubbles between the two electrodes will cause macroscale nonuniform current distribution on electrode surfaces.
3. Bubble formation, growth, and detachment from the electrode surface as well as ascending in the solution will cause natural convection in the cell, which will enhance mass and heat transfer. Some electrochemical processes (e.g., Krebs cell for chlorate synthesis) utilize gas lifting to achieve electrolyte solution circulation.

In summary, gassing has broad effects and cannot be neglected. To gain a better understanding of the gassing effect, we need to first examine the physics involved in gas bubble formation during electrolysis.

13.2.2.1 Bubble Formation During Electrolysis

The bubbling process includes nucleation, growth, and detachment.

The fundamental cause for bubble formation is the generation of a new phase, i.e., a new phase—gas—is formed at the interface of electrode (solid) and electrolyte solution (liquid). Early studies have tried to find the similarity between bubbling during electrolysis to boiling. However, due to microscale nonuniformity at the electrode surface (including surface energy caused by micro defects, potential and current distribution) and the complex processes, such as gas dissolving, supersaturation, and diffusion, involved, nucleation during electrolysis is much more complicated than boiling, and is still poorly understood.

Bubble growth initially is caused by transport of dissolving gas to the gas−liquid interface and the increase of pressure inside the bubble. However, the dominant mechanism for bubble growth is coalescing which includes:

1. Coalescing of small bubbles on electrode surface
2. Growing of mid-size bubbles by absorbing adjacent small bubbles
3. Moving coalescing, i.e., ascending of large bubbles on the electrode surface while continuing to absorb small bubbles.

Bubble detachment occurs when the buoyancy force is greater than the attachment force, and depends on not only the status of the electrode surface and electrochemical parameters but also the flow conditions of the electrolyte solution, such as flowing velocity.

13.2.2.2 Effect of Gassing on Electrolyte Solution Conductivity

The gassing process increases the effective electric resistivity of the electrolyte solution and hence the ohmic potential drop in electrochemical reactors. There are many formulas to evaluate the relation between gas holdup ε and electric conductivity κ, and the following three are most commonly used:

$$\text{Maxwell equation} \frac{\kappa}{\kappa_0} = \frac{1 - \varepsilon}{1 + \varepsilon/2} \tag{13.28}$$

$$\text{Bruggeman equation} \frac{\kappa}{\kappa_0} = (1 - \varepsilon)^{3/2} \tag{13.29}$$

$$\text{Prager equation} \frac{\kappa}{\kappa_0} = (1 - \varepsilon)\left(1 - \frac{\varepsilon}{2}\right) \tag{13.30}$$

$$\varepsilon = \frac{V_g}{V_g + V_l} \tag{13.31}$$

where V_g is the total volume of gas bubbles; V_l is the solution volume; and κ_0 is the solution conductivity without gas bubbles.

The estimated conductivity by these thee equations are listed in Table 13.3.

As shown in this table, estimated values by these three equations are very close to each other when ε is relatively low. Many commercial

TABLE 13.3 Estimated Conductivity Ratio κ/κ_0

ε	κ/κ_0			ε	κ/κ_0		
	Maxwell Equation	Bruggeman Equation	Prager Equation		Maxwell Equation	Bruggeman Equation	Prager Equation
0.1	0.8571	0.8538	0.8550	0.3	0.6087	0.5857	0.5950
0.2	0.7273	0.7155	0.7200	0.4	0.5000	0.4647	0.4800

electrochemical reactors use forced convection, which keeps ε low. The Bruggeman equation is more broadly used.

The main difficulty in using these equations is that it is difficult to obtain gas holdup.

Recent research indicates that the following factors contribute to the complex relationship between gas holdup and conductivity:

1. The gas holdup is not uniformly distributed and not stable. Therefore it is not easy to determine either local gas holdup or the overall average value.
2. The bubble size affects the $\varepsilon-\kappa$ relation. The smaller the bubble size, the stronger the effect of ε on κ. Therefore the above equations have to be modified to account for the effect.

13.2.2.3 Current Distribution of Gas-Evolving Electrodes

In an electrochemical reactor, gas evolving at the electrodes and nonuniform distribution will have an impact on not only the utilization of active electrode surface and electrode life but also the overall performance of the electrochemical reactor, such as higher cell voltage and DC power consumption.

As discussed above, there are two kinds of current distribution on a gas-evolving electrode surface, i.e., macrocurrent distribution caused by nonuniform distribution of gas bubbles between electrodes and microcurrent distribution caused by bubble attachment on the electrode surface.

Compared with other electrodes, current distribution on gas-evolving electrodes has the following characters:

1. The primary current distribution is not caused by system geometry, but bulk phase conductivity heterogeneity (gas—liquid system)
2. The secondary current distribution is affected by the bubble curtain on the electrode surface, leading to more pronounced electrochemical polarization
3. Its current distribution is closely related to the rate, velocity, and distribution of fluid flow
4. The structure of the reactor and electrodes, as well as the distance between electrodes (or between electrode and diaphragm), have stronger influences on the current distribution of gas-evolving electrodes.

Janssen, Nishiki, Martin, etc. conducted a series of research on the current distribution of gas-evolving electrodes, and their results revealed that higher current density, lower solution flow rate, and lower concentration of electrolyte solution led to less uniform current distribution on a gas-evolving electrode surface [2—4].

13.2.2.4 Effects of Gassing on Transport Processes

The effects of gassing on transport processes are important topics in electrochemical engineering. It is a widely accepted conclusion that gassing can

TABLE 13.4 i_d and δ Under Various Fluid Dynamic Conditions

Fluid Dynamic Conditions	i_d (A/m²)	δ (mm)	Fluid Dynamic Conditions	i_d (A/m²)	δ (mm)
Electrolysis with no convections	6.1	4.75	Rotating cylindrical electrode (3 rpm, $R = 0.05$ m)	810	0.036
Natural convection on vertical electrode surface (height $= 0.1$ m)	144	0.2	Cathode H_2 evolution (2.2 L/m²/s)	7200	0.004
Natural convection on horizontal electrode surface	365	0.08	Cathode H_2 evolution (0.17 L/m²/s)	1940	0.015
Laminar flow on plate electrode ($v = 0.25$ m/s)	300	0.10	Ultrasonic (7×10^{-4} W/m²)	5000	0.006
Turbulent flow on plate electrode ($v = 02$ m/s)	36,500	0.0008			

enhance transport processes, which has been confirmed by research results. The effects of gassing and other mechanical enhancement methods on diffusion-limited current density, i_d, and the thickness of diffusion layer δ in an electrochemical process of deposition of Cu from $CuSO_4$ solution are shown in Table 13.4.

Deposition conditions: 0.3 mol/L $CuSO_4$ solution, diffusivity is 10^{-9} m²/s, and kinematic viscosity is 10^{-6} m²/s.

Furthermore, the mass transfer coefficient, thickness of diffusion layer, and gas generation rate can be correlated with the following empirical equations:

$$K_m = \text{const}\left(\frac{V_G}{S}\right)^m \tag{13.32}$$

$$\delta = \text{const}\left(\frac{V_G}{S}\right)^{-m} \tag{13.33}$$

where:

K_m: mass transfer coefficient
V_G: gas generation rate
S: electrode surface area
δ: the thickness of diffusion layer and m: is a system-specific parameter, and its values in some systems are listed in Table 13.5.

TABLE 13.5 m in Equation (13.32)

Gas	Electrolyte	m	Researchers	Gas	Electrolyte	m	Researchers
H_2	Basic	0.43	Green and Robinson	H_2	Basic	0.36 (Hg)	Janssen and Hoogland
		0.29	Vondrsk and Balej			0.45	Kind
		0.36	Janssen and Hoogland	O_2	Basic	0.87/0.33	Janssen and Hoogland
		0.25	Fouad and Sedahmed			0.40	Fouad and Sedahmed
		0.65	Rousar et al.			0.583	Yanxi Chen et al.
		0.17~0.30	Jassen	O_2	Acidic	0.50	Beck
		0.287	Yanxi Chen et al.			0.40	Janssen and Hoogland
H_2	Acidic	0.5 (0.59)	Roald and Beck			0.60	Ibl
		0.525	Venczel			0.57	Janssen and Hoogland
		0.47	Janssen and Hoogland			0.66	Kind
		0.62(Pt)	Janssen and Hoogland	Cl_2	Acidic	0.51	Janssen and Hoogland

There are three models to explain the mechanism of gassing effects on transport processes.

1. Penetration model: the enhancement of mass transport process is caused by the intrusion of bulk solution into the space created by detachment of gas bubbles from the electrode surface. This model is most suitable for coalescing gases such as O_2
2. Hydrodynamic model: the upward migration of gas bubbles is the main mechanism for the enhancement of mass transfer. This model is suitable for noncoalescing gases such as H_2
3. Microconvection model: mass transport process is enhanced by local convection caused by growth gas bubbles.

13.2.3 Mass Transfer in Electrochemical Engineering

Mass transfer process is a critical step involved in electrode kinetics. When it becomes the rate-determining step, it will decide the overall rate and kinetic behaviors of the electrode reactions.

In addition to electrode kinetics, mass transfer also has an impact on other aspects of an electrochemical process:

1. It determines the productivity of electrochemical reactors (maximum current density). In addition, it also has great influences on operating cell voltage, current efficiency, space-time yield, and many other techno-economic parameters
2. It affects product quality. In metal electrochemical deposition processes, the product could have a rough surface, or even become powdery if the current density approaches its maximum limit. In electrochemical synthesis processes, the mass transfer limit causes elevated electrode voltage, which can induces side reactions and hence impairs product quality
3. Mass and heat transfer are closely correlated. Therefore, mass transfer conditions will also affect heat exchange, heat balance, and operating temperature
4. Requirements on mass transfer have to be considered in the design of electrochemical reactors, such as the composition, configuration, and control of electrolyte systems.

The three mass transfer mechanisms, i.e., convection, diffusion, and migration, can be represented by kinetic equations of electrode. However, in electrochemical engineering it is difficult to conduct quantitative calculation and theoretical analysis

In electrochemical engineering the mass transfer can be described using the mass transfer coefficient, i.e.:

$$j = k_m \Delta c$$

where:

j: mass transfer flux, mol/m$^3 \cdot$ s
k_m: mass transfer coefficient, m/s
Δc: concentration difference between the bulk phase and the interface, mol/m^3.

When studying the factors that affect the mass transfer coefficient in electrochemical engineering, conventional methods in chemical engineering are used, i.e., conducting experimental studies to develop correlations describing mass transfer behaviors, typically using dimensionless numbers. Of course, these dimensionless numbers are used to describe the impacts of various parameters on mass transfer inside an electrochemical reactor.

Some of the most widely used dimensionless numbers include:

1. Sherwood number (Sh)

$$Sh = \frac{k_m L}{D}$$

where:

k_m: mass transfer coefficient (m/s)

L: reactor character length (m)

D: diffusion coefficient (m²/s).

Since Sh is directly correlated to k_m, it can be used to characterize the mass transfer rate.

2. Reynold number (Re)

$$Re = \frac{uL}{v}$$

where:

u: flow velocity of electrolyte solution

L: reactor character length (m)

v: kinematic viscosity of the electrolyte solution (m²/s).

Re represents the ratio of inertia to viscous force of flow of the electrolyte solution.

3. Schmidt number (Sc)

$$Sc = \frac{v}{D}$$

Sc reflects the relation between convective and diffusive mass transfer.

4. Grashof number Gr

$$Gr = \frac{g\Delta\rho L^3}{v^2\rho}$$

where:

g: gravity (m²/s)

ρ: solution density (kg/m³)

$\Delta\rho$: density difference between electrolyte solution in the bulk phase and at electrode surface

v: kinematic viscosity

L: electrode length.

If the density difference $\Delta\rho$ is caused by concentration difference (Δc), densification coefficient (α) has been used to describe the correlation:

$$\alpha = \frac{\Delta\rho}{\rho\Delta c}$$

Then Gr can be rewritten as

$$Gr = \frac{ga\Delta c L^3}{v^2}$$

Gr describes mass transfer process induced by natural convection.

With these dimensionless numbers, the following correlation equation is generally applied to describe the mass transfer processes in electrochemical reactors,

$$Sh = k Re^a Sc^b \qquad (13.34)$$

Parameters in the above equation, k, a, and b, have to be estimated based on experimental data. And their values depend on experimental procedures and conditions. Table 13.6 gives some mass transfer correlations for various electrochemical reactors.

Sh number can be calculated using those correlations and other known conditions. Then the mass transfer coefficient k_m, and diffusion-limited current density (i_d) can be calculated as well, because $k_m = D/\delta$, $\delta = D/k_m$, and

$$i_d = \frac{nFDc_0}{\delta} = nFDc_0 \left(\frac{k_m}{D}\right) = nFk_mc_0 \qquad (13.35)$$

where c_0 is the concentration in bulk phase.

13.2.4   Heat Transfer and Balance in Electrochemical Engineering

Every electrochemical reactor has to be operated under a given temperature. Temperature has a direct impact on reaction rate, overpotential, and selectivity of the electrode reactions, as well as many technoeconomic performance factors of the electrochemical process, such as operating voltage current efficiency. Electrolyte corrosivity and stabilities of electrode and diaphragm materials also depend on temperature.

Selecting temperature for a electrochemical reaction depends on many factors. On the other hand, the operating temperature of an electrochemical reactor mainly depends on heat transfer and balance. Developing heat balance is an important task for designing an electrochemical reactor.

A general heat balance analysis can be expressed as:

Heat accumulation rate in the reactor = heat input rate by by inflow
+ heat generation rate in the reactor − heat removal rate by outflow
− heat loss rate to the environment ± heat exchange rate by heat exchanger

$$(13.36)$$

However, since rigorous theoretical analysis involves establishing and solving three-dimensional PDEs with complex boundary conditions, which is very difficult, a certain degree of simplification is preferred in electrochemical engineering.

TABLE 13.6 Mass Transfer Correlations for Different Electrochemical Reactors

No	Electrodes	Flow Pattern	Schematics	Conditions	Correlations	Comments
1	Parallel plate electrode	Forced convection, laminar flow			$Sh = 1.467\left(\frac{2}{1+r}\right)^{\frac{1}{3}}\left(Pe\frac{d_e}{L}\right)^{1/3}$	$Pe = Re \bullet Sc$ Laminar flow, $r = \frac{\text{Distance between electrodes}}{\text{Electrode width}}$ L: Electrode length d_e : equivalent diameter $de = 2BS/(B+S)$
2		Forced convection, laminar flow	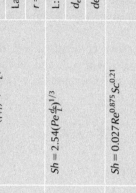	$10^5 \leq Pe\frac{d_e}{L} \leq 10^7$	$Sh = 2.54\left(Pe\frac{d_e}{L}\right)^{1/3}$	
3		Forced convection, turbulent flow		$10^4 < Re < 10^5$	$Sh = 0.027 Re^{0.875} Sc^{0.21}$	
4	Vertical plate electrode	Natural convection, laminar flow		$Gr \bullet Sc < 10^{12}$	$Sh = 0.45 Gr^{0.25} Sc^{0.25}$	
5		Natural convection, turbulent flow		$Gr \bullet Sc > 4 \times 10^{13}$	$Sh = 0.31 Gr^{0.28} Sc^{0.28}$	
6		Gas-evolving electrodes		$a = 1.38$ (spherical bubbles)	$Sh = a(1-\theta)^{0.5} Re^{0.5} Sc^{0.5}$	$l = L$
				$a = 1.74$ (half-spherical bubbles)		$Re = \frac{Lv_g}{\gamma}$

#	Electrode type		Conditions	Sh	Character length
7	Horizontal parallel electrodes	Channel flow, fully developed laminar flow	$Re < 2000, B > SL/d_e < 35$	$Sh = 1.85Re^{0.33}Sc^{0.33}(d_e/L)^{0.33}$	$l = d_e = 2BS/(B+S)$
8		Turbulent flow	$Re < 2300$ $\frac{L}{d_e} > 10$	$Sh = 0.023Re^{0.8}Sc^{0.33}$	$l = L$
9	Rotating cylindrical electrodes	Laminar flow	$10^2 < Re < 10^4$	$Sh = 0.62Re^{0.5}Sc^{0.33}$	$Re = \frac{r^2\omega}{\gamma}$
10		Turbulent flow	$Re > 10^6$	$Sh = 0.011Re^{0.87}Sc^{0.33}$	$Re = \frac{r^2\omega}{\gamma}$
11	Coaxial cylindrical electrodes	Laminar flow $(\omega = 0)$	$Re < 2000$	$Sh = 1.61\phi Re^{0.33}Sc^{0.33}(d_e/L)^{0.33}$	$r = r_i/r_0$ $L = 2(r_0 - r_i)$
12		Turbulent flow $(\omega = 0)$	$Re > 2000$	$Sh = 0.023Re^{0.8}Sc^{0.33}$	$Re = \frac{2r_i\omega}{\gamma}$
13		Rotating cylinder $(v = 0)$	$100 < Re < 1.6 \times 10^6$	$Sh = 0.079Re^{0.7}Sc^{0.36}$	$L = 2r_i$
14	Packed bed electrodes			$Sh = 1.52Re^{0.55}Sc^{0.33}$	l is average particle size ε is bed void fraction
15	Fluidized bed electrodes			$Sh = \frac{(1-\varepsilon)^{0.5}}{\varepsilon} Re^{0.5}Sc^{0.33}$	

Note: l: character length; θ: coverage of electrode surface; v_g: gas generation rate; ω: angular velocity; v: flow velocity; and $\phi = \frac{r-1}{r}\left(\frac{0.5 + \left[\frac{r^2}{1-r^2}\right]\ln(r)}{1 + \left[\frac{r^2}{1-r^2}\right]\ln(r)}\right)$.

Items in Eq. (13.36) include:

1. Heat carried in (out) by inflow (outflow) per unit time (J/s)

$$Q_1 = S \sum J_1 M_i C_{p,i} T \tag{13.37}$$

where:

    S: cross section area of fluid flow

    J: flow rate (positive for inflow and negative for outflow)

    M_i: molecular weight of component i

    T: temperature (K)

    $C_{p,i}$: specific heat capacity of component i under constant pressure.

2. Heat generation rate by electrochemical processes per unit time (J/s)

$$Q_2 = I \left(V - \frac{\Delta H}{nF} \right) \tag{13.38}$$

where:

    I: current

    V: electrode operating voltage

    ΔH: reaction enthalpy.

3. Heat exchange with the environment (J/s)

$$Q_3 = \sum_j U_j S_j \Delta T_j \tag{13.39}$$

Eq. (13.39) accounts for heat exchange through all heat exchanging surfaces of the reactor. U_j is the overall heat exchange coefficient of surface j, S_j is surface area for heat exchange, and ΔT_j is the temperature difference between the reactor and the environment.

4. Heat brought in/out by heat exchangers per unit time Q_4 (J/s)

Eq. (13.36) can be written as:

$$m c_p \frac{dT}{dt} = I \left(V - \frac{\Delta H}{nF} \right) + S \sum_i j_i M_i C_{p,i} T - \sum_j U_j S_j \Delta T_j \pm Q_4 \tag{13.40}$$

Eq. (13.40) can be used for:

a. Determining equilibrium temperature. When a system reaches steady state, temperature will not change with time, i.e., $dT/dt = 0$, and Eq. (13.40) becomes

$$I \left(V - \frac{\Delta H}{nF} \right) + S \sum_i j_i M_i C_{p,i} T = \sum_j U_j S_j \Delta T_j + Q_4 \tag{13.41}$$

If the temperature difference between the reactor and the environment is known, the current and heat transfer rate of heat exchangers can be calculated using Eq. (13.41).

b. Estimating the time needed for an electrochemical system to reach certain temperature. According to Eq. (13.40):

$$
t = \int_{T_0}^{T} \frac{mc_p}{I\left(V - \left(\dfrac{\Delta H}{nF}\right)\right) + S\sum_i j_i M_i C_{p,i} T - \sum_j U_j S_j \Delta T \pm Q_4} \, dt \quad (13.42)
$$

In Eq. (13.40), heat carried by inflow and outflow (Q_1), heat loss (Q_3), and heat exchange (Q_4) are relatively easy to understand for readers familiar with transport phenomena. Heat generated by the electrochemical reactions (Q_2) has to be addressed separately, since it is a special issue for electrochemical reactors. The operating voltage of an electrochemical reactor can be expressed as:

$$
V = -\frac{\Delta G}{nF} + \sum |\Delta \varphi| + \sum IR \quad (13.43)
$$

Therefore,

$$
Q_2 = I\left(V - \frac{\Delta H}{nF}\right) = I\left(\sum |\Delta \varphi| + \sum IR - \frac{T\Delta S}{nF}\right) + 2IE \quad (13.44)
$$

Eq. (13.44) indicates that the heat generated in an electrochemical reactor Q_2, is determined by not only ohmic resistivity of all components (electrode, diaphragm, electrolyte solution) in the reactor but also thermodynamic properties of the electrochemical reaction (ΔH, ΔS, etc.) and the irreversibility of reaction kinetics (i.e., various overpotentials $\Delta \varphi$).

13.3   ELECTROCHEMICAL REACTORS

An electrochemical reactor is a piece of equipment used for electrochemical reactions. In three major areas of electrochemical engineering, i.e., electrolysis, chemical power resource, and electroplating, electrochemical reactors include electrolytic cells, electroplating cells, primary batteries, secondary batteries, and fuel cells. These electrochemical reactors differ significantly in structure, size, function, etc., but share two fundamental characters:

1. All electrochemical reactors consist of electrodes (usually made of electronic conductors) and an electrolyte solution (Ionic conductor)
2. Overall, all electrochemical reactors can be grouped into two categories: (i) consuming electric energy, i.e., electrolytic reactor which consume electrical energy to drive with electrochemical reactions at electrode-electrolyte interfaces and (ii) galvanic cells which generate electrical energy through spontaneous electrochemical reactions at the electrode–electrolyte interfaces.

13.3.1 Types of Electrochemical Reactors

According to their structures, electrochemical reactors can be categorized into three major types:

1. Tank cells

   This type is the most widely used. They are usually rectangular in shape, and the scales in the three dimensions (length, width, height) are different. The electrodes are usually plates, and vertically installed in the reactors, parallel to each other.
2. Filter-press or plate-frame cells

   This type of reactor usually includes multiple identical cell units, which are pressed and sealed together. Each unit consists of electrodes, plates, frames, and diaphragms.
3. Specially designed reactors

   Electrochemical reactors can be customer designed in order to increase their specific electrode areas, enhance mass transfer processes, and improve productivity.

13.3.1.1 Tank Cells

Box-shaped reactors are usually operated in batch or semibatch modes. Secondary batteries are typical batch reactors with electrodes and electrolytes installed and sealed in the cell. Secondary batteries can be operated in recharging or discharging modes. In electroplating processes, open cells are typically used so that the parts and finished products can be periodically placed in or removed from the electroplating baths. Obviously, this is a batch operation. In many traditional electrolysis industries such as the electrochemical production of aluminum and fluoride, semibatch operation is most commonly used. In most tank cells, anodes and cathodes are vertically installed next to each other in order to reduce the distance and enhance the productivity. However, the reduction of electrode distance is limited by some other factors. For example, the growth of metal crystal may cause short circuits in electrometallurgy processes. In electrolytic synthesis processes the mixing of products from the two electrodes has to be minimized to eliminate side reactions. Therefore, a diaphragm is installed between the electrodes. Natural convection is usually the main mechanism for enhancement of mass transfer processes in a tank cell, and it is not common to introduce forced convection. For example, natural convection induced by gas bubble lifting in electrolysis processes can effectively enhance mass transfer.

Monopolar type circuit connections are typically used in tank cells. However, bipolar connections may also be used after necessary modifications.

Tank cells are widely used because of their structure and design, which makes their manufacture and maintenance easy. But they suffer the disadvantages of low productivity, and are not suitable for large-scale continuous operation and for industrial processes that require strict control of mass transfer. Fig. 13.9 shows a monopolar electrolyzer for water electrolysis.

13.3.1.2 Filter-Press or Plate-Frame Electrochemical Reactor

As indicated above, this type of reactor usually contains several identical units, which are pressed and sealed together. Each unit consists of electrodes, plates, frames, and diaphragm. Electrodes can be placed vertically or horizontally, with electrolyte solution flowing through them, and there is no need of reactor body, as shown in Fig. 13.10. The number of units in one filter-press cell could reach more than 100.

The advantages of filter-press electrochemical reactors include:

1. The structures of unit cells can be simple and standardized to allow mass production and easy replacement
2. Materials of electrodes and diaphragms can be selected to meet various requirements

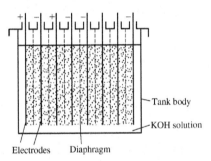

FIGURE 13.9   Monopolar electrolyzer for water electrolysis.

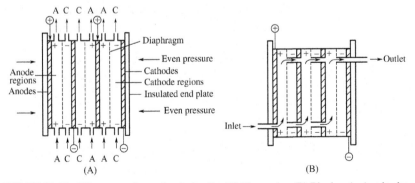

FIGURE 13.10   Filter-press electrochemical cells. (A) Monopolar. (B) Bipolar. A: Anode electrolyte; C: Cathode electrolyte.

TABLE 13.7 Key Parameters of a Series of Electrochemical Cells by ElectroCell

Parameter	Micro Flow Cell	Electro MP Cell	Electro Syn Cell	Electro Prod Cell
Electrode area (m^2)	0.001	$0.01 \sim 0.2$	$0.04 \sim 1.04$	$0.4 \sim 16$
Current density(kA/m^2)	< 4	< 4	< 4	< 4
Electrode gap (mm)	$3 \sim 6$	$6 \sim 12$	5	$0.5 \sim 4$
Electrolyte flow per cell (L/min)	$0.18 \sim 1.5$	$1 \sim 5$	$5 \sim 15$	$10 \sim 30$
Flow rate in each cell (m/s)	$0.05 \sim 0.4$	$0.03 \sim 0.3$	$0.2 \sim 0.6$	$0.15 \sim 0.45$
Number of electrode pairs	1	$1 \sim 20$	$1 \sim 26$	$1 \sim 40$

3. Distributions of electric current and potential are relatively uniform
4. Many different types of turbulence promoter can be installed to enhance mass transfer and control electrolyte velocity
5. The productivity can be readily adjusted by changing electrode surface areas in unit cells or number of units to meet needs of different customers. Table 13.7 shows some parameters of one series (ElectroCell) of filter-press electrochemical cells
6. Bipolar circuit configuration can be used (to decrease potential drop, save material and improve current distribution).

Filter-press cells can be built with unit cells with special configurations, such as unit cells with a heat exchanger or electrodialyzer, porous electrodes and electrodes with complex 3D structure.

Increasing the surface area of a single electrode, in addition to enhancing the productivity, can improve diaphragm utilization, reduce maintenance cost, and reduce space requirement. For example, for filter-press membrane electrolyzers widely used in chlor-alkali industry, the electrode surface area is $0.2-3$ m^2 in monopolar reactors and $1-5.4$ m^2 in bipolar reactors.

The plates and frames used in filter-press type cells can be made of diverse materials, including nonmetal (rubber, plastic, etc.) and metal materials. Nonmetal materials are cheap, but have shorter life, and their maintenance and replacement could be time-consuming; the metal ones are more durable, but more expensive.

In the electrochemical industry, filter-press type reactors have been successfully applied in water electrolysis, the chlor-alkali industry, organic synthesis (e.g., adiponitrile electrosynthesis), and galvanic cells (e.g., fuel cells and 1 layer-built cells).

13.3.1.3   Specially Designed Electrochemical Reactors

As discussed in previous sections, the space-time yield is an important parameter to assess the performance of a electrochemical reactor. According to Eq. (13.21)

$$Y_{ST} = \frac{G}{tV_R} = \frac{iStK\eta_1}{tV_R} = i\left(\frac{S}{V_R}\right) \cdot K\eta_1 = i \cdot A_S K\eta_1 = A_S \cdot i(K\eta_1) \qquad (13.45)$$

For a given reaction system, $K\eta_1$ is approximately a constant, then Y_{ST} basically depends on the product of A_s (the specific electrode surface area m^2/m^3) and current density i (A/m^2). The dimension of $A_s \cdot i$ is A/m^3, which is the same as that of the volumetric current density.

For a reaction system with very low electric conductivity or very low reactant concentration, it is not feasible to significantly increase current density for the electrode reactions because of mass transfer limitation. Therefore, A_s has to be increased to enhance the space-time yield. Many electrochemical reactors with specially designed structures have been developed, and Table 13.8 summarizes the main characters of some of such specially designed electrochemical reactors.

Figs. 13.11 and Figs. 13.12 illustrate the dependence of specific electrode surface area, A_s, and volumetric current density (i_v) on characteristic lengths of the reactor (electrode distance or particle size).

13.3.2   Operation Characters of Electrochemical Reactors

As indicated previously. there are three operation modes for electrochemical reactors, and the characters of each mode can be described using the mathematical models that describe reactant concentration changes.

13.3.2.1   Simple Batch Reactor

This type of reactor is widely used. Reactant concentrations inside the reactor tend to be uniform when stirring is sufficient. Assuming that the initial reactant concentration is $c(0)$, and after reaction time t the concentration decreases to $c(1)$. If the reaction is first order, the concentration rate can be described as:

$$\frac{dc(t)}{dt} = kc(t) \qquad (13.46)$$

For an electrochemical system, reactant concentration change can be calculated by Faraday's Law,

$$\frac{-dc(t)}{dt} = \frac{I(t)}{nFV_R} \qquad (13.47)$$

TABLE 13.8 Characters of Specially Designed Electrochemical Reactors

Electrochemical Cells	Ohmic Potential Drop	Mass Transfer Rate	Specific Electrode Area	Electrochemical Cells	Ohmic Potential Drop	Mass Transfer Rate	Specific Electrode Area
Thin file cell of Trickle tower cell	High	High	Large	Fluidized bed cell	Moderately high	Moderately low	Very Large
Capillary gap cell	Low	Low	Small	Swiss-role cell/ Sandwich and roll cell	Low	High	Small
Rotating electrode cell	High	High	Small	Zero-gap cell	Low	Low	Small
Pump cell	Low	High	Small	Solid polymer electrolyte (SPE) cell	Low	Low	Small
Packed bed cell	Moderately high	Moderately low	Very Large	Cells with turbulence promoters	High	High	Small

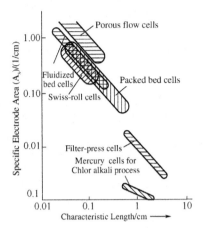

FIGURE 13.11   Specific electrode area versus characteristic length.

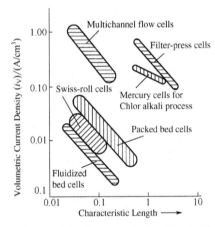

FIGURE 13.12   Volumetric current density versus characteristic length.

where V_R is reactor volume and $I(t)$ is the instantaneous current at time t. If the electrode reaction is controlled by diffusion

$$I(t) = I_d = k_m SnFc(t) \tag{13.48}$$

where:

I_d: diffusion-limited current
k_m: mass transfer coefficient
S: electrode surface area.

Substituting Eq. (13.48) into Eq. (13.47) gives

$$\frac{-dc(t)}{dt} = \frac{k_m Sc(t)}{V_R} \tag{13.49}$$

Comparing Eq. (13.49) and (13.46), we can obtain

$$k = \frac{k_m S}{V_R} \qquad (13.50)$$

Integrating Eq. (13.49) gives

$$c(t) = c(0)\exp\left(-\frac{k_m S}{V_R}t\right) \qquad (13.51)$$

This equation illustrates the relationship between reactant concentration and time and the factors that will influence the concentration change. When mass transfer coefficient or electrode surface area increases, reactant concentration drops faster. The ratio S/V_R is the specific electrode surface area introduced in Section 13.2.1, which is defined as active electrode surface area per unit reactor volume and expressed as A_S in this section. Therefore Eq. (13.51) can be written as

$$c(t) = c(0)\exp(-k_m A_S t) \qquad (13.52)$$

The conversion in the reactor (X_A) can be expressed as:

$$X_A = 1 - \frac{c(t)}{c(0)} = 1 - \exp(-k_m A_S t) \qquad (13.53)$$

We can see that batch cells are operated under unsteady state and concentrations of both the reactant and the product change with time.

In actual applications, there are some modifications. For example, electrolyte solution can be added periodically. In some applications product can be removed continuously to avoid redissolution, plugging, or short-circuiting.

13.3.2.2   Plug Flow Reactors

When electrolyte solution flows through the reactor at a constant flow rate, reactant concentrations at reactor inlet and outlet will not change with time under steady state, as shown in Fig. 13.13.

For an electrochemical reactor, mass conservation can be expressed as:

Materials that entered the reactor−Material that left the reactor=Materials consumed by the electrochemical reactions.

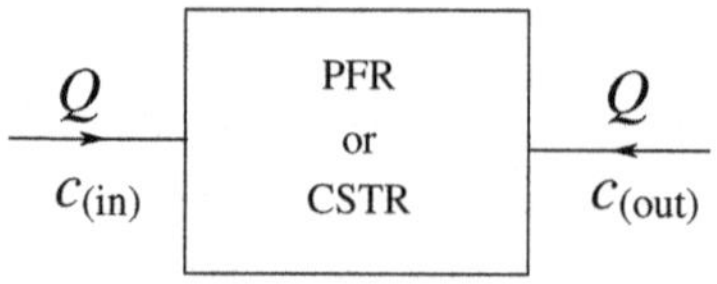

FIGURE 13.13   Plug flow or CSTR reactor.

According to Faraday's law:

$$Qc_{(in)} - Qc_{(out)} = \frac{1}{nF} \tag{13.54}$$

where Q is the volumetric flow rate, m^3/s.
Eq. (13.54) can be converted to

$$\Delta c = c_{(in)} - c_{(out)} = \frac{1}{nFQ} \tag{13.55}$$

In a plug flow reactor, the reactant concentration decreases along electrode direction (x direction, $x = 0$ at inlet), and total current over the electrode surface can be calculated by integration:

$$I = \int_0^x I(x)dx \tag{13.56}$$

If the electrode reaction process is controlled by diffusion, the current density at a specific location depends on the concentration at that point:

$$i_x = i_d = k_m nFc(x) \tag{13.57}$$

If diffusion resistance can be ignored, concentration gradient along x direction is

$$\frac{-dc(x)}{dx} = \frac{I(x)S'}{nFQ} \tag{13.58}$$

where S' is the electrode surface area per unit length.
Replacing $I(x)$ with I_d, Eq. (13.58) becomes:

$$\frac{-dc(x)}{dx} = \frac{k_m S'}{Q}c(x) \tag{13.59}$$

Integrating along the x direction, the concentration at outlet can be obtained as a function of inlet concentration and other parameters:

$$c_{(out)} = c_{(in)}\exp\left(-\frac{k_m S'}{Q}\right) \tag{13.60}$$

Obviously, if the flow rate and concentration at inlet remain constant, increasing the mass transfer coefficient and electrode surface area will lower the reactant concentration at outlet.
Similarly, conversion at outlet is

$$X_A = 1 - \frac{c_{(out)}}{c_{(in)}} = 1 - \exp\left(-\frac{k_m S'}{Q}\right) \tag{13.61}$$

The diffusion-limited current can be expressed as a function of concentration at the inlet by combining Eqs. (13.55) and Eq. (13.60):

$$I_d = nFQc_{(in)}X_A = nFQc_{(in)}\left[1 - \exp\left(-\frac{k_m S}{Q}\right)\right] \qquad (13.62)$$

The residence time $\tau = V_R/Q$, so Eq. (13.60) can be rewritten as:

$$c_{(out)} = c_{(in)}\exp\left(-\frac{k_m S}{V_R}\tau\right) \qquad (13.63)$$

If $t = \tau_m$ this equation and Eq. (13.51) are identical. The conversion can be written as:

$$X_A = 1 - \exp\left(-\frac{k_m S}{V_R}\tau\right) \qquad (13.64)$$

Therefore, for given vales of k_m, S, and V_R, the conversion of a batch reactor is equal to that of a plug flow reactor if the reaction time in the batch reactor is equal to the residence time in the plug flow reactor. On the other hand, if the conversions in these two reactors are equal, k_m, S, and V_R of these reactors should be the same as well.

13.3.2.3 Continuous Flow Stirred Tank Cells

One important feature of this type of reactor is that reactant concentration inside the cell is equal to that at the outlet due to ideal mixing conditions. As a result, the diffusion-limited current and mass transfer do not change with either time or location.

$$I_d = k_m nFSc_{(out)} \qquad (13.65)$$

The mass balance equation of this type of reactor is:

$$\Delta c = c_{(in)} - c_{(out)} = \frac{k_m S}{Q}c_{(out)} \qquad (13.66)$$

Or

$$c_{(out)} = \frac{c_{(in)}}{1 + k_m S/Q} \qquad (13.67)$$

And the conversion is

$$X_A = 1 - \frac{c_{(out)}}{c_{(in)}} = 1 - \frac{1}{1 + k_m S/Q} \qquad (13.68)$$

Increasing k_m and S can improve conversion. According to Eqs. (13.68) and (13.55), the diffusion-limited current can be correlated to inlet concentration:

$$I_d = nFQc_{(in)} \left[- \frac{1}{1 + k_m S/Q} \right] \tag{13.69}$$

For a given set values of k_m, S, and Q, the conversion in a continuous stirred tank reactor (CSTR) is lower than that in a plug flow reactor, However, when the stirring is intensive enough, k_m is almost independent of Q. Under such conditions the performance of a continuous flow stirred tanks cells is very similar to that of an electrochemical reactor with rotating cylinder electrodes under turbulent flow condition.

In fact, both reactors are so-called single pass reactors, and their conversion is limited. Reducing the flow rate of electrolyte solution can increase conversion but it also causes a decrease of k_m and accumulations of products and reaction heat, which are undesirable. Recycling operation mode (as shown in Fig. 13.14) can be used to enhance conversion. A more effective operation mode is batch recycle, i.e., the outflow of a reactor flows through a reservoir and then recycled back into the reactor (as shown in Fig. 13.15). Another option is using multiple reactors in series.

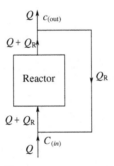

FIGURE 13.14   Reactor with recycle.

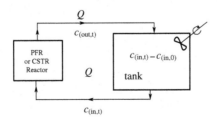

FIGURE 13.15   Batch recycle.

13.3.2.4 Batch Recycle

An electrochemical reactor and a reservoir or another reactor are required for batch recycle operation, as shown in Fig. 13.15. This operation mode has been applied in organic eletrosynthesis processes.

This operation mode is flexible and convenient, and the functions of the reservoirs or reactors include:

1. Increasing electrolyte solution capacity
2. Adjusting pH value of the electrolyte solution
3. Facilitating heat exchange and stabilizing temperature
4. Allowing secondary homogeneous reactions to take place to make the final product
5. Providing space for reactant mixing and pretreatment
6. Achieving gas−liquid or liquid−solid separations
7. Allowing sampling for analysis.

If the volume of reservoir is significantly larger than that of the reactor ($V_T \gg V_R$), and the residence time (τ_T) is long enough, this system can be approximately considered as a CSTR. However, its reactant concentration at both inlet and outlet varies with time, which is different from a single pass reactor.

If the reservoir is well mixed, based on mass balance of reactor and reservoir:

$$V_T \frac{dc_{(in)}}{dt} = Q(c_{(out)} - c_{(in)}) \tag{13.70}$$

Substituting c_{out} from Eq. (13.60) into Eq. (13.70) and integrating:

$$c_{(in,t)} = c_{(in,0)} \left\{ -\frac{t}{\tau_1} \left[1 - \exp\left(-\frac{k_m S}{Q} \right) \right] \right\} \tag{13.71}$$

Or

$$c_{(in,t)} = c_{(in,0)} \exp\left(-\frac{t}{\tau_T} \theta_A^{PFR} \right) \tag{13.72}$$

where θ_A^{PFR} is the conversion of a single pass PFR, the overall conversion of the reactor system is:

$$\theta_{A,t}^{PFR} = 1 - \left[\frac{c_{(in,t)}}{c_{(in,0)}} \right] = 1 - \exp\left(-\frac{t}{\tau_T} \theta_A^{PFR} \right) \tag{13.73}$$

For the continuous flow reactor in the batch recycle reactor system, the following equation can be derived by using Eqs. (13.67) and Eqs. (13.70):

$$c_{(in,t)} = c_{(in,0)} \exp\left\{ -\frac{1}{\tau_T} \left[1 - \frac{1}{1 + \left(\frac{k_m S}{Q} \right)} \right] \right\} \tag{13.74}$$

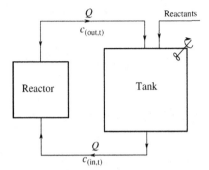

FIGURE 13.16 Semibatch reactor with recycle.

Using single pass conversion from Eq. (13.68):

$$c_{(in,t)} = c_{(in,0)} \exp\left(-\frac{t}{\tau} \theta_A^{CSTR} \right) \tag{13.75}$$

And the overall conversion for the continuous flow reactor of the batch recycle system is:

$$\theta_{A,t}^{CSTR} = 1 - \frac{c_{(in,t)}}{c_{(in,0)}} = 1 - \exp\left(-\frac{t}{\tau_T} \theta_A^{CSTR} \right) \tag{13.76}$$

A batch recycle operation can be achieved using different configurations. Fig. 13.16 shows an example. For a batch operation, reactants can be added in continuously or intermittently.

Example 13.3

The overall reaction of electrochemical production of Zn is:

$$ZnSO_4 + H_2O \rightarrow Zn + H_2SO_4 + \frac{1}{2}O^2$$

The current efficiency of anode reaction is 96%, current density is 450 A/m², and cathode area is 40 m². The deposit layer (Zn) on cathode surface is removed once its thickness reaches 3.5 mm. The reactor is operated continuously (335 days/year). How many times does the Zn deposit layer have to be removed per month (30 days)? How many reactors are needed to produce 16,000 t of Zn per year?

Solution

1. The anode reaction is: $Zn^{2+} + 2e \rightarrow Zn$. Therefore for every mol of Zn the electricity needed is 2F. Therefore the electrochemical equivalent for Zn deposition can be calculated using Faraday's Law

$$K_{Zn} = \frac{65.38}{2 \times 26.8} = 1.219 \text{ g/Ah}$$

(In the above equation: 26.8 Ah = 1F electricity)

(Continued)

(Continued)

2. Assume the thickness of Zn deposition is δ (cm) and the length of each deposition period is τ (h). The density of Zn is 7.14 g/cm³, so the quantity of the chemical reaction is:

$$G = \delta d = i\tau K\eta_1$$

where G is the weight of product (g) per unit electrode surface area (cm²) Therefore:

$$\tau = \frac{\delta \cdot d}{iK\eta_1} = \frac{0.35 \times 7.14}{0.045 \times 1.219 \times 0.96} = 47.45 \text{ h} \approx 2 \text{days}$$

Every month the Zn deposition has to be removed $n = 30/2 = 15$ times.

3. The number of reactors (m) needed to achieved the required productivity is:

$$m = \frac{G_{total}}{iSTK\eta_1} = \frac{1.6 \times 10^{10}}{450 \times 40 \times 24 \times 335 \times 1.219 \times 0.95} = 95$$

13.3.2.5 Operation Characters of Electrochemical Reactors

The kinetic behavior of industrial electrochemical reactors can be demonstrated by discussing a simple batch reactor.

If the current is constant during an electrochemical reaction process, reactant concentration changes can be calculated using Faraday's Law:

$$\Delta c = c_{(0)} - c_{(t)} = \frac{It\eta_1}{nFV_S} \tag{13.77}$$

where η is the average current efficiency up to reaction time t and V_s is solution volume.

This equation can be rewritten as

$$c(t) = c(0) - \frac{It\eta_1}{nFV_S} \tag{13.78}$$

Or

$$\eta_1 = (c(t) - c(0))\frac{nFV_S}{It} \tag{13.79}$$

Take natural logarithm on both sides of Eq. (13.79)

$$\ln(c(t)) = \ln\left(c(0) - \frac{It\eta_1}{nFV_S}\right) \tag{13.80}$$

We can see that for an electrochemical reactor operated under constant current, its kinetics is closely correlated to electric current. Depending on mass transfer situations, the reaction kinetics can be classified as current control or diffusion control.

1. Current control: When reactant concentrations are high enough, actual reaction current is far below the diffusion-limited current (I_d). Then the consuming rate of reactants can be considered as a constant, and the corresponding current efficiency is η_I'. Under such conditions Eq. (13.78) can be written as:

$$c(t) = c(0) - \frac{It\eta_I'}{nFV_S} \tag{13.81}$$

$$\eta_I' = (c(0) - c(t))\frac{nFV_S}{It} \tag{13.82}$$

$$\ln c(t) = \ln\left(c(0) - \frac{It\eta_I'}{nFV_S}\right) \tag{13.83}$$

And the conversion is

$$\theta_A = \frac{c(0) - c(t)}{c(0)} = \frac{It\eta_I'}{nFc_0} \tag{13.84}$$

Since η_I', I, and $c(0)$ can be considered as constants, θ_A increases linearly with time t. The space-time yield is:

$$Y_{ST} = \frac{M(c(0) - c(t))}{t} \tag{13.85}$$

where M is molecular weight, kg/mol.
Using Eq. (13.82), one can obtain:

$$Y_{ST} = \frac{MI\eta_I'}{nFV_s} \tag{13.86}$$

We can see that the space-time yield is also a constant.
2. Diffusion control: After a reaction proceeds for a certain period of time ($t>t'$), due to the decrease of reactant concentration ($c<c'$), operation current reaches or even becomes greater than the diffusion-limited current ($I>I_d$). Under such conditions, current efficiency will decrease with time and the reaction rate becomes diffusion controlled. The concentration decrease is proportional to the logarithm of time. If the reactant concentration corresponded to I_d is c', then:
3.

$$c(t) = c'\exp\left[-\frac{k_m S}{V_s}(t-t')\right] = \frac{I_d}{nFSk_m}\exp\left[-\frac{k_m S}{V_s}(t-t')\right] \tag{13.87}$$

The current efficiency at this time is:

$$\eta_I = \frac{k_m SnFc(t)}{I_d} \tag{13.88}$$

Substituting Eq. (13.87) into (13.88) gives

$$\eta_l = \frac{k_m SnF}{I_d} c' \exp\left[-\frac{k_m S}{V_s}(t - t')\right] \tag{13.89}$$

Taking logarithm on both sides gives

$$\ln c(t) - \ln c' - \left[\frac{k_m S}{V_s}(t - t')\right] \tag{13.90}$$

So, when $t > t'$ (i.e., system is under diffusion control), the conversion is

$$\theta_A = 1 - \frac{c_t}{c'} = 1 - \exp\left[-\frac{k_{\bar{m}}A}{V_s}(t - t')\right] \tag{13.91}$$

i.e., the relation between θ_A and time is nonlinear.
Similarly, the space-time yield after $t > t'$

$$Y_{ST} = \frac{M(c' - c(t))}{t} \tag{13.92}$$

Or

$$Y_{ST} = \frac{IM\eta_l}{nFV_s} \tag{13.93}$$

It should be noted that the above discussions are very general. During the transition from electron exchange control to diffusion control, there will be a region where both processes are important, i.e., the transitions will occur gradually. Meanwhile, there will be many conversion steps (e.g., adsorption and surface film formation) on electrode surfaces, and there might be homogeneous reactions in the bulk phase, all of which make electrode processes more complex. Therefore, it is very challenging to quantitatively describe the operation characters of a electrochemical reactor. However, the above discussions are still beneficial, since they provide a methodology to analyze and estimate the conversion, space-time yield, current efficiency, as well as the correlation between reactant concentration and reaction time of electrochemical reactors.

Example 13.4
An electrochemical synthesis reaction is proceeded at a current density of 680 A/m². The operating voltage of the reactor is 5.4V, current efficiency is 85%, and daily productivity is 1200 kg. Calculate daily DC electricity consumption. If the relation between operating voltage and current density can be described by an empirical correlation $\Delta V = B\Delta i$, $B = 5.5 \times 10^{-4}$V, calculate the daily DC electricity consumption if productivity is 1500 kg/day. Assume
(Continued)

(Continued)

current efficiency remains constant and electrochemical equivalent of the product is 0.72 g/Ah.

Solution

1. Based on electrochemical equivalent (K) the theoretical electricity consumption (k) can be calculated:

$$k = \frac{1}{K} = \frac{1}{0.72} = 1.389 \text{ Ah/g}$$

DC electricity consumption per kg of product can be calculated using Eq. (13.16):

$$w = \frac{kV}{\eta_I} = \frac{1.389 \times 5.4}{0.85} = 8.824 \text{ kWh/kg}$$

DC electricity consumption per day is:

$$W_{total} = GW = 1200 \times 8.824 = 10,588.8 \text{ kWh}$$

2. If the current efficient is a constant, productivity will increase with current density. The percent of productivity increase required is:

$$\frac{1500 - 1200}{1200} = 25\% = 0.25$$

So the current density should be increased to $i_2 = 1.25 i_1 = 1.25 \times 680 = 850 \text{ A/m}^2$
i.e., $\Delta i = i_2 - i_1 = 850 - 680 = 170 \text{ A/m}^2$
The corresponding operating voltage for the reactor is:

$$\Delta V = B\Delta i = 5.5 \times 10^{-4} \times 170 = 935 \times 10^{-4} \text{V}$$

Therefore, $V_2 5.4 + 0.0935 = 5.4935 \text{V}$
Total electricity consumption after productivity increase is:

$$W_{Total} = GW_2 = 1500 \times \frac{1.389 \times 5.4935}{0.85} = 13,465.5 \text{ kWh}$$

13.3.3 Connections and Combination of Electrochemical Reactors

Although the productivity of unit reactors used in the modern electrochemical industry keep increasing, their structures and performance have been improved and enhanced, and operation current density is increased, the capacity of unit electrochemical reactors is relatively small when compared with typical chemical and metallurgical equipment. As a result, it is very common that multiple electrochemical reactors are required to meet production needs. Therefore, the common problems in electrochemical engineering are how to connect and combine multiple electrochemical reactors.

Connections of electrochemical reactors include both electric circuit and liquid flow paths. The electrical connections can further include connections between electrodes inside a reactor and those between reactors.

13.3.3.1 Circuit Connection of Electrochemical Reactors

1. Connection of electrodes in reactor

The connection between electrodes inside a reactor can be classified as monopolar and bipolar connections, and the corresponding reactors are referred to as monopolar cells and bipolar cells, respectively, as shown in Fig. 13.17.

In monopolar electrochemical reactors, every electrode is connected to one pole of an electric source, and the polarity of both sides of an electrode is the same, acting as an anode or a cathode. In bipolar electrochemical reactors, only the two electrodes at both ends are connected to the two poles of electric source. The polarity of the two sides of an electrode is opposite, i.e., one side is the anode and the other is the cathode. Some characters of these types of electrochemical reactors are listed in Table 13.9.

Fig. 13.18 shows the potential distribution in a bipolar electrochemical reactor. It contains five unit reactors and six electrodes.

The following two issues have to be considered when operating a bipolar reactor:

i. Avoid bypass and electric leakage, as shown in Fig. 13.17.

The electrolyte solution between two adjacent unit reactors is connected, and can allow electric current between electrodes of two unit reactors (instead of the two electrodes in the same unit reactor), which will lead to lower current efficiency and even corrosions of electrodes in the middle.

ii. Only certain types of electrodes are suitable for bipolar connection. The two surfaces of bipolar electrodes act as anode and cathode, and each surface must have electrocatalytic activities for the corresponding

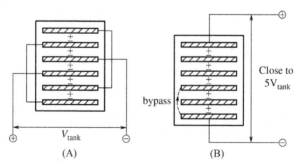

FIGURE 13.17 Circuit connection of monopolar (A) and bipolar (B) type reactors.

TABLE 13.9 Comparison of Monopolar and Bipolar Electrochemical Reactors

Characters	Monopolar Electrochemical Reactors	Bipolar Electrochemical Reactors
Polarity on two sides of the electrode	Same	Different
Electrode processes	Only one type of electrode reaction takes place (Anode reaction or cathode reaction)	Anode reaction on one side of the electrode and cathode reaction on the other side
Electrodes in the cell	In parallel	In series
Current	High ($I_{total} = \sum I_i$)	Low ($I_{total} = I_i$)
Cell voltage	Low ($V_{total} = V_i$)	High ($V_{total} = \sum V_i$)
Requirements for DC power supply	Low voltage and high current, more expensive	High voltage and high current, less expensive
Maintenance	Easy, and less impact on production	More complicated, production has to stop
Safety	Safer	Less safe
Design and manufacture	Simple	More complicated
Materials loading and unloading	Simple	More complicated
Space requirement	Larger	Smaller due to more compact structure
Material and installation cost	Higher	Lower
Ohmic potential drop per unit cell	Higher	Lower
Current distribution on electrodes	Less uniform	More uniform
More suitable for	Tank cells	Filter-Press cells

reactions. One electrode material usually cannot meet such requirements. Therefore, the two surfaces of the electrode in bipolar reactors are typically made of different materials or are specially treated (e.g., coating, plating, different layers of electrocatalyst).

2. Circuit connection between electrochemical reactors

The main consideration on connections between electrochemical reactors is the demand on DC power supply. A silicon rectifier is most

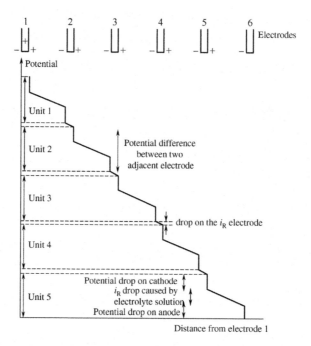

FIGURE 13.18 Electric potential distribution in a bipolar reactor.

commonly used for the DC power supply. When its DC voltage is in the range of 200–700V, AC efficiency could reach 95%. Therefore, when connecting multiple electrochemical reactors we should control the overall voltage in this range to maintain efficiency. For example, if the overall voltage is 450V, the voltage at the two end electrodes are +225V and −225, when the grounding is in the middle. The DC current could be controlled by properly selecting the capacity of the rectifier and connecting the reactors in parallel to meet productivity requirements.

Series connections is more commonly used for monopolar electrochemical reactors due to the low operating voltage and high current that are the major characters of this type of reactors. On the other hand, bipolar electrochemical reactors are usually operated at high operating voltage and low current, and hence are more suitable to be connected in parallel.

13.3.3.2 Flow Path Connections of Electrochemical Reactors

Similarly, both series and parallel configurations can be used to connect liquid flow path for multiple electrochemical reactors, as shown in Figure 11.19.

When high conversion is required, such as electrosynthesis, series connection is usually adopted. For plug flow reactors, according to Eq. (13.60)

$$c_{(out)} = c_{(in)}\exp\left(-\frac{k_m S}{Q}\right)$$

When n reactors are connected in series, reactant concentration at outlet is

$$c_{(out, n)} = c_{(in)}\exp\left(-\frac{nk_m S}{Q}\right) \tag{13.94}$$

And conversion is

$$X_{A,n} = 1 - \frac{c_{(out,n)}}{c_{(in)}} = 1 - \exp\left(\frac{-nk_m S}{Q}\right) \tag{13.95}$$

For CSTR, Eq. (13.68) can be used, and for n CSTRs in series, reactant concentration at outlet is

$$c_{(out,n)} = c_{(in)}\left(\frac{1+k_m A}{Q}\right)^n \tag{13.96}$$

The conversion is

$$X_{A,n} = 1 - \frac{c_{(in)}}{c_{(out,n)}} = 1 - \frac{1}{(1+k_m S/Q)^n} \tag{13.97}$$

The above discussions are for two ideal flow situations. In reality the flow pattern in the reactors is likely to be somewhere in between these two ideal situations. Many nonideal situations could exist, such as insufficient mixing and bypass flowing. And actual connecting configurations could be much more complex. There could be complete or partial recycle, or a combination of series and parallel connections, all depending on the purposes of electrolyte solution recycle and flow, such as reactant and product transport, heat exchange, maintaining and adjusting concentration and temperature distribution. It is worthy of note that the requirements on an electrolyte solution system, including equipment design, pump and reservoir's configurations and control, of these two connection configurations are different. For example, in order to maintain constant flow rate and velocity in each unit reactor, the total liquid flow rate under parallel configuration is much higher, and ensuring uniform distribution of liquid flow to each unit reactor is critical. On the other hand, for a series connected reactor system, the flow path is long and flow resistance is greater, which will require a pump that can provide higher pressure. Meanwhile the temperature increase is more significant and needs to be addressed properly.

13.3.3.3 Combination of Electrochemical Reactors

In order to meet certain productivity requirement, operating multiple (tens and even hundreds) electrochemical reactors simultaneously is a common scenario. There are two main reasons to use multiple electrochemical reactors: (1) limited capacity and productivity of a single reactor and (2) power supply and circuit connection consideration. Each electrochemical reactor has to be operated at a certain voltage and their combination has to meet the requirement of electrical connection between electrochemical cells discussed above in order to maintain overall DC voltage in the region to achieve high AC efficiency.

In industrial electrolysis processes, productivity depends on the capacity of each electrochemical reactor and numbers of reactors. However, a given productivity can be satisfied by various combinations of reactor capacities and numbers.

One way to choose the reactor capacity and number is to analyze the impact of reactor capacities and numbers on the capital investment and operation cost, and then select the reactor numbers to minimize overall production cost.

$$M_c = F_c + O_c \qquad (13.98)$$

where M_c is the total cost, F_c is the capital investment, and O_c is the operation cost.

It is usually very difficult to find a quantitative correlation between the total cost and so many affecting factors (such as reactor capacities and numbers, and current density). Although some literatures propose using general mathematical methods to find an optimal value of n, e.g., by solving Eq. (13.99), it is not always feasible.

$$\frac{\partial M_c}{\partial n} = 0 \qquad (13.99)$$

When selecting the number of reactors and capacity, in addition to considering the availability of electrochemical reactors and rectifiers on the market, their reliability and maintenance convenience have to be considered.

REFERENCES

[1] Ibl N. Comprehensive treatise of electrochemistry, Vol. 6. New York: Plenum Press; 1983. p. 239–315.

[2] Janssen JJ, Barendrecht E. Mass transfer at gas evolving electrodes. J Appl Electrochem 1985;15(4):549–55.

[3] Martin AD, Wragg AA. The vertical distribution of current in a gas-evolving membrane cell. J Appl Electrochem 1989;19(5):657–67.

[4] Yoshinori N, Koichi A, Koichi T, Hiroaki M. Effect of gas evolution on current distribution and ohmic resistance in a vertical cell under forced convection conditions. J Appl Electrochem 1986;16(4):615–25.

FURTHER READING

Chen Y. Electrolysis engineering. Tianjin: Tianjin Science and Technology Press; 1993 (In Chinese).

Pletcher D. Industrial electrochemistry. 2nd ed. London: Chapman and Hall; 1990.

Fahidy TZ. Principles of electrochemical reactors analysis. Amsterdam: Elsevier; 1985.

Pickett D. Electrochemical reactor design. 2nd ed. New York: Elsevier; 1979.

Scott K. Electrochemical reaction engineering. London: Academic Press; 1991.

PROBLEMS

13.1 NaCl aqueous solution is electrolyzed to produce chlorine. What is the highest O_2 concentration in anode gas in order to maintain current efficiency greater than 95%?

13.2 KOH solution is electrolyzed to produce oxygen and hydrogen. The solution electric conductivity is $120 \text{ m}^{-1}/\Omega$, electrode surface area is 2.4 m^2. The distance from anode to the diaphragm is 6 mm, and gas holdup in the region is 0.22; the distance from cathode to the diaphragm is 5 mm, and gas holdup in the region is 0.3. The thickness of the diaphragm is 2 mm, whose electric conductivity is $32 \text{ m}^{-1}/\Omega$. The theoretical decomposition voltage (E_0) is 1.9V, and the overpotential can be estimated according to current density:

$$\Delta \varphi_A = 0.23 + 0.08 \lg i$$
$$\Delta \varphi_k = 0.06 - 0.12 \lg i$$

(i is current density, in the unit of A/m^2)

Calculate the operating voltage.

13.3 In the electrochemical reactor where MnO_2 is produced from $MnSO_4$, the overall reaction is $MnSO_4 + 2 \text{ H}_2O \rightarrow MnO_2 + H_2SO_4 + H_2$. The operating voltage is 2.5V, and current efficiency is 85%. Calculate the DC power consumption.

13.4 The reaction for producing fluorine by electrochemical oxidization of HF is $2HF \rightarrow H_2 + F_2$. DC power consumption is 15 W h/kg, the current efficiency can be kept at 95%, and theoretical decomposition voltage is 2.9V. Please estimate the range of overpotential.

Index

Note: Page numbers followed by "*f*" and "*t*" refer to figures and tables, respectively.

Printed in the United States
By Bookmasters